THE CYBERNETIC BRAIN

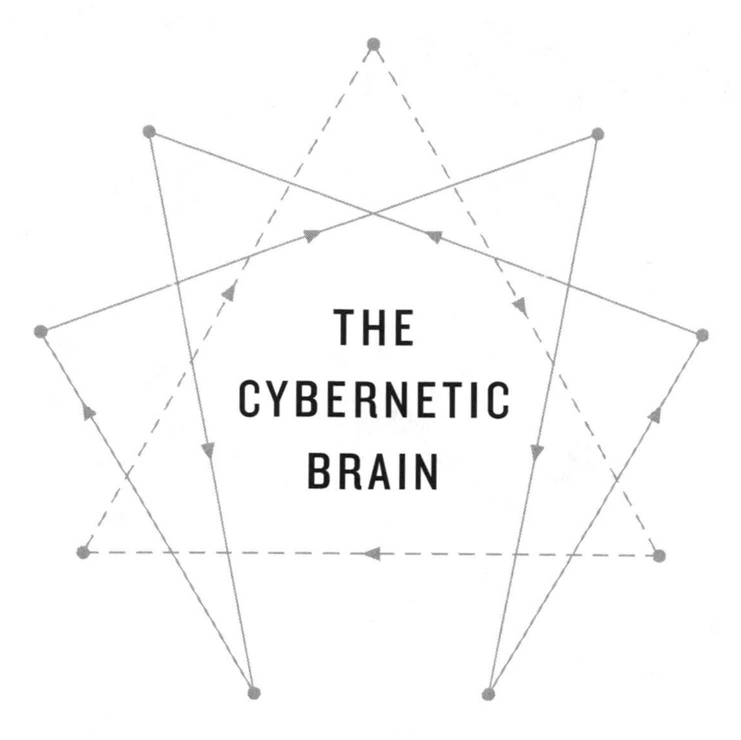

THE CYBERNETIC BRAIN

SKETCHES OF ANOTHER FUTURE

Andrew Pickering

THE UNIVERSITY OF CHICAGO PRESS
CHICAGO AND LONDON

THE UNIVERSITY OF CHICAGO PRESS, CHICAGO 60637
THE UNIVERSITY OF CHICAGO PRESS, LTD., LONDON

PAPERBACK EDITION 2011
PRINTED IN THE UNITED STATES OF AMERICA

20 19 18 17 16 15 14 13 12 11 2 3 4 5 6

ISBN-13: 978-0-226-66789-8 (CLOTH)
ISBN-13: 978-0-226-66790-4 (PAPER)
ISBN-10: 0-226-66789-8 (CLOTH)
ISBN-10: 0-226-66790-1 (PAPER)

Library of Congress Cataloging-in-Publication Data

Pickering, Andrew.
The cybernetic brain : sketches of another future / Andrew Pickering.
p. cm.
Includes bibliographical references and index.
ISBN-13: 978-0-226-66789-8 (cloth : alk. paper)
ISBN-10: 0-226-66789-8 (cloth : alk. paper) 1. Cybernetics. 2. Cybernetics—History. 3. Brain. 4. Self-organizing systems. I. Title.
Q310.P53 2010
003'.5—dc22

2009023367

⊗ THIS PAPER MEETS THE REQUIREMENTS OF ANSI/NISO Z39.48-1992 (PERMANENCE OF PAPER).

DEDICATION

For Jane F.

CONTENTS

PART 2: BEYOND THE BRAIN

ACKNOWLEDGMENTS

This project began in fall 1998, with the support of fellowships from the John Simon Guggenheim Foundation and the Dibner Institute for the History of Science and Technology at MIT. A lengthy first draft was written during a sabbatical at the Science Studies Unit of the University of Edinburgh in 2002–3, supported in part by a grant from the National Science Foundation. In 2005–6 I enjoyed a fellowship from the Illinois Program for Research in the Humanities at the University of Illinois at Urbana-Champaign, and the final draft of the book was completed while I was a fellow at the Center for Advanced Study in the Behavioral Sciences at Stanford in 2006–7. Along the way, visits of a month or two to the Max Planck Institute for the History of Science in Berlin made possible bursts of sustained research and writing. I am immensely grateful to all of these institutions, and to David Bloor as director of the Science Studies Unit and Ursula Klein and Hans-Jörg Rheinberger, whose research groups I joined in Berlin.

The book could not have been written without the active assistance of many cyberneticians, colleagues, friends and family members of the book's principal characters, and various others whose work I have referred to and discussed: Jill Ashby, John Ashby, Mick Ashby, Sally Bannister, Ian Beer, Stafford Beer, Vanilla Beer, Rodney Brooks, Peter Cariani, Raul Espejo, John Frazer, Ranulph Glanville, Nick Green, Amanda Heitler, Garnet Hertz, Stewart Kauffman, Allenna Leonard, Paul Pangaro, the late Elizabeth Pask, Simon Penny, Ruth Pettit, Jasia Reichardt, Bernard Scott, Yolanda Sonnabend, Joe

Truss, David Whittaker, and Stephen Wolfram. My thanks to all of them, especially perhaps to Paul Pangaro, who introduced me to many of the others at an early stage in my research and was always ready with information, feedback, and encouragement.

Among my colleagues and friends, frequent conversations, critiques, and encouragement from the likes of Peter Asaro, Geof Bowker, Bruce Lambert, and Fernando Elichirigoity were at different times and in different ways very important to me. Gordon Belot, Michael Friedman, Eden Medina, Laura Ruetsche, and Fred Turner read and commented on the entire manuscript; Erica Goode, Peter Harries-Jones, Malcolm Nicolson, Dave Perkins, and Henning Schmidgen on various chapters. Others who have contributed constructively and critically include Ian Carthy, Adrian Cussins, Manuel DeLanda, Otniel Dror, John Geiger, Slava Gerovitch, Rhodri Hayward, David Hopping, Sharon Irish, Evelyn Fox Keller, Vera Ketelboeter, Jeffrey Kripal, Julia Kursell, David Lambert, Mike Lynch, Brian Marick, Clark McPhail, Diana Mincyte, Anya Pantuyeva, Jan Nederveen Pieterse, Judith Pintar, Amit Prasad, Carol Steiner, Steve Sturdy, Lucy Suchman, and Norton Wise. My thanks go to all of them, too.

I thank Jane Flaxington for her help in the endless quest for permission to reproduce figures. At the University of Chicago Press, I thank Douglas Mitchell and Timothy McGovern for their editorial care and assistance. Erik Carlson's copyediting was exemplary, educational, and fun.

Finally, my love to Jane, Lucy, Thomas, and Alex for putting up with my eccentricities, now including this book.

Exeter, April 10, 2008

1

THE ADAPTIVE BRAIN

THE MAKING OF A SYNTHETIC BRAIN REQUIRES NOW LITTLE MORE THAN TIME AND LABOUR. . . . SUCH A MACHINE MIGHT BE USED IN THE DISTANT FUTURE . . . TO EXPLORE REGIONS OF INTELLECTUAL SUBTLETY AND COMPLEXITY AT PRESENT BEYOND THE HUMAN POWERS. . . . HOW WILL IT END? I SUGGEST THAT THE SIMPLEST WAY TO FIND OUT IS TO MAKE THE THING AND SEE.

ROSS ASHBY, "DESIGN FOR A BRAIN" (1948, 382–83)

On 13 December 1948, the *Daily Herald* carried a front-page article entitled "The Clicking Brain Is Cleverer Than Man's," featuring a machine called the homeostat built by W. Ross Ashby. Soon the rest of the press in Britain and around the world followed suit. In the United States, an article in *Time* magazine, "The Thinking Machine," appeared on 24 January 1949 (p. 66), and by 8 March 1949 Ashby was holding forth on BBC radio on "imitating the brain." At much the same time, W. Grey Walter appeared on BBC television showing off a couple of small robots he had built, Elmer and Elsie, the first examples of his robot "tortoises," or, more pretentiously, of a new inorganic species, *Machina speculatrix*. One appeared in a family photo in *Time* (fig. 1.1). In 1952, Gordon Pask began work on his Musicolour machine—an electromechanical device that collaborated in obscure ways with a musician to generate a synesthetic light show. Soon he was also experimenting with

Figure 1.1. The cyborg family. Source: de Latil 1956, facing p. 34.

quasi-biological electrochemical computers that could evolve new senses, and within a decade he was designing buildings that could reconfigure themselves in "conversation" with their users. In 1959 Stafford Beer published a book imagining an automated factory controlled by a biological computer—perhaps a colony of insects or perhaps a complex ecosystem such as a pond. By the early 1970s, he was redesigning the "nervous system" of the Chilean economy at the invitation of the socialist government of Salvador Allende.

Examples like these convey some of the flavor of the history explored in the following chapters. In this chapter and the next I want to discuss more generally what cybernetics is, or was, and why it interests me. (The tense is difficult; cybernetics as a field is alive today, but the main characters of this book are all now dead. I will tend therefore to speak of cybernetics in the past tense, as referring to a historical body of work.)

— — — — —

SOME PEOPLE THINK THAT CYBERNETICS IS ANOTHER WORD FOR AUTOMATION; SOME THAT IT CONCERNS EXPERIMENTS WITH RATS; SOME THAT IT IS A BRANCH OF MATHEMATICS; OTHERS THAT IT WANTS TO BUILD A COMPUTER CAPABLE OF RUNNING THE COUNTRY. MY HOPE IS THAT . . . PEOPLE WILL UNDERSTAND BOTH HOW THESE WONDERFULLY DIFFERENT NOTIONS CAN BE SIMULTANEOUSLY CURRENT, AND ALSO WHY NONE OF THEM IS MUCH TO THE POINT.

STAFFORD BEER, *CYBERNETICS AND MANAGEMENT* (1959, VI)

TO SPEAK OF A HISTORY, ANY HISTORY, AS THOUGH THERE WAS BUT ONE SOMEHOW CANONICAL HISTORY . . . IS MISLEADING. . . . ANY ENTITY, CULTURE OR CIVILISATION . . . CARRIES INNUMERABLE, IN SOME WAYS DIFFERING, HISTORIES.

GORDON PASK, "INTERACTIONS OF ACTORS" (1992, 11)

The word "cybernetics" was coined in 1947 by the eminent American mathematician Norbert Wiener and his friends to name the kind of science they were discussing at the famous Macy conferences held between 1946 and 1953.[1] It was derived from the Greek word *kybernetes* (Latin equivalent, *gubernator*) meaning "governor" in the sense of "steersman," so one could read "cybernetics" as "the science of steersmanship"—and this is, as it happens, a good definition as far as this book is concerned. The matter was made more interesting and complicated, however, by Wiener's 1948 book which put the word into circulation, *Cybernetics; or, Control and Communication in the Animal and the Machine*. There Wiener tried to tie together all sorts of more or less independent lines of scientific development: digital electronic computing (then still novel), information theory, early work on neural networks, the theory of servomechanisms and feedback systems, and work in psychology, psychiatry, decision theory, and the social sciences. There are many stories to be told of the evolution, the comings together, and the driftings apart of these threads, only a few of which have so far attracted the attention of scholars.[2] One can almost say that everyone can have their own history of cybernetics.

In this book I do not attempt a panoptic survey of everything that could be plausibly described as cybernetic. I focus on the strand of cybernetics that interests me most, which turns out to mean the work of a largely forgotten group of British cyberneticians, active from the end of World War II almost to the present. Even to develop an overview of British cybernetics would require several books, so I focus instead on a few leading lights of the field, the ones mentioned already: Grey Walter (1910–77), Ross Ashby (1903–72), Stafford Beer (1926–2002), and Gordon Pask (1928–96), with a substantial detour through the work of Gregory Bateson and R. D. Laing. And even with this editorial principle, I have to recognize that each of my four easily warrants his own biography, which I have not attempted to write. So what follows is very much my own history of cybernetics in Britain—not a comprehensive survey, but the story of a set of scientific, technological, and social developments that speak to me for reasons I will explain and that I hope will interest others.

A further principle of selection is also in play. Most accounts of the history of cybernetics are in the mode of a history of ideas; they concentrate on grasping the key ideas that differentiate cybernetics from other sciences. I am not uninterested in ideas, but I am interested in ideas as engaged in practice, and at the heart of this book is a series of real-world projects encompassing all sorts of strange machines and artifacts, material and social. I want to document what cybernetics looked like when people did it, rather than just thought it. That is why the opening paragraph ran from artificial brains to the Chilean economy, rather than offering an abstract discussion of the notion of "feedback" or whatever.

The choice of principals for this study makes sense sociologically inasmuch as my four cyberneticians interacted strongly with one another. Walter and Ashby were first-generation cyberneticians, active in the area that became known as cybernetics during and even before World War II, and were leading members of the first protocybernetic organization in Britain, the so-called Ratio Club, which met between 1949 and 1958 (Alan Turing was the best-known recruit). They never collaborated in research, but they knew, took account of, and commented on each other's work, though relations became strained in 1959 when Ashby briefly became Walter's boss. Beer and Pask were second-generation cyberneticians, coming onto the scene in the 1950s after the foundations of the field had been laid. They were lifelong friends, and Beer became almost the social secretary of the British branch of cybernetics, with strong personal ties not only to Walter, Ashby, and Pask and but also to Wiener and to Warren McCulloch, the guiding spirit of cybernetics in the United States. But what about the technical content of British cybernetics? Is there any unity there?

The standard origin story has it that cybernetics evolved out of the intersection of mathematics and engineering in U.S. military research in World War II, and this is certainly a good description of Wiener's trajectory (Galison

Figure 1.2. The four pioneers of cybernetics (*left to right*): Ross Ashby, Warren McCulloch, Grey Walter, and Norbert Wiener. Source: de Latil 1956, facing p. 53.

1994). But figure 1.2, a photograph taken in the early 1950s, originally appeared with the not unreasonable caption "The Four Pioneers of Cybernetics," and what I find striking is that, with Wiener as the exception, three of the four—Ashby, Walter, and McCulloch—spent much or all of their professional careers in research on the human brain, often in psychiatric milieus.[3] We can explore the specifically psychiatric origins of cybernetics in detail in chapters 3 and 4, but for the moment it is enough to note that the distinctive object of British cybernetics was *the brain*, itself understood in a distinctive way. This requires some explanation now, since it is a way into all that follows.

To put it very crudely, there are two ways to think about the brain and what it does. The way that comes naturally to me is to think of the brain as an organ of *knowledge*. My brain contains representations, stories, memories, pictures of the world, people and things, myself in it, and so on. If I know something, I have my brain (and not my kidneys, say) to thank for it. Of course, I did not get this image of the brain from nowhere. It is certainly congenial to us academics, professional knowers, and it (or an equivalent image of mind) has been stock-in-trade for philosophy for centuries and for philosophy of science throughout the twentieth century. From the mid-1950s onward this image has been instantiated and highly elaborated in the branch of computer science concerned with artificial intelligence (AI). AI—or, at least, the approach to AI that has become known as GOFAI: good, old-fashioned AI—just is traditional philosophy of science implemented as a set of computer algorithms. The key point that needs to be grasped is that the British cyberneticians' image of the brain was *not* this representational one.

What else could a brain be, other than our organ of representation? This question once baffled me, but the cyberneticians (let me take the qualifier "British" for granted from now on unless needed) had a different answer. As Ashby put it in 1948, "To some, the critical test of whether a machine is or is not a 'brain' would be whether it can or cannot 'think.' But to the biologist the brain is not a thinking machine, it is an *acting* machine; it gets information and then it does something about it" (Ashby 1948, 379). The cyberneticians, then, conceived of the brain as an immediately embodied organ, intrinsically tied into bodily performances. And beyond that, they understood the brain's special role to be that of adaptation. The brain is what helps us to get along and come to terms with, and survive in, situations and environments we have never encountered before. Undoubtedly, knowledge helps us get along and adapt to the unknown, and we will have to come back to that, but this simple contrast (still evident in competing approaches to robotics today) is what we need for now: the cybernetic brain was not representational but *performative*, as I shall say, and its role in performance was *adaptation*.

As a preliminary definition, then, we can regard cybernetics as a postwar science of the adaptive brain, and the question then becomes: What did cybernetics look like in practice? Just how did the cyberneticians attack the adaptive brain? The answer is, in the first instance, by building electromechanical devices that were themselves adaptive and which could thus be understood as perspicuous and suggestive models for understanding the brain itself. The simplest such model was the servomechanism—an engineering device that reacts to fluctuations in its environment in such a way as to cancel them out. A domestic thermostat is a servomechanism; so was the nineteenth-century steam-engine "governor" which led Wiener to the word "cybernetics." Working with servomechanisms in the war was, in fact, what led Wiener into the field he subsequently named. Walter's robot tortoises and Ashby's homeostat were more striking and original examples of adaptive mechanisms, and they were at the forefront of "brain science" in the late 1940s and throughout the 1950s. A phrase of Warren McCulloch's comes to mind. Speaking of another British protocybernetician, the experimental psychologist Kenneth Craik, McCulloch remarked that Craik always wanted to understand "the go of it"—meaning, to grasp the specific mechanical or quasi-mechanical connections that linked inputs and outputs in complex systems like the brain.[4] Cybernetic devices like tortoises and homeostats aimed precisely to illuminate the go of the adaptive brain.

There is something strange and striking about adaptive mechanisms. Most of the examples of engineering that come to mind are not adaptive. Bridges and buildings, lathes and power presses, cars, televisions, computers, are all designed to be *indifferent* to their environment, to withstand fluctuations, not to adapt to them. The best bridge is one that just stands there, whatever the weather. Cybernetic devices, in contrast, explicitly aimed to be sensitive and responsive to changes in the world around them, and this endowed them with a disconcerting, quasi-magical, disturbingly lifelike quality. Wiener himself was well aware of this, and his writings are dotted with references to the Sorcerer's Apprentice (who casts a magical spell that sets matter in motion and cannot be undone) and the Golem of Prague (magically animated clay). Walter likewise spoke of "the totems of primitive man" and invoked the figure of Frankenstein's monster (1953, 113, 115). This sense of mystery and transgression has always attached to cybernetics, and accounts, I think, for much of its *glamour*—the spell it casts over people, including myself.

— — — — —

I need to say more about cybernetics, the brain, and psychiatry. The early cybernetics of Walter and Ashby directly concerned the brain as an anatomical organ. The tortoise and the homeostat were intended as electromechanical models of the physiological brain, normal and pathological, with the latter providing a direct link to the brutal approaches to psychiatry that were dominant from the 1930s to the 1950s, chemical and electrical shock therapies and lobotomy. In the 1950s and 1960s, however, a different form of cybernetic psychiatry emerged, often, though somewhat misleadingly, labeled "antipsychiatry" for its opposition to violent interventions in mental illness (and, indeed, for its opposition to the concept of mental illness). I associate this latter form of cybernetic psychiatry with the work of the expatriate Englishman Gregory Bateson (1904–80) and, in the 1960s, with the radical therapeutic experiments of the Scottish psychiatrist R. D. Laing (1927–89).

Unlike my four principals, Bateson and Laing are relatively well known to scholars, the subject of several book-length studies, so I will not discuss their work here to the same depth as the others. But I include a chapter on them for three reasons. First, because Bateson's approach to psychiatry exemplifies a move in cybernetics beyond a concern with the physiological brain and toward something less biologically specified. If Walter and Ashby focused on the adaptive brain, Bateson was concerned with something less precise and less structured, the adaptive subject or self, and how that could be disrupted by what he called double binds. Laing, from this perspective, played out what Batesonian psychiatry might look like in practice. Second, simply to emphasize that cybernetics was not forever irrevocably locked into the world of electroshock. And third, continuing that line of thought, because there is an important sense in which Bateson and Laing were *more cybernetic* than Walter and Ashby. Laing's psychiatry took seriously, as Walter and Ashby's did not, the idea that we are all adaptive systems, psychiatrists and schizophrenics alike. I am interested to follow the practical and institutional ramifications of this move here.

These features of Bateson and Laing's work—looking beyond the biological brain, and an extension of cybernetics into the field of the self and social relations—move us to another theme of this book, namely, the multiplicity of cybernetics, its protean quality. I began by defining cybernetics as the science of the adaptive brain, but even the earliest manifestations of cybernetics ran in several directions. Tortoises and homeostats could be understood as "brain science" in the sense of trying to explicate the functioning of the normal brain as a complex adaptive system—a holistic counterpoint to reductive neurophysiology, say. At the same time, as I just mentioned, tortoises and homeostats could also simulate the abnormal, pathological brain—madness—and hence stand as a contribution to psychiatry. Furthermore, these cybernetic devices did not have to be seen in relation to the brain at all, but could also be seen as things in themselves. Walter's tortoises, for example, were foundational to approaches to robotics that are very influential today—the situated robotics that I associate with the work of Rodney Brooks, and extremely interesting related work in biologically inspired robotics. From a different angle again, although Ashby's work from the 1930s onward has to be understood as attempting to shed light on the brain, by the 1950s he had begun to see his cybernetics as a general theory, applicable to all sorts of complex systems besides the brain: adaptive autopilots, the British economy, the evolution of species.

The brain, one might say, could not contain cybernetics; cybernetics spilled out all over the disciplinary and professional map. It was a strongly interdisciplinary field, or, better, an antidisciplinary one: it did not aggregate disciplinary perspectives; it rode roughshod over disciplinary boundaries—which also contributes to its glamour. Bateson and Laing, as I said, went beyond the narrow focus of cybernetics on the biological brain to a wider focus on intrinsically social selves, and if we add in Beer and Pask the picture gets still richer. On the one hand, these two second-generation cyberneticians followed Ashby and Walter in the pursuit of material models of the adaptive brain, but in extremely original ways. Beer's experiments with *Daphnia* and ponds and Pask's with electrochemical "threads" were precisely attempts to "grow" adaptive brains—nondigital and nonrepresentational, biological or quasi-biological computers. This is some of the most striking and visionary work I have come across in the history of science and engineering. On the other hand, much of Beer and Pask's work can be seen as extending the achievements of the first generation, especially Ashby's, into new spaces, while echoing the social concerns of Bateson and Laing beyond the realm of psychiatry. Beer drew heavily upon Ashby's work on the homeostat (as well as upon contemporary neurophysiology) in developing his "management cybernetics" in the late 1950s, which later evolved into his viable system model of social organizations and his "team syntegrity" approach to collective decision making. Beer also extended his cybernetics beyond organizations into politics, up to the level of world politics, and even into the spiritual domain, entirely beyond the mundane world. Pask's elaboration of cybernetics started in the world of entertainment with the Musicolour machine and ran, in one direction, into the development of cybernetic trainers and teaching machines and, in another, into robot artworks, interactive theater, and adaptive architecture.

The world of cybernetics was, then, very rich. Cybernetic practices and artifacts first emerged as brain science and psychiatry, but quickly and distinctively spread to all the fields I have just mentioned (and more): robotics, engineering, a science of general systems with applications in many fields, biological computing, management, politics, spirituality (if that is a field), entertainment, the arts, theater and architecture (music, too), education. Unlike more familiar sciences such as physics, which remain tied to specific academic departments and scholarly modes of transmission, cybernetics is better seen as *a form of life*, a way of going on in the world, even an attitude, that can be, and was, instantiated both within and beyond academic departments, mental institutions, businesses, political organizations, churches, concert halls, theaters, and art museums. This is to put the case positively. But from another angle, we should note the continuing *marginality* of cybernetics to established institutions.

— — — — —

I am struck, first, by the profound amateurism of British cybernetics. Key contributions often had an almost hobbyist character: Walter built his first tortoises at home in his spare time; so did Ashby his homeostat (at least, in the apocryphal version of the story); likewise Beer and Pask's experimentation with biological and chemical computers; Bateson never in his life had a steady job; Laing's experiments in psychiatry took place beyond the established institutional framework. Cybernetics welled up outside the usual channels, and it found little support within those channels. One might have expected the universities to be the natural home for such a field, and, indeed, Beer and Pask did hold a variety of part-time academic positions, but only a handful of academic units devoted to the production and transmission of cybernetic knowledge appeared in the West, and then only over finite time spans. One thinks principally of Warren McCulloch's group at MIT's Research Laboratory of Electronics (1952–69), Heinz von Foerster's Biological Computer Laboratory at the University of Illinois (1958–75) (where Ashby was a professor for the ten years before his retirement), and, in Britain, the Cybernetics Department at Brunel (1969–85).[5] (Interestingly, various versions of cybernetics were institutionalized in the post-Stalinist Soviet Union. To follow that would take us too far afield, but see Gerovitch 2002.)

Conferences and less formal gatherings instead constituted scholarly centers of gravity for the field: the Macy conferences in the United States; the Ratio Club, a self-selected dining club, in Britain (1949–58); and in Europe a series of international conferences held at Namur in Belgium from 1958 onward. Our cyberneticians were thus left to improvise opportunistically a social basis for their work. After graduating from Cambridge in 1952, Pask, for example, set up his own research and consulting company, System Research, and looked for contracts wherever he could find them; in 1970 Beer gave up a successful career in management to become an independent consultant. And along with this instability of the social basis of cybernetics went a very chancy mode of transmission and elaboration of the field. Thus, quasi-popular books were very important to the propagation of cybernetics in a way that one does not find in better-established fields. Norbert Wiener's *Cybernetics* (1948) was enormously important in crystallizing the existence of cybernetics as a field and in giving definition to the ambitions of its readers. Grey Walter's

The Living Brain (1953) found an active readership diverse enough to span protoroboticists and the Beat writers and artists. It was a turning point in his musical career when Brian Eno's mother-in-law lent him a copy of Stafford Beer's book, *Brain of the Firm*, in 1974.

Sociologically, then, cybernetics wandered around as it evolved, and I should emphasize that an undisciplined wandering of its subject matter was a corollary of that. If PhD programs keep the academic disciplines focused and on the rails, chance encounters maintained the openness of cybernetics. *Brain of the Firm* is a dense book on the cybernetics of management, and music appears nowhere in it, but no one had the power to stop Eno developing Beer's cybernetics however he liked. Ashby's first book, *Design for a Brain* (1952), was all about building synthetic brains, but Christopher Alexander made it the basis for his first book on architecture, *Notes on the Synthesis of Form* (1964). A quick glance at *Naked Lunch* (1959) reveals that William Burroughs was an attentive reader of *The Living Brain*, but Burroughs took cybernetics in directions that would have occurred to no one else.

Cybernetics was thus a strange field sociologically as well as substantively. We might think of the cyberneticians as nomads, and of cybernetics as a nomad science, perpetually wandering and never finding a stable home. For readers of Gilles Deleuze and Félix Guattari's *A Thousand Plateaus* (1987), the phrase "nomad science" has a special resonance in its contrast with "royal science." The royal sciences are the modern sciences, which function as part of a stable social and political order—which prop up the state. The nomad sciences, on Deleuze and Guattari's reading, are a different kind of science, one which wanders in from the steppes to undermine stability. We can come back to this thought from time to time.

— — — — —

THE STUDY OF THINKING MACHINES TEACHES US MORE ABOUT THE BRAIN THAN WE CAN LEARN BY INTROSPECTIVE METHODS. WESTERN MAN IS EXTERNALIZING HIMSELF IN THE FORM OF GADGETS. EVER POP COKE IN THE MAINLINE? IT HITS YOU RIGHT IN THE BRAIN, ACTIVATING CONNECTIONS OF PURE PLEASURE. . . . C PLEASURE COULD BE FELT BY A THINKING MACHINE, THE FIRST STIRRINGS OF HIDEOUS INSECT LIFE.

WILLIAM BURROUGHS, *NAKED LUNCH* (2001 [1959], 22)

As John Geiger (2003) discovered, if you look at the works of Aldous Huxley or Timothy Leary or William Burroughs and the Beats, you find Grey Walter.

You also find yourself at one of the origins of the psychedelic sixties. From a different angle, if you are interested in the radical critique of psychiatry that was so important in the late 1960s, you could start with its high priest in Britain, R. D. Laing, and behind him you would find Gregory Bateson and, again, Walter. If you were interested in intersections between the sixties and Eastern spirituality, you might well come across Stafford Beer, as well as experimentation with sensory deprivation tanks and, once more, Bateson. In 1960, Ross Ashby lectured at the Institute for Contemporary Arts in London, the hub of the British art scene, on "art and communication theory," and, at the ICA's 1968 *Cybernetic Serendipity* exhibition, Gordon Pask displayed his Colloquy of Mobiles—an array of interacting robots that engaged in uncertain matings with one another—alongside Beer's Statistical Analogue Machine, SAM. Pask's "conversation" metaphor for cybernetics, in turn, gets you pretty close to the underground "antiuniversity" of the sixties.

What should we make of this? One might continue Deleuze and Guattari's line of thought and say that the sixties were the decade when popular culture was overrun by not one but two bands of nomads. On the one hand, the sixties were the heyday of cybernetics, the period when this marginal and antidisciplinary field made its greatest inroads into general awareness. On the other hand, the sixties can be almost defined as the period when a countercultural lifestyle erupted from the margins to threaten the state—"the Establishment." Given more space and time, this book might have been the place for an extended examination of the counterculture, but to keep it within bounds I will content myself with exploring specific crossovers from cybernetics to the sixties as they come up in the chapters to follow. I want to show that some specific strands of the sixties were in much the same space as cybernetics—that they can be seen as continuations of cybernetics further into the social fabric. This extends the discussion of the protean quality of cybernetics and of the sense in which it can be seen as an interesting and distinctive form of life.

— — — — —

Two more, possibly surprising, strands in the history of cybernetics are worth noting. First, as we go on we will repeatedly encounter affinities between cybernetics and Eastern philosophy and spirituality. Stafford Beer is the extreme example: he both practiced and taught tantric yoga in his later years. There is, I think, no necessary connection between cybernetics and the East; many cyberneticians evince no interest whatsoever in Eastern spirituality. Nevertheless, it is worth exploring this connection where it arises (not least, as a site of interchange between cybernetics and the sixties counterculture).

In the next chapter I will outline the peculiar ontology that I associate with cybernetics—a nonmodern ontology, as I call it, that goes with a performative understanding of the brain, mind and self, and which undoes the familiar Western dualism of mind and matter, resonating instead with many Eastern traditions.

Second, cyberneticians have shown a persistent interest in what I call strange performances and altered states. This, too, grows out of an understanding of the brain, mind, and self as performative. One might imagine the representational brain to be immediately available for inspection. Formal education largely amounts to acquiring, manipulating, and being examined on representational knowledge. Such activities are very familiar to us. But the performative brain remains opaque and mysterious—who knows what a performative brain can do? There is something to be curious about here, and this curiosity is a subtheme of what follows. As I said, early cybernetics grew out of psychiatry, and the topics of psychiatry are nothing more than altered states—odd, unpleasant, and puzzling ways to be relative to some norm. We will see, however, that cybernetics quickly went beyond any preoccupation with mental illness. Grey Walter, for example, did research on "flicker": it turns out that exposure to strobe lights can induce, on the one hand, symptoms of epilepsy, but also, on the other, surprising visions and hallucinations. I think of flicker as a peculiar sort of *technology of the self*—a technique for producing states of being that depart from the everyday—and we can explore several of them, material and social, and their associated states as we go along. Walter also offered cybernetic analyses of yogic feats and the achievement of nirvana. All of this research makes sense if one thinks of the brain as performative, and it connects, in ways that we can explore further, both to the spiritual dimension of cybernetics and to the sixties.

— — — — —

I have been trying to indicate why we might find it historically and anthropologically interesting to explore the history and substance of cybernetics, but my own interest also has a political dimension. The subtitle of this book—*Sketches of Another Future*—is meant to suggest that we might learn something from the history of cybernetics for how we conduct ourselves in the present, and that the projects we will be examining in later chapters might serve as models for future practice and forms of life. I postpone further development of this thought to the next chapter, where the overall picture should become clearer, but for now I want to come at it from the opposite angle. I need to confront the fact that cybernetics has a bad reputation in some quarters. Some people think of it as the most despicable of the sciences. Why is that? I do not have a panoptic grasp of the reasons for this antipathy, and it is hard to find any canonical examples of the critique, but I can speak to some of the concerns.[6]

One critique bears particularly on the work of Walter and Ashby. The idea is that tortoises and homeostats in fact fail to model the human brain in important respects, and that, to the degree that we accept them as brain models, we demean key aspects of our humanity (see, e.g., Suchman 2005). The simplest response to this is that neither Walter nor Ashby claimed actually to have modelled anything approaching the real human brain. In 1999, Rodney Brooks gave his book on neo-Walterian robotics the appropriately modest title of *Cambrian Intelligence*, referring to his idea that we should start at the bottom of the evolutionary ladder (not the top, as in symbolic AI, where the critique has more force). On the other hand, Ashby, in particular, was not shy in his speculations about human intelligence and even genius, and here the critique does find some purchase. His combinatoric conception of intelligence is, I believe, inadequate, and we can explore this further in chapter 4.

A second line of critique has to do with cybernetics' origins in Wiener's wartime work; cybernetics is often thought of as a militarist science. This view is not entirely misleading. The descendants of the autonomous antiaircraft guns that Wiener worked on (unsuccessfully) in World War II (Galison 1994) are today's cruise missiles. But first, I think the doctrine of original sin is a mistake—sciences are not tainted forever by the moral circumstances of their birth—and second, I have already noted that Ashby and Walter's cybernetics grew largely from a different matrix, psychiatry. One can disapprove of that, too, but the discussion of Bateson and Laing's "antipsychiatry" challenges the doctrine of original sin here as well.

Another line of critique has to do with the workplace and social inequality. As Wiener himself pointed out, cybernetics can be associated with the postwar automation of production via the feedback loops and servomechanisms that are crucial to the functioning of industrial robots. The sense of "cybernetics" is often also broadened to include anything to do with computerization and the "rationalization" of the factory floor. The ugly word "cybernation" found its way into popular discourse in the 1960s as part of the critique of intensified control of workers by management. Again, there is something to this critique (see Noble 1986), but I do not think that such guilt by association should lead us to condemn cybernetics out of hand.[7] We will, in fact, have the opportunity to examine Stafford Beer's management cybernetics at length. The force of the critique turns out to be unclear, to say the least, and we will see how, in Beer's hands, management cybernetics ran into a form of politics that the critics would probably find congenial. And I should reemphasize that my concern here is with the whole range of cybernetic projects. In our world, *any* form of knowledge and practice that looks remotely useful is liable to taken up by the military and capital for their own ends, but by the end of this book it should be abundantly clear that military and industrial applications come nowhere close to exhausting the range of cybernetics.

Finally, there is a critique pitched at a more general level and directed at cybernetics' concern with "control." From a political angle, this is the key topic we need to think about, and also the least well understood aspect of the branch of cybernetics that this book is about. To get to grips with it properly requires a discussion of the peculiar ontological vision of the world that I associate with cybernetics. This is the topic of the next chapter, at the end of which we can return to the question of the political valence of cybernetics and of why this book has the subtitle it does.

— — — — —

The rest of the book goes as follows: Chapter 2 is a second introductory chapter, exploring the strange ontology that British cybernetics played out, and concluding, as just mentioned, with a discussion of the way in which we can see this ontology as political, in a very general sense.

Chapters 3–7 are the empirical heart of the book. The chapters that make up part 1—on Walter, Ashby, Bateson, and Laing—are centrally concerned with the brain, the self, and psychiatry, though they shoot off in many other directions too. Part 2 comprises chapters on Beer and Pask and the directions in which their work carried them beyond the brain. The main concern of each of these chapters is with the work of the named individuals, but each chapter also includes some discussion of related projects that serve to broaden the field of exploration. One rationale for this is that the book is intended more as an exploration of cybernetics in action than as collective biography, and I am interested in perspicuous instances wherever I can find them. Some of these instances serve to thicken up the connections between cybernetics and the sixties that I talked about above. Others connect historical work in cybernetics to important developments in the present in a whole variety of fields. One object here is to answer the question: what happened to cybernetics? The field is not much discussed these days, and the temptation is to assume that it died of some fatal flaw. In fact, it is alive and well and living under a lot of other names. This is important to me. My interest in cybernetics is not purely historical. As I said, I am inclined to see the projects discussed here as models for future practice, and, though they may be odd, it is nice to be reassured that they are not a priori ridiculous. Also, unlike their cybernetic predecessors, the contemporary projects we will be looking at are fragmented; their interrelations are not obvious, even to their practitioners. Aligning them with a cybernetic lineage is a way of trying to foreground such interrelations in the present—to produce a world.

The last chapter, chapter 8, seeks to summarize what has gone before in a novel way, by pulling together various cross-cutting themes that surface in different ways in some or all of the preceding chapters. More important, it takes further the thought that the history of cybernetics might help us imagine a future different from the grim visions of today.

2

ONTOLOGICAL THEATER

OUR TERRESTRIAL WORLD IS GROSSLY BIMODAL IN ITS FORMS: EITHER THE FORMS IN IT ARE EXTREMELY SIMPLE, LIKE THE RUN-DOWN CLOCK, SO THAT WE DISMISS THEM CONTEMPTUOUSLY, OR THEY ARE EXTREMELY COMPLEX, SO THAT WE THINK OF THEM AS BEING QUITE DIFFERENT, AND SAY THEY HAVE LIFE.

ROSS ASHBY, *DESIGN FOR A BRAIN* (1960, 231–32)

In the previous chapter I approached cybernetics from an anthropological angle—sketching out some features of a strange tribe and its interesting practices and projects, close to us in time and space yet somehow different and largely forgotten. The following chapters can likewise be read in an anthropological spirit, as filling in more of this picture. It is, I hope, a good story. But more can be said about the substance of cybernetics before we get into details. I have so far described cybernetics as a science of the adaptive brain, which is right but not enough. To set the scene for what follows we need a broader perspective if we are to see how the different pieces fit together and what they add up to. To provide that, I want to talk now about *ontology*: questions of what the world is like, what sort of entities populate it, how they engage with one another. What I want to suggest is that the ontology of cybernetics is a strange and unfamiliar one, very different from that of the modern sciences. I also want to suggest that *ontology makes a difference*—that the strangeness of specific cybernetic projects hangs together with the strangeness of its ontology.[1]

A good place to start is with Bruno Latour's (1993) schematic but insightful story of modernity. His argument is that modernity is coextensive with a certain *dualism* of people and things; that key features of the modern West can be traced back to dichotomous patterns of thought which are now institutionalized in our schools and universities. The natural sciences speak of a world of things (such as chemical elements and quarks) from which people are absent, while the social sciences speak of a distinctly human realm in which objects, if not entirely absent, are at least marginalized (one speaks of the "meaning" of "quarks" rather than quarks in themselves). Our key institutions for the production and transmission of knowledge thus stage for us a dualist ontology: they teach us how to think of the world that way, and also provide us with the resources for acting as if the world were that way.[2]

Against this backdrop, cybernetics inevitably appears odd and *nonmodern*, to use Latour's word. At the most obvious level, synthetic brains—machines like the tortoise and the homeostat—threaten the modern boundary between mind and matter, creating a breach in which engineering, say, can spill over into psychology, and vice versa. Cybernetics thus stages for us a nonmodern ontology in which people and things are not so different after all. The subtitle of Wiener's foundational book, *Control and Communication in the Animal and the Machine*, already moves in this direction, and much of the fascination with cybernetics derives from this challenge to modernity. In the academic world, it is precisely scholars who feel the shortcomings of the modern disciplines who are attracted most to the image of the "cyborg"—the cybernetic organism—as a nonmodern unit of analysis (with Haraway 1985 as a key text).

This nonmodern, nondualist quality of cybernetics will be evident in the pages to follow, but it is not the only aspect of the unfamiliarity of cybernetic ontology that we need to pay attention to. Another comes under the heading of time and temporality. One could crudely say that the modern sciences are sciences of pushes and pulls: something already identifiably present *causes* things to happen this way or that in the natural or social world. Less crudely, perhaps, the ambition is one of *prediction*—the achievement of general knowledge that will enable us to calculate (or, retrospectively, explain) why things in the world go this way or that. As we will see, however, the cybernetic vision was not one of pushes and pulls; it was, instead, of forward-looking search. What determined the behavior of a tortoise when set down in the world was not any presently existing cause; it was whatever the tortoise found there. So cybernetics stages for us a vision not of a world characterized by graspable causes, but rather of one in which reality is always "in the making," to borrow a phrase from William James.

We could say, then, that the ontology of cybernetics was nonmodern in two ways: in its refusal of a dualist split between people and things, and in an evolutionary, rather than causal and calculable, grasp of temporal process. But we can go still further into this question of ontology. My own curiosity about such matters grew out of my book *The Mangle of Practice* (1995). The analysis of scientific practice that I developed there itself pointed to the strange ontological features just mentioned: I argued that we needed a nondualist analysis of scientific practice ("posthumanist" was the word I used); that the picture should be a forward-looking evolutionary one ("temporal emergence"); and that, in fact, one should understand these two features as constitutively intertwined: the reciprocal coupling of people and things happens in time, in a process that I called, for want of a better word, "mangling." But upstream of those ideas, so to speak, was a contrast between what I called the representational and performative idioms for thinking about science. The former understands science as, above all, a body of representations of reality, while the latter, for which I argued in *The Mangle*, suggests that we should start from an understanding of science as a mode of performative engagement with the world. Developing this thought will help us see more clearly how cybernetics departed from the modern sciences.[3]

— — — — —

WHAT IS BEING SUGGESTED NOW IS NOT THAT BLACK BOXES BEHAVE SOMEWHAT LIKE REAL OBJECTS BUT THAT THE REAL OBJECTS ARE IN FACT ALL BLACK BOXES, AND THAT WE HAVE IN FACT BEEN OPERATING WITH BLACK BOXES ALL OUR LIVES.

ROSS ASHBY, *AN INTRODUCTION TO CYBERNETICS* (1956, 110)

Ross Ashby devoted the longest chapter of his 1956 textbook, *An Introduction to Cybernetics*, to "the Black Box" (chap. 6), on which he had this to say (86): "The problem of the Black Box arose in electrical engineering. The engineer is given a sealed box that has terminals for input, to which he may bring any voltages, shocks, or other disturbances, he pleases, and terminals for output from which he may observe what he can." The Black Box was a key concept in the early development of cybernetics, and much of what I need to say here can be articulated in relation to it. The first point to note is that Ashby emphasized the ubiquity of such entities. This passage continues with a list of examples of people trying to get to grips with Black Boxes: an engineer faced with "a secret and sealed bomb-sight" that is not working properly, a clinician studying a brain-damaged patient; a psychologist studying a rat in a maze. Ashby then remarks, "I need not give further examples as they are to be found everywhere. . . . Black Box theory is, however, even wider in its application than these professional studies," and he gives a deliberately mundane example: "The child who tries to open a door has to manipulate the handle (the input) so as to produce the desired movement at the latch (the output); and he has to learn how to control the one by the other without being able to see the internal mechanism that links them. In our daily lives we are confronted at every turn with systems whose internal mechanisms are not fully open to inspection, and which must be treated by the methods appropriate to the Black Box" (Ashby 1956, 86). On Ashby's account, then, Black Boxes are a ubiquitous and even universal feature of the makeup of the world. We could say that his cybernetics assumed and elaborated a *Black Box ontology*, and this is what we need to explore further.

Next we can note that Black Box ontology is a performative image of the world. A Black Box is something that *does something*, that one does something to, and that does something back—a partner in, as I would say, a dance of agency (Pickering 1995). Knowledge of its workings, on the other hand, is *not* intrinsic to the conception of a Black Box—it is something that may (or may not) grow out of our performative experience of the box. We could also note that there is something right about this ontology. We are indeed enveloped by lively systems that act and react to our doings, ranging from our fellow humans through plants and animals to machines and inanimate matter, and one can readily reverse the order of this list and say that inanimate matter is itself also enveloped by lively systems, some human but most nonhuman. The world just is that way.

A Black Box ontology thus seems entirely reasonable. But having recognized this, at least two *stances* in the world of Black Boxes, ways of going on in the world, become apparent. One is the stance of modern science, namely, a refusal to take Black Boxes for what they are, a determination to strip away their casings and to understand their inner workings in a representational fashion. All of the scientist's laws of nature aim to make this or that Black Box (or class of Black Boxes) transparent to our understanding. This stance is so familiar that I, at least, used to find it impossible to imagine any alternative to it. And yet, as will become clear, from the perspective of cybernetics it can be seen as entailing a *detour*, away from performance and through the space of representation, which has the effect of *veiling* the world of performance from us. The modern sciences invite us to imagine that our relation to the world is basically a cognitive one—we act in the world through our knowledge of it—and that, conversely, the world is just such a place that can be known through the methods and in the idiom of the modern sciences. One could say that the modern sciences stage for us a modern ontology of the world as a knowable and representable place. And, at the same time, the product of the modern sciences, scientific knowledge itself, enforces this vision. Theoretical physics tells us about the unvarying properties of hidden entities like quarks or strings and is silent about the performances of scientists, instruments, and nature from which such representations emerge. This is what I mean by veiling: the performative aspects of our being are unrepresentable in the idiom of the modern sciences.[4]

The force of these remarks should be clearer if we turn to cybernetics. Though I will qualify this remark below, I can say for the moment that the hallmark of cybernetics was a refusal of the detour through knowledge—or, to put it another way, a conviction that in important instances such a detour would be mistaken, unnecessary, or impossible in principle. The stance of cybernetics was a concern with performance *as performance*, not as a pale shadow of representation. And to see what this means, it is perhaps simplest to think about early cybernetic machines. One could, for example, imagine a highly sophisticated thermostat that integrated sensor readings to form a representation of the thermal environment and then transmitted instructions to the heating system based upon computational transformations of that representation (in fact, Ashby indeed imagined such a device: see chap. 4). But my thermostat at home does no such thing. It simply reacts directly and performatively to its own ambient temperature, turning the heat down if the temperature goes up and vice versa.[5] And the same can be said about more sophisticated cybernetic devices. The tortoises engaged directly, performatively and nonrepresentationally, with the environments in which they found themselves, and so did the homeostat. Hence the idea expressed in the previous chapter, that tortoises and homeostats modelled the performative rather than the cognitive brain.

So what? I want to say that cybernetics drew back the veil the modern sciences cast over the performative aspects of the world, including our own being. Early cybernetic machines confront us, instead, with interesting and engaged material performances that do not entail a detour through knowledge. The phrase that runs through my mind at this point is *ontological theater*. I want to say that cybernetics *staged* a nonmodern ontology for us in a double sense. Contemplation of thermostats, tortoises, and homeostats helps us, first, to grasp the ontological vision more generally, a vision of the world as a place of continuing interlinked performances. We could think of the tortoise, say, exploring its world as a little model of what the world is like in general, an *ontological icon*. Going in the other direction, if one grasps this ontological vision, then building tortoises and homeostats stages for us examples of how it might be brought down to earth and played out in practice, as robotics, brain science, psychiatry, and so on. The many cybernetic projects we will examine can all stand as ontological theater in this double sense: as aids to our ontological imagination, and as instances of the sort of endeavors that might go with a nonmodern imagining of the world.[6]

This modern/nonmodern contrast is a key point for all that follows. I want in particular to show that the consistent thread that ran through the history of British cybernetics was the nonmodern performative ontology I have just sketched out. All of the oddity and fascination of this work hangs together with this unfamiliar vision of the sort of place the world is. And I can immediately add a corollary to that observation. In what follows, I am interested in cybernetics as ontological theater in both of the senses just laid out—as both an aid to our imaginations and as exemplification of the fact that, as I said earlier, ontology makes a difference. I want to show that how we imagine the world and how we act in it reciprocally inform one another. Cybernetic projects, in whatever field, look very different from their modern cognates.

From here we can proceed in several directions. I turn first to the "so what?" question; then we can go into some important nuances; finally, we can go back to the critique of cybernetics and the politics of ontology.

THE ESSENCE OF LIFE IS ITS CONTINUOUSLY CHANGING CHARACTER; BUT OUR CONCEPTS ARE ALL DISCONTINUOUS AND FIXED, . . . AND YOU CAN NO MORE DIP UP THE SUBSTANCE OF REALITY WITH THEM THAN YOU CAN DIP UP WATER WITH A NET, HOWEVER FINELY MESHED.

WILLIAM JAMES, "BERGSON AND INTELLECTUALISM" (1943 [1909, 1912], 253)

Why should we be interested in cybernetics? Haven't modern science and engineering served us well enough over the past few hundred years? Of course, their achievements have been prodigious. But I can still think of a few reasons why it might be interesting and useful to try understanding the world in a different way:

1. It is an exercise in mental gymnastics: the White Queen (or whoever it was) imagining a dozen impossible things before breakfast. Some of us find it

fun to find new ways to think, and sometimes it leads somewhere (Feyerabend 1993).

2. Perhaps modern science has succeeded too well. It has become difficult for us to recognize that much of our being does not have a cognitive and representational aspect. I suppose I could figure out how my doorknob works, but I don't need to. I established a satisfactory performative relation with doorknobs long before I started trying to figure out mechanisms. A science that helped us thematize performance as prior to representation might help us get those aspects of our being into focus. And, of course, beyond the human realm, most of what exists does not have the cognitive detour as an option. It would be good to be able to think explicitly about performative relations between things, too.

3. Perhaps there would be positive fruits from this move beyond the representationalism of modern science. In engineering, the thermostat, the tortoise, the homeostat, and the other nonmodern cybernetic projects we will be looking at all point in this direction.

4. Perhaps in succeeding too well, modern science has, in effect, blinded us to all of those aspects of the world which it fails to get much grip upon. I remember as a physicist trying to figure out why quarks always remained bound to one another and reflecting at the same time that none of us could calculate in any detail how water flowed out of a tap. Contemporary complexity theorists like to argue that the methods of modern science work nicely for a finite class of "linear" systems but fail for "nonlinear" systems—and that actually the latter are in some sense most of the world. Stafford Beer foreshadowed this argument in his first book, *Cybernetics and Management*, where he argued that we could think of the world as built from three different kinds of entities or systems (Beer 1959, 18). We can go into this in more detail in chapter 6, but, briefly, Beer referred to these as "simple," "complex," and "exceedingly complex" systems. The first two kinds, according to Beer, are in principle knowable and predictable and thus susceptible to the methods of modern science and engineering. Exceedingly complex systems, however, are not. They are systems that are so complex that we can never fully grasp them representationally and that change in time, so that present knowledge is anyway no guarantee of future behavior. Cybernetics, on Beer's definition, was the science of exceedingly complex systems that modern science can never quite grasp.

I will come back repeatedly to Beer's idea of exceedingly complex systems as we go along, and try to put more flesh on it. This is the aspect of cybernetics that interests me most: the aspect that assumes an *ontology of unknowability*, as one might call it, and tries to address the problematic of getting along performatively with systems that can always surprise us (and this takes us back to the adaptive brain, and, again, to nonhuman systems that do not have the option of the cognitive detour). If there are examples of Beer's exceedingly complex systems to be found in the world, then a nonmodern approach that recognizes this (rather than, or as well as, a modern one that denies it) might be valuable. It is not easy, of course, to say where the dividing line between aspects of the world that are "exceedingly complex" rather than just very complicated is to be drawn. Modern science implicitly assumes that everything in the world will eventually be assimilated to its representational schema, but the time horizon is infinite. Here and now, therefore, a cybernetic stance might be appropriate in many instances. This is where the intellectual gymnastics get serious, and where the history of cybernetics might be needed most as an aid to the imagination.

5. I may as well note that my interest in cybernetics stems originally from a conviction that there is indeed something *right* about its ontology, especially the ontology of unknowability just mentioned. As I said earlier, I arrived at something very like it through my empirical studies in the history of modern science, though the substance of scientific knowledge speaks to us of a different ontology. I lacked the vocabulary, but I might have described modern science as a complex adaptive system, performatively coming to terms with an always-surprising world. At the time, I thought of this as a purely theoretical conclusion. When pressed about its practical implications, I could not find much to say: modern science seems to get on pretty well, even as it obscures (to my way of thinking) its own ontological condition.[7] The history of cybernetics, however, has helped me to see that theory, even at the level of ontology, can return to earth. Cybernetic projects point to the possibility of novel and distinctive constructive work that takes seriously a nonmodern ontology in all sorts of fields. They show, from my perspective, where the mangle might take us. And one further remark is worth making. Theory is not enough. One cannot *deduce* the homeostat, or Laing's psychiatry, or Pask's Musicolour machine from the cybernetic ontology or the mangle. The specific projects are not somehow already present in the ontological vision. In each instance creative work is needed; something has to be *added* to the ontological vision to specify it and pin it down. That is why we need to be interested in particular manifestations of cybernetics as well as ontological imaginings. That is how, from my point of view, cybernetics carries us beyond the mangle.[8]

Now for the nuances. First, knowledge. The discussion thus far has emphasized the performative aspect of cybernetics, but it is important to recognize that cybernetics was not simply and straightforwardly antirepresentational. Representational models of a firm's economic environment, for example, were a key part of Beer's viable system model (VSM) of the organization. Once one sees that, the clean split I have made between cybernetics and modern science threatens to blur, but I think it is worth maintaining. On the one hand, I want to note that many cybernetic projects did not have this representational aspect. The great advantage that Beer saw in biological computing was that it was immediately performative, involving no detours through the space of representation. On the other hand, when representations did appear in cybernetic projects, as in the VSM, they figured as immediately geared into performance, as revisable guides to future performance rather than as ends in themselves. Beer valued knowledge, but he was also intensely suspicious of it—especially of our tendency to mistake representations for the world, and to cling to specific representations at the expense of performance. We might thus think of cybernetics as staging for us a *performative epistemology*, directly engaged with its performative ontology—a vision of knowledge as *part of* performance rather than as an external controller of it. This is also, as it happens, the appreciation of knowledge that I documented and argued for in *The Mangle*.

— — — — —

Now that we have these two philosophical terms on the table—ontology and epistemology—I can say more about my own role in this history. In chapter 1 I said that anyone can have their own history of cybernetics, and this one is mine. I picked the cast of characters and which aspects of their work to dwell upon. But beyond that, the emphasis on ontology is more mine than the cyberneticians'. It is the best way I have found to grasp what is most unfamiliar and valuable about cybernetics, but the fact is that the word "ontology" does not figure prominently in the cybernetic literature. What I call the cybernetic ontology tends to be simply taken for granted in the literature or not labeled as such, while "epistemology" is often explicitly discussed and has come increasingly to the fore over time. Contemporary cyberneticians usually make a distinction between "first-order" cybernetics (Walter and Ashby, say) and "second-order" cybernetics (Bateson, Beer, and Pask), which is often phrased as the difference between the cybernetics of "observed" and "observing" systems, respectively. Second-order cybernetics, that is, seeks to recognize that the scientific observer is part of the system to be studied, and this in turn leads to a recognition that the observer is *situated* and sees the world from a certain perspective, rather than achieving a detached and omniscient "view from nowhere." Situated knowledge is a puzzling and difficult concept, and hence follows an intensified interest in the problematic of knowledge and epistemology.

What should I say about this? First, I take the cybernetic emphasis on epistemology to be a symptom of the dominance of specifically epistemological inquiry in philosophy of science in the second half of the twentieth century, associated with the so-called linguistic turn in the humanities and social sciences, a dualist insistence that while we have access to our own words, language, and representations, we have no access to things in themselves. Cybernetics thus grew up in a world where epistemology was the thing, and ontology talk was verboten. Second, my own field, science studies, grew in that same matrix, but my own research in science studies has convinced me that we need to undo the linguistic turn and all its works. The shift from a representational to a performative idiom for thinking about science, and from epistemology alone to ontology as well, is the best way I have found to get to grips with the problematic of situated knowledge (and much else).

So I think that second-order cybernetics has talked itself into a corner in its intensified emphasis on epistemology, and this book could therefore be read as an attempt to talk my way out of the trap. Again, of course, the "so what?" question comes up. Words are cheap; what does it matter if I use the word "ontology" more than the cyberneticians? Actually—though it is not my reason for writing the book—something might be at stake. Like ontology itself, ontology talk might make a difference. How one conceives a field hangs together with its research agendas. To see cybernetics as being primarily about epistemology is to invite endless agonizing about the observer's personal responsibility for his or her knowledge claims. Fine. But the other side of this is the disappearance of the performative materiality of the field. All of those wonderful machines, instruments, and artifacts get marginalized if one takes cybernetics to be primarily about knowledge and the situatedness of the observer. Tortoises, homeostats, biological computers, Musicolour machines, adaptive architecture—all of these are just history as far as second-order cybernetics is concerned. We used to do things like that in our youth; now we do serious epistemology.

Evidently, I think this position is a mistake. I am interested in cybernetics as the field that brought nonmodern ontology down to earth, and played it out and staged it for us in real projects. I think we need more of this kind of thing, not less. I did not make the history up; I don't have enough imagination; it has taken me years to find it out and struggle with it. But the chapters that follow invite, in effect, a redirection of cybernetics. I think the field might be far more lively and important in the future if it paid attention to my description of its past.

— — — — —

Now for the trickiest point in this chapter. I began with Black Boxes and the differing stances toward them of modern science and cybernetics: the former seeking to open them up; the latter imagining a world of performances in which they remained closed. This distinction works nicely if we want to think about the work of the second-generation cyberneticians, Beer and Pask, and also Bateson and Laing. Nothing more needs to be said here to introduce them. But it works less well for the first generation, Walter and Ashby, and this point needs some clarification.

I quoted Ashby earlier defining the problematic of the Black Box in terms of an engineer probing the Box with electrical inputs and and observing its outputs. Unfortunately for the simplicity of my story, the quotation continues: "He is to deduce what he can of its contents." This "deduction" is, needless to say, the hallmark of the modern scientific stance, the impulse to open the box, and a whole wing of Ashby's cybernetics (and that of his students at Illinois in the 1960s) circled around this problematic. Here I am tempted to invoke the author's privilege and say that I am not going to go into this work in any detail in what follows. While technically fascinating, it does not engage much with the ontological concerns which inform this book. But it is not so easy to get off this hook. Besides a general interest in opening Black Boxes, Ashby (and Walter) wanted to open up one particular Black Box, the brain, and it is impossible to avoid a discussion of that specific project here—it was too central to the development of cybernetics.[9] I need to observe the following:

Seen from one angle, the tortoise and the homeostat function well as nonmodern ontological theater. These machines interacted with and adapted to their worlds performatively, without any representational detours; their worlds remained unknowable Black Boxes to the machines. This is the picture I want to contemplate. But from another angle, Walter and Ashby remained securely within the space of modern science. As brain scientists, they wanted to open up the brain to our representational understanding by a classically scientific maneuver—building models of its interior. These models were unusual in that they took the form of machines rather than equations on paper, but their impulse was the same: precisely to get inside the Black Box and to illuminate the inner go of the adaptive brain.

What should we make of this? Clearly, this branch of cybernetics was a hybrid of the modern and the nonmodern, staging very different acts of ontological theater depending on the angle one watched them from. I could therefore say that the invitation in what follows is to look at them from the nonmodern angle, since this is the aspect of our imagination most in need of stimulation. But, as we will see in more detail later, it is, in fact, also instructive to look more closely at them from the modern angle too. We can distinguish at least three aspects in which Walter and Ashby's cybernetics in fact departed from the paradigms of modern science.

First, sciences like physics describe a homogeneous field of entities and forces that lacks any outside—a cosmos of point masses interacting via an inverse-square law, say. Cybernetic brain modelling, in contrast, immediately entailed an external other—the unknown world to which the brain adapts. So even if early cybernetic brain models can be placed in a modern lineage, they necessarily carried with them this reference to performative engagement with the unknown, and this is what I will focus on in the following chapters.

Second, we can think not about the outside but about the inside of cybernetic brain models. The tortoise and the homeostat were instances of what theoretical biologist Stuart Kauffman (1971) called "articulation of parts explanation."[10] Kauffman's examples of this were taken from work in developmental biology in which one appeals to the properties of single cells, say, to explain morphogenesis at a higher level of cellular aggregation. Ashby's and Walter's brain models had just this quality, integrating atomic parts—valves, capacitors, and so on—to achieve higher-level behavior: adaptation. This is a very different style of explanation from that of modern physics, which aims at a calculable representation of some uniform domain—charged particles responding identically to an electric field, for example. And it is worth noting that articulation of parts explanation immediately thematizes performance. One is more concerned with what entities *do* than what they *are*. Ashby and Walter were not exploring the properties of relays and triodes; they were interested in how they would behave in combination. From this angle, too, cybernetic brain modelling once more dramatized performative engagement, now within the brain.

And third, we can take this line of thought further. This is the place to mention what I think of as a cybernetic *discovery of complexity*. At an "atomic" level, Walter and Ashby understood their machines very well. The individual components were simple and well-understood circuit elements—resistors, capacitors, valves, relays, some wires to make the connections. But the discovery of complexity was that such knowledge is not enough when it comes to understanding aggregate behavior; that explanation by articulation of parts is not as straightforward as one might imagine; that especially—and in contrast to paradigmatic instances of modern science—*prediction* of overall performance on the basis of an atomic understanding can be difficult to the point of impossibility. Walter reported that he was surprised by the behavior of his tortoises. Ashby was baffled and frustrated by the homeostat's successor—a machine called DAMS—so much so that he eventually abandoned the DAMS project as a failure. We could say, therefore, that Walter and Ashby both discovered in their scientific attack on the brain that even rather simple systems can be, at the same time, exceedingly complex systems in Beer's terms. Beer's favorite examples of such systems were the brain itself, the firm, and the economy, but even Ashby's and Walter's little models of the brain fell into this class, too.

Two observations follow. First, despite the modern scientific impulse behind their construction, we could take the tortoise and, especially, DAMS as themselves instances of ontological theater, in a somewhat different sense from that laid out above. We could, that is, try to imagine the world as populated by entities like the tortoise and DAMS, whose behavior we can never fully predict. This is another way in which the modern scientific approach to the brain of Walter and Ashby in effect turns back into a further elaboration of the nonmodern ontology that this book focuses upon. It is also a rephrasing of my earlier remark on hybridity. Seen from one end of the telescope, the cybernetic brain models shed genuinely scientific light on the brain—in adapting to their environment, they represented an advance in getting to grips with the inner go of the brain itself. But seen from the other end, they help us imagine what an exceedingly complex system is. If "toys" like these, to borrow Walter's description of them, can surprise us, the cybernetic ontology of unknowability seems less mysterious, and cybernetic projects make more sense.

Continuing with this line of thought, in chapter 4 we can follow one line of Ashby's work into the mathematical researches of Stuart Kauffman and Stephen Wolfram. I just mentioned some important philosophical work by Kauffman, but at issue here is another aspect of his theoretical biology. In computer simulations of complex systems in the late 1960s, Kauffman came across the emergence of simple structures having their own dynamics, which he could interfere with but not control. These systems, too, might help give substance to our ontological imaginations. In understanding the work of Bateson, Laing, Beer, and Pask, the idea of performative interaction with systems that are not just unknowable but that also have their own inner dynamics—that go their own way—is crucial. Wiener derived the word "cybernetics" from the Greek for "steersman"; Pask once compared managing a factory with sailing a ship (chap. 7); and the sense of sailing we will need in later chapters is just that of participating performatively in (rather than representationally computing) the dynamics of sails, winds, rudders, tides, waves, and what have you.

The motto of Wolfram's *New Kind of Science* (2002) is "extremely complex behaviour from extremely simple systems," and this is precisely the phrase that goes with the earlier cybernetic discovery of complexity. Whereas the cyberneticians built machines, Wolfram's work derives from experimentation with very simple formal mathematical systems called cellular automata (CAs). And Wolfram's discovery has been that under the simplest of rules, the time evolution of CAs can be ungraspably complex—the only way to know what such a system will do is set it in motion and watch. Again we have the idea of an unpredictable endogenous dynamics, and Wolfram's CAs can thus also function as ontological theater for us in what follows—little models of the fundamental entities of a cybernetic ontology. In their brute unpredictability, they conjure up for us what one might call an *ontology of becoming*. Much of the work to be discussed here had as its problematic questions of how to go on in such a world.

Again, in the case of Kauffman and Wolfram, a certain ontological hybridity is evident. In classically modern fashion, Wolfram would like to know which CA the world is running. The recommendation here is to look through the other end of the telescope—or pick up the other end of the stick—and focus on the literally unpredictable properties of mathematical systems like these as a way of imagining more generally how the world is.[11]

THE FACT IS THAT OUR WHOLE CONCEPT OF CONTROL IS NAIVE, PRIMITIVE AND RIDDEN WITH AN ALMOST RETRIBUTIVE IDEA OF CAUSALITY. CONTROL TO MOST PEOPLE (AND WHAT A REFLECTION THIS IS UPON A SOPHISTICATED SOCIETY!) IS A CRUDE PROCESS OF COERCION.

STAFFORD BEER, *CYBERNETICS AND MANAGEMENT* (1959, 21)

MODERN SCIENCE'S WAY OF REPRESENTING PURSUES AND ENTRAPS NATURE AS A CALCULABLE COHERENCE OF FORCES. . . . PHYSICS. . . SETS NATURE UP TO EXHIBIT ITSELF AS A COHERENCE OF FORCES CALCULABLE IN ADVANCE.

MARTIN HEIDEGGER, "THE QUESTION CONCERNING TECHNOLOGY" (1976 [1954], 302-3)

WE HAVE TO LEARN TO LIVE ON PLANETARY SURFACES AND BEND WHAT WE FIND THERE TO OUR WILL.

NASA ADMINISTRATOR, *NEW YORK TIMES*, 10 DECEMBER 2006

I want to conclude this chapter by thinking about cybernetics as politics, and to do so we can pick up a thread that I left hanging in the previous chapter. There I ran through some of the critiques of cybernetics and indicated lines of possible response. We are now in a position to consider one final example. Beyond the specifics of its historical applications, much of the suspicion of cybernetics seems to center on just one word: "control." Wiener defined the field as the science of "control and communication," the word "control" is everywhere in the cybernetics literature, and those of us who have a fondness for human liberty react against that. There are more than enough controls imposed on us already; we don't want a science to back them up and make them more effective.

The cyberneticians, especially Stafford Beer, struggled with this moral and political condemnation of their science, and I can indicate the line of response. We need to think about possible meanings of "control." The objectionable sense is surely that of control as *domination*—the specter of Big Brother watching and controlling one's every move—people reduced to automata. Actually, if this vision of control can be associated with any of the sciences, it should be the modern ones. Though the word is not much used there, these are Deleuze and Guattari's royal sciences, aligned with the established order, that aspire to grasp the inner workings of the world through knowledge and thus to dominate it and put it entirely at our disposal. Beyond the natural sciences, an explicit ambition of much U.S. social science throughout the twentieth century was "social engineering." Heidegger's (1976 [1954]) understanding of the sciences as integral to a project of *enframing* and subjugation comes to mind. And the point I need to stress is that the cybernetic image of control was *not like that*.

Just as Laingian psychiatry was sometimes described as antipsychiatry, the British cyberneticians, at least, might have been rhetorically well advised to describe themselves as being in the business of *anticontrol*. And to see what that means, we have only to refer back to the preceding discussion of ontology. If cybernetics staged an ontology in which the fundamental entities were dynamic systems evolving and becoming in unpredictable ways, it could hardly have been in the business of Big Brother–style domination and enframing. It follows immediately from this vision of the world that enframing will fail. The entire task of cybernetics was to figure out how to get along in a world that was not enframable, that could not be subjugated to human designs—how to build machines and construct systems that could adapt performatively to whatever happened to come their way. A key aspect of many of the examples we will examine was that of open-ended search—of systems that would explore their world to see what it had to offer, good and bad. This, to borrow another word from Heidegger, is a stance of *revealing* rather than enframing—of openness to possibility, rather than a closed determination to achieve some preconceived object, come what may (though obviously this assertion will need to be nuanced as we go along). This is the ontological sense in which cybernetics appears as one of Deleuze and Guattari's nomad sciences that upset established orders.

One theme that will emerge from the chapter on Ashby onward, for example, is that of a distinctly cybernetic notion of *design*, very different from that more familiar in modern science and engineering. If our usual notion of design entails the formulation of a plan which is then imposed upon matter, the cybernetic approach entailed instead a continuing interaction with materials, human and nonhuman, to explore what might be achieved—what one might call an *evolutionary* approach to design, that necessarily entailed a degree of *respect* for the other.

Readers can decide for themselves, but my feeling is, therefore, that the critique of cybernetics that centers on the word "control" is importantly misdirected. British cybernetics was not a scientized adjunct of Big Brother. In fact, as I said, the critique might be better redirected toward modernity rather than cybernetics, and this brings us to the question of ontological politics. The period in which I have been writing this book has not been a happy one, and the future looks increasingly grim. In our dealings with nature, 150 years of the enframing of the Mississippi by the U.S. Army Corps of Engineers came to a (temporary) end in 2005 with Hurricane Katrina, the flooding of New Orleans, many deaths, massive destruction of property, and the displacement of hundreds of thousands of people.[12] In our dealings with each other, the United States's attempt to enframe Iraq—the installation of "freedom and democracy"—became another continuing disaster of murder, mayhem, and torture.

In one of his last public appearances, Stafford Beer (2004 [2001], 853) argued, "Last month [September 2001], the tragic events in New York, as cybernetically interpreted, look quite different from the interpretation supplied by world leaders—and therefore the strategies now pursued are quite mistaken in cybernetic eyes." Perhaps we have gone a bit overboard with the modern idea that we can understand and enframe the world. Perhaps we could do with a few examples before our eyes that could help us imagine and act in the world differently. Such examples are what the following chapters offer. They demonstrate concretely and very variously the possibility of a nonmodern stance in the world, a stance of revealing rather than enframing, that hangs together with an ontology of unknowability and becoming. Hence the invitation to see the following scenes from the history of cybernetics as sketches of another future, models for another way to go on, an invitation elaborated further in chapter 8.

This book is not an argument that modernity must be smashed or that science as we know it should be abandoned. But my hope is that it might do something to weaken the spell that modernity casts over us—to question its hegemony, to destabilize the idea that there is no alternative. Ontological monotheism is not turning out to be a pretty sight.

PART ONE

PSYCHIATRY TO CYBERNETICS

3

GREY WALTER

FROM ELECTROSHOCK TO THE PSYCHEDELIC SIXTIES

THE BRUTE POINT IS THAT A WORKING GOLEM IS . . . PREFERABLE TO TOTAL IGNORANCE. . . . IT IS CLEAR BY NOW THAT THE IMMEDIATE FUTURE OF STUDY IN MODELLING THE BRAIN LIES WITH THE SYNTHESIS OF GADGETS MORE THAN WITH THE ANALYSIS OF DATA.

JEROME LETTVIN, *EMBODIMENTS OF MIND* (1988, VI, VII)

In an obituary for his long-standing friend and colleague, H. W. Shipton described Grey Walter as, "in every sense of the phrase a free thinker [with] contempt for those who followed well paved paths. He was flamboyant, persuasive, iconoclastic and a great admirer of beauty in art, literature, science, and not least, woman" (1977, iii). The historian of science Rhodri Hayward remarks on Walter's "swashbuckling image" as an "emotional adventurer," and on his popular and academic reputation, which ranged from "robotics pioneer, home guard explosives expert, wife swapper, t.v.-pundit, experimental drugs user and skin diver to anarcho-syndicalist champion of leucotomy and electro-convulsive therapy" (2001a, 616). I am interested in Walter the cybernetician, so the swashbuckling will get short shrift, alas.[1]

After an outline of Walter's life and career, I turn to robot-tortoises, exploring their contribution to a science of the performative brain while also showing the ways in which they went beyond that. I discuss the tortoises as ontological

Figure 3.1. Grey Walter. Reproduced from *The Burden: Fifty Years of Clinical and Experimental Neuroscience at the Burden Neurological Institute*, by R. Cooper and J. Bird (Bristol: White Tree Books, 1989), 50. (By permission of White Tree Books, Bristol.)

theater and then explore the social basis of Walter's cybernetics and its modes of transmission. Here we can look toward the present and contemporary work in biologically inspired robotics. A discussion of CORA, a learning module that Walter added to the tortoises, moves the chapter in two directions. One adds epistemology to the ontological picture; the other points to the brutal psychiatric milieu that was a surface of emergence for Walter's cybernetics. The chapter concludes with Walter's interest in strange performances and altered states, and the technologies of the self that elicit them, including flicker and biofeedback. Here we can begin our exploration of crossovers and resonances between cybernetics and the sixties, with reference to William Burroughs, the Beats, and "brainwave music." I also discuss the hylozoist quality of the latter, a theme that reappears in different guises throughout the book.

The ontological hybridity of first-generation cybernetics will be apparent. While we can read Walter's work as thematizing a performative vision of ourselves and the world, the impulse to open up the Black Box of the brain will also be evident. Cybernetics was born in the matrix of modern science, and we can explore that too.

— — — — —

William Grey Walter was born in Kansas City, Missouri, in 1910.[2] His parents were journalists, his father English, his mother Italian-American. The family moved to Britain in 1915, and Walter remained there for the rest of his life. At some stage, in a remarkable coincidence with Ashby, Beer, and Pask, Walter stopped using his first name and was generally known as Grey (some people understood him to have a double-barreled surname: Grey-Walter). He was educated at Westminster School in London and then at King's College Cambridge, where he gained an honors degree in physiology in 1931 and stayed on for four years' postgraduate research on nerve physiology and conditioned reflexes, gaining his MA degree for his dissertation, "Conduction in Nerve and Muscle." His ambition was to obtain a college fellowship, but he failed in that and instead took up a position in the Central Pathological Laboratory of the Maudsley mental hospital in London in 1935, at the invitation of Frederick Golla, the laboratory's director, and with the support of a fellowship from the Rockefeller Foundation.[3]

Golla encouraged Walter to get into the very new field of electroencephalography (EEG), the technique of detecting the electrical activity of the brain, brainwaves, using electrodes attached to the scalp. The possibility of detecting these waves had first been shown by the Jena psychiatrist Hans Berger in 1928 (Borck 2001) but the existence of such phenomena was only demonstrated in Britain in 1934 by Cambridge neurophysiologists E. D. Adrian and B. H. C. Matthews. Adrian and Matthews confirmed the existence of what they called the Berger rhythm, which later became known as the alpha rhythm: an oscillation at around ten cycles per second in electrical potentials within the brain, displayed by all the subjects they examined. The most striking feature of these waves was that they appeared in the brain when the subjects' eyes were shut, but vanished when their eyes were opened (fig. 3.2). Beyond that, Adrian and Matthews found that "the Berger rhythm is disappointingly constant" (Adrian and Matthews 1934, 382). But Walter found ways to take EEG research further. He was something of an electrical engineering genius, designing and building EEG apparatus and frequency analyzers and collaborating with the Ediswan company in the production of commercial equipment, and he quickly made some notable clinical achievements, including the first diagnosis and localization

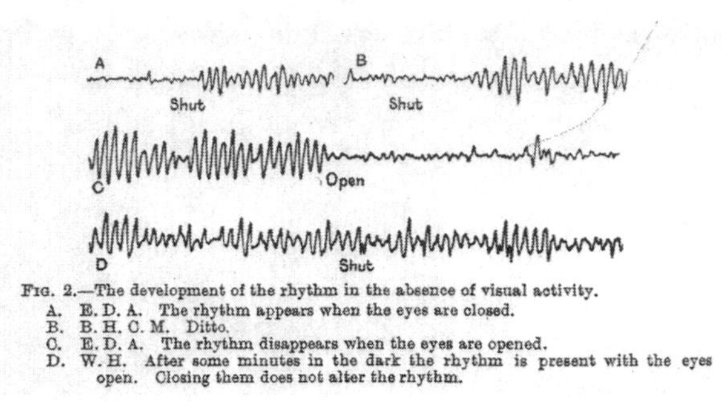

Figure 3.2. Alpha rhythms in the brain, showing the effect of opening and closing the eyes. Source: E. D. Adrian and B. H. C. Matthews, "The Berger Rhythm: Potential Changes from the Occipital Lobes in Man," *Brain*, 57 (1934), 355–85. (By permission of Oxford University Press.)

of a cerebral tumor by EEG, the discovery that a significant proportion of epileptics show unusual brainwaves even between fits, and intervention in a famous murder case (Hayward 2001a, 620).[4] Following his pioneering work, EEG was at the center of Walter's career for the rest of his life. In 1949 he was a cofounder and coeditor of the felicitously titled journal *Electroencephalography and Clinical Neurophysiology* (self-described on its title page as "The EEG Journal") and from 1953 to 1957 he was president of the International Federation of EEG Societies.[5]

In 1939 Walter and Golla moved together to the newly established Burden Neurological Institute near Bristol, with Golla as its first director and Walter as director of its Physiology Department (at annual salaries of £1,500 and £800, respectively). The Burden was a small, private institution devoted to "clinical and experimental neuroscience" (Cooper and Bird 1989), and Walter remained there for the rest of his working life, building a reputation as one of the world's leaders in EEG research and later in research using electrodes implanted in the brain (rather than attached to the scalp).[6] Walter's best-recognized and most lasting contribution to brain science was his discovery in the 1960s of contingent negative variation, the "expectancy wave," a shift in the electrical potential of the brain that precedes the performance of intentional actions. He was awarded the degree of ScD by Cambridge in 1947 and an honorary MD degree by the University of Aix-Marseilles in 1949.

Besides his technical work, in 1953 Walter published an influential popular book on the brain, *The Living Brain*, with a second edition in 1961, and in 1956 he published a novel, *Further Outlook*, retitled *The Curve of the Snowflake* in the United States.[7] He was married twice, from 1934 to 1947 to Katherine

Ratcliffe, with whom he had two children, and from 1947 to 1960 to Vivian Dovey, a radiographer and scientific officer at the Burden, with whom he coauthored papers and had a son. From 1960 to 1974 he lived with Lorraine Aldridge in the wife swap mentioned above (R. Cooper 1993; Hayward 2001a, 628). In 1970 Walter's research career came to an end when he suffered a serious head injury as a result of a collision between the scooter he was riding (at the age of sixty, let us recall) and a runaway horse. He was in a coma for a week, suffered serious brain damage, and never fully recovered. He returned to work at the Burden as a consultant from 1971 until his retirement in 1975 and died suddenly of a heart attack in 1976 (Cooper and Bird 1989, 60).

Walter's most distinctive contribution to cybernetics came in 1948, with the construction of the first of his robot tortoises. He was one of the founders of the Ratio Club, the key social venue for the British cyberneticians, which met from 1949 until 1955 (Clark 2002, chap. 3, app. A1). He was an invited guest at the tenth and last of the U.S. Macy cybernetics conferences in 1953 (Heims 1991, 286), and he was a member of the four-man scientific committee of the first meeting of the European counterpart of the Macys, the 1956 Namur conference—the First International Congress on Cybernetics—where he presided over section IV, devoted to "cybernetics and life."

The Tortoise and the Brain

How might one study the brain? At different stages of his career, Walter pursued three lines of attack. One was a classically reductionist approach, looking at the brain's individual components. Working within a well-established research tradition, in his postgraduate research at Cambridge he explored the electrical properties of individual neurons which together make up the brain. One can indeed make progress this way. It turns out, for example, that neurons have a digital character, firing electrical signals in spikes rather than continuously; they have a certain unresponsive "dead time" after firing; they have a threshold below which they do not respond to incoming spikes; they combine inputs in various ways. But if one is interested in the properties of whole brains, this kind of understanding does not get one very far. A crude estimate would be that the brain contains 10^{10} neurons and many, many more interconnections between them, and no one, even today, knows how to sum the properties of that many elements to understand the behavior of the whole. As Walter put it, "One took an anatomical glance at the brain, and turned away in despair" (1953, 50). We could see this as a simple instance of the problem of complexity which will appear in various guises in this chapter and the next: there exist systems for which an atomic understanding fails to translate into a global one. This is the sense in which the brain counted for Stafford Beer as an exemplary "exceedingly complex system."

Walter's second line of attack emerged on his move to London. His EEG work aimed at mapping the properties of the brain. What does the brain *do*? Well, it emits small but complicated electrical signals that are detectable by sensitive electronic apparatus. Such signals, both oscillatory (waves) and singular, were what Walter devoted his life to studying. This proved to be difficult. Other rhythms of electrical activity—the so-called beta, theta, and delta bands of brainwaves at frequencies both above and below the alphas—were discovered, but EEG readouts revealed the brain to be very noisy, and distinguishing correlations between inputs and outputs was problematic. Echoing the findings of Adrian and Matthews in 1934, Walter (1953, 90) observed that "very few of the factors affecting the spontaneous rhythms were under the observation or control of experimenter or subject. Usually only the effects of opening or closing the eyes, of doing mental arithmetic, of overbreathing and of changes in the blood sugar could be investigated. . . . The range and variety of methods were not comparable with the scope and sensitivity of the organ studied, and the information obtained by them was patchy in the extreme." The electric brain, one could say, proved more complex than the variables in terms of which researchers might hope to map it.[8]

We can return to Walter's EEG work at various points as we go along, but I can enter a couple of preliminary comments on it here. As ontological theater, it evidently stages for us a vision of the brain as a performative organ rather than a cognitive one—an organ that acts (here, emitting electrical signals) rather than thinks. Equally evidently, such a conception of the brain destabilizes any clean dualist split between people and things: the performative brain as just one Black Box to be studied among many.[9] At the same time, though, as we will see shortly, Walter's ambition was always to open up the Black Box, in pursuit of its inner go. This is what I mean by referring to the hybrid quality of his cybernetics.

Walter's third line of attack on the brain was the one that I have talked about before: the classically scientific tactic of building models of the brain. The logic here is simple: if a model can emulate some feature of the system modelled, one has learned something, if only tentatively, about the go of the latter, its inner workings. As Roberto Cordeschi (2002) has shown, one can trace the lineage of this approach in experimental psychology back to the early years of the twentieth century, including, for example, the construction of a phototropic electric dog in 1915. The early years of cybernetics were marked by a proliferation of such models, including the maze-learning robots built by Claude Shannon and R. A. Wallace—which Walter liked to call *Machina labyrinthea*—and Ashby's homeostat (*Machina sopora*) (Walter 1953, 122–23), but we need to focus on the tortoise.[10]

The tortoises (or "turtles") were small electromechanical robots, which Walter also referred to as members of a new inorganic species, *Machina speculatrix*. He built the first two, named Elsie and Elmer, at home in his spare time between Easter of 1948 and Christmas of 1949. In 1951, a technician at the Burden, W. J. Warren—known as Bunny, of course—built six more, to a higher engineering standard (Holland 1996, 2003). The tortoises had two back wheels and one front (fig. 3.3). A battery-powered electric motor drove the front wheel, causing the tortoise to move forward; another motor caused the front forks to rotate on their axis, so the basic state of the tortoise was a kind of cycloidal wandering. If the tortoise hit an obstacle, a contact switch on the body would set the machine into a back and forth oscillation which would usually be enough to get it back into the open. Mounted on the front fork was a photocell. When this detected a source of illumination, the rotation of the front fork would be cut off, so the machine would head toward the light. Above a certain intensity of illumination, however, the rotation of the forks would normally be switched back on, so the life of the tortoise was one of perpetual wanderings up to and away from lights (fig. 3.4). When their batteries were low, however, the tortoises would not lose interest in light sources; instead, they would enter their illuminated hutches and recharge themselves.

The tortoises also executed more complex forms of behavior which derived from the fact that each carried a running light that came on when the tortoise was in search mode and went off when it locked onto a light. The running lights were originally intended simply to signal that a given tortoise was working properly, but they bestowed upon the tortoise an interesting sensitivity to its own kind. It turned out, for example, that a tortoise passing a mirror would be attracted to the reflection of its own light, which light would then be extinguished as the tortoise locked onto its image; the light would then reappear as the scanning rotation of the front wheel set back in, attracting the tortoise's attention again, and so on (fig. 3.5). The tortoise would thus execute a kind of mirror dance, "flickering, twittering and jigging," in front of the mirror, "like a clumsy Narcissus." Likewise, two tortoises encountering one another would repetitively lock onto and then lose interest in one another, executing a mating dance (fig. 3.6) in which "the machines cannot escape from one another; but nor can they ever consummate their 'desire' " (Walter 1953, 128, 129).

Figure 3.3. Anatomy of a tortoise. Source: de Latil 1956, facing p. 50.

So much for the behaviors of the tortoises; now to connect them to the brain. One can analogize a tortoise to a living organism by distinguishing its motor organs (the power supply, motors, and wheels), its senses (the contact switch and the photocell), and its brain (connected to the motor organs and senses by nerves: electrical wiring). The brain itself was a relatively simple piece of circuitry consisting of just two "neurons," as Walter (1950a, 42) put it, each consisting of an electronic valve, a capacitor, and a relay switch (fig. 3.7). In response to different inputs, the relays would switch between different modes of behavior: the basic wandering pattern, locking onto a light, oscillating back and forth after hitting an obstacle, and so on.

What can we say about the tortoise as brain science? First, that it modelled a certain form of adaptive behavior. The tortoise explored its environment and reacted to what it found there, just as all sorts of organisms do—the title of Walter's first publication on the tortoises was "An Imitation of Life" (1950a). The suggestion was thus that the organic brain might contain similar structures to the tortoise's—not valves and relays, of course, but something functionally equivalent. Perhaps, therefore, it might not be necessary to descend to the level of individual neurons to understand the aggregate properties of the brain. This is the sense in which Jerome Lettvin (once a collaborator of Warren McCulloch) could write in 1988 that "a working golem is . . . preferable to total ignorance" (1988, vi). But the tortoises also had another significance for Walter.

The tortoise's method of finding its targets—the continual swiveling of the photocell through 360 degrees—was novel. Walter referred to this as *scanning*, and scanning was, in fact, a topic of great cybernetic interest at the time. The

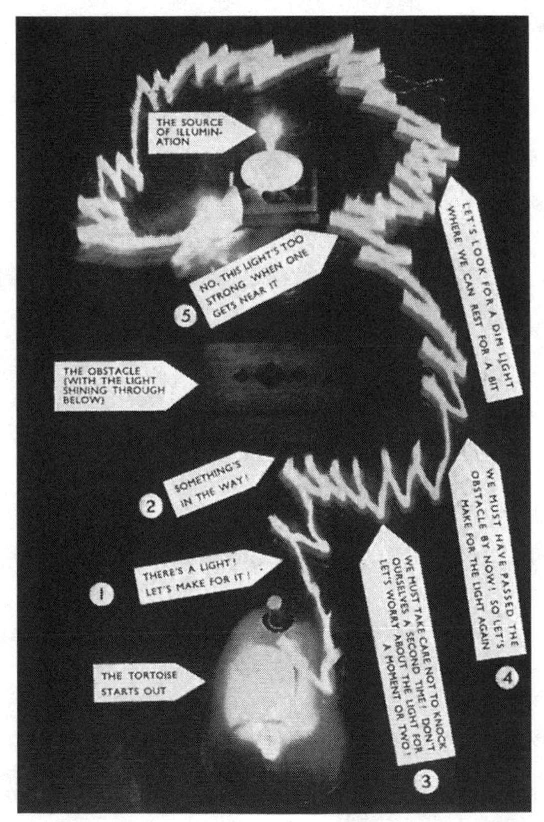

Figure 3.4. The tortoise in action. Source: de Latil 1956, facing p. 275.

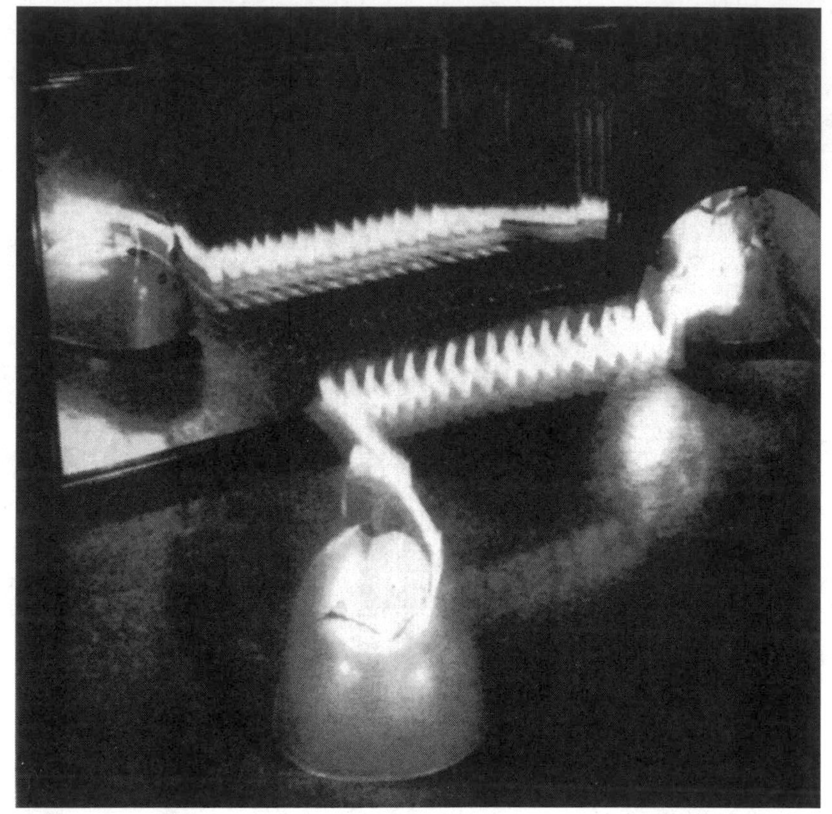

Figure 3.5. The mirror dance. Source: Holland 1996.

central question addressed here was how the brain goes from atomistic sensory impressions to a more holistic awareness of the world. In the United States in 1947 Walter Pitts and Warren McCulloch published an influential paper, "How We Know Universals," which aimed to explain pattern recognition—for example, recognizing individual letters of the alphabet independently of their size and orientation—in terms of a scanning mechanism. More relevant to Walter, in his 1943 book *The Nature of Explanation*, Kenneth Craik (1943, 74), the British experimental psychologist, speculated about the existence of some cerebral scanning mechanism, always, it seems, explained by an analogy with TV. "The most familiar example of such a mechanism is in television, where a space-pattern is most economically converted for transmission into a time sequence of impulses by the scanning mechanism of the camera" (Walter 1953,

108). The basic idea was that the brain contains some such scanning mechanism, which continually scans over its sensory inputs for features of interest, objects, or patterns in the world or in configurations internal to itself.[11]

One of the tortoise's most striking features, the rotation of the front forks and the photocell, was thus an implementation of this cybernetic notion of scanning. And beyond that, scanning had a further degree of significance for Walter. Craik visited Walter in the summer of 1944 to use the Burden's automatic frequency analyzers, and from that time onward both of them were drawn to the idea that the brainwaves recorded in Walter's EEGs were somehow integral to the brain's scanning mechanism (Hayward 2001b, 302). The basic alpha rhythm, for example, which stopped when the eyes were opened, could be interpreted as a search for visual information, a search "which relaxes when a pattern is found," just as the tortoise's photocell stopped going

Figure 3.6. The mating dance. Source: Holland 1996.

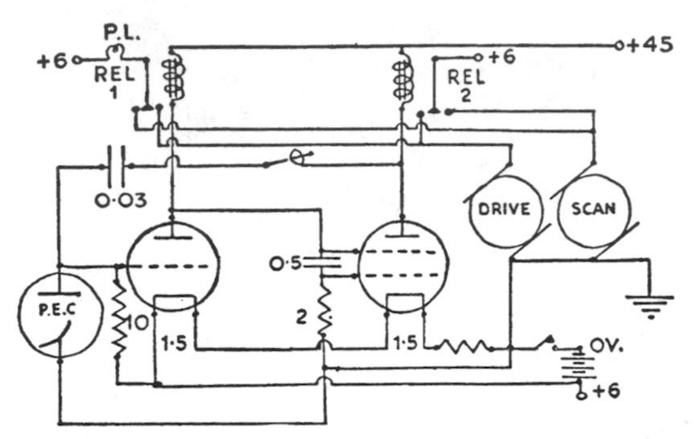

Figure 3.7. The brain of the tortoise. Source: Walter 1953, 289, fig. 22.

around when it picked up a light.[12] This interpretation found some empirical support. As Walter noted (1953, 109), "There was the curious coincidence between the frequency of the alpha rhythms and the period of visual persistence. This can be shown by trying how many words can be read in ten seconds. It will be found that the number is about one hundred—that is, ten per second, the average frequency of the alpha rhythms" (Walter 1953, 109). He also mentioned the visual illusion of movement when one of a pair of lights is turned off shortly after the other. Such data were at least consistent with the idea of a brain that lives not quite in the instantaneous present, but instead scans its environment ten times a second to keep track of what is going on.[13]

From a scientific perspective, then, the tortoise was a model of the brain which illuminated the go of adaptation to an unknown environment—how it might be done—while triangulating between knowledge of the brain emanating from EEG research and ideas about scanning.

Tortoise Ontology

We can leave the technicalities of the tortoise for a while and think about ontology. I do not want to read too much into the tortoise—later machines and systems, especially Ashby's homeostat and its descendants, are more ontologically interesting—but several points are worth making. First, the assertion that the tortoise, manifestly a machine, had a "brain," and that the functioning of its machine brain somehow shed light on the functioning of the human brain, challenged the modern distinction between the human and the nonhuman, between people and animals, machines and things. This is the most obvious sense in which Walter's cybernetics, like cybernetics more broadly, staged a nonmodern ontology.[14] Second, we should reflect on the way the tortoise's brain latched onto its world. The tortoise is our first instantiation of the performative perspective on the brain that I introduced in chapter 1, the view of the brain as an "acting machine" rather than a "thinking machine," as Ashby put it. The tortoise did not construct and process representations of its environment (à la AI robotics); it did things and responded to whatever turned up (cycloidal wandering, locking onto lights, negotiating obstacles). The tortoise thus serves to bring the notion of a performative brain down to earth. In turn, this takes us back to the notion of Black Box ontology that I introduced in chapter 2. The tortoise engaged with its environment as if the latter were a Black Box, in Ashby's original sense of this word—a system to be performatively explored.[15] As ontological theater, the tortoise staged a version of this Black Box ontology, helping us to grasp it and, conversely, exemplifying a sort of robotic brain science that might go with such an ontology.

Now we come to the complication I mentioned in chapter 2. In one sense the tortoise staged a nonmodern Black Box ontology, but in another it did not. For Walter, the point of the exercise was to open up one of these boxes, the brain, and to explore the inner go of it in the mode of modern science. How should we think about that? We could start by remembering that in Walter's work the world—the tortoise's environment—remained a Black Box. In this sense, Walter's cybernetics had a hybrid character: nonmodern, in its thematization of the world as a performative Black Box; but also modern, in its representational approach to the inner workings of the brain. My recommendation would then be to pay attention to the nonmodern facet of this hybrid, as the unfamiliar ontology that cybernetics can help us imagine. But there is more to think about here. The question concerns the extent to which Walter's brain science in fact conformed to the stereotype of modern science. As I mentioned in chapter 2, cybernetic brain science was an odd sort of science in several ways. First, the scientifically understood brain had as its necessary counterpart the world as an unopened Black Box, so that the modern and the nonmodern aspects of this branch of cybernetics were two sides of a single coin. Second, the style of scientific explanation here is what I called "explanation by articulation of parts." Walter's brain science did not emulate physics, say, in exploring the properties of the fundamental units of the brain (neurons or their electromechanical analogues); instead, it aimed to show that when simple units were interconnected in a certain way, their aggregate performance had a certain character (being able to adapt to the unknown). Again, this sort of science thematizes performance rather than knowledge of individual parts. And third, this style of explanation had a tendency to undermine its own modern impulse in what I call the cybernetic discovery of complexity, to which we can now turn.

— — — — —

IT IS ONE OF THE INTERESTING CONSEQUENCES OF THIS KIND OF MODEL-MAKING—THOUGH I ONLY REALISED IT AFTER I STARTED MAKING THESE TOYS—THAT A VERY SMALL NUMBER OF NERVE ELEMENTS WOULD PROVIDE FOR AN EXTREMELY RICH LIFE.

GREY WALTER, "PRESENTATION" (1971 [1954], 29)

The tortoises were simple and comprehensible artifacts. Anyone could understand how their two-neuron brains worked—at least, anyone familiar with the relay and triode circuits of the time. But, as Walter argued, "the variation of behaviour patterns exhibited even with such economy of structure are complex and unpredictable" (1953, 126). He noted, for example, that he had been taken by surprise by the tortoises' mirror and mating dances (1953, 130). The tortoises engaged with their environments in unexpected ways, displaying *emergent properties* relative to what Walter had designed into them. After the fact, of course, Walter explained such performances in terms of the tortoises' running lights, as mentioned above. But it is worth recognizing that such interpretations were themselves not beyond dispute. On the basis of his own tortoise reconstructions, Owen Holland (2003, 2101–8) was led to challenge Walter's interpretation of the source of these dances, arguing that they are a function of the oscillatory behavior set in motion by physical contact, rather than anything to do with the running lights. Here it begins to become clear that the tortoises remained mini–Black Boxes. As Walter put it, "Even in the simple models of behaviour we have described, it is often quite impossible to decide whether what the model is doing is the result of its design or its experience" (1953, 271).[16]

The tortoise thus again appears as ontological theater, but in a different sense from that discussed above. As a piece of engineering, it displayed the fact that a reductive knowledge of components does not necessarily translate into a predictive understanding of aggregate performance—one still has to run the machine and find out what it will do. As I said in chapter 2, I find this ontologically instructive too. Many people, including me, tend to think that the world has some determinate structure that is, in principle, fully comprehensible. What the tortoise stages us for us is that, even if that were true, we might still have to find out about the world in real-time performative interaction. For such people, it might be helpful to start by imagining the world as full of tortoiselike entities—unknowable in any predictive sense and always capable of surprising us, as the tortoise proved to be. This is another way to begin getting the hang of the ontology of cybernetics.

In his first publication on the tortoises, in *Scientific American* in May 1950, Walter (1950a, 44) emphasized this discovery of complexity in a striking extrapolation beyond the two-neuron tortoise brain: "It is unlikely that the number of perceptible functional elements in the brain is anything like the total number of the nerve cells; it is more likely to be of the order of 1,000. But even if it were only 10, this number of elements could provide enough variety for a lifetime of experience for all the men who ever lived or will be born if mankind survives a thousand million years." At stake here are not Walter's precise numbers (see Walter 1953, 118–20, for the calculation)—though cybernetic combinatorics readily generates enormous numbers, as we will see later. Walter was *not* suggesting that given ten elements he could predict the future of the human race in classically scientific fashion. His point concerned, rather, I think, the unimaginable richness of performance that could be generated by a few simple parts articulated with one another. Even if we knew what the ten functional elements of the brain are and how they are interconnected, we would not be able to "solve" the system and thus calculate and predict all possible forms of human behavior over the next billion years. We would just have to build the system and run it, like the tortoise, to see what it would do—or we could just let history run its course and find out. In general, even if we know all that there is to know about the primitive components of a Black Box, we might still not know anything about how the ensemble will perform. At this level of aggregation, the box remains black, and this is what Walter learned from his tortoises.

— — — — —

Thus my sense of the tortoise as ontological theater—as variously conjuring up and playing out an ontological vision of performance and unknowability. We will see this ontology elaborated in all sorts of ways in the pages to follow. But here I should note two qualifications concerning just how much the tortoise can enlighten us. First, the tortoise was indeed adaptive, but only to a degree. Especially, it had *fixed goals* hard wired into it, such as pursuing lights. The tortoise did not evolve new goals as it went along in the world. This fixity of goals was a common feature of early cybernetic engineering, going back all the way to the steam-engine governor (which sought to keep the engine speed constant) and beyond. As ontological theater this has to be seen as a shortcoming. There is no reason to think that human beings, for example, are characterized by a fixity of goals, and every reason, in fact, to argue against it (Pickering 1995). From this angle too, then, we should see the tortoise as staging a hybrid ontology, part adaptive and part not.[17] As I have said before, the adaptive aspects of cybernetics are what I want most to get into focus here, as pointing toward the unfamiliar aspects of nonmodern ontology.

The tortoise's *world* also left something to be desired. It was a world that, to a first approximation, never changed, a fixed array of lights and obstacles. The tortoise adapted to its environment, but the environment did nothing in response.[18] There was no place for a dance of agency between the tortoise and its world. This has to be regarded as another shortcoming of Walter's cybernetics as ontological theater, and we can see in later chapters how other cybernetic systems, beginning with Ashby's homeostat, transcended this limitation.

Tortoises as Not-Brains

IT IS UP TO M. WALTER TO EXPLAIN THE IMPORTANCE OF HIS MODELS FOR PHYSIOLOGY. THE ENGINEER IS INTERESTED IN THE MACHINE THAT IMITATES SENSE ORGANS AND THE MACHINE THAT LEARNS. ONE CAN IMAGINE A DAY WHEN MACHINES THAT LEARN WOULD HAVE A GENERAL IMPORTANCE IN INDUSTRY. THAT IS WHY WE HAVE REPEATED HIS APPROACH.

HEINZ ZEMANEK, "LA TORTUE DE VIENNE ET LES AUTRES TRAVAUX CYBERNÉTIQUES" (ZEMANEK 1958, 772, MY TRANSLATION)

In the opening chapter I mentioned the protean quality of cybernetics, that although the brain was its original referent, the brain could not contain it, and I can elaborate on that remark now. I have shown how the tortoise took shape as a model of the brain and as a contribution to brain science; I will shortly explore its specific connection to psychiatry. But one did not have to see a brain when contemplating a tortoise. One could simply see a machine, an interesting example of a particular style of adaptive engineering, a robot. Here is Walter's own account of the origins of the tortoise from *The Living Brain* (1953, 125): "The first notion of constructing a free goal-seeking mechanism goes back to a wartime talk with the psychologist, Kenneth Craik. . . . When he was engaged on a war job for the Government, he came to get the help of our automatic [EEG] analyser with some very complicated curves he had obtained, curves relating to the aiming errors of air gunners. Goal-seeking missiles were literally much in the air in those days; so, in our minds, were scanning mechanisms. . . . The two ideas, goal-seeking and scanning, . . . combined as the essential mechanical conception of a working model that would behave like a very simple animal." Craik was a young experimental psychologist and protocybernetican, who died at the age of thirty-one in a bicycle accident in Cambridge on 18 May 1945, the last day of World War II in Europe. He was very much the British Wiener, even more heavily involved in military research into gun aiming and the like during the war, and there are clear echoes of Wiener's wartime work on autonomous weapons systems in this quotation from Walter.[19] And though there is no evidence that Walter ever sought to develop the tortoise for such purposes, if one wanted to find a *use* for it, an obvious thing to do would be to fix a gun next to the guiding photocell or fill its body with explosives detonated by the contact switch. And Walter was certainly well aware of such possibilities. At the end of his technical description of tortoise construction, he stated that "the model may be made into a better 'self-directing missile' by using two photocells in the usual way" (1953, 291–92).[20]

Walter's contribution to brain science was thus also a contribution to the history of engineering and robotics (on which more below). And beyond the technical realms of brain science and robotics, the tortoises also found a place in popular culture. They were not simply technical devices. Walter showed them off and people liked them. He demonstrated the first two tortoises, Elmer and Elsie, in public in 1949, though "they were rather unreliable and required frequent attention." Three of the tortoises built by Bunny Warren were exhibited at the Festival of Britain in 1951; others were demonstrated in public regularly throughout the 1950s. They appeared on BBC television (Holland 2003, 2090–91, gives an account and analysis of a 1950 BBC newsreel on the tortoises). Walter set them loose at a meeting of the British Association for the Advancement of Science, where they displayed a lively interest in women's legs (presumably attracted to the light-reflecting qualities of nylon stockings: Hayward, 2001b).

This popular appeal, in turn, manifested itself in at least two lines of subsequent development. One was an embryonic eruption into the toy market: a tortoise was sent to the United States after the Festival of Britain as the prototype for a line of transistorized children's toys—which never went into production, alas (Holland 1996, n.d.; Hayward 2001b). One can now, however, buy construction kits for devices which are clearly versions of the tortoise. Along another axis, the tortoise entered the world of science fiction and popular entertainment. In the BBC's long-running *Doctor Who* TV series, I find it hard to doubt that the tortoise was the model for K-9, the Doctor's robot dog (which looked just like a tortoise, with a small tail attached). One thinks also of the Daleks, with their sinister optical scanner, and my recollection is that the Daleks were first seen in an electronic readout from a human brain which itself took the form of a TV image—another imaginative version of the cybernetic notion of scanning. What should we make of this popular appeal? It derived, I assume, from the quasi-magical properties of tortoises I mentioned in chapter 1, as mechanical devices that behaved as if they were alive. We are back in the territory of the Golem and the Sorcerer's Apprentice, and a fascination with transgression of the boundary between the animate and the inanimate. This animation of the inanimate hangs together, of course, with the implementation of the cybernetic ontology just discussed: the tortoises appeared so lively just because of their autonomy and sensitivity to their environment.

Brain science, psychiatry, robotics, toys, TV sci-fi: these are some of the areas that the tortoises contributed to. This list starts to establish what I mean by the protean quality of cybernetics, and as the book goes on, we can extend it.

The Social Basis of Cybernetics

THE MECHANICAL DESIGN [OF A TORTOISE] IS USUALLY MORE OF A PROBLEM THAN THE ELECTRICAL. . . . THERE IS NOT A GREAT CHOICE OF MOTORS; THOSE USED FOR DRIVING SMALL HOME-CONSTRUCTED MODELS ARE ADEQUATE BUT NOT EFFICIENT. . . . IT IS OFTEN ADVISABLE TO RE-BUSH THE BEARINGS. . . . THE GEAR TRAINS TO THE DRIVING AND SCANNING SHAFTS ARE THE MOST AWKWARD PARTS FOR THE AMATEUR CONSTRUCTOR. THE FIRST MODEL OF THIS SPECIES WAS FURNISHED WITH PINIONS FROM OLD CLOCKS AND GAS-METERS.

GREY WALTER, *THE LIVING BRAIN* (1953, 290–91)

SO MANY DISCOVERIES HAVE BEEN MADE BY AMATEURS THAT THERE MUST BE A SPECIAL STATE OF MIND AND A PHASE OF SCIENTIFIC EVOLUTION WHEN TOO MUCH KNOWLEDGE IS A DANGEROUS THING. COULD ONE SAY THAT AN AMATEUR IS ONE WHO DOES NOT KNOW HIS OWN IMPOTENCE?

GREY WALTER, "TRAPS, TRICKS AND TRIUMPHS IN E.E.G." (1966, 9)

I mentioned in the opening chapter that cybernetics had an unconventional social basis as well as an unfamiliar ontology, and here we can begin the investigation of the former. One point to bear in mind is that Walter did have a steady job throughout his working life, spending the thirty-one years prior to his scooter accident at the Burden Neurological Institute. As I said, however, his career there revolved around his EEG work and electrophysiological research more generally, and the question that I want to focus on here concerns the social basis for his cybernetics as exemplified by the tortoises.

In the quotation above on Craik and the origins of the tortoise, I skipped over a phrase, "long before the home study was turned into a workshop," which precedes "the two ideas, goal-seeking and scanning, had combined." Walter built the first tortoises at home, in his spare time.[21] Hence, for example, the practical advice to readers on tortoise construction just quoted. Walter's key contribution to cybernetics was, then, the work of an amateur, a hobbyist. And, as we will see, this was true of all four of our principals. In this sense, then, we can say that at its origins British cybernetics had *no social basis*. It emerged from nowhere as far as established fields and career paths were concerned. The cyberneticians and their projects were *outsiders* to established fields of endeavor.

Some discussion is appropriate here. First, it is worth emphasizing that the amateur and hobbyist roots of British cybernetics are a marker of its oddity: there was no obvious field for it to grow from. Perhaps the most likely matrix would have been experimental psychology (one thinks of Kenneth Craik) but in fact cybernetics did not originate there. Second, we should go back to the standard origin story of cybernetics, connecting it to Norbert Wiener's military research. There is, as I said in chapter 1, a contrast here between British and American cybernetics. As I have already indicated, the primary referent of Walter's tortoise work was not some piece of military technology such as Wiener's antiaircraft predictor; it was the brain. Walter always presented the tortoise precisely as a model brain, and though I just quoted him on the tortoise as a self-guided missile, this was a passing remark. And, of course, it makes sense that a brain researcher working at a neurological institute would have the brain rather than weapons systems on his mind.[22]

This, then, is the other origin story of cybernetics that I can develop further as we go on, the story of cybernetics as emerging from and as brain science rather than military research. This story requires some nuance, needless to say. Little research in the 1940s and 1950s was immune to military influence, and it was Craik, the British Wiener, who gave Walter the idea of scanning. Nevertheless, it would be misleading to try to center the story of British cybernetics on war; it is much more illuminating to focus on the brain.[23] That said, there is another connection to warfare that is worth mentioning, which in fact deepens the contrast with Wiener.

In the second Grey Walter Memorial Lecture, the veteran EEG researcher W. A. Cobb told a story of wartime shortages of equipment and of how he eventually obtained a special timer from the wreckage of a crashed Spitfire (Cobb 1981, 61). We can take this as iconic of the conditions under which British cybernetics developed. Wiener worked on a well-funded military project at the cutting edge of research at MIT, the very heart of the U.S. military-academic complex; like Cobb, Walter and the other British cyberneticians cobbled together their creations from the detritus of war and a couple of centuries of industrialization.[24] The electronic components of machines like the tortoise were availably cheaply as war surplus (Hayward 2001b, 300), and, as Walter said, other parts were salvaged from old clocks and gas meters. If Wiener's cybernetics grew directly out of a military project, Walter's was instead improvised in a material culture left over from the war.

One last remark on the origins of British cybernetics. Inescapably associated with the notions of the amateur and the hobbyist are notions of sheer pleasure and fun. Just as there is no reason to doubt that Walter intended the tortoises as a serious contribution to brain science, there is no reason to doubt that he had fun building them and watching them perform. This theme of having fun is another that runs through the history of British cybernetics and again presents a stark contrast with that of cybernetics in the United States, where the only fun one senses in reading the proceedings of the Macy Conferences is the familiar and rather grim academic pleasure of the cut and thrust of scholarly debate. The chairman of the meetings, Warren McCulloch (2004, 356), recalled: "We were unable to behave in a familiar, friendly or even civil manner. The first five meetings were intolerable. Some participants left in tears, never to return. We tried some sessions with and some without recording, but nothing was printable. The smoke, the noise, the smell of battle are not printable." Of the many conventional boundaries and dichotomies that British cybernetics undermined, that between work and fun was not the least.

We can turn from the origins of British cybernetics to its propagation. Walter made no secret of his hobby; quite the reverse: he publicized the tortoises widely, engaging with at least three rather distinct audiences which we can discuss in turn. The first audience was the general public. According to Owen Holland (2003, 2090), "by late 1949, Grey Walter was demonstrating Elmer and Elsie, the first two tortoises, to the press, with all the showmanship that some held against him," and the first major press report appeared in the *Daily Express* on 13 December 1949, written by Chapman Pincher. The BBC TV newsreel mentioned above followed in 1950, and so on. Outside the world of journalism, Walter himself wrote for a popular readership. The principal published technical sources on the tortoises are Walter's two articles in *Scientific American* in May 1950 and August 1951 and his 1953 popular book *The Living Brain*. These contributed greatly to the public visibility of Walter and the tortoises, but let me postpone discussion of substantive outcomes of this publicity for a while.

As an academic myself, I have tended to assume that the proper readership and home for a field like cybernetics would be a scholarly one. Walter did not publish any detailed accounts of the tortoises alone in the scholarly literature, but in the early 1950s they often featured as parts of his emerging account of the brain as otherwise explored in EEG research. A lecture delivered to a psychiatric audience published in January 1950, for example, began with a discussion of the tortoises (not named as such), their complexity of behavior, and the significance of scanning, before plunging into the details of EEG findings and their interpretation (Walter 1950b, 3–6). But it is also safe to say that the major impact of cybernetics was not centered on any established field. Historical overviews of twentieth-century psychiatry, for example (on which more shortly), make little or no mention of cybernetics (e.g., Valenstein 1986; Shorter 1997).[25] And one can see why this should have been. The combination of brain science and engineering made concrete in the tortoises was a strange one, both to the sciences of the brain (neurophysiology, EEG research, psychology, psychiatry) and, from the other direction, to engineering. To *do* any of these disciplines on the model of Walter and the tortoises would have required drastic shifts in practice, which are much harder to make than any simple shift in the realm of ideas.

This brings us to the third community with which Walter engaged, the nascent community of cyberneticians in Britain. The 1948 publication of Wiener's *Cybernetics* both put the word itself into circulation in Britain and helped crystallize the formation of a self-consciously cybernetic community there. On 27 July 1949 John Bates of the Neurological Research Institute of the National Hospital in London wrote to Walter as follows:

> Dear Grey,
>
> I have been having a lot of "Cybernetic" discussions during the past few weeks here and in Cambridge during a Symposium on Animal Behaviour Mechanisms, and it is quite clear that there is a need for the creation of an environment in which these subjects can be discussed freely. It seems that the essentials are a closed and limited membership and a post-prandial situation, in fact a dining-club in which conventional scientific criteria are eschewed. I know personally about 15 people who had Wiener's ideas before Wiener's book

appeared and who are more or less concerned with them in their present work and who I think would come. . . .

Besides yourself, Ashby [see the next chapter] and Shipton [Walter's colleague and collaborator at the Burden], and Dawson and Morton from here, I suggest the following:-

Mackay	-	computing machines, Kings Coll. Strand.
Barlow	-	sensory physiologist—Adrian's lab.
Hick	-	Psychological lab. Camb.
Scholl	-	Statistical Neurohistologist—U.C. Anat. lab.
Uttley	-	ex Psychologist, Radar etc. T.R.E.
Gold	-	ex radar zoologists at Cambridge
Pringle		

I could suggest others but this makes 13. I would suggest a few more non neurophysiologists communications or servo folk of the right sort to complete the party but those I know well are a little too senior and serious for the sort of gathering I have in mind.

We might meet say once a quarter and limit the inclusive cost to 5/- less drinks. Have you any reactions? I have approached all the above list save Uttley so far, and they support the general idea.

Walter replied the next day to this "exciting letter"—"We also have been having some pretty free CYBERNETIC discussions and your notion of a sort of Dining Club attracts me very much. I agree that it will be nice to keep the gathering rather small, about the size of a witches coven owing to the shortage of broomsticks." Walter also mentioned that Warren McCulloch was visiting Britain in September 1949 and suggested that this would provide a good occasion for the first meeting of the group.[26] And thus it came to pass. McCulloch addressed the first meeting of the Ratio Club on 14 September 1949 on the topic of "Finality and Form in Nervous Activity." Sixteen member were present, including Ashby but not Walter, "owing to the delivery of a male homeostat which I was anxious to get into commission as soon as possible." Expenditure on food was £1-4-0; on beer and wine, £7. Thereafter, the club met at least thirty-four times up to 1955 (with decreasing frequency after 1952) before being wound up at a reunion meeting on 27 November 1958.[27]

There is much that might be said on the Ratio Club, its membership, and their doings, but this would easily carry us too far afield, and I will confine myself to a few observations.[28] We should note first Ratio's interdisciplinarity. Bates described its proposed membership as "half primarily physiologists though with 'electrical leanings' and half communication theory and ex-radar folk with biological leanings" and, later, to Turing, as "half biologists—(mostly neurophysiologists) and half engineers and mathematicians," while remarking to himself that the club was "incomplete—no sociologists, northeners, professors" (Clark 2002, 78–80).[29] But beyond that, Ratio was interinstitutional, as one might say. It did not simply elide disciplinary boundaries within the university; it brought together representatives from different sorts of institutions: people from the universities, but also medical men and physiologists based in hospitals and research institutes, including Walter and Ashby, and workers in government laboratories (Albert Uttley at the Telecommunications Research Establishment, the TRE).[30] The Ratio Club was the center of gravity for work in cybernetics in Britain from 1949 to the mid-1950s, and it existed *transversely*, or orthogonally, to the usual institutions for the production of knowledge, cutting across not just academic disciplinary boundaries, but also across the usual institutional classifications, too. And this transversality continued to be a conspicuous feature of British and European cybernetics after the demise of Ratio, when the series of Namur conferences became the key institutional venue from 1956 onward.[31]

Two observations follow. First, ontology and sociology were entangled here. This transverse crystallization had the character of a purification that was at once social and ontological. From the side of traditional fields of practice, it would be a mistake to think that an interest in the adaptive brain was actively excluded. But the formation of first the Ratio Club and then the Namur conference series attests to a perceived marginality of the cyberneticians in their own fields, and a perceived closeness to workers in other fields with similar interests. From the other side, the shared interest in the adaptive brain came to center precisely on transverse institutions like the Ratio Club. Ratio—rather than their home disciplines and institutions—was where people like Walter found an active and engaged audience for their cybernetics. And, as we will see later, much of the propagation of cybernetics up the present has continued to be located in such strange antidisciplinary and interinstitutional spaces, even as the range of cybernetics has gone far beyond the brain.

My second observation is this. The Ratio Club and its successor institutions were undoubtedly successful in maintaining the postwar cybernetic ferment, but they were conspicuously lacking in the means of social reproduction. The Ratio Club had no mechanism for training students: a dining club does not grant PhD's. Among our cyberneticians, only Stafford Beer in the second generation seems to have taken this problem seriously, but we can note now that this ad hoc organization contributed importantly to the way cybernetics evolved. Academic disciplines are very good at holding neophytes to specific disciplinary agendas, and it was both a strength and a weakness of cybernetics that it could not do this—a strength, inasmuch as cybernetics retained an undisciplined and open-ended vitality, an ability to sprout off in all sorts of new directions, that the established disciplines often lack; a weakness, as an inability both to impose standards on research and to establish career paths for new cyberneticians left enthusiasts to improvise careers much as did the founders.

These remarks return us to a topic broached above. Popular writing and, in Walter's case especially, public performances assumed an importance in the propagation of cybernetics that one does not find in established fields. In doing the research for this book I have been surprised to discover just how many first and consequential contacts with cybernetics have been with popular books, articles and performances. We just saw that Wiener's *Cybernetics* was central to the crystallization of the British cybernetics community, and Beer fell into cybernetics after reading the same book. Walter's cybernetics traveled and mutated along the same lines. In chapter 7 we can discuss the adaptive architecture of John Frazer, who tried to build his own robots after seeing a display of the tortoises as a schoolboy, before falling in with Pask (who declared himself a cybernetician after meeting Wiener in person as an undergraduate). Later in this chapter, we can see how William Burroughs laundered elements of cybernetics into the counterculture after reading *The Living Brain*. And in the following section I want to bring the discussion of robotics up to the present by focusing on another *Living Brain* reader, Rodney Brooks.[32] The general point to note here, however, is that the propagation of cybernetics was indeed both unsystematic and undisciplined. Walter's cybernetics was addressed to the brain, but Brooks understood it as robotics, Frazer took it into architecture, and Burroughs transplanted it into the domain of altered states and that classic sixties project, the exploration of consciousness. Hence the protean quality of cybernetics, with individuals free to adapt it to their own interests and obsessions, unconstrained by disciplinary policing.[33]

Rodney Brooks and Robotics

Rodney Brooks is currently director of the MIT Computer Science and Artificial Intelligence Laboratory, Panasonic Professor of Robotics at MIT, and past chairman and now chief technical officer of iRobot Corporation.[34] Brooks began his career in robotics as a schoolboy in Australia when "I came across a Pelican edition of Grey Walter's book, and tried to build my own version of *Machina Speculatrix*, using transistor technology rather than vacuum tubes. . . . The subtleties of the original electronics were a little beyond me, but I did manage to get my first robot, Norman, to the point where it could wander around the floor, respond to light, and bumble its way around obstacles" (Brooks 2002, 27). From Australia he moved to the United States, completed a PhD in computer science at Stanford University in 1981, and held postdoctoral positions at Carnegie Mellon University and MIT and a faculty position at Stanford, before rejoining MIT as an assistant professor in 1984. The first machine that Brooks and a few collaborators then constructed was a robot called Allen, which made Brooks's reputation, in certain quarters at least, and began his rise to his current position as leader of one of the most important computer science and AI laboratories in the world. And the point to grasp is that Allen was very much an updated version of the tortoise. Using a ring of twelve sonar range detectors in place of the tortoise's photocell and contact switch, and solid-state logic elements instead of electronic valves, Allen would explore its environment, pursuing goals (such as finding and trying to get to the most distant part of the room) and avoiding obstacles along the way. Even Brooks's construction strategy, which he called a "subsumption architecture," getting different layers of the control system working one after the other, mirrored Walter's transit from the tortoise itself to CORA (see below).[35]

So, if one is looking for a "weighty" answer to the question, what happened to cybernetics? one answer would be: it is alive and well in Brooks's lab at MIT. But then another question arises. How on earth could one make a reputation in computer science by building an updated tortoise thirty-six years after Walter? Of course, Brooks displayed his own originality, but some important history is needed here, which I want just to mention without going into detail. In the opening chapter I contrasted the performative brain of cybernetics with the representational one of AI, and I need to say a little about the development of the latter field.

The canonical history of AI dates its self-conscious inception to the six-week workshop "Artificial Intelligence" at Dartmouth College organized by John McCarthy in 1956 (Brooks 2002, 21–31). Many of the principals of the nascent field were present, and what followed was a rapid purification, as I called it above, but going in the opposite direction. From World War II to the mid-1950s speculation about the mind in terms of machine models was an exceptionally rich, diverse, and fascinating field, in which cybernetics in many ways took the lead. From the mid-1950s onward a representationalist strand of AI came to the fore, and it achieved institutional dominance within the space of about ten years. In GOFAI—good, old-fashioned AI—the aim was to mimic mental performances. Alan Newell and Herbert Simon's Logic Theorist program was an early landmark, and it was a program that mimicked the proofs to be found in Bertrand Russell and Alfred North Whitehead's *Principia Mathematica*. In robotics this translated into the problematic of generating computer representations (maps, models) of environments and operating on them to execute plans, such as moving from A to B while avoiding obstacles. This style of AI and robotics, then, can stand as a piece of ontological theater for the other ontology from that of cybernetics, the modern ontology of knowability. AI robots sought to know their worlds substantively, and to accomplish their goals through that knowledge. AI robotics was the other to Walter-style robotics.

Historically, representational, or symbolic, AI quickly became the dominant paradigm in the universities, largely displacing cybernetics from its already tenuous foothold, not only from computer science departments and their ilk, but from social science departments, too, in the so-called cognitive revolution, in which human mental powers were conceived by analogy to digital computers as information processors (Gardner 1987). Of course, the rise of AI and the associated "cognitive sciences" is an immense historical story in itself, but let me just comment briefly. How did AI come to exert such a fascination over the academic and popular imagination? Part of the answer must lie in its very familiar ontology. It is easy to think of the brain and mind as the organ of knowledge, and AI thus conceived presents a straightforward problem of mimicking very familiar (especially to academics) mental performances. At the same time, AI was uniquely associated with digital computers and their programming and thus fitted very naturally into the agenda of novel postwar departments of computer science (unlike the odd machines of Walter et al.). And third, the military bought it. Almost all the funding for AI research was provided by the U.S. military, and almost all of that went to research in symbolic AI (Edwards 1996).[36]

Cybernetics thus lost much of its social basis in the universities from the mid-1950s onward; the cyberneticians became even more marginal there than they had been before—which is another kind of answer to the question, what happened to cybernetics? But this gets us back to the story of Rodney Brooks. In robotics, symbolic AI promised much but never quite delivered. Machines were never quite fast enough to accomplish real-time control.[37] In his first years in the United States, Brooks worked within this tradition, focusing on computer models of environments, but became increasingly frustrated with it. In the late 1970s at Stanford, he helped Hans Moravec, a future leader in

AI-style robotics, on a robot which moved so slowly (due to the time taken for computation) that, outdoors, the movement of sun and shadows would confuse its internal repesentations (Brooks 2002, 30):

> Despite the serious intent of the project, I could not but help feeling disappointed. Grey Walter had been able to get his tortoises to operate autonomously for hours on end, moving about and interacting with a dynamically changing world and with each other. His robots were constructed from parts costing a few tens of dollars. Here at the center of high technology, a robot relying on millions of dollars of equipment did not appear to operate nearly as well. Internally it was doing much more than Grey Walter's tortoises had ever done—it was building accurate three-dimensional models of the world and formulating detailed plans within those models. But to an external observer all that internal cogitation was hardly worth it.

It was against this background that Brooks's 1985 robot, Allen, stood out as a revolutionary alternative. Allen dispensed with the "cognitive box" (Brooks 2002, 36) that was the hallmark and center of attention in contemporary robotics in favor of the performative and adaptive engagement with the environment that was the hallmark of the tortoises.[38] This, of course, put him on the wrong side of the law as far as the academic establishment was concerned, and he has repeatedly told the story of how, during his first scholarly presentation of his new approach, one senior computer scientist whispered to another, "Why is this young man throwing away his career?" Three referees unanimously recommended rejection of his first paper on this approach—though it was published anyway (Brooks 1999 [1986]) and went on to become "one of the most highly cited papers in all of robotics and computer science" (Brooks 2002, 43). In the event, though the "arguments . . . continue to today" (Brooks 2002, 43), Brooks's approach did succeed in redirecting the work of a substantial fraction of the robotic community back into Walterian, cybernetic channels. One token of this success came in 2002, with the organization of a major international conference, "Biologically-Inspired Robotics," held at the Hewlett-Packard Laboratory near Bristol and close to the Burden Institute. Marking the twenty-fifth anniversary of Walter's death, the subtitle of the conference was simply "The Legacy of W. Grey Walter." Many of the principals of this "new field" gave invited addresses, and graduate students presented an impressive array of talks.[39]

After a decades-long hiatus, then, at Brooks's lab at MIT, and many other academic centers, too, the robotic wing of cybernetics finally gained what it conspicuously lacked in its formative years, a solid institutional base not only for research but also for social reproduction, the training of graduate students as future researchers with a prospect of recognizable career paths in the field.[40] And one concluding remark is worth making for future reference. In the following chapters we will encounter many imaginative initiatives in cybernetics which eventually fizzled out, and one inevitably wonders whether this points to some essential flaw in cybernetics itself. I think we should remember that Walter's robotics once fizzled out, but that in retrospect it is clear that the fizzling had more to do with the lack of an institutional base and support than any inherent flaws.[41]

CORA and *Machina docilis*

AFTER FOUR YEARS [IN CAMBRIDGE] SPENT LITERALLY IN A CAGE AND CHAINED BY THE ANKLE—NOT FOR PUNISHMENT BUT FOR ELECTRICAL SCREENING—. . . IMAGINE, THEN, HOW REFRESHING AND TANTALIZING WERE THE RESULTS FROM PAVLOV'S LABORATORY IN LENINGRAD TO THOSE ENGAGED ON THE METICULOUS DISSECTION OF INVISIBLE NERVE TENDRILS AND THE ANALYSIS OF THE IMPULSES WHICH WE INDUCED THEM TO TRANSMIT.

GREY WALTER, *THE LIVING BRAIN* (1953, 51)

The tortoise served as a model of the adaptive brain, but only a primitive one. It lived in real time, reacting to environmental cues (lights, contacts) as they happened to it and never learning anything from its experience. Walter quickly sought to go beyond this limitation by building in a second layer of adaptability, and he concluded his first publication on the tortoise by mentioning that the "more complex models that we are now constructing have memory circuits" (1950a, 45). These more complex models entailed two modifications to the basic tortoise. One was to equip it with more senses—wiring a microphone into its circuits, for example, to give it a sensitivity to sound as well as light. The other was the addition of a clever circuit called CORA, for conditioned reflex analogue (figs. 3.8, 3.9). Wired into the tortoise, CORA converted *Machina speculatrix* to *Machina docilis*, as Walter called it—the easily taught machine. CORA detected repeated coincident inputs in different sensory channels and, when a certain threshold of repetition was reached, opened up a link from one sense to the other—so that the tortoise would become "conditioned" to react to a sound, say, in the way that it had hitherto reacted to the contact switch on its body.[42]

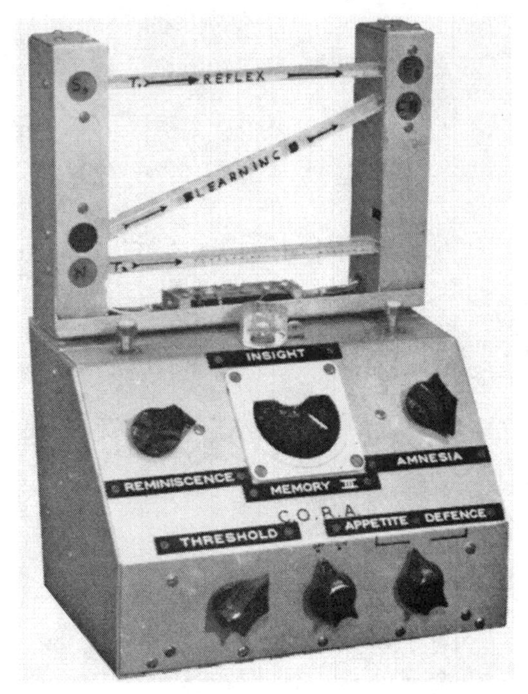

Figure 3.8. CORA. Source: de Latil 1956, facing p. 51.

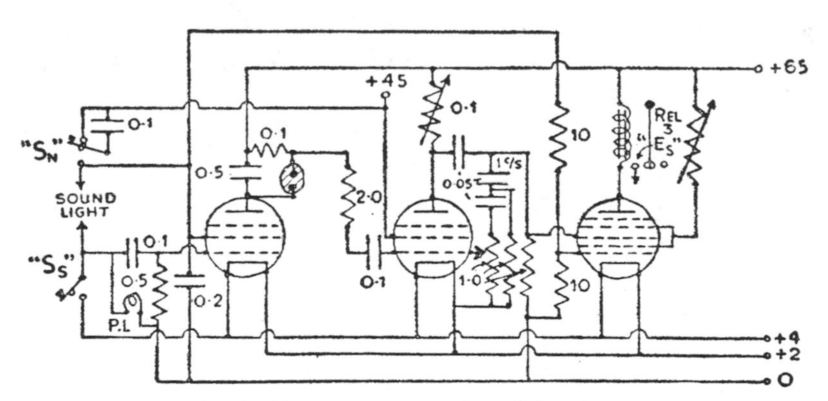

Figure 3.9. CORA: circuit diagram. Source: Walter 1953, 295.

What can we say about CORA? As its name implies, CORA was intended to emulate classical Pavlovian conditioning in animals. As a student at Cambridge, Walter had worked with a student of Pavlov for more than a year to set up a conditioning laboratory, "mastering the technique and improving it by the introduction of certain electronic devices." When "the great" Pavlov himself visited England, Walter, "as the English exponent of his work . . . had the privilege of discussing it with him on familiar terms. . . . I asked him if he saw any relation between the two methods of observing cerebral activity, his and Berger's [EEG readings]. . . . But Pavlov showed no desire to look behind the scenes. He was not in the least interested in cerebral mechanisms" (Walter 1953, 51–52).[43] CORA, in contrast, was explicitly a further scientific attempt to look behind the scenes, to open up the Black Box of the adaptive brain by building a model that could mimic its performances, just like the tortoises before that.

CORA did, indeed, make a connection between the electrical rhythms of the brain and conditioned learning. The key element in connecting one sense to another was precisely the build up of an oscillating voltage in CORA, and Walter laid much store by this, even arguing that CORA displayed the contingent negative variation in electrical potential which was his most important contribution to neurophysiology (1966, 13), but I cannot explore this further here.[44] Instead, I want to comment on CORA as brain science from several angles before connecting it to Walter's psychiatric milieu.

Reversing the earlier order, I begin with a quick comment on the relation of CORA to the social basis of cybernetics. CORA was a virtuoso piece of electrical engineering, in both its design and construction. The tortoise was imitable—by Frazer, Brooks, and many others—but CORA was inimitable. I know of no attempts to replicate it, never mind take the development of the tortoise beyond it.[45] Even Walter discontinued his robotics after CORA. *Machina speculatrix* pointed to a difficult but, to some—odd schoolboys like Frazer and Brooks; contributors to the Namur conferences—manageable synthesis of brain science and engineering. *Machina docilis* upped the ante too far. Nothing grew specifically from it in the cybernetic tradition. In the late 1980s revival of Walter-style robotics, machines that learned were indeed built, but that learning was based on neural networks, not CORA-style electronics, and the oscillatory features that intrigued Walter were lost (at least, temporarily). In that sense, CORA remains an unexploited resource in the history of cybernetics.

— — — — —

CORA also invites us to extend the discussion of cybernetic ontology to encompass epistemology. The great novelty of *M. docilis* was that it acquired a sort of knowledge about the world: it learned what to associate with what. How should we think about that? The point I would emphasize is that the knowledge of *M. docilis* waxed and waned with its performance, integrating over its experience—associations between stimuli would be lost if the robot's expectations did not continue to be reinforced—and thus functioned as a heuristic guide, emerging from and returning to performance, not as any kind of controlling center. *Docilis* thus offers us a vision of knowledge as engaged in, and as part of, performance, rather than as a thing itself or as some sort of external determinant of action—a vision of knowledge as being *in the plane of practice*, as I put it in *The Mangle of Practice*, not above it. Much as *speculatrix* acted out for us a performative ontology, then, *docilis* also staged a performative epistemology, as I called it in chapter 2—an appreciation of knowledge not as a hopefully definitive mapping of the world, but as another component of performance. This is the vision of knowledge that goes with the cybernetic ontology, and that we will see elaborated in succeeding chapters.[46]

Cybernetics and Madness

PSYCHIATRISTS USED TO BE REPROACHED WITH THEIR LACK OF THERAPEUTIC ZEAL; IT WAS SAID THEY WERE RESIGNED WHEN THEY SHOULD HAVE BEEN HOPEFUL AND ENTERPRISING, AND TORPID WHEN ENERGY WAS CALLED FOR. WHETHER THE REPROACH WAS DESERVED OR NOT IT SEEMS TO HAVE STUNG THEM INTO THERAPEUTIC FURY. CONTINUOUS NARCOSIS, INSULIN SHOCK AND CARDIAZOL FITS HAVE PROVED THAT THE PSYCHIATRIST IS AT LEAST AS DARING AS THE SURGEON NOWADAYS. THE PAPERS BY KALINOWSKY IN OUR ISSUE OF DEC. 9, AND BY FLEMING, GOLLA AND WALTER IN THIS RECORD ANOTHER BOLD STEP. . . . BUT WE MUST NOT LET UNCONSCIOUS ASSOCIATIONS WITH WHAT IS DONE PERIODICALLY IN A ROOM AT SING SING PREJUDICE US AGAINST WHAT MAY WELL TURN OUT TO BE A VALUABLE STEP FORWARD.

EDITORIAL, "MORE SHOCKS," *LANCET, 234*, NO. 6070 (30 DECEMBER 1939): 1373

It is tempting to think of the tortoises and CORA as freestanding scientific and engineering projects, divorced from mundane concerns. Walter may have mentioned the tortoise's potential as an autonomous weapons system, but he did nothing to pursue it. On the other hand, one of the first things he did with CORA was drive his tortoises mad. This points to connections between his cybernetics and psychiatry that we can explore.

If the CORA-equipped tortoise could be understood as a model of a normal brain, Walter was keen to show that it was a model for the pathological brain too. In his first article on the tortoise, Walter (1950a, 45) noted that, with

'Prefrontal leucotomy, Sir . . .'

Figure 3.10. "Prefrontal lobotomy, Sir . . ." Source: Beer 1994a, 162. (Courtesy of Cwarel Isaf Institute and Malik Management Zentrum St. Gallen [www.management.kybernetik.com, www.malik-mzsg.ch].)

CORA, "the possibility of a conflict neurosis immediately appears," and in a follow-up article in August 1951 he observed that (63) "it becomes only too easy to establish an experimental neurosis. Thus if the arrangement is such that the sound becomes positively associated both with the attracting light and with the withdrawal from an obstacle, it is possible for both a light and a sound to set up a paradoxical withdrawal. The 'instinctive' attraction to a light is abolished and the model can no longer approach its source of nourishment. This state seems remarkably similar to the neurotic behavior produced in human beings by exposure to conflicting influences or inconsistent education." Or, as he put it more poetically in *The Living Brain* (1953, 183), "in trying, as it were, to sort out the implications of its dilemma, the model ends up, 'sicklied o'er with the pale cast of thought,' by losing all power of action."[47]

The idea that mental problems might be precipitated by conflicting patterns of conditioning was not original to Walter. As he acknowledged, its history went back to the induction of "experimental neuroses" in dogs by clashing conditioning regimes, carried out in the laboratory of "the master," Pavlov himself, in 1921 (Gray 1979, 119; Pavlov 1927).[48] And in March 1950, for example, two months before Walter's first tortoise article appeared, *Scientific American* featured an article entitled "Experimental Neuroses" by Jules Masserman, a psychiatrist at Northwestern University, which discussed the induction of pathological symptoms by cross-conditioning in cats. Drawing upon Auguste Comte's typology, Masserman argued that the experimentalization of neuroses moved psychiatry from the "mystic" and "taxonomic" stages into the ultimate "dynamic" phase of "science" (Masserman 1950). Walter could have made the same point about his experiments with CORA. And one could say he had gone one step beyond Masserman. Not content with simply demonstrating that cross-conditioning could produce pathological behavior, he had, again, produced an electromechanical model which enabled one to grasp the go of this process at the hardware level.

It is thus revealing to think of cybernetics as a science of psychiatry, not in the sense that it could be reduced to psychiatry—even with the tortoise it already overflowed the bounds of the brain—but in the sense that psychiatry was a surface of emergence (Pickering 2005b) for cybernetics: Walter's cybernetics (and Ashby's) grew out of the phenomena and problematics of his psychiatric milieu. And we can take this line of thought further in a couple of directions. One is to note that after driving his tortoises mad, Walter cured them (1953, 184): "When a complex learning model develops an excess of depression or excitement, there are three ways of promoting recovery. After a time the conflicting memories may die away—except in obsessional states. . . . Switching off all circuits and switching on again clears all lines and provides, as it were, a new deal for all hands. Very often it has been necessary to disconnect a circuit altogether—to simplify the whole arrangement." And in case his readers missed the point, Walter went on to analogize these electromechanical procedures to those of psychiatric therapy, adding his cybernetic apologia for the latter:

> Psychiatrists also resort to these stratagems—sleep [leaving the machine alone for a long time], shock [switching it off and on again] and surgery [disconnecting electrical circuits within it]. To some people the first seems natural, the second repulsive, and the third abhorrent. Everyone knows the benison of sleep, and many have been shocked into sanity or good sense, but the notion that a mental disorder could be put right by cutting out or isolating a part of the brain was an innovation which roused as much indignation and dispute as any development in mental science. There are volumes of expert testimony from every point of view, but our simple models would indicate that, insofar as the power to learn implies the danger of breakdown, simplification by direct attack may well and truly arrest the accumulation of self-sustaining antagonism and "raze out the written troubles of the brain."

So cybernetics was a science of psychiatry in a double sense, addressing the go of both mental disorder and psychiatric therapy and offering a legitimation of the latter along the way. And since Walter does not use the usual terms for the therapies he mentions, we should note that we are plunged here into the

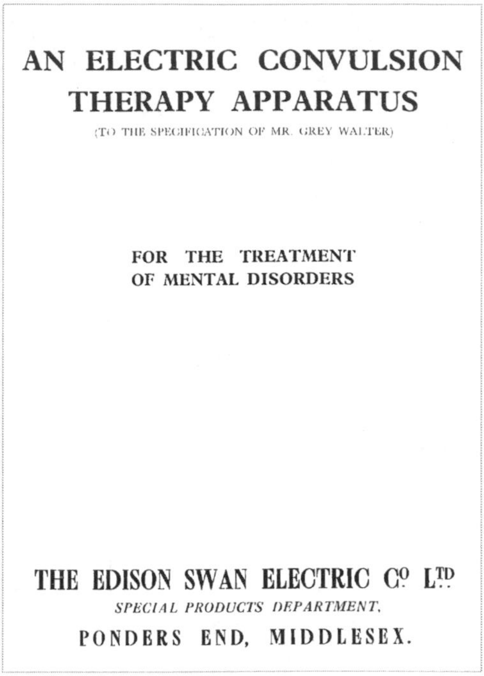

AN ELECTRIC CONVULSION
THERAPY APPARATUS
(TO THE SPECIFICATION OF MR. GREY WALTER)

FOR THE TREATMENT
OF MENTAL DISORDERS

THE EDISON SWAN ELECTRIC Co. LTD
SPECIAL PRODUCTS DEPARTMENT,
PONDERS END, MIDDLESEX.

Figure 3.11. ECT brochure, cover page. Source: Science Museum, London, BNI archive.

"great and desperate cures"—insulin shock, chemical shock, electroshock (electroconvulsive therapy—ECT), and lobotomy—that arose in psychiatry in the 1930s and had their heyday in the 1940s and early 1950s, the same period as the first flush of cybernetics.[49]

And to put some more flesh on this connection, we should note that despite his evident desire to "do science" as he had in his student days at Cambridge, Walter continually found himself entangled with the concerns of the clinic. During his brief stint in London, he wrote in a 1938 report to the Rockefeller Foundation on his EEG work (1938, 16) that "the volume of clinical work which I have been asked to undertake has grown to embarrassing proportions. . . . These examinations are, of course, undertaken most willingly . . . but the clerical and other routine work, to say nothing of the maintenance of apparatus . . . take up so much time that little is left for breaking new ground."[50] Walter's later work on flicker (see below) also had a significant clinical element. But more directly to the point here is that the newly established Burden Neurological Institute took the lead in transplanting the new approaches to psychiatry to Britain, claiming an impressive list of firsts, including the first use of ECT in Britain in 1939 and the first prefrontal leucotomy in 1940 (Cooper and Bird 1989). And Walter was very much involved in these achievements. His technical skill and creativity were such that he had a standing relationship with the Ediswan Company in the development of commercial apparatus, and Britain's first commercial ECT machine was developed by Ediswan and carried on the title page of its brochure the statement that it was built "to the specification of Mr. Grey Walter" (fig. 3.11).[51] Walter was one of the authors of the first three papers to appear from the Burden on ECT. The earliest of these appeared in the *Lancet* in December 1939 (Fleming, Golla, and Walter 1939), describes ECT as EEG in reverse, and includes EEG readings of two post-ECT patients.[52] During World War II Walter himself performed ECT treatments on American soldiers suffering from "battle fatigue."[53]

Walter's interest in mental pathologies and therapies was thus by no means that of a detached observer, and if one wanted to identify the worldly matrix from which his cybernetics emerged, it would have to be psychiatry; more specifically the psychiatry of the great and desperate cures; and more specifically still the world of electroshock, electroconvulsive therapy, ECT.[54]

— — — — —

Two remarks to end this section. In chapter 1 I said that it was interesting to think of cybernetics as one of Deleuze and Guattari's "nomad sciences" that destabilize the state, and we can come back to that thought now. Earlier in the present chapter I described the nomadic wandering of Walter's cybernetics through inter- and antidisciplinary worlds such as the Ratio Club and the Namur conferences, and yet when it came to real-world psychiatry Walter's work was evidently no threat to the established order. What should we make of this? The obvious remark is that Walter's cybernetics was adjusted to his professional career in a double way: its radical aspects flourished outside psychiatry's gates, and within those gates it was domesticated to conform to the status quo. There is no need to be cynical about this: in the forties and early fifties it was possible to be optimistic about the great and desperate psychiatric cures (compared with what had gone before), and there is no reason to doubt that Walter (and Ashby) were genuinely optimistic. Nevertheless, we can also see how ontology and sociology were correlated here. As will become clearer in chapter 5 when we discuss Bateson and Laing, it was possible to go much further than Walter in developing the cybernetic theme of adaptation in psychiatry, but the price of this was a transformation of the social relations of doctors and patients that did, in a Deleuzian fashion, threaten the established order.

Second, it is interesting to ask where this line of first-generation cybernetic psychiatry went. The answer is: nowhere. In Britain, Walter and Ashby were the leading theorists of mental pathology and therapy in the forties and fifties, with their models offering a new understanding of the brain, madness, and its treatment, but histories of twentieth-century psychiatry give them hardly a mention (try Valenstein 1986 and Shorter 1997). And one can think of several reasons why this should be. The first takes us back to the sheer oddity of cybernetics. Walter remained to a significant degree an outsider to psychiatry in his specifically cybernetic work, and it was also the case that Walter's cybernetics made little constructive contribution to psychiatric therapy—it offered an explanation of the mechanisms of ECT and lobotomy without suggesting new therapeutic approaches. Again, one might imagine that Walter had got as far as he could with CORA. It is not clear what his next step might have been in developing this line of research further, or where he could have found support for what would probably have been a significant engineering effort.

But beyond all that, we need to think about two broader developments bearing on psychiatry as a clinical field. The first was the introduction in the 1950s of psychoactive drugs that proved effective in controlling the symptoms of mental disorder, beginning with chlorpromazine (Shorter 1997, chap. 7; Rose 2003). These drugs had their own unfortunate side effects but did not entail the violence and irreversibility of lobotomy, which went into rapid decline in the mid-1950s. ECT enjoyed a longer history, up to the present, but its use also declined in the face of drugs, and the technique lost its cutting-edge, if I can say that, status as the most advanced form of psychiatric therapy.[55] Cybernetics was thus left high and dry in the later 1950s, as a science of clinical practices which were, if not entirely extinct, at least less prevalent than they had been in the 1940s. It is hardly surprising, then, that Walter found other lines of research more attractive from the mid-1950s onward. Ashby continued developing his own cybernetics as a science of psychiatry into the later 1950s, but, as we shall see, he too abandoned his psychiatrically oriented research from around 1960.

The other development we need to think about here is a growing critique in the 1950s of violent psychiatric therapies and even of the use of antipsychotic drugs, a critique which burst into popular consciousness in the 1960s as what was often called the antipsychiatry movement. This movement was not, despite the name, pure critique. It entailed a different way of conceptualizing and acting upon mental disorders, which was, as it happens, itself cybernetic. This gets us back to Bateson and Laing, and on to chapter 5.

Strange Performances

For the remainder of this chapter I want to come at Walter's work from a different angle. The tortoises and CORA were Walter's most distinctive contribution to the early development of cybernetics, but they occupied him only in the late 1940s and early 1950s, and here we can examine some of his more enduring concerns with the brain—and how they crossed over into the counterculture of the sixties.

As I mentioned earlier, Walter's research career was centered on EEG work, and this, like the tortoises though in a different register, again thematized the brain as a performative organ. And the point we need to dwell on now is that, as I remarked in chapter 1, one can be *curious* about the performative brain in a way that a cognitive conception hardly invites. If one thinks about conscious mental operations, as in mainstream AI and the cognitive sciences, there is not much to be curious about. The task for AI is thus to model on a computer familiar cognitive feats like playing chess, solving equations, or logical deduction. In contrast, the performative brain is more of a challenge. We have little conscious access to processes of adaptation, for example. Who knows what a performative brain can do? This is a sense in which the brain appears as one of Beer's exceedingly complex systems, endlessly explorable. Finding out what the brain can do was a central aspect of Walter's research throughout his career, and we can examine some interesting aspects of that here.[56]

Walter's 1953 book *The Living Brain* is largely devoted to the science of the normal brain and its pathologies, epilepsy and mental illness. But in different passages it also goes beyond the pathological to include a whole range of what one might call altered states and strange performances: dreams, visions, synesthesia, hallucination, hypnotic trance, extrasensory perception, the achievement of nirvana and the weird abilities of Eastern yogis and fakirs—the "strange feats" of "grotesque cults" (1953, 148) such as suspending breathing and the heartbeat and tolerating intense pain.[57] What should we make of this?

1. It exemplifies the sort of curiosity about the performative brain that I just mentioned—this is a list of odd things that brains, according to Walter, can do.[58]

2. It conjures up an understanding of the brain as an active participant in the world. Even in the field of perception and representation, phenomena such as dreams and hallucinations might be taken to indicate that the brain does not copy the world but assimilates sensory inputs to a rich inner dynamics. The tortoise did not thematize this aspect of the brain (except, to

a limited degree, in its scanning mechanism), but it is part of what I tried to get at in chapter 2 by mentioning the work of Kauffman and Wolfram on the endogenous dynamics of complex systems, which we will see elaborated in in the following chapters.[59]

3. It is clear that Walter spoke with personal authority about some items on his list of strange performances, while others were abstracted from a more general awareness of other cultures, especially of the East, with India never all that far away in the British imagination. What strikes me about all of the items on the list is that they refer to aspects of the self that are devalued in modernity. We could think of the paradigmatic modern self in terms of the self-contained individual, dualistically opposed to other selves and the material world, a center of reason, calculation, planning, and agency; and measured against such a yardstick dreamers and madmen are defective selves. Or, to put the point more positively, it appears almost inevitable that curiosity about the performative brain is liable to lead one to a nonmodern conception of the self, different from and more expansive than the modern. We might see yogic feats, for instance, as another example of ontological theater—pointing to an understanding of the brain and self as endlessly explorable, exceedingly complex systems and, at the same time, pointing to the sort of performances one might attempt given such a conception of the brain (but that one might never imagine in relation to the representational brain). We can also note that a certain nonmodern spirituality begins to surface here in association with the nonmodern self—a species of earthy spirituality that goes with embodied yogic performances, say, rather than the purified spirituality and the "crossed-out God" of Christianity that Latour (1993) characterizes as part of the "modern settlement." This form of spirituality will also reappear in the following chapters.[60]

4. Walter associated particular altered states and strange performances with specific *technologies of the self*, as I will call them, following Michel Foucault (1988). We have already encountered several examples of these—the specific material setups that Walter used to drive his robots mad (contradictory conditioning across different sensory channels), his techniques for restoring them to sanity (leaving them alone for extended periods, switching them on and off, disconnecting circuits), and their presumptive equivalents in the human world—and we can examine more of them as we go on. But now I should note that the technologies that will concern us are not substantively the same ones that interested Foucault. Foucault's concern was with the histories of specific techniques of *self-control*, aimed at forming specific variants of the autonomous freestanding individual, of the modern self as I just de-

fined it. The technologies that we need to explore, in contrast, undermine the modern duality of people and things by foregrounding couplings of self and others—another instance of ontological theater. On Walter's account, inner states of the brain and, by extension, the self were not to be ascribed to pure inner causes, but to intersections with the nonself, to external configurations like the cross-conditioning setups associated with madness. To emphasize this, I will refer to such techniques as technologies of the nonmodern self. From this angle, too, we see how a conception of the performative brain can lead to a nonmodern decentering of the self—a theme that will come back repeatedly in the following chapters.[61]

5. *The Living Brain* did not simply offer a catalog of altered states and technologies of the self. In more or less detail, Walter also sought to sketch out the mechanisms that connected them. His most detailed accounts were of the go of madness, along the lines sketched out above, and epilepsy (see below). But he also argued that CORA could be taken to illuminate conditioning mechanisms by which Eastern yogis acquired their odd powers over otherwise autonomous bodily functions; that nirvana—"the peace that passeth understanding, the derided 'happiness that lies within' "—could be understood as "the experience of homeostasis" (1953, 39; more on homeostasis in the next chapter); and so on. Again, cybernetics as brain science appears here as the other side of a performative brain that inhabits spaces of ecstasy and madness as well as the everyday world.

6. If Walter's list of strange performances and altered states seems odd and wild, it is because the marginalization of many of its entries has been central to the constitution of modernity and the conception of the dualist, freestanding modern self. The East, with its yogis and fakirs, is the other to modern science, the modern self, and the modern West. Dreams and visions are, shall we say, at the edges of modern consciousness.[62] This is the nonmodernity of cybernetics, once more. But . . .

7. There was a time when the list appeared less wild: the sixties. Madness and ecstasy, the East and Eastern spirituality, strange performances, altered states, explorations of consciousness—these were some trademark preoccupations and practices of the sixties counterculture. We can examine below a couple of direct crossovers from Walter and *The Living Brain* to the sixties, but to make the connection another way, we could think of the work of a canonical sixties author, Aldous Huxley. Huxley's visionary account of his first experience of mescaline in *The Doors of Perception* (1954) became required reading in the sixties, along with its sequel, *Heaven and Hell* (1956; published as a single volume in 1963). And what interests me here is that *Heaven and*

Hell is isomorphous with *The Living Brain* in the respects now under discussion. It, too, offers a long catalog of altered states running from madness to ecstasy and enlightenment, coupled both with an exegesis in terms of Eastern spirituality (specifically Buddhism) and with scientific explanations of the origins of such states. This isomorphism between Walter and Huxley points, I think, to a commonality between cybernetics and the sixties, precisely in a shared interest in the performative brain, a curiosity about what it can do, and, in general, a fascination with nonmodern selves.[63] We can return to the sixties in a moment, but first we need to examine another aspect of Walter's technical EEG research.

Flicker

"Flicker" is a long-standing term of art in experimental psychology, referring to visual effects induced by flickering lights (Geiger 2003, 12–15). A spinning top with black and white bands induces perceptions of color, for example. Walter became interested in flicker and incorporated it into his EEG research in 1945, when he came across a new piece of technology that had become available during the war, an electronic stroboscope. Staring at the machine through closed eyelids, he reported, "I remember vividly the peculiar sensation of light-headedness I felt at flash rates between 6 and 20 [per second] and I thought at once 'Is this how one feels in a petit mal attack?—Of course this could be how one can *induce* a petit mal attack" (Walter 1966, 8).[64] And, indeed, when he experimented with a strobe on an epileptic patient, "within a few seconds a typical wave-&-spike discharge developed as predicted." The quotation continues: "This was enormously exciting because I think it was the first time that a little theory [in EEG research] based on empirical observation had actually been confirmed by experiment. This meant that there might be some hope of reinstating the EEG as a scientific rather than merely utilitarian pursuit. . . . This was one of the critical turning points in our history." The scientific import of flicker in EEG research was thus that it offered a new purchase on the performative brain, and a new neurophysiological and clinical research program opened up here, pursuing the effects of "photic driving" at different frequencies with different subjects. Walter and his colleagues at the Burden, including his wife, Vivian Dovey, experimented on nonepileptic as well as epileptic subjects and found that (Walter 1953, 97) "epileptic seizures are not the exclusive property of the clinically epileptic brain. . . . We examined several hundred 'control' subjects—schoolchildren, students, various groups of adults. In three or four percent of these, carefully adjusted flicker evoked responses indistinguishable from those previously regarded as 'diagnostic' of clinical epilepsy. When these responses appeared, the subjects would exclaim at the 'strange feelings,' the faintness or swimming in the head; some became unresponsive or unconscious for a few moments; in some the limbs jerked in rhythm with the flashes of light." It turned out the optimal flicker frequency for the induction of such effects was often hard to find, and at the Burden Harold "Shippy" Shipton built a feedback apparatus (Walter 1953, 99) "in the form of a trigger circuit, the flash being fired by the brain rhythms themselves. . . . With this instrument the effects of flicker are even more drastic than when the stimulus rate is fixed by the operator. The most significant observation is that in more than 50 per cent of young normal adult subjects, the first exposure to feedback flicker evokes transient paroxysmal discharges of the type seen so often in epileptics" (fig. 3.12).

To follow the details of this research would take us too far afield, so let me make a few comments on it before going back to the sixties.[65] First, Walter's work here exemplifies my earlier remarks about the possibility of being curious about the performative brain. If our capacity for cognitive tasks is immediately before us—I already know that I can do crosswords and sudoku puzzles—the epileptic response to flicker was, in contrast, a surprise, a discovery about what the performative brain can do. Second, this research points again to the psychiatric matrix in which Walter's cybernetics developed. Third, experiments aimed at inducing quasi-epilieptic fits in schoolchildren should only make us grateful for the controls on human-subjects experimentation that have since been introduced.[66] Fourth, flicker is a nice exemplification of my notion of a technology of the self, a material technology for the production of altered states. If you want a paradigmatic example of a technology of the nonmodern self, think of flicker. Fifth and finally, Shippy's feedback circuit deserves some reflection. In the basic flicker setup the brain was pinned down in a linear relation to the technology. The technology did something—flickered—and the brain did something in response—exhibited epileptic symptoms. This counts as a piece of ontological theater inasmuch as it thematizes the performative brain, the brain that acts rather than thinks. But it does not thematize the adaptive brain, the key referent of cybernetics per se: there is no reciprocal back-and-forth between the brain and its environment. Feedback flicker, in contrast, staged a vision of the adaptive brain, albeit in a rather horrifying way. The strobe stimulated the brain, the emergent brainwaves stimulated the feedback circuit, the circuit controlled the strobe, which stimulated the brain, and so on around the loop. We could say that the brain explored the performative potential of the material technology

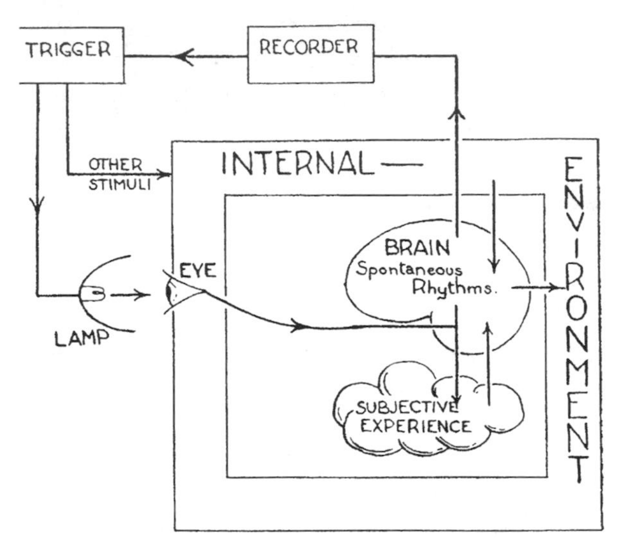

Figure 3.12. Feedback-controlled flicker. Source: V. J. Walter and W. G. Walter, "The Central Effects of Rhythmic Sensory Stimulation," *Electroencephalography and Clinical Neurophysiology,* **1** (1949), 57–86, p. 84, fig. 18.

(in an entirely nonvoluntary, nonmodern fashion), while the technology explored the space of brain performance. I suggested earlier that the tortoise was unsatisfactory as ontological theater inasmuch as its world was largely passive and unresponsive, and I therefore want to note that feedback flicker offers us a more symmetric ontological spectacle, lively on both sides—a dance of agency between the human and the nonhuman. What acted in these experiments was genuinely a cyborg, a lively, decentered combination of human and machine.

We can come back to this below in a discussion of the history of biofeedback, and at a more general level in the following chapter on Ashby's cybernetics.

Flicker and the Sixties

Walter and his colleagues experimented with strobes not only on laboratory subjects but also on themselves, and (Walter 1953, 101) "we all noticed a peculiar effect . . . a vivid illusion of moving patterns whenever one closed one's eyes and allowed the flicker to shine through the eyelids. The illusion . . . takes a variety of forms. Usually it is a sort of pulsating check or mosaic, often in bright colours. At certain frequencies—around 10 per second—some subjects see whirling spirals, whirlpools, explosions, Catherine wheels." Again we can understand these observations as a discovery about the performative brain, continuing a longer tradition of research into such effects in experimental psychology. Walter (1953, 107–13) in fact conjectured that the moving patterns were related to the scanning function of the alpha waves (as materialized in the tortoise): since there is no motion in the strobe light, perhaps the pulsation and whirling in the visual effects comes from the scanning mechanism itself, somehow traveling around the brain. But the language itself is interesting. This passage continues: "A vivid description is given by Margiad Evans in 'A Ray of Darkness': 'I lay there holding the green thumbless hand of the leaf. . . . Lights like comets dangled before me, slow at first and then gaining a fury of speed and change, whirling colour into colour, angle into angle. They were all pure unearthly colours, mental colours, not deep visual ones. There was no glow in them but only activity and revolution.'"[67] What should we make of a passage like that? The word that came to my mind when I first read it was "psychedelic." And I immediately thought of some key texts that were required reading in the sixties, especially Huxley's *The Doors of Perception*. Then I was fortunate enough to obtain a copy of a wonderful recent book by John Geiger called *Chapel of Extreme Experience* (2003).[68] Geiger traces out beautifully how Walter's work on flicker entered into sixties culture. I have little substance to add to Geiger's account, but I want to review his story, since it adds importantly to our topic.

We need to think of three lines of development. First and most conventionally, Walter's observations on flicker fed into a distinctive branch of work in experimental psychology aimed at elucidating its properties, exploring, for example, the kinds of images and visions that flicker produced, and into philosophical reflections on the same. Interestingly, these explorations of flicker were typically entwined with explorations of the effects of psychoactive drugs such as mescaline and LSD. It turned out that the hallucinogenic effects of these drugs are intensified by flicker and vice versa. These fascinating branches of psychological and philosophical research on the performative brain flourished in the 1950s and 1960s but seem since to have been largely forgotten—no doubt due to the criminalization of the drugs.[69] Of more direct interest to the student of popular culture is that Aldous Huxley indeed appears in this story. His 1956 book *Heaven and Hell* indeed includes flicker, experienced on its own or in conjunction with LSD, in its catalog of technologies of the nonmodern self (A. Huxley 1956, 113–14).

At the wildest end of the spectrum, in the late 1950s flicker came to the attention of the group of writers and artists that centered on William Burroughs and Allen Ginsberg, often to be found in Tangiers, where Paul Bowles was a key figure, or staying at the Beat Hotel, 9 rue Git le Coeur in Paris. As I mentioned earlier, the Beats' connection to Walter was textual, chancy, and undisciplined, going via *The Living Brain*. Burroughs read it and was fascinated to find that "consciousness expanding experience has been produced by flicker."[70] For the Beats also, flicker and drugs ran together. In 1959, when Ginsberg took acid for the first time at the Mental Research Institute in Palo Alto, it was in the framework of a typical Grey Walter setup: "Burroughs suggested he did so in concert with a stroboscope. The researchers . . . connected the flicker machine to an EEG, so that Ginsberg's own alpha waves would trigger the flashes." I mentioned earlier the strikingly cyborg aspect of such a configuration, and interestingly, Ginsberg experienced it as such (quoted by Geiger 2003, 47): "It was like watching my own inner organism. There was no distinction between inner and outer. Suddenly I got this uncanny sense that I was really no different than all of this mechanical machinery all around me. I began thinking that if I let this go on, something awful would happen. I would be absorbed into the electrical grid of the entire nation. Then I began feeling a slight crackling along the hemispheres of my skull. I felt my soul being sucked out through the light into the wall socket." Burroughs also gave a copy of *The Living Brain* to another of the Beats, the writer and artist Brion Gysin, who recognized in Walter's description of flicker a quasi-mystical experience he had once had on a bus, induced by sunlight flashing through the trees. Gysin in turn discussed flicker with another member of Burroughs's circle, Ian Sommerville, a mathematics student at Cambridge, and in early 1960 Sommerville built the first do-it-yourself flicker machine—a cylinder with slots around its circumference, standing on a 78 rpm turntable with a 100 watt lightbulb in the middle (fig. 3.13). It turned out that fancy and expensive stroboscopes were not necessary to induce the sought-after effects—this cheap and simple Dream Machine (or Dreamachine), as Gysin called it, was quite enough (Geiger 2003, 48–49).[71]

From here one can trace the cultural trajectory of flicker in several directions. Burroughs both referred to flicker in his writing and built it into his prose style in his "cut-up" experiments (Geiger 2003, 52–53).[72] Gysin and Sommerville published essays on and construction details for their Dream Machine in the journal *Olympia* in February 1962 (Geiger 2003, 62). Timothy Leary, ex-Harvard psychologist and acid guru, was one of the Beats' suppliers of drugs and learned from them of flicker, which he began to discuss, along

Figure 3.13. Brion Gysin and the Dream Machine. Source: Geiger 2003, 50. (Copyright © John Geiger from *Chapel of Extreme Experience: A Short History of Stroboscopic Light and the Dream Machine*. Reprinted by permission of Counterpoint. Photograph copyright © 2000 by Harold Chapman.)

with Grey Walter, in his own writings.[73] Gysin displayed Dream Machines as art objects in a series of exhibitions and argued that they marked a break into a new kind of art that should displace all that had gone before: "What is art? What is color? What is vision? These old questions demand new answers when, in the light of the Dream Machine, one sees all ancient and modern abstract art with eyes closed" (Gysin quoted by Geiger 2003, 62).[74]

Gysin was also taken with the idea of the Dream Machine as a drug-free point of access to transcendental states, and had plans to develop it as a commercial proposition, something to replace the television in people's living rooms, but all his efforts in that direction failed (Geiger 2003, 66 & passim). And in the end, the flicker technology that entered popular culture was not the cheap Dream Machine but the hi-tech strobe light.[75] As Geiger puts it (2003, 82–83): "By 1968 . . . stroboscopic lights were flashing everywhere. They . . . had been taken up by the drug culture. Ken Kesey featured strobe lights in his 'Acid Tests'—parties where he served guests LSD-laced Kool-Aid to the music of the Grateful Dead. . . . Tom Wolfe wrote in *The Electric Kool-Aid Acid Test*: 'The strobe has certain magical properties in the world of acid heads. At certain speeds stroboscopic lights are so synched in with the pattern of brain waves that they can throw epileptics into a seizure. Heads discovered that strobes could project them into many of the sensations of an LSD experience without taking LSD.'" Flicker, then, was an axis along which Walter's cybernetics played into the distinctive culture of the high 1960s.[76] And Walter himself was happy to claim a share of the credit. In a 1968 talk he remarked,

"Illusory experiences produced by flashing lights . . . nowadays are used as a standard method of stimulation in some subcultures. I should be paid a royalty because I was the first to describe these effects" (quoted by Geiger 2003, 83).

This is as far as I can take the story of flicker and the sixties, and the key points to note are, first, that this cultural crossover from Walter's cybernetics to the drug culture and the Beats indeed took place and, second, that the crossover is easy to understand ontologically.[77] In different ways, the sixties and cybernetics shared an interest in the performative brain, with technologies of the decentered self as a point of exchange. The sixties were the heroic era of explorations of consciousness, and flicker joined a whole armory of such sixties technologies: psychedelic drugs, as already mentioned, meditation, sensory deprivation tanks, as pioneered by John Lilly (1972), and even trepanning.[78] In the next section we can take a quick look at yet another such technology: biofeedback. For now, three remarks are in order.

First, just as I conceive of cybernetics as ontology in action, playing out the sort of inquiries that one might associate with a performative understanding of the brain, one can equally see the sixties as a form of ontological theater staging the same concerns, not in brain science but in unconventional forms of daily life.

Second, I want to emphasize the sheer oddity of Gysin's Dream Machines, their discordant relation to everyday objects and the traditions in which they are embedded. In the field of art, it is probably sufficient to quote Gysin himself, who justifiably described the Dream Machine as "the first artwork in history made to be viewed with closed eyes" (Geiger 2003, 54). As a commercial proposition, the Dream Machine was just as problematic. In December 1964, Gysin showed a version to representatives from Columbia Records, Pocketbooks, and Random House, and "all present were soon trying to understand what they had and how to market it. Was it something that could be sold in book form with cut-outs, or was it something that could be sold with LPs? Columbia Records' advertising director Alvin Goldstein suggested the Dream Machine would make a great lamp. Someone said they could be used in window displays" (Geiger 2003, 69). In its unclassifiability, the Dream machine exemplifies in the realm of material technology my thesis that ontology makes a difference.

Finally, I should return to the question of the social transmission of cybernetics. Just as we saw earlier in the history of robotics, flicker's crossover from cybernetics to the Beats took place via a popular book, *The Living Brain*, and thus outside any disciplined form of social transmission. The focus of Walter's book is resolutely on the human brain; it is not a book about art or living-room furniture. But Gysin read "half a sentence," and "I said, 'Oh, wow, that's it!'" (quoted in Geiger 2003, 49). Although not evident in the story of the Walter-Brooks connection in robotics, a corollary of the chancy mode in which cybernetics was transmitted was, as I said earlier, the opportunity for wild mutation—the transmutation of brain science into art objects and psychedelic replacements for the TV.

Biofeedback and New Music

THE SOUNDS THAT ARE "ALLOWED TO BE THEMSELVES" IN LUCIER'S WORK HAVE ALWAYS HAD A MYSTERIOUSLY "EXPRESSIVE" QUALITY. SOMETIMES I THINK IT IS INARTICULATE NATURE SPEAKING TO US HERE.

JAMES TENNEY, "THE ELOQUENT VOICE OF NATURE" (1995)

"Biofeedback" refers to another set of technologies of the nonmodern self, techniques for reading out "autonomous" bodily parameters such as brain rhythms and displaying them to subjects, thus making them potentially subject to purposeful intervention. Shipton's flicker-feedback circuit might be described as such a device, except that there was no choice in the matter: the circuit locked onto the subject's brainwaves and fed them back as flicker whether the subject liked it or not. Walter describes a more voluntary biofeedback arrangement in *The Living Brain* (1953, 240). The onset of sleep and anger is marked by an increase in low-frequency theta rhythms in the brain, and Walter imagines an EEG setup in which this increase flashes a light or rings a bell: "Worn by hard-driving motorists, theta warning-sets would probably save more lives than do motor horns, and they might assist self-knowledge and self-control."[79] In the 1960s, biofeedback came to refer to a species of self-training, in which subjects learned to control aspects of their EEG spectrum (without ever being able to articulate how they did it).[80]

We could follow the history of biofeedback in several directions. Going back to our earlier clinical concerns, Jim Robbins (2000) offers a popular account of the history of biofeedback in psychiatry and of present-day uses in the treatment of a whole range of disorders including epilepsy, learning disabilities, autism, and PTSD.[81] He notes, however, that biofeedback was also taken up by the sixties counterculture in pursuit of alpha-wave-dominated states that had become identified with transcendental experiences (fig. 3.14). The first meeting of biofeedback professionals took place at Snowmass, Colorado, in 1968, and the attendees were "a mixture of uptight scientific types . . . and people barefooted, wearing white robes, with long hair. It attracted the

Figure 3.14. EEG biofeedback. The photograph ran with an article entitled "What a Sexy Brainwave" and had a caption reading, "Georgina Boyle calls up those no-worry waves." Source: *Sunday Mirror* (London), 12 December 1971, 22.

heads to a tremendous extent" (Robbins 2000, 65, quoting Joe Kamiya, a pioneer in the scientific exploration of biofeedback). David Rorvik (1970) elaborates on this in much the same terms as were applied to flicker: "Now, with the dawning of the cybernetic seventies, it is not too surprising that LSD and the other hallucinogens of the sixties are about to be eclipsed, in a sense, by an electronic successor: BFT. Bio-Feedback Training, or 'electronic yoga' as it has been called, puts you in touch with inner space, just like LSD but, unlike acid, leaves you in full control of your senses. And, unlike meditation, it doesn't take years of sitting on mountaintops to master. . . . There are those who believe that biofeedback training may not only illuminate the myriad workings of the mind but may even fling open the doors to entirely new kinds of experience, extending the inner dimensions of the emergent cybernetic man" (1970, 175–76).

Here, though, I want to explore another intersection of cybernetics and the arts. If flicker was a distinctive and paradoxical contribution to the visual arts, biofeedback in turn fed into the New Music of the 1960s, usually associated with names like John Cage and David Tudor.[82] The idea was simple enough. In a basic brainwave biofeedback setup, a light comes on when the subject's alpha output, say, exceeds some level, and by focusing on keeping the light lit, subjccts somehow learn to boost their alpha level at will. To go from this setup to music, all that was required was to substitute sound for the visual element of the feedback circuit. The difficulty was, in the first instance, that alpha frequencies are below the range of hearing, and one solution, used from time to time, was to record brain activity to tape and then play it back in speeded-up form, thus making it audible. The drawback to such a solution was that it blocked the possibility of any real-time feedback coupling between performer and performance, and the first recognized EEG music event followed a different route. First performed live in 1965, Alvin Lucier's *Music for Solo Performer* fed the EEG readout directly into loudpeakers whenever the alpha rhythms were above the threshold, generating an audible output by putting the speakers next to or in contact with "gongs, timpani, bass drums, anything that loudspeakers could vibrate sympathetically" (Lucier 1995, 50)—even a metal dustbin (fig. 3.15).[83]

Several points are worth noting about this style of alpha music. Most evidently, like feedback-controlled flicker, it brings us face to face with a form of decentering of the self into a technosocial apparatus. Any given performance of *Music for Solo Performer* was not the work of a solo performer: it was the work of a human plus EEG electrodes, amplifiers and signal analyzers, switches, loudspeakers, and sound generating devices of all sorts. Second, even with extensive biofeedback training, in such a setup the performer does not exercise absolute control over the performance. From one angle, the sounds themselves are what enable the performer to tune into the generation of alpha waves—that is the principle of biofeedback. Nevertheless, "although theoretically it [the alpha rhythm] is a continual pattern of ten hertz, it never comes out that way because it stops when your eyelids flutter or you visualise a little and it tends to drift down a bit if you get bored or sleepy" (Lucier 1995, 58). One has the sense, then, of a reciprocal and open-ended interplay between the performer and the performance, with each both stimulating and interfering with the other—a kind of reciprocal steersmanship, in the sense discussed in chapter 2. We can go into this further in chapter 6, on Brian Eno's music, and chapter 7, on Pask's cybernetic aesthetics, but I want to suggest here that biofeedback music can stand as another and very nice example of

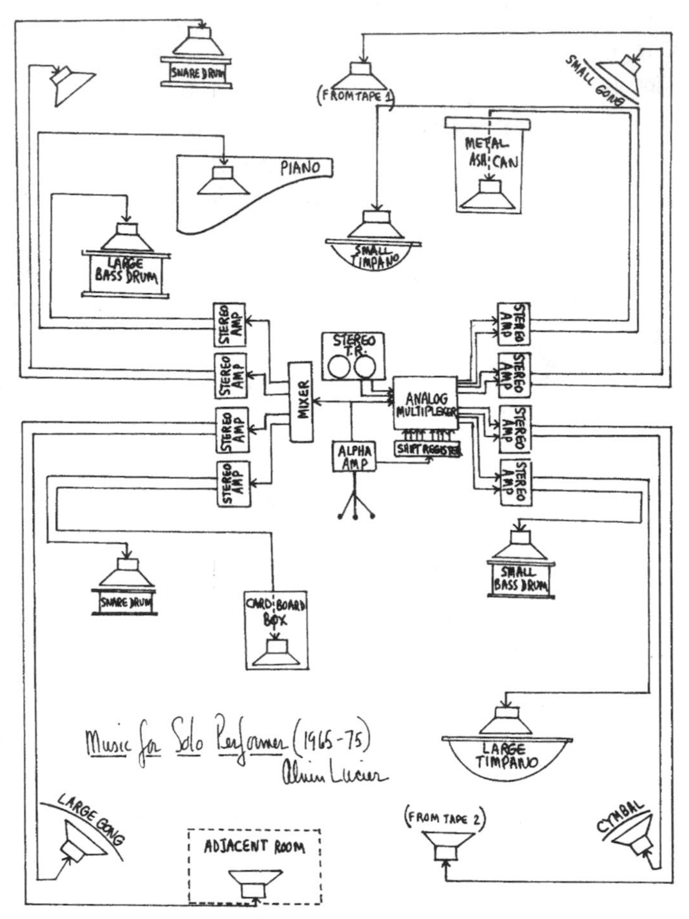

Figure 3.15. Music for solo performer. Source: A. Lucier, *Reflections: Interviews, Scores, Writings*, edited by G. Gronemeyer and R. Oehlschlägel (Köln: MusikTexte, 1995), 54.

ontological theater—of an open-ended and performative interplay between agents that are not capable of dominating each other. Second, I do not need to labor the point that here again ontology makes a difference—*Music for Solo Performer* is self-evidently different from mainstream notions of music. As James Tenney (1995, 12) put it, "Before [the first performance of *Music for a Solo Performer*] no one would have thought it necessary to define the word 'music' in a way which allowed for such a manifestation; afterwards some definition could not be avoided." Third, we can note that we are once more back on the terrain of altered states (and, literally, strange performances!). Lucier speaks of a "perfectly meditative alpha state" (1995, 56), and, in this sense, the decentered quality of the musical performance hung together with a decentered, nonmodern subject position of the performer. Fourth, I want to comment on what I think of as the *hylozoism* of this sort of music, but to get clear on that it helps to refer to the work of another pioneer in this field, Richard Teitelbaum.

Teitelbaum was yet another person who had a transformative encounter with Walter's writings. In 1966, "by chance, I found a copy of W. Grey Walter's pioneering work *The Living Brain* in Rome. Studying it thoroughly, I was particularly interested in the sections on flicker and alpha feedback, and by descriptions of the hallucinatory experiences reported by some subjects" (Teitelbaum 1974, 55). Having learned of Lucier's work, Teitelbaum hit upon the idea of using EEG readouts to control the electronic synthesizers then being developed in the United States by Robert Moog (on which see Pinch and Trocco 2002), which led to the first performance of a work called *Spacecraft* by the Musica Elettronica Viva Group on a tour of Europe in autumn 1967 (Teitelbaum 1974, 57). On the experience of performing in *Spacecraft*, Teitelbaum recalled that (59)

> the unusual sensations of body transcendence and ego-loss that occurred in this music—and in related biofeedback experiences—seemed aptly described . . . in the Jewish mystical texts of the Kabbalah: in the state of ecstacy a man "suddenly sees the shape of his self before him talking to him and he forgets his self and it is disengaged from him and he sees the shape of his self before him talking to him and predicting the future." With five musicians simultaneously engaged in the same activities—electronically mixing, inter-modulating with each other and issuing from the same loudspeakers—a process of non-ordinary communication developed, guiding individual into collective consciousness, merging the many into one.

By the slippery word "hylozoism" I want to refer to a spiritually charged awe at the performative powers of nature that seems to inhabit this quotation: the idea evident in Teitelbaum's and Lucier's work (and in the New Music of the sixties more generally) that, so to speak, it's all there in nature already, that the classically modern detour through human creativity and design is just that, a detour that we could dispense with in favor of making nature itself—here the alpha rhythms of the brain—audible (or visible).[84] Let me just note for the moment that this idea goes very well with the cybernetic ontology of performative interaction. Again we can understand Teitelbaum's work as cybernetic ontological theater—an approach to music that at once conjures

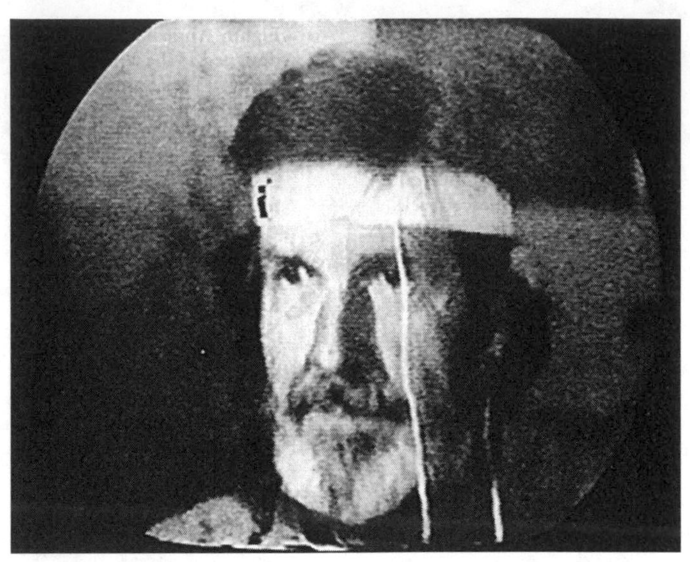

Figure 3.16. Still from a video of John Cage during alpha feedback. Source: Teitelbaum 1974, 68.

up the overall ontological vision and exemplifies how that vision might be distinctively instantiated and developed in real-world practice. The topic of hylozoism recurs in the following chapters in various guises, at greatest length in chapter 6, on Stafford Beer. We can pick up the related question of a distinctively cybernetic stance on design in the next chapter, on Ross Ashby.[85]

— — — — —

This is the end of our first close encounter with British cybernetics. In terms of material technologies, I described Walter's tortoises as specimens of ontological theater, contemplation of which helps one to grasp the performative and adaptive ontology of cybernetics, and as ontology in action, an instance of how one might proceed in brain science (and other fields) if one takes that ontology seriously. The contrast between Walter's robotics and that associated with AI illustrates my idea that ontology makes a difference—that very different practices can hang together with different understandings of what the world is like. From the tortoises we moved on to CORA, which staged for us a performative epistemology, directly geared into the performative ontology staged by the naked tortoise, and which also made the connection between Walter's cybernetics and the psychiatric milieu from which it emerged. Finally, the discussion of flicker and biofeedback touched on other lines of inquiry into the performative brain and crossovers between cybernetics and the psychedelic sixties, with the sixties, too, graspable as ontological theater and ontology in action.

At the same time, the history of Walter's cybernetics begins an exemplification of what I called the protean quality of cybernetics, with the tortoises spanning the worlds of brain science, psychiatry, robotics, and entertainment—and we can now add to this list the Dream Machine and biofeedback setups as pivots to the wider culture and the arts. This multiplicity can be associated with the lack of any stable institutional basis for cybernetics, with first the Ratio Club and then the Namur conferences as key nexuses in Britain and Europe; and with the disorganized, undisciplined mode of cybernetic transmission and the possibilities for mutation that went with that.

Next we come to Walter's contemporary in the first generation of British cyberneticians, Ross Ashby. As we shall see, Ashby's cybernetics grew around a notion of adaptation that was different from and richer than Walter's, and it was, in fact, Ashby's vision of adaptation (shared by Gregory Bateson) that informed the work of the second generation, Stafford Beer and Gordon Pask.

4

ROSS ASHBY

PSYCHIATRY, SYNTHETIC BRAINS, AND CYBERNETICS

HAVING DECIDED (HEAVEN FORGIVE ME, BUT IT IS MY CONVICTION) TO FOLLOW IN DARWIN'S FOOTSTEPS, I BOUGHT HIS AUTOBIOGRAPHY TO GET SOME HINTS ON HOW TO DO IT.

ROSS ASHBY, JOURNAL ENTRY, 29 JUNE 1945 (ASHBY 1951–57, P. 1956)

William Ross Ashby (fig. 4.1), always known as Ross, was born in London on 6 September 1903.[1] After failing the entrance exam for the City of London School, he finished his schooling at the Edinburgh Academy between 1917 and 1921 and then graduated from Sidney Sussex College, Cambridge, with a BA in zoology in 1924. He was an unhappy child, incapable of living up to the expectations of a demanding father, and this unhappiness remained with him for many years.[2] Ashby's father wanted him to pursue a career in either medicine or the law and, opting for the former, on leaving Cambridge Ashby trained at St. Bartholomew's Hospital, receiving the M.B. and B.Ch. degrees in 1928 (qualifying him to practice as a doctor) and the M.D. degree in 1935, both from Cambridge. In 1931 he was awarded a diploma in psychological medicine by the Royal College of Physicians and Surgeons. From 1930 to 1936 he was employed by London County Council as a clinical psychiatrist at Leavesden Mental Hospital in Hertfordshire. In 1931 Ashby married Elsie Maud Thorne—known to her intimates as Rosebud; Mrs. Ashby to others; born in 1908; employed at that point in the Millinery Department at Liberty's

Figure 4.1. W. Ross Ashby. (By permission of Jill Ashby, Sally Bannister, and Ruth Pettit.)

on Regent Street—and between 1932 and 1935 they had three daughters, Jill, Sally, and Ruth.

From 1936 to 1947 Ashby was a research pathologist at St. Andrew's mental hospital in Northampton, an appointment he continued to hold while serving from 1945 until 1947 as a specialist pathologist in the Royal Army Medical Corps with the rank of lieutenant and later major. From June 1945 until May 1946 he was posted to India, in Poona and Bangalore. Returning to England, he became director of research at another mental institution, Barnwood House in Gloucester, in 1947 and remained there until 1959, when he was appointed director of the Burden Neurological Institute in Bristol, succeeding Frederick Golla and becoming Grey Walter's boss. In January 1961, after just a year at the Burden, Ashby moved to the United States to join the University of Illinois (Urbana-Champaign) as a professor in the Department of Electrical Engineering, primarily associated with Heinz von Foerster's Biological Computer Laboratory (BCL) but with a joint appointment in biophysics. He remained at the BCL until his retirement as an emeritus professor in 1970, when he returned to Britain as an honorary professorial fellow at the University of Wales, Cardiff. He died of a brain tumor shortly afterward, on 15 November 1972, after five months' illness.

Ashby's first recognizably cybernetic publication, *avant la lettre*, appeared in 1940. In the mid-1940s he began to make contact with other protocyberneticians, and in 1948 at Barnwood House he built the cybernetic machine for which he is best remembered, the homeostat, described by Norbert Wiener (1967 [1950], 54) as "one of the great philosophical contributions of the present day." The concept of adaptation staged by the homeostat, different from Walter's, will echo through the following chapters. Over the course of his career, Ashby published more than 150 technical papers as well as two enormously influential books: *Design for a Brain* in 1952 and *An Introduction to Cybernetics* in 1956, both translated into many languages. From the homeostat onward, Ashby was one of the leaders of the international cybernetics community—a founding member of the Ratio Club in Britain, an invitee to the 1952 Macy cybernetics conference in the United States, and, reflecting his stature in the wider world of scholarship, an invited fellow at the newly established Center for Advanced Study in the Behavioral Sciences in Palo Alto, California, in 1955–56. After moving to Illinois, he was awarded a Guggenheim Fellowship in 1964–65, which he spent back in England as a visiting research fellow at Bristol University.[3]

Ashby's contributions to cybernetics were many and various, and I am not going to attempt to cover them all here. Speaking very crudely, one can distinguish three series of publications in Ashby's oeuvre: (1) publications relating to the brain that one can describe as distinctly cybernetic, running up to and beyond *Design for a Brain*; (2) distinctly medical publications in the same period having to do with mental pathology; and (3) more general publications on complex systems having no especial reference to the brain, running roughly from the publication of *An Introduction to Cybernetics* and characterizing Ashby's later work at Illinois. My principle of selection is to focus mostly on the first and second series and their intertwining, because I want to explore how Ashby's cybernetics, like Walter's, developed as brain science in a psychiatric milieu. I will explore the third series only as it relates to the "instability of the referent" of the first series: although Ashby's earlier work always aimed to elucidate the functioning of the brain, normal and pathological, he developed, almost despite himself, a very general theory of machines. My object here is thus to explore the way that Ashby's cybernetics erupted along this line into a whole variety of fields, but I am not going to follow in any detail his later articulation of cybernetics as a general science of complex systems. This later work is certainly interesting as theory, but, as I have said before, I am most interested in what cybernetics looked like when put into practice in real-world projects, and here the natural trajectory runs from Ashby's cybernetic brain not into his own work on systems but into Stafford Beer's management cybernetics—the topic of the next chapter.

The skeleton of what follows is this. I begin with a brief discussion of Ashby's distinctly clinical research. Then I embark on a discussion of the development of his cybernetics, running through the homeostat and *Design for a Brain* up to the homeostat's failed successor, DAMS. Then I seek to reunite these two threads in an exploration of the relation between Ashby's cybernetics and his clinical work up the late 1950s. After that, we can pick up the third thread just mentioned, and look at the extensions of Ashby's research beyond the brain. Finally, I discuss echoes of Ashby's work up to the present, in fields as diverse as architecture, theoretical biology and cellular automata studies. Throughout, I draw heavily upon Ashby's handwritten private journal that he kept throughout his adult life and various notebooks, now available at the British Library in London.[4]

The Pathological Brain

When one reads Ashby's canonical works in cybernetics it is easy to imagine that they have little to do with his professional life in medicine and psychiatry. It is certainly the case that in following the trajectory of his distinctive contributions to cybernetics, psychiatry recedes into the shadows. Nevertheless, as I will try to show later, these two strands of Ashby's research were intimately connected, and, indeed, the concern with insanity came first. To emphasize this, I begin with some remarks on his medical career.

Overall, it is important to remember that Ashby spent his entire working life in Britain in mental institutions; it would be surprising if that milieu had nothing to do with his cybernetic vision of the brain. More specifically, it is clear that Ashby, like Walter, belonged to a very materialist school of psychiatry led in Britain by Frederick Golla. Though I have been unable to determine when Ashby first met Golla and Walter, all three men moved in the same psychiatric circles in London in the mid-1930s, and it is probably best to think of them as a group.[5] It is clear, in any event, that from an early date Ashby shared with the others a conviction that all mental phenomena have a physical basis in the brain and a concomitant concern to understand the go of the brain, just how the brain turned specific inputs into specific outputs. And this concern is manifest in Ashby's earliest publications. At the start of his career, in London between 1930 and 1936, he published seventeen research papers in medical journals, seeking in different ways to explore connections between mental problems and physical characteristics of the brain, often based on postmortem dissections. Such writings include his very first publication, "The Physiological Basis of the Neuroses," and a three-part series, "The Brain of the Mental Defective," as well as his 1935 Cambridge MA thesis, "The Thickness of the Cerebral Cortex and Its Layers in the Mental Defective" (Ashby 1933, 1935; Ashby and Stewart 1934–35).

Such research was by no means untypical of this period, but it appears to have led nowhere. No systematic physiological differences betwen normal and pathological brains were convincingly identified, and Ashby did not publish in this area after 1937.[6] After his move to St. Andrew's Hospital in 1936, Ashby's research into insanity moved in several directions.[7] The January 1937 annual report from the hospital mentions a survey of "the incidence of various mental and neurological abnormalities in the general population, so that this incidence could be compared with the incidence in the relatives of those suffering from mental or neurological disorders. . . . Dr. Ashby's work strongly suggests that heredity cannot be so important a factor as has sometimes been maintained" (Ashby 1937a). The report also mentions that Ashby and R. M. Stewart had studied the brain of one of Stewart's patients who had suffered from a rare form of brain disease (Ashby, Stewart, and Watkin 1937), and that Ashby had begun looking into tissue culture methods for the investigation of brain chemistry (Ashby 1937b). Ashby's pathological work continued to feature in the January 1938 report, as well as the fact that "Dr. Ashby has also commenced a study on the theory of organisation as applied to the nervous system. It appears to be likely to yield interesting information about the fundamental processes of the brain, and to give more information about the ways in which these processes may become deranged"—this was the beginning of Ashby's cybernetics, the topic of the next section.

According to the St. Andrew's report from January 1941, "Various lines of research have been undertaken in connection with Hypoglycaemic Therapy. Drs. Ashby and Gibson have studied the effects of Insulin as a conditioned stimulus. Their results have been completed and form the basis of a paper awaiting publication. They are actively engaged also in studying various metabolic responses before and after treatment by Insulin and Cardiazol. The complications arising from treatment by these methods are being fully investigated and their subsequent effects, if any, carefully observed. It is hoped to publish our observations at an early date." Here we are back in the realm of the great and desperate psychiatric cures discussed in the previous chapter. Insulin and cardiazol were used to induce supposedly therapeutic convulsions in

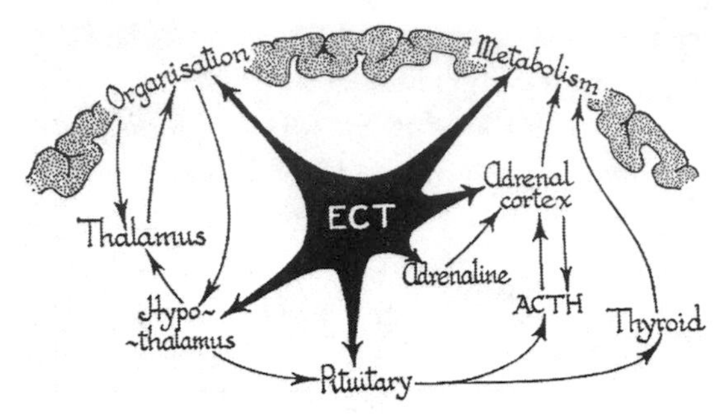

Figure 4.2. "The most important variables affected by E.C.T." Reproduced with permission from W. R. Ashby, "The Mode of Action of Electro-convulsive Therapy," *Journal of Mental Science, 99* (1953), 203, fig. 1. (© 1953 The Royal College of Psychiatrists.)

mental patients, and we can note that in this work Ashby had moved from his earlier interest in the pathological brain per se to the biological mechanisms of psychiatric treatment.

This shift in focus intensified after Ashby's move to Barnwood House in 1947. Not far from the Burden Neurological Institute, Barnwood House was at the epicenter of radical psychiatric cures in Britain. Its director, G. W. T. H. Fleming, was the first author listed, with Golla and Walter, on the first published report on the use of electroconvulsive therapy in Britain (Fleming, Golla, and Walter 1939, discussed in the previous chapter). Ashby had no doubts about the efficacy of ECT: "Electroshock therapy . . . has long passed its period of probation and is now universally accepted as active and effective." "Yet," he wrote, "its mode of action is still unknown." From its introduction there had been speculation that ECT achieved its ends not directly, via the shock itself, but by inducing some therapeutic change in the chemistry of the brain, and this was what Ashby sought to elucidate at Barnwood House, most notably in a long essay on his empirical research published in 1949, which won a prize—the £100 Burlingame Prize awarded by the Royal Medico-Psychological Association. There, Ashby reported on his own observations on fourteen mental patients who had been subjected to ECT and concluded, "The usual effect of convulsive therapy is to cause a brisk outpouring of adrenal chemical steroids during the first few days of the treatment. . . . There is evidence that [this] outpouring . . . is associated with a greater tendency to clinical recovery" (Ashby 1949a, 275, 321). Again, we see the characteristic concern to illuminate the material "go of it"—now to spell out the beginning of a chain of effects leading from the administration of electroshock to modified mental performances. And Ashby followed this up in, for example, a 1953 paper entitled "The Mode of Action of Electro-convulsive Therapy," in which he reported his own research on rats subjected to ECT, using an assay of his own devising to explore ECT's effects on the "adenohypophyseal-adrenocortical system" (Ashby 1953a; see also Ashby 1949b for earlier rat experiments on this topic).

It is clear, then, that Ashby was actively involved in a certain kind of clinical psychiatric research well into his fifties, trying to understand the material peculiarities of pathological brains and how therapeutic interventions worked. This was his professional life until he left Britain in 1961, and I will come back to it. Now, however, we can move to a more rarefied plane and explore the development of Ashby's distinctive cybernetic understanding of the brain.

Ashby's Hobby

Shortly after Ashby's death, his wife wrote to Mai von Foerster, Heinz's wife and a family friend at the University of Illinois:

> I came across a very private notebook the other day written in 1951. In it Ross wrote: After I qualified, work on the brain, of the type recorded in my notebooks, was to me merely a delightful amusement, a hobby I could retreat to, a world where I could weave complex and delightful patterns of pure thought, untroubled by social, financial or other distractions. So the work which I had treated for years only as a hobby began to arouse interest. I was asked to broadcast about it in March, 1949. My fear is now that I may become conspicuous, for a book of mine is in the press. For this sort of success I have no liking. My ambitions are vague—someday to produce something faultless.[8]

The notebook in question is "Passing through Nature," Ashby's biographical notebook, written between 1951 and 1957 (see note 4).[9] The broadcast Ashby referred to was a thirty-minute program on BBC radio, "Imitating the Brain," transmitted on 8 March 1949, for which he was paid twenty-six pounds and five shillings (i.e., twenty-five guineas) plus fifteen shillings and threepence rail fare; the book is *Design for a Brain*, which appeared in 1952.[10] My aim now is to trace out the evolution of the strand of Ashby's early work that led up to and included *Design*. I am interested in its substance and how it emerged from the hobbyist shadows to establish Ashby's reputation as one of the world's leading cyberneticians. In a biographical note from 1962 Ashby wrote that "since 1928 Ashby has given most of his attention to the problem: How can the brain be at once mechanistic and adaptive? He obtained the solution in

1941, but it was not until 1948 that the Homeostat was built to embody the special process. . . . Since then he has worked to make the theory of brainlike mechanisms clearer" (Ashby 1962, 452). I will not try to trace out the evolution of his thinking from 1928 onward; instead, I want to pick up the historical story with Ashby's first protocybernetic publication. As I said, Ashby's clinical concerns are very much marginalized in his key cybernetic works, which focus on the normal rather than the pathological brain, but we can explore the interconnections later.

Ashby's first step in translating his hobbyist concerns into public discourse was a 1940 essay entitled "Adaptiveness and Equilibrium" published in the *Journal of Mental Science*. In a journal normally devoted to reports of mental illness and therapies, this paper introduced in very general terms a dynamic notion of equilibrium drawn from physics and engineering. A cube lying on one of its faces, to mention Ashby's simplest example, is in a state of dynamic equilibrium inasmuch as if one tilts it, it will fall back to its initial position. Likewise, Ashby noted, if the temperature of a chicken incubator is perturbed, its thermostat will tend to return it to its desired value. In both cases, any disturbance from the equilibrium position calls forth opposing forces that restore the system to its initial state. One can thus say that these systems are able to *adapt* to fluctuations in their environment, in the sense of being able to cope with them, whatever they turn out to be. Much elaborated, this notion of adaptation ran through all of Ashby's later work on cybernetics as brain science, and we can note here that it is a different notion from the one I associated with Walter and the tortoise in the previous chapter. There "adaptation" referred to a sensitive spatial engagement with the environment, while for Ashby the defining feature of adaptation was finding and maintaining a relation of dynamic equilibrium with the world. This divergence lay at the heart of their different contributions to cybernetics.

Why should the readers of the *Journal of Mental Science* be interested in all this? Ashby's idea unfolded in two steps. One was to explain that dynamic equilibrium was a key feature of life. A tendency for certain "essential variables" to remain close to some constant equilibrium value in the face of environmental fluctuations was recognized to be a feature of many organisms; Ashby referred to the pH and sugar levels of the blood and the diameter of the pupil of the eye as familiar examples. Tilted cubes and thermostats could thus be seen as formal models for real organic adaptive processes—the mechanisms of *homeostasis*, as it was called, though Ashby did not use that word at this point. And Ashby's second step was to assert that "in psychiatry its importance [i.e., the importance of adaptiveness] is central, for it is precisely the loss of this 'adaptiveness' which is the reason for certification [i.e., forcible confinement to a mental institution]" (478). Here he tied his essay into a venerable tradition in psychiatry going back at least to the early twentieth century, namely, that madness and mental illness pointed to a failure to adapt—an inappropriate mental fixity in the face of the flux of events (Pressman 1998, chap. 2). As we saw, Walter's *M. docilis* likewise lost its adaptivity when driven mad.

Ashby's first cybernetic paper, then, discussed some very simple instances of dynamic equilibrium and portrayed them as models of the brain. One is reminded here of Wiener's cybernetics, in which feedback systems stood in as model of the brain, and indeed the thermostat as discussed by Ashby was none other than such a system. And two points are worth noting here. First, a historical point: Ashby's essay appeared in print three years before Arturo Rosenblueth, Wiener, and Julian Bigelow's classic article connecting servomechanisms and the brain, usually regarded as the founding text of cybernetics. And second, while Rosenblueth, Wiener, and Bigelow (1943) thought of servomechanisms as models for purposive action in animals and machines, Ashby's examples of homeostatic mechanisms operated below the level of conscious purpose. The brain adumbrated in Ashby's paper was thus unequivocally a performative and precognitive one.

— — — — —

I quoted Ashby as saying that he solved the problem of how the brain can be at once mechanistic and adaptive in 1941, and his major achievement of that year is indeed recorded in a notebook entitled "The Origin of Adaptation," dated 19 November 1941, though his first publication on this work came in an essay submitted in 1943 and only published in 1945, delayed, no doubt, by the exigencies of war (Ashby 1945a). The problematic of both the notebook and the 1945 publication is this: Some of our biological homeostatic mechanisms might be given genetically, but others are clearly acquired in interaction with the world. One of Ashby's favorite adages was, The burned kitten fears the fire. The kitten *learns* to maintain a certain distance from the fire—close enough to keep warm, but far away enough not get to burned again, depending, of course, on how hot the fire is. And the question Ashby now addressed himself to was how such learning could be understood mechanistically—what could be the go of it? As we have seen, Walter later addressed himself to the question of learning with his conditioned reflex analogue, CORA. But Ashby found a different solution, which was his first great contribution to brain science and cybernetics.

The 1945 essay was entitled "The Physical Origin of Adaptation by Trial and Error," and its centerpiece was a strange imaginary machine: "a frame with a number of heavy beads on it, the beads being joined together by elastic strands to form an irregular network." We are invited to think of the positions and velocities of the beads as the variables which characterize the evolution of this system in time, and we are invited also to pay attention to "the constants of the network: the masses of the beads, the lengths of the strands, their arrangement, etc. . . . These constants are the 'organization' [of the machine] by definition. Any change of them would mean, really, a different network, and a change of organization." And it is important to note that in Ashby's conception the "constants" can change; the elastic breaks if stretched too far (Ashby 1945a, 15–16).[11]

The essay then focuses on the properties of this machine. Suppose we start it by grabbing one of the beads, pulling it against the elastic, and letting go; what will happen? There are two possibilities. One is that the whole system of beads and elastic will twang around happily, eventually coming to a stop. In that case we can say that the system is in a state of dynamic equilibrium, as defined in the 1940 essay, at least in relation to the initial pull. The system is already adapted, as one might say, to that kind of pull; it can cope with it.

But now comes the clever move, which required Ashby's odd conception of this machine in the first place. After we let go of the bead and everything starts to twang around, one of the strands of elastic might get stretched too far and break. On the above definition, the machine would thus change to a different state of organization, in which it might again be either stable or unstable. In the latter case, more strands would break, and more changes of organization would take place. And, Ashby observed, this process can continue indefinitely (given enough beads and elastic) until the machine reaches a condition of stable equilibrium, when the process will stop. None of the individual breaks are "adaptive" in the sense of necessarily leading to equilibrium; they might just as well lead to new unstable organizations. In this sense, they are *random*—a kind of nonvolitional trial-and-error process on the part of the machine. Nevertheless, the machine is *ultrastable*—a technical term that Ashby subsequently introduced—inasmuch as it tends inexorably to stable equilibrium and a state of adaptedness to the kinds of pull that initially set it in motion. "The machine finds this organization automatically if it is allowed to break freely" (1945a, 18).

Here, then, Ashby had gone beyond his earlier conception of a servomechanism as a model for an adaptive system. He had found the solution to the question of how a machine might become a servo relative to a particular stimulus, how it could learn to cope with its environment, just as the burned kitten learns to avoid the fire. He had thus arrived at a far more sophisticated model for the adaptive and performative brain than anyone else at that time.

The Homeostat

The bead-and-elastic machine just discussed was imaginary, but on 19 November 1946 Ashby began a long journal entry with the words "I have been trying to develope [*sic*] further principles for my machine to illustrate stability, & to develope ultrastability." There followed eight pages of notes, logic diagrams and circuit diagrams for the machine that he subsequently called the homeostat and that made him famous. The next entry was dated 25 November 1946 and began: "Started my first experiment! How I hate them! Started by making a Unit of a very unsatisfactory type, merely to make a start."[12] He then proceeded to work his way through a series of possible designs, and the first working homeostat was publicly demonstrated at Barnwood House in May 1947; a further variant was demonstrated at a meeting of the Electroencephalographic Society at the Burden Neurological Institute in May 1948.[13] This machine became the centerpiece of Ashby's cybernetics for the next few years. His first published account of the homeostat appeared in the December 1948 issue of the journal *Electronic Engineering* under the memorable title "Design for a Brain," and the same machine went on to feature in the book of the same name in 1952. I therefore want to spend some time discussing it.

The homeostat was a somewhat baroque electromechanical device, but I will try to bring out its key features. Figure 4.4a in fact shows four identical homeostat units which are all electrically connected to one another. The interconnections cannot be seen in the photograph, but they are indicated in the circuit diagram of a single unit, figure 4.4c, where it is shown that each unit was a device that converted electrical inputs (from other units, on the left of the diagram, plus itself, at the bottom) into electrical outputs (on the right). Ashby understood these currents as the homeostat's essential variables, electrical analogues of blood temperature or acidity or whatever, which it sought to keep within bounds—hence its name—in a way that I can now describe.

The inputs to each unit were fed into a set of coils (*A*, *B*, *C*, *D*), producing a magnetic field which caused a bar magnet (*M*) to pivot about a vertical axis. Figure 4.4b is a detail of the top of a homeostat, and shows the coils as a flattened oval within a Perspex housing, with the right-hand end of the bar magnet just protruding from them into the light. Attached to the magnet and rotating with it was a metal vane—the uppermost element in figures 4.4b and 4.4c—which was bent at the tip so as to dip into a trough of water—the

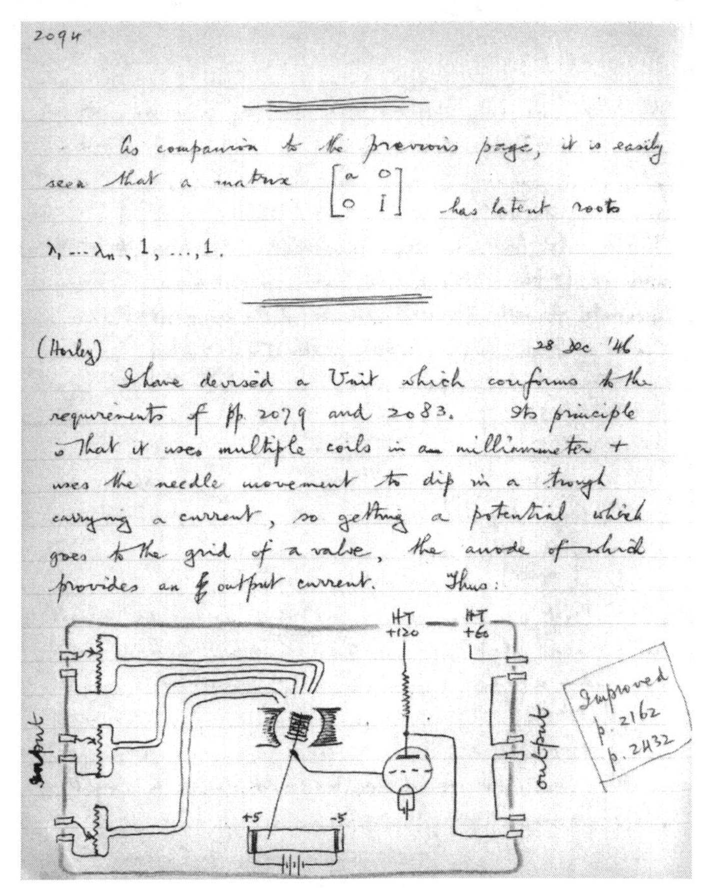
2094

As companion to the previous page, it is easily seen that a matrix $\begin{bmatrix} a & 0 \\ 0 & I \end{bmatrix}$ has latent roots $\lambda_1, \ldots \lambda_n, 1, \ldots, 1$.

(Henley) 28 Dec '46

I have devised a Unit which conforms to the requirements of pp. 2079 and 2083. Its principle is that it uses multiple coils in a milliammeter + uses the needle movement to dip in a trough carrying a current, so getting a potential which goes to the grid of a valve, the anode of which provides an output current. Thus:

Figure 4.3. Page from Ashby's journal, including his first sketch of the homeostat wiring diagram. Source: Journal entry dated 28 December 1946 (p. 2094). (By permission of Jill Ashby, Sally Bannister, and Ruth Pettit.)

curved Perspex dish at the front of figure 4.4b, the arc at the top of figure 4.4c. As indicated in figure 4.4c, an electrical potential was maintained across this trough, so that the tip of the vane picked up a voltage dependent on its position, and this voltage then controlled the potential of the grid of a triode valve (unlabeled: the collection of elements enclosed in a circle just below and to the right of *M* in figure 4.4c; the grid is the vertical dashed line through the circle), which, in turn, controlled the output currents.

Thus the input-output relations of the homeostat except for one further layer of complication. As shown in figure 4.4c, each unit could operate in one of two modes, according to the setting of the switches marked *S*, the lower row of switches on the front of the homeostat's body in figure 4.4a. For one

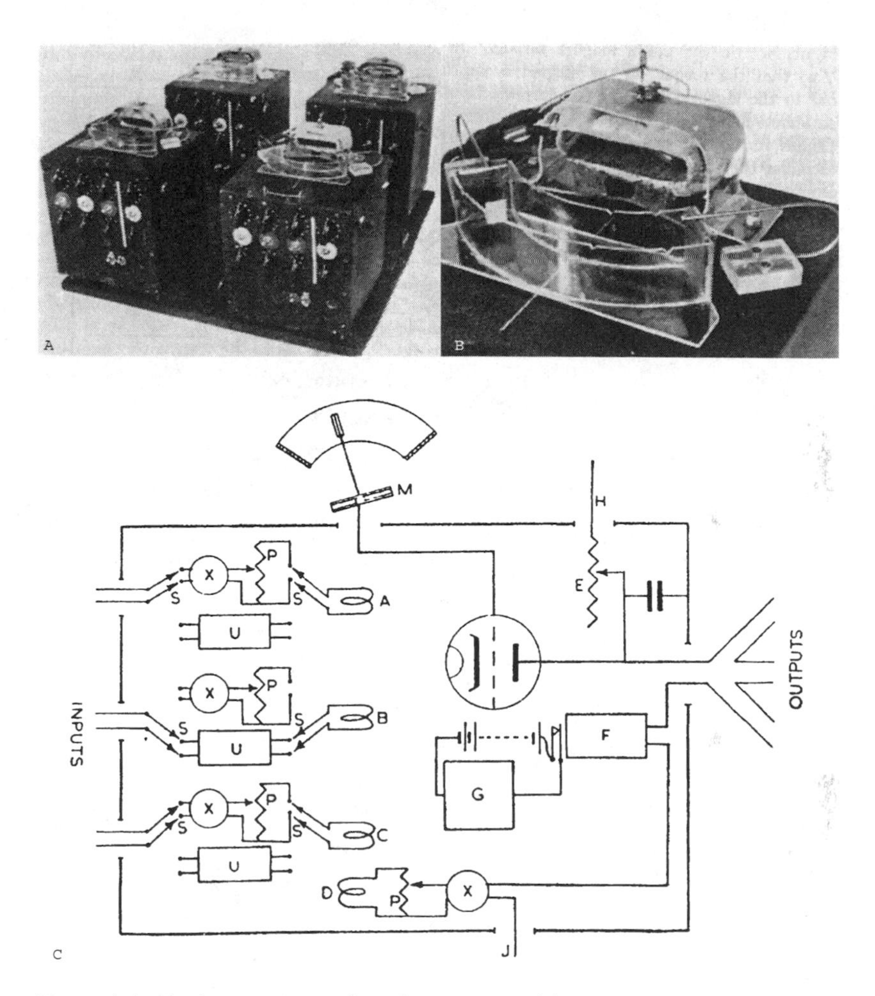

Figure 4.4. The homeostat: *a*, four interconnected homeostats; *b*, detail of the top of a homeostat unit, showing the rotating needle; *c*, circuit diagram. Source: W. R. Ashby, "Design for a Brain," *Electronic Engineering, 20* (December 1948), 380, figs. 1, 2. (With kind permission from Springer Science and Business Media.)

setting, the input current traveled to the magnet coil through a commutator, *X*, which reversed the polarity of the input according to its setting, and through a potentiometer, *P*, which scaled the current according to its setting. The settings for *P* and *X* were fixed by hand, using the upper and middle set of knobs on the front of the homeostat in figure 4.4a. More interesting, the switch *S* could also be set to route the input current through a "uniselector" or "stepping switch"—*U* in figure 4.4c. Each of these uniselectors had twenty-five positions, and each position inserted a specific resistor into the input circuit, with the different values of the twenty-five resistances being "deliberately randomised, the actual numerical values being taken from a published table of random numbers" (Ashby 1948, 381). Unlike the potentiometers and commutators, these uniselectors were not set by hand. They were controlled instead by the internal behavior of the homeostat. When the output current of the unit rose beyond some preset limit, relay *F* in figure 4.4c would close, driving the uniselector (via the coil marked *G*) to its next setting, thus replacing the resistor in the input circuit by another randomly related to it.

So what? The first point to bear in mind is that any single homeostat unit was quite inert: it did nothing by itself. On the other hand, when two or more units were interconnected, dynamic feedback interrelations were set up between them, as the outputs of each unit fed as input to the others and thence returned, transformed, as input to the first, on and on, endlessly around the loop. And to get to grips with the behavior of the whole ensemble it helps to specialize the discussion a bit. Consider a four-homeostat setup as shown in figure 4.4a, and suppose that for one of the units—call it homeostat 1—the switch *S* brings a uniselector into the input circuit, while for the three remaining homeostats the switches *S* are set to route the input currents through the manually set potentiometers and commutators. These latter three, then, have fixed properties, while the properties of homeostat 1 vary with its uniselector setting.

When this combination is switched on, homeostat 1 can find itself in one of two conditions. It might be, as Ashby would say, in a condition of *stable* equilibrium, meaning that the vane on top of the unit would come to rest in the middle of its range, corresponding by design to zero electrical output from the unit, and return there whenever any of the vanes on any of the units was given a small push. Or the unit might be *unstable*, meaning that its vane would be driven toward the limits of its range. In that event, the key bit of the homeostat's circuitry would come into play. As the electrical output of the unit increased above some preset value, the relay would close and drive the uniselector to its next position. This, in effect, would change the electrical properties of homeostat 1, and then we can see how it goes. The unit might again find itself in one of two conditions, either stable or unstable. If the latter, the relay would again drive the uniselector to its next position, inserting a new resistance in the circuit, and so on and so on, until homeostat 1 *found* a condition of stable equilibrium in which its vane gravitated to the center of its range.

This is the key point about the homeostat: it was a real ultrastable machine of the kind that Ashby had only imagined back in 1941. The uniselectors took the place of the bands that broke in the fantasy machine of his 1945 publication (with the added advantage that the uniselectors were always capable of moving to another position, unlike elastic bands, which never recover from breaking). Started off in any configuration, the homeostat would *randomly reorganize* itself to find a condition of dynamic equilibrium with its environment, without any external intervention.

The homeostat was, then, a major milestone in Ashby's twenty-year quest to understand the brain as a machine. Now he had a real electromechanical device that could serve in understanding the go of the adaptive brain. It was also a major development in the overall cybernetic tradition then crystallizing around Wiener's *Cybernetics*, also published in 1948.[14] I want to pause, therefore, to enter some commentary before returning to the historical narrative—first on ontology, then on the social basis of Ashby's cybernetics.

The Homeostat as Ontological Theater

ASHBY'S BRILLIANT IDEA OF THE UNPURPOSEFUL RANDOM MECHANISM WHICH SEEKS FOR ITS OWN PURPOSE THROUGH A PROCESS OF LEARNING IS . . . ONE OF THE GREAT PHILOSOPHICAL CONTRIBUTIONS OF THE PRESENT DAY.

NORBERT WIENER, *THE HUMAN USE OF HUMAN BEINGS*, 2ND ED. (1967 [1950]), 54

THERE CAN'T BE A PROPER THEORY OF THE BRAIN UNTIL THERE IS A PROPER THEORY OF THE ENVIRONMENT AS WELL. . . . THE SUBJECT HAS BEEN HAMPERED BY OUR NOT PAYING SUFFICIENTLY SERIOUS ATTENTION TO THE ENVIRONMENTAL HALF OF THE PROCESS. . . . THE "PSYCHOLOGY" OF THE ENVIRONMENT WILL HAVE TO BE GIVEN ALMOST AS MUCH THOUGHT AS THE PSYCHOLOGY OF THE NERVE NETWORK ITSELF.

ROSS ASHBY, DISCUSSION AT THE 1952 MACY CONFERENCE (ASHBY 1953B, 86-87)

My ontological commentary on the homeostat can follow much the same lines as that on the tortoise, though I also want to mark important differences. First, like the tortoise, the homeostat stages for us an image of an immediately performative engagement of the brain and the world, a little model of a performative ontology more generally. Again, at the heart of this engagement was a process of random, trial-and-error search. The tortoise physically explored its environment, finding out about distributions of lights and obstacles; the homeostat instead searched its inner being, running through the possibilities of its inner circuitry until it found a configuration that could come into dynamic equilibrium with its environment.

Next we need to think about Ashby's modelling not of the brain but of the world.[15] The world of the tortoise was largely static and unresponsive—a given field of lights and obstacles—but the homeostat's world was lively and dynamic: it was, as we have seen, more homeostats! If in a multiunit setup homeostat 1 could be regarded as a model brain, then homeostats 2, 3, and 4 constituted homeostat 1's world. Homeostat 1 perturbed its world dynamically, emitting currents, which the other homeostats processed through their circuits and responded to accordingly, emitting their own currents back, and so on around the loop of brain and world. This symmetric image, of a lively and responsive world to be explored by a lively and adaptive brain, was, I would say, echoing Wiener, the great philosophical novelty of Ashby's early cybernetics, its key feature.

As ontological theater, then, a multihomeostat setup stages for us a vision of the world in which fluid and dynamic entities evolve together in a decentered fashion, exploring each other's properties in a performative back-and-forth dance of agency. Contemplation of such a setup helps us to imagine the world more generally as being like that; conversely, such a setup instantiates a way of bringing that ontological vision down to earth as a contribution to the science of the brain. This is the ontology that we will see imaginatively elaborated and played out in all sorts of ways in the subsequent history of cybernetics.[16] Biographically, this is where I came in. In *The Mangle of Practice* I argued that scientific research has just this quality of an emergent and performative dance of agency between scientists and nature and their instruments and machines, and despite some evident limitations mentioned below, a multihomeostat setup is a very nice starting point for thinking about the ontological picture I tried to draw there. It was when I realized this that I became seriously interested in the history of cybernetics as elaborating and bringing that ontological picture down to earth.

Three further remarks on homeostat ontology might be useful. First, I want simply to emphasize that relations between homeostats were entirely noncognitive and nonrepresentational. The homeostats did not seek to know one another and predict each other's behavior. In this sense, each homeostat was unknowable to the others, and a multihomeostat assemblage thus staged what I called before an *ontology of unknowability*. Second, as discussed in chapter 2, paradigmatic modern sciences like physics describe a world of fixed entities subject to given forces and causes. The homeostat instead staged a vision of fluid, ever-changing entities engaged in trial-and-error search processes. And a point to note now is that such processes are intrinsically *temporal*. Adaptation happens, if it happens at all, in time, as the upshot of a temporally extended process, trying this, then that, and so on. This is the sense in which the homeostat adumbrates, at least, an *ontology of becoming* in which nothing present in advance determines what entities will turn out to be in the future. This is another angle from which we can appreciate the nonmodernity of cybernetics. Third, we could notice that the brain/world symmetry of Ashby's setups in fact problematized their specific reference to the brain. We can explore Ashby's response to this later, but to put the point positively I could say now that this symmetry indexes the potential *generality* of the homeostat as ontological theater. If the phototropism and object avoidance of the tortoise tied the tortoise to a certain sort of brainlike sensing entity, very little tied the homeostat to the brain (or any other specific sort of entity). A multihomeostat configuration could easily be regarded as a model of a world built from any kind of performatively responsive entities, possibly including brains but possibly also not. Here, at the level of ontological theater, we again find cybernetics about to overflow its banks.

— — — — —

So much for the general ontological significance of the homeostat. As in the previous chapter, however, we should confront the point that Ashby, like Walter, aimed at a distinctly modern understanding of the brain: neither of them was content to leave the brain untouched as one of Beer's exceedingly complex systems; both of them wanted to open up the Black Box and grasp the brain's inner workings. Ashby's argument was that the homeostat was a positive contribution to knowledge of how the performative brain adapts. What should we make of that? As before, the answer depends upon the angle from which one looks. From one angle, Ashby's argument was certainly correct: it makes sense to see the homeostat's adaptive structure as a model for how the brain works. From another angle, however, we can see how, even as modern science, the homeostat throws us back into the world of exceedingly complex systems rather than allowing us to escape from it.

The first point to note is, again, that Ashby's science had a rather different quality from that of the classical modern sciences. It was another instance of explanation by articulation of parts (chap. 2): if you put together some valves and relays and uniselectors *this* way, then the whole assemblage can adapt performatively. Ashby's science thus again thematized performance, at the level of parts as well as wholes. Second, and again like Walter's, Ashby's science was a science of a heterogeneous universe: on the one hand, the brain, which Ashby sought to understand; on the other, an unknown and cognitively unknowable (to the homeostat) world. Performative interaction with the unknowable was thus a necessary constituent of Ashby's science, and in this sense the homeostat returns us to an ontology of unknowability. And, third, a discovery of complexity also appears within Ashby's cybernetics, though this again requires more discussion.

In chapter 3 we saw that despite its simplicity the tortoise remained, to a degree, a Black Box, capable of surprising Walter with its behavior. The modern impulse somehow undid itself here, in an instance where an atomic understanding of parts failed to translate into a predictive overview of the performance of the whole. What about the homeostat? In one sense, the homeostat did not display similarly emergent properties. In his published works and his private journals, Ashby always discussed the homeostat as a demonstration device that displayed the adaptive properties he had already imagined in the early 1940s and first discussed in print in his 1945 publication on the bead-and-elastic machine.

Nevertheless, *combinations* of homeostats quickly presented analytically insoluble problems. Ashby was interested, for example, in estimating the probability that a set of randomly interconnected homeostats with fixed internal settings would turn out to be stable. In a 1950 essay, he explored this topic from all sorts of interesting and insightful angles before remarking that, even with simplifying assumptions, "the problem is one of great [mathematical] difficulty and, so far as I can discover, has not yet been solved. My own investigations have only convinced me of its difficulty. That being so we must collect evidence as best we can" (Ashby 1950a, 478). Mathematics having failed him, Ashby turned instead to his machines, fixing their parameters and interconnections at random in combinations of two, three, or four units and simply recording whether the needles settled down in the middle of their ranges or were driven to their limits. His conclusion was that the probability of finding a stable combination probably fell off as $(1/2)^n$, where n was the number of units to be interconnected, but, rather than that specific result, what I want to stress is that here we have another discovery of complexity, now in the analytic opacity of multihomeostat setups. Ashby's atomic knowledge of the individual components of his machines and their interconnections again failed to translate into an ability to predict how aggregated assemblages of them would perform. Ashby just had to put the units together and see what they did.

As in the previous chapter, then, we see here how the modern impulse of early cybernetics bounced back into the cybernetic ontology of unknowability. While illuminating the inner go of the brain, homeostat assemblages of the kind discussed here turned out to remain, in another sense, mini–Black Boxes, themselves resistant to a classically scientific understanding, which we can read again as suggestive icons for a performative ontology. Imagine the world in general as built from elements like these opaque dynamic assemblages, is the suggestion. We can go further with this thought when we come to DAMS, the homeostat's successor.

Making much same point, the following quotation is from a passage in *Design for a Brain* in which Ashby is discussing interconnected units which have just two possible states, described mathematically by a "step-function" and corresponding to the shift in a uniselector from one position to the next (1952, 129): "If there are *n* step-functions [in the brain], each capable of taking two values, the total number of fields available will be 2^n. . . . The number of fields is moderate when *n* is moderate, but rapidly becomes exceedingly large when *n* increases. . . . If a man used fields at the rate of ten a second day and night during his whole life of seventy years, and if no field were ever repeated, how many two-valued step-functions would be necessary to provide them? Would the reader like to guess? The answer is that thirty-five would be ample!" One is reminded here of Walter's estimate that ten functional elements in the brain could generate a sufficient variety of behaviors to cover the entire experience of the human race over a period of a thousand million years. What the early cyberneticians discovered was just how complex (in aggregate behavior) even rather simple (in atomic structure) systems can be.

— — — — —

The homeostat is highly instructive as ontological theater, but I should also note its shortcomings. First, like all of the early cybernetic machines including the tortoise, the homeostat had a *fixed goal*: to keep its output current within predetermined limits. This was the unvarying principle of its engagement with the world. But, as I said about the tortoise, I do not think that this is a general feature of our world—in many ways, for example, human goals emerge and are liable to transformation in practice. At the same time, we might note an important difference between the homeostat's goals and, say, the tortoise's. The latter's goals referred to states of the outer world—finding and pursuing lights. The homeostat's goals instead referred inward, to its internal states. One might therefore imagine an indefinite number of worldly projects as bearing on those inner states, all of them obliquely structured by the pursuit of inner equilibrium. This is certainly a step in the right ontological direction beyond the tortoise.

Second, I described the homeostat as exploring its environment open-endedly, but this is not strictly true. My understanding of open-endedness includes an indefinitely large range of possibilities, whereas the homeostat had precisely twenty-five options—the number of positions of its uniselector. A four-homeostat setup could take on $25^4 = 390{,}625$ different states in all.[17] This is a large number, but still distinctly finite. As ontological theater, therefore, we should think of the homeostat as pointing in the direction of open-ended adaptation, without quite getting there.

Third, and most important, as the word "uniselector" suggests, adaptation in the homeostat amounted to the *selection* of an appropriate state by a process of trial and error within a combinatoric space of possibilities. This notion of selection appears over and over again in Ashby's writings, and, at least from an ontological point of view, there is something wrong with it. It leaves no room for creativity, the appearance of genuine novelty in the world; it thus erases what I take to be a key feature of open-endedness. It is easiest to see what is at stake here when we think about genuinely cognitive phenomena, so I will come back to this point later. For the moment, let me just register my conviction that as models of the brain and as ontological theater more generally, Ashby's homeostats were deficient in just this respect.

— — — — —

One final line of thought can round off this section. It is interesting to examine how Ashby's cybernetics informed his understanding of himself. As mentioned above, a multihomeostat assemblage foregrounded the role of time—adaptation as necessarily happening in time. And here is an extract from Ashby's autobiographical notebook, "Passing through Nature" (Ashby 1951–57), from September 1952 (pp. 36–39):

> For forty years [until the mid-1940s—the first blossoming of his cybernetics] I hated change of all sorts, wanting only to stay where I was. I didn't want to grow up, didn't want to leave my mother, didn't want to go from school to Cambridge, didn't want to go to hospital, and so on. I was unwilling at every step.
>
> Now I seem to be changed to the opposite: my only aim is to press on. The march of time is, in my scientific theorising, the only thing that matters. Every thing, I hold, must go on: if human destiny is to go on and destroy itself with an atomic explosion, well then, let us get on with it, and make the biggest explosion ever!

> I am now, in other words a Time-worshipper, seized with the extra fervour of the convert. I mean this more or less seriously. "Time" seems to me to be big enough, impersonal enough, to be a possible object of veneration—the old man of the Bible with his whims & bargains, & his impotence over evil, and his son killing, has always seemed to me to be entirely inadequate as the Spirit of All Existent, if not downright contemptible. But Time has possibilities. As a variable it is utterly different from all others, for they exist in it as a fish lives in the ocean: so immersed that its absence is inconceivable. My aim at the moment is to reduce all adaptation to its operation, to show that if only Time will operate, whether over the geological periods on an earth or over a childhood in an individual, then adaptation will inevitably emerge. This gives to time a position of the greatest importance, equalled only by that "factor" that called space & matter into existence.

This passage is interesting in a couple of respects. On the one hand, Ashby records a change in his perspective on time and change (in himself and the world) that is nicely correlated with the flourishing of his cybernetics. On the other, this passage returns us to the relation between cybernetics and spirituality that surfaced in the last chapter and runs through those that follow. Walter made the connection via his discussion of the strange performances associated with Eastern spirituality, which he assimilated to his understanding of the performative brain and technologies of the self. There are also definite echoes of the East in this passage from Ashby—one thinks of Shiva indifferently dancing the cosmos into and out of existence—though now the bridge from cybernetics to spirituality goes via time and adaptation, the key themes of Ashby's cybernetics as exemplified in the homeostat, rather than technologies of the self.[18]

The self does, however, reappear in a different guise in this passage. "The old man of the Bible with his whims & bargains" is the very paradigm of the modern, self-determined, centered, human subject writ as large as possible. And it is interesting to note that Ashby's rejection of this image of the Christian God went with a nonmodern conception of himself. Just as a multihomeostat setup dramatized a decentered self, not fully in control and constitutively plunged into its environment, so "Passing through Nature" begins (Ashby 1951–57, pp. 1–3) with the story of a meeting in January 1951 at which Warren McCulloch was present. Realizing how important McCulloch was to his career as a cybernetician, Ashby took the initiative and shook hands with him, but then immediately found himself going back to a conversation with someone of "negligible . . . professional importance." "What I want to make clear is that I had no power in the matter. The series of events ran with perfect smoothness and quite irresistibly, taking not the slightest notice of whatever conscious views I may have had. Others may talk of freewill and the individual's power to direct his life's story. My personal experience has convinced me over and over again that my power of control is great—where it doesn't matter: but at the important times, in the words of Freud, I do not live but 'am lived.'"

By the early 1950s, then, Ashby's understanding of himself and God and his cybernetics all hung together, with questions of time and change as their pivot. I take this as another instance of the fact that ontology makes a difference—here in the realm of spirituality and self-understanding, as well as brain science and much else: time worship and "I am lived" as an ontology of performative becoming in action.[19]

The Social Basis of Ashby's Cybernetics

Turning from ontology to sociology, it is evident already that there are again clear parallels between Ashby and Walter. Ashby was telling no more than the truth when he described his early work—up to 1940, say—as having no social basis, as "a hobby I could retreat to": something pursued outside his professional life, for his own enjoyment. Even after 1940, when he began to publish, his work for a long time retained this extraprofessional, hobbyist quality, very largely carried on in the privacy of his journals. In an obituary, his student Roger Conant (1974, 4) speaks of Ashby building the homeostat "of old RAF parts on Mrs Ashby's kitchen table" and of writing his two books "in Dr. Ashby's private padded cell" at Barnwood House.[20]

When he did begin to publish his protocybernetic theorizing, Ashby submitted his work initially to the journals in which his earlier distinctively psychiatric papers had appeared. His very first paper in this series (Ashby 1940) appeared in the leading British journal for research on mental pathologies, the *Journal of Mental Science*. It appears that there was no great response to Ashby's work within this field, outside the narrow but important circle defined by himself, Grey Walter, Frederick Golla, and G. W. T. H. Fleming, the editor of the journal in question. And one can understand why this might have been: clinical psychiatrists and psychologists were concerned with the practical problems of mental illness, and, besides its oddity as engineering, Ashby's work sticks out like a sore thumb in the pages of the psychiatric journals—his theoretical work offered little constructive input to psychiatric practice (though more on this below).

Conversely, in seeking to create a community of interest for his work, Ashby, like Walter, systematically looked beyond his profession. A journal entry from early June 1944 (p. 1666) records that "several of my papers have been returned recently & it seems that there is going to be considerable difficulty in floating this ship."[21] At this point he began writing to other scholars with whom he appears to have had no prior contact about his and their work, and it is notable that none of the people he addressed shared his profession. Thus, the small existing collection of Ashby's correspondence from this period includes letters to or from the experimental psychologists Kenneth Craik and E. Thorndike in 1944, and in 1946 the anthropologist-turned-cybernetician Gregory Bateson, the eminent neurophysiologist E. D. Adrian, the doyen of American cybernetics, Warren McCulloch, the British mathematician Alan Turing, and Norbert Wiener himself. In most cases it is clear that Ashby was writing out of the blue, and that he identified this extraprofessional and protocybernetic community from his reading of the literature. Through these contacts, and also by virtue of something of an explosion in his publication record—around twenty cybernetic essays appeared in various journals between 1945 and 1952—Ashby quickly assumed a leading position in the nascent cybernetic community, though, as we saw in the previous chapter, this was itself located outside the usual social structures of knowledge production. In Britain, its heart was the Ratio Club, the dining club of which Ashby was a founder member; Ashby was an invited speaker at the 1952 Macy cybernetics conference in the United States, and he regularly gave papers at the Namur cybernetics conferences in Europe. As far as knowledge dissemination was concerned, Ashby's route into the wider social consciousness was, like Walter's and Wiener's, via the popular success of his books.

Ashby's cybernetics largely existed, then, in a different world from his professional life, though that situation began to change in the late 1950s. Through what appears to be a certain amount of chicanery on the part of G. W. T. H. Fleming, who was chairman of the trustees of the Burden Neurological Institute as well as director of Barnwood House, where Ashby then worked, Ashby was appointed in 1959 to succeed Golla as the director of the Burden. His ineptitude in that position—including trying to purge the library of outdated books, setting exams for all the staff, and setting private detectives on Grey Walter—remains legendary in British psychiatric circles, and Ashby was saved from a disastrous situation by the opportunity to flee to the United States (Cooper and Bird 1989, 15–18). Stafford Beer's diary for 1960 records the circumstances of an offer from Heinz von Foerster to join the faculty of the University of Illinois, made while Beer, Pask, and Ashby were all on campus for a conference on self-organization—an offer which Ashby understandably accepted without hesitation (Beer 1994 [1960], 299–301).

At Illinois, Ashby's formal position was that of professor in the Department of Electrical Engineering with an associated position on the biophysics committee. His primary affiliation was to von Foerster's Biological Computer Laboratory, the BCL. The BCL was an independently funded operation housed within the Electrical Engineering Department and was, during the period of its existence, 1958–75, the primary institutional basis for cybernetics in the capitalist world.[22] At the BCL Ashby became the only one of our cyberneticians to enjoy full-time institutional support for his work, both in research and teaching. Ashby retired from the BCL in 1970 at the age of sixty-seven and returned to England, and Conant (1974, 4) records that "the decade spent in the United States resulted in a host of publications and was in his own estimation the most fruitful period of his career." It seems clear that this time of singular alignment between paid work and hobby was also one of the happiest periods of Ashby's life, in which he could collaborate with many graduate students on topics close to his heart, and for which he is remembered fondly in the United States (unlike the Burden) as "an honest and meticulous scholar . . . a warm-hearted, thoughtful, and generous person, eager to pass to his students the credit for ideas he had germinated himself" (Conant 1974, 5).

Most of Ashby's cybernetic career thus displayed the usual social as well as ontological mismatch with established institutions, finding its home in improvised social relations and temporary associations lacking the usual means of reproducing themselves. In this respect, of course, his time at the BCL is anomalous, an apparent counterinstance to the correlation of the ontological and the social, but this instance is, in fact, deceptive. The BCL was itself an anomalous and marginal institution, only temporarily lodged within the academic body. It was brought into existence in the late 1950s by the energies of von Foerster, a charming and energetic Austrian postwar emigré, with powerful friends and sponsors, especially Warren McCulloch, and ready access to the seemingly inexhaustible research funding available from U.S. military agencies in the decades following World War II. When such funding became progressively harder to find as the sixties went on, the BCL contracted, and it closed down when von Foerster retired in 1975. A few years later its existence had been all but forgotten, even at the University of Illinois. The closure of the BCL—rather than, say, its incorporation within the Electrical Engineering Department—once again illustrates the social mismatch of cybernetics with existing academic structures.[23]

Design for a Brain

We can return to the technicalities of Ashby's cybernetics. The homeostat was the centerpiece of his first book, *Design for a Brain*, which was published in 1952 (and, much revised, in a second edition, in 1960). I want to discuss some of the principal features of the book, as a way both to clarify the substance of Ashby's work in this period and to point the way to subsequent developments.

First, we should note that Ashby had developed an entire mathematical apparatus for the analysis of complex systems, and, as he put it, "the thesis [of the book] is stated twice: at first in plain words and then in mathematical form" (1952, vi). The mathematics is, in fact, relegated to a forty-eight-page appendix at the end of the book, and, following Ashby's lead, I, too, postpone discussion of it to a later section. The remainder of the book, however, is not just "plain words." The text is accompanied by a distinctive repertoire of diagrams aimed to assist Ashby and the reader in thinking about the behavior of complex systems. Let me discuss just one diagram to convey something of the flavor of Ashby's approach.

In figure 4.5 Ashby schematizes the behavior of a system characterized by just two variables, labeled *A* and *B*. Any state of the system can thus be denoted by a "representative point," indicated by a black dot, in the *A-B* plane, and the arrows in the plane denote how the system will change with time after finding itself at one point or another. In the unshaded central portions of the plane, the essential variables of the system are supposed to be within their assigned limits; in the outer shaded portions, they travel beyond those limits. Thus, in panel I, Ashby imagines that the system starts with its representative point at *X* and travels to point *Y*, where the essential variables exceed their limits. At this point, the parameters of the system change discontinuously in a "step-function"—think of a band breaking in the bead-and-elastic machine of 1943, or a uniselector moving to its next position in the homeostat—and the "field" of system behavior thus itself changes discontinuously to that shown in panel II. In this new field, the state of the system is again shown as point *Y*, and it is then swept along the trajectory that leads to *Z*, followed by another reconfiguration leading to state field III. Here the system has a chance of reaching equilibrium: there are trajectories within field III that swirl into a "stable state," denoted by the dot on which the arrows converge. But Ashby imagines that the system in question lies on a trajectory that again sweeps into the forbidden margin at *Z*. The system then transmogrifies again into state IV and at last ceases its development, since all the trajectories in that field configuration converge on the central dot in a region where the essential variables are within their limits.

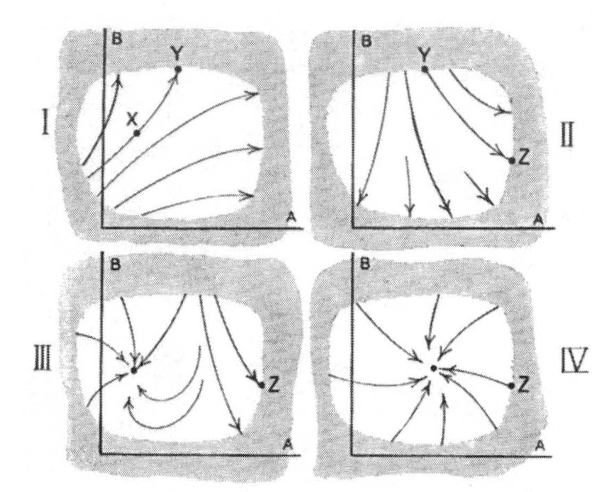

Figure 4.5. Changes of field in an ultrastable system. Source: W. R. Ashby, *Design for a Brain* (London: Chapman & Hall, 1952), 92, fig. 8/7/1. (With kind permission from Springer Science and Business Media.)

Figure 4.5 is, then, an abstract diagram of how an ultrastable system such as a homeostat finds its way to state of equilibrium in a process of trial and error, and I want to make two comments on it. The first is ontological. The basic conceptual elements of Ashby's cybernetics were those of the sort analyzed in this figure, and they were *dynamic*—systems that change in time. Any trace of stability and time independence in these basic units had to do with the specifics of the system's situation and the special circumstance of having arrived at a stable state. Ashby's world, one can say, was built from such intrinsically dynamic elements, in contrast to the modern ontology of objects carrying unvarying properties (electrons, quarks). My second comment is historical but forward looking. In *Design for a Brain*, one can see Ashby laboriously assembling the technical elements of what we now call complex systems theory. For those who know the jargon, I can say that Ashby already calls diagrams like those of figure 4.5 "phase-space diagrams"; the points at which the arrows converge in panels III and IV are what we now call "attractors" (including, in Ashby's diagrams, both point and cyclical attractors, but not "strange" ones); and the unshaded area within panel IV is evidently the "basin of attraction" for the central attractor. Stuart Kauffman and Stephen Wolfram, discussed at the end of this chapter, are among the leaders of present-day work on complexity.

Now for matters of substance. Following Ashby, I have so far described the possible relation of the homeostat to the brain in abstract terms, as both being adaptive systems. In *Design for a Brain*, however, Ashby sought to evoke more substantial connections. One approach was to point to real biological examples of putatively homeostatic adaptation. Here are a couple of the more horrible of them (Ashby 1952, 117–18):

> Over thirty years ago, Marina severed the attachments of the internal and external recti muscles of a monkey's eyeball and re-attached them in crossed position so that a contraction of the external rectus would cause the eyeball to turn not outwards but inwards. When the wound had healed, he was surprised to discover that the two eyeballs still moved together, so that binocular vision was preserved.
>
> More recently Sperry severed the nerves supplying the flexor and extensor muscles in the arm of the spider monkey, and re-joined them in crossed position. After the nerves had regenerated, the animal's arm movements were at first grossly inco-ordinated, but improved until an essentially normal mode of progression was re-established.

And, of course, as Ashby pointed out, the homeostat showed just this sort of adaptive behavior. The commutators, X, precisely reverse the polarities of the homeostat's currents, and a uniselector-controlled homeostat can cope with such reversals by reconfiguring itself until it returns to equilibrium. A very similar example concerns rats placed in an electrified box: after some random leaping about, they learn to put their foot on a pedal which stops the shocks (1952, 106–8). Quite clearly, the brain being modelled by the homeostat here is not the cognitive brain of AI; it is the performative brain, the Ur-referent of cybernetics: "excitations in the motor cortex [which] certainly control the rat's bodily movements" (1952, 107). In the second edition of *Design for a Brain*, Ashby added some less brutal examples of training animals to perform in specified ways, culminating with a discussion of training a "house-dog" not to jump on chairs (1960, 113): "Suppose then that jumping into a chair always results in the dog's sensory receptors being excessively stimulated [by physical punishment, which drives some essential variable beyond its limits]. As an ultrastable system, step-function values which lead to jumps into chairs will be followed by stimulations likely to cause them to change value. But on the occurrence of a set of step-function values leading to a remaining on the ground, excessive stimulation will not occur, and the values will remain." He then goes on to show that similar training by punishment can be demonstrated on the homeostat. He discusses a set up in which just three units were connected with inputs running 1→2→3→1, where the trainer, Ashby, insisted that an equilibrium should be reached in which a small forced movement of the needle on 1 was met by the opposite movement of the needle on 2. If the

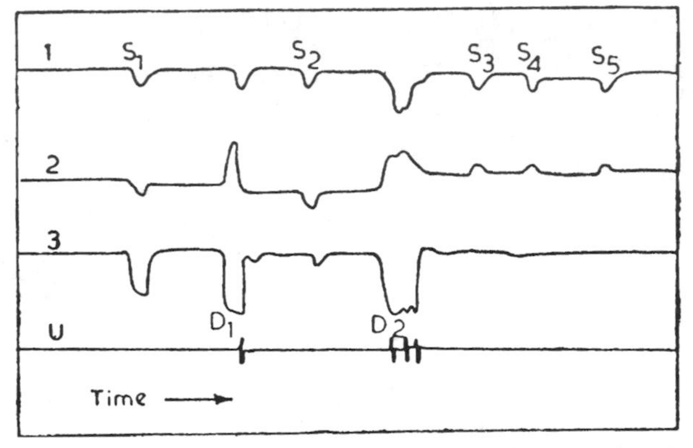

Figure 4.6. Training a three-homeostat system. The lines running from left to right indicate the positions of the needles on the tops of units 1, 2, and 3. The punishments administered to unit 3 are marked D_1 and D_2. The shifts in the uniselectors are marked as vertical blips on the bottom line, *U*. Note that after the second punishment a downward displacement of needle 1 evokes an upward displacement of needle 2, as desired. Source: W. R. Ashby, *Design for a Brain: The Origin of Adaptive Behaviour*, 2nd ed. (London: Chapman & Hall, 1960), 114, fig. 8/9/1. (With kind permission from Springer Science and Business Media.)

system fell into an equilibrium in which the correlation between the needles 1 and 2 was the wrong way around, Ashby would punish homeostat 3 by pushing its needle to the end of its range, causing its uniselector to trip, until the right kind of equilibrium for the entire system, with an anticorrelation of needles 1 and 2, was achieved. Figure 4.6 shows readouts of needle positions from such a training session.

Ashby thus sought to establish an equation between his general analysis of ultrastable systems and brains by setting out a range of exemplary applications to the latter. Think of the response of animals to surgery, and then think about it *this* way. Think about training animals; then think about it *this* way. In these ways, Ashby tried to train his readers to make this specific analogical leap to the brain.

But something is evidently lacking in this rhetoric. One might be willing to follow Ashby some of the way, but just what are these step mechanisms that enable animals to cope with perverse surgery or training? Having warned that "we have practically no idea of where to look [for them], nor what to look for [and] in these matters we must be vary careful to avoid making asssumptions unwittingly, for the possibilities are very wide" (1960, 123), Ashby proceeds to sketch out some suggestions.

One is to note that "every cell contains many variables that might change in a way approximating to the step-function form. . . . Monomolecular films,

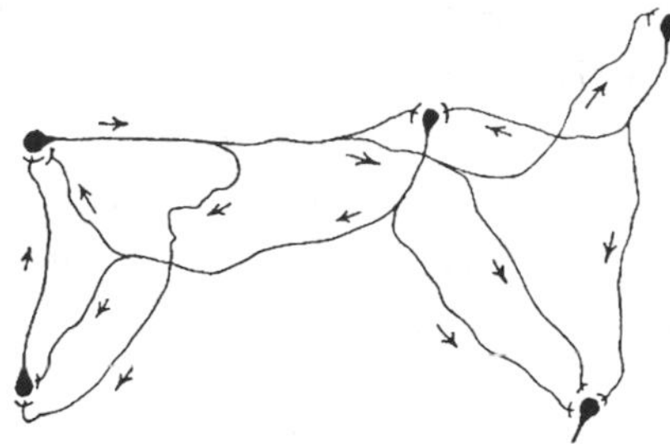

Figure 4.7. Interconnected circuit of neurons. Source: W. R. Ashby, *Design for a Brain* (London: Chapman & Hall, 1952), 128, fig. 10/5/1. (With kind permission from Springer Science and Business Media.)

protein solutions, enzyme systems, concentrations of hydrogen and other ions, oxidation-reduction potentials, adsorbed layers, and many other constituents or processes might act as step-mechanisms" (1952, 125). A second suggestion is that neurons are "amoeboid, so that their processes could make or break contact with other cells" (126). And third, Ashby reviews an idea he associates with Rafael Lorente de Nó and Warren McCulloch, that the brain contains interconnected circuits of neurons (fig. 4.7), on which he observes that "a simple circuit, if excited, would tend either to sink back to zero excitation, if the amplification factor was less than unity, or to rise to the maximal excitation if it was greater than unity." Such a circuit would thus jump discontinuously from one state to another and "its critical states would be the smallest excitation capable of raising it to full activity, and the smallest inhibition capable of stopping it" (128). Here, then, were three suggestions for the go of it—plausible biological mechanisms that might account for the brain's homeostatic adaptability.

— — — — —

The homeostat appears midway through *Design for a Brain*. The preceding chapters prepare the way for it. Then its properties are reviewed. And then, in the book's concluding chapters, Ashby looks toward the future. "My aim," he says, with a strange kind of modesty, "is simply to copy the living brain" (1952, 130). Clearly, a single homeostat was hardly comparable in its abilities to the brain of a simple organism, never mind the human brain—it was "too larval" (Ashby 1948, 343)—and the obvious next step was to contemplate a multiplication of such units. Perhaps the brain was made up of a large number of ultrastable units, biological homeostats. And the question Ashby then asked was one of speed or efficiency: how long would it take such an assembly to come into equilibrium with its environment?

Here, some back-of-an-envelope calculations produced interesting results. Suppose that any individual unit had a probablity *p* of finding an equilibrium state in one second. Then the time for such a unit to reach equilibrium would be of the order of $1/p$. And if one had a large number of units, *N* of them, acting quite independently of one another, the time to equilibrium for the whole assemblage would still be $1/p$. But what if the units were fully interconnected with one another, like the four units in the prototypical four-homeostat setup? Then each of the units would have to find an equilibrium state in the same trial as all the others, otherwise the nonequilibrium homeostats would keep changing state and thus upsetting the homeostats that had been fortunate enough already to reach equilibrium. In this configuration, the time to equilibrium would be of the order of $1/p^N$. Ashby also considered an intermediate case in which the units were interconnected, but in which it was possible for them to come into equilibrium sequentially: once unit 1 had found an equilibrium condition it would stay there, while 2 hunted around for the same, and so on. In this case, the time to equilibrium would be N/p.

Ashby then put some numbers in: $p = 1/2$; $N = 1{,}000$ units. This leads to the following estimates for *T*, the time for whole system to adapt (1952, 142):

> for the fully interconnected network: $T_1 = 2^{1000}$ seconds;
> for interconnected but sequentially adapting units, $T_2 = 2{,}000$ seconds;
> for the system of entirely independent units, $T_3 = 2$ seconds.[24]

Two seconds or 2,000 seconds are plausible figures for biological adaptation. According to Ashby, 2^{1000} seconds is 3×10^{291} centuries, a number vastly greater than the age of the universe. This last hyperastronomical number was crucial to Ashby's subsequent thinking on the brain and how to go beyond the homeostat, and the conclusion he drew was that if the brain were composed of many ultrastable units, they had better be only *sparsely connected* to one another if adaptation were going to take a realistic time. At this point he began the construction of a new machine, but before we come to that, let me note again the ontological dimension of Ashby's cybernetics.

The brain that adapted fastest would be composed of fully independent units, but Ashby noted that such a brain "cannot represent a complex biological system" (1952, 144). Our brains do not have completely autonomous subsystems each set to adapt to a single feature of the world we inhabit, on the one hand; the neurons of the brain are observably very densely interconnected, on the other. The question of achieving a reasonable speed of adaptation thus resolved itself, for Ashby, into the question of whether some kind of serial adaptation was possible, and he was very clear that this depended not just on how the brain functioned but also on what the world was like. Thus, he was led to distinguish between "easy" environments "that consist of a few variables, independent of each other," and "difficult" ones "that contain many variables richly cross-linked to form a complex whole" (1952, 132). There is a sort of micro-macro correspondence at issue here. If the world were too lively—if every environmental variable one acted on had a serious impact on many others—a sparsely interconnected brain could never get to grips with it. If when I cleaned my teeth the cat turned into a dog, the rules of mathematics changed and the planets reversed their courses through the heavens, it would be impossible for me to grasp the world piecemeal; I would have to come to terms with all of it in one go, and that would get us back to the ridiculous time scale of T_1.[25]

In contrast, of course, Ashby pointed out that not all environmental variables are strongly interconnected with one another, and thus that sequential adaptation within the brain is, in principle, a viable strategy. In a long chapter on "Serial Adaptation" he first discusses "an hour in the life of *Paramecium*," traveling from a body of water to its surface, where the dynamics are different (due to surface tension), from bodies of water with normal oxygen concentration to those where the oxygen level is depleted, from cold to warm, from pure water to nutrient-rich regions, occasionally bumping into stones, and so on (1952, 180–81). The idea is that each circumstance represents a different environment to which *Paramecium* can adapt in turn and more or less independently. He then discusses the business of learning to drive a car, where one can try to master steering on a straight road, then the accelerator, then changing gears (in the days before automatics, at least in Britain)—though he notes that at the start these tend to be tangled up together, which is why learning to drive can be difficult (181–82). "A puppy can learn how to catch rabbits only after it has learned to run; the environment does not allow the two reactions to be learned in the opposite order. . . . Thus, the learner can proceed in the order 'Addition, long multiplication, . . .' but not in the order 'Long multiplication, addition, . . .' Our present knowledge of mathematics has in fact been reached only because the subject contains such stage-by-stage routes" (185).[26] There follows a long description of the steps in training falcons to hunt (186), and so on.

So, in thinking through what the brain must be like as a mechanism, Ashby also further elaborated a vision of the world in which an alchemical correspondence held between the two terms: the microcosm (the brain) and the macrocosm (the world) mirrored and echoed one another inasmuch as both were sparsely connected systems, not "fully joined," as Ashby put it. We can follow this thread of the story below, into the fields of architecture and theoretical biology as well as Ashby's next project after the homeostat, DAMS. But I can finish this section with a further reflection.

Warren McCulloch (1988) notably described his cybernetics as "experimental epistemology," meaning the pursuit of a theory of knowledge via empirical and theoretical analysis of how the brain actually represents and knows the world. We could likewise think of Ashby's cybernetics as *experimental ontology*. I noted earlier that the general performative vision of the world does not imply any specific cybernetic project; that such projects necessarily *add* something to the vision, both pinning it down and vivifying it by specifying it in this way or that. The homeostat can certainly be seen as such a specification, in the construction of a definite mechanism. But in Ashby's reflections on time to equilibrium, this specification reacted back upon the general vision, further specifying that. If one recognizes the homeostat as a good model for adaptation, then these reflections imply something, not just about the brain but about the world at large as well: both must consist of sparsely connected dynamic entities.

We are back to the idea that ontology makes a difference, but with a twist. My argument so far has been that the nonmodern quality of cybernetic projects can be seen as the counterpart of a nonmodern ontology. Here we have an example in which one of these projects fed back as a fascinating ontological conclusion about the coupling of entities in the world. It is hard to see how one could arrive at a similar conclusion within the framework of the modern sciences.[27]

DAMS

AS A SYMBOL OF HIS INTEREST IN RELATIONS HE CARRIED A CHAIN CONSTRUCTED OF THREE SIMPLER CHAINS INTERLOCKED IN PARALLEL; HE ENJOYED WATCHING MICROSCOPIC ECOSYSTEMS (CAPTURED WITH FISHPOLE AND BOTTLE FROM THE BONEYARD CREEK IN URBANA) FOR THE RICHNESS OF INTERACTION THEY DISPLAYED, AND HE BUILT A SEMI-RANDOM ELECTRONIC CONTRAPTION WITH 100 DOUBLE TRIODES AND WATCHED IT FOR TWO YEARS BEFORE ADMITTING DEFEAT IN THE FACE OF ITS INCOMPREHENSIBLY COMPLEX BEHAVIOR.

ROGER CONANT, "W. ROSS ASHBY (1903-1972)" (1974, 4)

The 1952 first printing of *Design for a Brain* included just one footnote: on page 171 Ashby revealed that he was building a machine called DAMS. In the 1954

Figure 4.8. Photograph of DAMS. (By permission of Jill Ashby, Sally Bannister, and Ruth Pettit.)

second printing of the first edition the footnote was removed, though the entry for DAMS could still be found in the index and a citation remained on page 199 to the only publication in which Ashby described this device, the paper "Statistical Machinery" in the French journal *Thalès* (Ashby 1951). In the second edition, of 1960, both the index entry and the citation also disappeared: DAMS had been purged from history. Despite the obscurity to which Ashby was evidently determined to consign it, his journal in the 1950s, especially from 1950 to 1952, is full of notes on this machine. It would be a fascinating but terribly demanding project to reconstruct the history of DAMS in its entirety; I will discuss only some salient features.

I opened the book with Ashby's suggestion that "the making of a synthetic brain requires now little more than time and labour" (1948, 382), and he evidently meant what he said. DAMS was to be the next step after the homeostat. Its name was an acronym for dispersive and multistable system. A *multistable* system he defined as one made up of many interconnected ultrastable systems. A *dispersive* system was one in which different signals might flow down different pathways (Ashby 1952, 172). This gets us back to the above discussion of times to reach equilibrium. Ashby conceived DAMS as a system in which the ultrastable components were linked by switches, which, depending on conditions, would either isolate components from one another or transmit signals between them. In this way, the assemblage could split into smaller subassemblies appropriate to some adaptive task without the patterns of splitting having to be hard wired in advance. DAMS would thus turn itself into a sparsely connected system that could accumulate adaptations to differing stimuli in a finite time (without disturbing adaptive patterns that had already been established within it).

At the hardware level, DAMS was an assemblage of electronic valves, as in a multihomeostat setup, but now linked not by simple wiring but by neon lamps. The key property of these lamps was that below some threshold voltage they were inert and nonconducting, so that they in fact isolated the valves that they stood between. Above that threshold however, they flashed on and became conducting, actively joining the same valves, putting the valves in communication with one another. According to the state of the neons, then, parts of DAMS would be isolated from other parts by nonconducting neons, "walls of constancy," as Ashby put it (1952, 173), and those parts could adapt independently of one another at a reasonable, rather than hyperastronomical, speed.

Not to leave the reader in undue suspense, I can say now that DAMS never worked as Ashby had hoped, and some trace of this failure is evident in the much-revised second edition of *Design for a Brain*. There Ashby presents it as a rigorous deduction from the phenomenon of cumulative adaptation to different stimuli, P_1, P_2, and so on, that the step mechanisms (uniselectors in the homeostat, neon tubes in DAMS) "*must* be divisible into non-overlapping sets, that the reactions to P_1 and P_2 must each be due to their particular sets, and that the presentation of the problem (i.e., the value of P) must determine which set is to be brought into functional connexion, the remainder being left in functional isolation" (1960, 143). One can see how this solves the problem of accumulating adaptations, but how is it to be achieved? At this point, Ashby wheels on his deus ex machina, a "gating mechanism," Γ, shown in figure 4.9. This picks up the state of the environmental stimulus P via the reaction R of the organism to it and switches in the appropriate bank of uniselectors, neons, or whatever that the essential variables (the dial on the right) can trigger, if necessary, to preserve the equilibrium of the system. But then the reader is left hanging: What is the go of this gating mechanism? How does it do its job? Almost at the end of the book, eighty-four pages later, Ashby acknowledges that "it was shown that . . . a certain *gating-mechanism* was necessary; but nothing was said about how the organism should acquire one" (1960, 227). Two pages later, Ashby fills in this silence, after a fashion (1960, 229–30): "The biologist, of course, can answer the question at once; for the work of the last century . . . has demonstrated that natural, Darwinian,

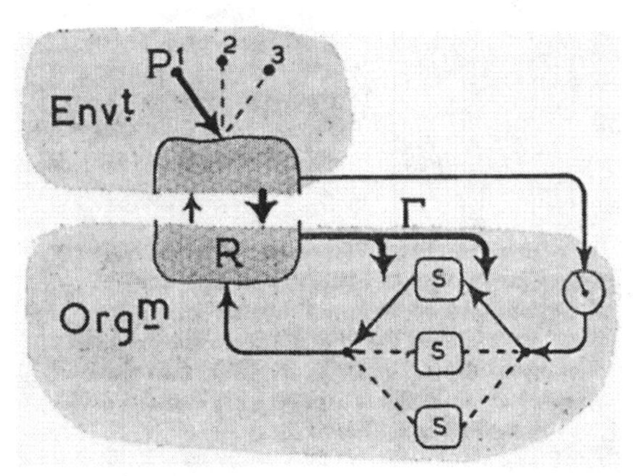

Figure 4.9. The gating mechanism. Source: W. R. Ashby, *Design for a Brain: The Origin of Adaptive Behaviour*, 2nd ed. (London: Chapman & Hall, 1960), 144, fig. 10/9/1. (With kind permission from Springer Science and Business Media.)

selection is responsible for all the selections shown so abundantly in the biological world. Ultimately, therefore, these ancillary mechanisms [the gating mechanism and, in fact, some others] are to be attributed to natural selection. They will, therefore, come to the individual (to our kitten perhaps) either by the individual's gene-pattern or they develop under an ultrastability of their own. There is no other source." Within the general framework of Ashby's approach to the brain and adaptation, these remarks make sense. We need a gating mechanism if multiple adaptations are to be achieved in a finite time; we do adapt; therefore evolution must have equipped us with such a mechanism. But what Ashby had been after with DAMS was the go of multiple adaptation. What he wanted was that DAMS should evolve its own gating mechanism in interacting with its environment, and it is clear that it never did so. To put the point the other way around, what he had discovered was that *the structure of the brain matters*—that, from Ashby's perspective, a key level of organization had to be built in genetically and could not be achieved by the sort of trial-and-error self-organization performed by DAMS.[28]

— — — — —

Though DAMS failed, Ashby's struggles with it undoubtedly informed his understanding of complex mechanisms and the subsequent development of his cybernetics, so I want to pursue these struggles a little further here.[29] First, I want to emphasize just how damnably complicated these struggles were. DAMS first appeared in Ashby's journal on 11 August 1950 (pp. 2953–54) with the words "First, I might as well record my first idea for a new homeostat [and, in the margin] found a month ago." The next note, also dated 11 August 1950, runs for twenty pages (pp. 2955–74) and reveals some of the problems that Ashby had already run into. It begins, "For a time the construction of the new machine (see previous page) went well. Then it forced me to realise that my theory had a yawning hole in it" (p. 2955).

This yawning hole had to do with DAMS's essential variables, the parameters it should control. In the original homeostat setups all of the currents were essential variables, capable of triggering discontinuous changes of state via the relays and uniselectors. But there was no reason why all of the currents in DAMS should be essential variables. Some of them should be, but others would have simply to do with making or breaking connections. Thus, a new problem arose: how the environment should be supposed to connect to DAMS's essential variables, and how those variables might act back onto the environment.[30] The homeostat offered no guidance on this, and the remainder of this entry is filled with Ashby's thoughts on this new problem. It contains many subsequently added references to later pages which develop these early ideas further. In a passage on page 2967, for example, one thought is linked by an asterisk to a note at the bottom of the page which says, "May '51. Undoubtedly sound in aim, but wrong in the particular development used here," while in the margin is a note in black ink, "Killed on p. 2974," and then another note, "Resurrected p. 3829," in red. The next paragraph then begins, "This was the point I reached before I returned to the designing of the electrical machine, but, as usual, the designing forced a number of purely psychological problems into the open. I found my paragraph (2) (above) [i.e., the one just discussed here] was much too vague to give a decisive guide." The penultimate paragraph of the entire note ends (p. 2974), "I see no future this way. The idea of p. 2967 (middle) [i.e., again the one under discussion here] seems to be quite killed by this last figure." But then a marginal note again says, "Resurrected p. 3829" (i.e., 17 May 1952).

The substantial point to take from all this is that the construction of DAMS posed a new set of problems for Ashby, largely having to do with the specification of its essential variables and their relation to the environment, and it was by no means clear to him how to solve them.[31] And what interests me most here is that in response to this difficulty, Ashby, if only in the privacy of his journal, articulated an original *philosophy of design*.

"The relation of the essential variables to a system of part-functions [e.g., the neon tubes] is still not clear, though p. 3074 helps. Start again from first principles," Ashby instructed himself on 28 January 1951, but a second note dated the same day recorded that DAMS was "going to be born any time" (pp. 3087–8). Six weeks later Ashby recorded that "DAMS has reached the size of ten valves, and," he added, "has proved exceedingly difficult to understand." He continued (14 March 1951, pp. 3148–51),

> But while casting around for some way of grasping it I came across a new idea. Why not make the developent of DAMS follow in the footsteps marked out by evolution, by making its variations struggle for existence? We measure in some way its chance of "survival," and judge the values of all proposed developments by their effects on this chance. We know what "survival" means in the homeostat: we must apply the same concept to DAMS. . . .
>
> The method deserves some comment. First notice that it totally abandons any pretence to "understand" the assembly in the "blue-print" sense. When the system becomes highly developed the constructor will be quite unable to give a simple and coherent account of why it does as it does. . . . Obviously in these circumstances the words "understand" and "explain" have to receive new meanings.
>
> This rejection of the "blue-print" attitude corresponds to the rejection of the "blue-print" method in the machine itself. One is almost tempted to dogmatise that the Darwinian machine is to be developed only by the Darwinian process! (there may be more in this apothegm than a jest). After all, every new development in science needs its own new techniques. Nearly always, the new technique seems insufficient or hap-hazard or plain crazy to those accustomed to the old techniques.
>
> If I can, by this method, develop a machine that imitates advanced brain activities without my being able to say how the activities have arisen, I shall be like the African explorer who, having heard of Lake Chad, and having sought it over many months, stood at last with it at his feet and yet, having long since lost his bearings, could not say for the life of him where in Africa Lake Chad was to be found.

This is a remarkable passage of ontological reflection, which gets us back to the cybernetic discovery of complexity from a new angle. Like Walter's tortoise, the homeostat had been designed in detail from the ground up—the blueprint attitude—and this approach had been sufficient, inasmuch as the two machines did simulate performances of the adaptive brain. My argument was, however, that when constructed, they remained to a degree impermeable Black Boxes, displaying emergent properties not designed into them (the tortoise), or otherwise opaque to analysis (the multihomeostat setup). But it was only with DAMS that Ashby had to confront this discovery of complexity head-on. And in this passage, he takes this discovery to what might be its logical conclusion. If, beyond a certain degree of complexity, the performance of a machine could not be predicted from a knowledge of its elementary parts, as proved to be the case with DAMS, then one would have to abandon the modern engineering paradigm of knowledge-based design in favor of evolutionary tinkering—messing around with the configuration of DAMS and retaining any steps in the desired direction.[32] The scientific detour away from and then back to performance fails for systems like these.

The blueprint attitude evidently goes with the modern ontological stance that presumes a knowable and cognitively disposable world, and Ashby's thoughts here on going beyond design in a world of mechanisms evolving quasi-organically once more make the point that ontology makes a difference, now at the level of engineering method. We can come back to this point in later chapters.

Ashby never reached the shores of Lake Chad, but one feature of DAMS's performance did become important to his thinking: a behavior called "habituation." In his only published discussion of DAMS, after a discussion of DAMS itself, Ashby turns to a theoretical argument, soon to appear in *Design for a Brain*, that he claims is generally applicable to any "self-switching network, cortex or D. A. M. S. or other, . . . no matter in what random pattern the parts are joined together and no matter in what state its 'memories' have been left by previous activities." This argument has two parts: first, that a system like DAMS will naturally split itself up into subsystems that "*tend to be many and small rather than few and large*"; and second, that such a system becomes habituated to a repeated stimulus, inamsuch as "*it will tend to set its switches so that it is less, rather than more, disturbed by it*." Then Ashby returns to his machines, noting first that the latter effect had been demonstrated on the homeostat, where, indeed, it is true almost by definition: the first application of any stimulus was liable to provoke a large response—the tripping of the unselectors—while once the homeostat had found an equilibrium configuration, its response to the same stimulus would be small: a damped oscillation returning to the equilibrium state. By 1951, Ashby could also remark that this property "is already showing on the partly-constructed D. A. M. S." (1951, 4, 5; Ashby's italics).

Ashby regarded habituation in his machines as support for his general approach to the brain. "In the cerebral cortex this phenomenon [of diminishing response to a stimulus] has long been known as 'habituation.' It is in fact not restricted to the cerebral cortex but can be observed in every tissue that is capable of learning. Humphrey considers it to be the most fundamental form of learning" (1951, 5). But, as Ashby put it in *Design for a Brain*, "The nature of habituation has been obscure, and no explanation has yet received general approval. The results of this chapter suggest that it is simply a consequence of the organism's ultra-stability, a by-product of its method of adaptation" (1952, 152).[33] The significance of this observation is that Ashby had gone beyond the simple mimicry of adaptation to a novel result—discovering the go of a phenomenon that hitherto remained mysterious.[34] And in his journals, Ashby took this line of thought still further. Reflecting on DAMS on 22 May 1952 (p. 3829), he arrived at an analysis of "dis-inhibition" (he writes it in quotes): "The intervention of a second stimulus will, in fact, restore the δ-response to its original size. This is a most powerful support to my theory. All other theories, as far as I know, have to postulate some special mechanism simply to get dis-inhibition."[35]

If DAMS never reached the promised land and Ashby never quite reached Lake Chad, then, certainly the DAMS project led to this one substantive result: an understanding of habituation and how it could be undone in ultrastable machines. We can come back to this result when we return to Ashby's psychiatric concerns.

I can add something on the social basis of Ashby's research in the DAMS era and its relation to the trajectory of his research. In the early 1950s, Pierre de Latil visited the leading cyberneticians of the day, including Walter as well as Ashby, and wrote up a report on the state of play as a book, *Thinking by Machine: A Study of Cybernetics*, which appeared in French in 1953 and in English in 1956, translated by Frederick Golla's daughter, Yolande. De Latil recorded that "Ashby already considers that the present DAMS machine is too simple and is planning another with even more complex action. Unfortunately, its construction would be an extremely complex undertaking and is not to be envisaged for the present" (de Latil 1956, 310). I do not know where the money came from for the first versions of DAMS, but evidently cost became a problem as Ashby began to aim at larger versions of it. On an ill-starred Friday the 13th in September 1957, Ashby noted to himself, "As the RMPA [Royal Medico-Psychological Association] are coming to B. H. [Barnwood House] in May 1960 I have decided to get on with making a DAMS for the occasion, doing as well as I can on the money available. By building to a shoddiness that no commercial builder would consider, I can probably do it for far less than a commercial firm would estimate it at." Clearly, by this time Ashby's hobby was turning into a habit he could ill afford and remained a hobby only for lack of institutional support.[36] That his work on DAMS had lapsed for some time by 1957 is evident in the continuation of the note: "In addition, my theoretical grasp is slowly getting bogged down for lack of real contact with real things. And the deadline of May 1960 will force me to develop the practical & immediate" (p. 5747).

Ashby's strained optimism of 1957 was misplaced. A year later, on 29 September 1958, we find him writing (pp. 6058–60): "The new DAMS . . . having fizzled out, a new idea occurs to me today—why not make a small DAMS, not for experimental purposes but purely for demonstration. . . . The basic conception is that all proofs are elsewhere, in print probably; the machine is intended purely to enable the by-stander to see what the print means & to get some intuitive, physical, material feeling for what it is about. (Its chief virtue will be that it will teach me, by letting me see something actually do the things I think about.) Summary: Build devices for demonstration." The drift in this passage from DAMS to demonstration machines is significant. After a break, the same journal entry continues poignantly: "The atmosphere at Namur (Internatl. Assoc. for Cybs., 2–9 Sep.) showed me that I am now regarded more as a teacher than as a research worker. The world wants to hear what I have found out, & is little interested in future developments. Demonstration should therefore be my line, rather than exploration. In this connexion it occurs to me that building many small machines, each to show just one point, may be easier (being reducible) than building a single machine that includes the lot. Summary: Build small specialist machines, each devised to show one fact with perfect clarity." A formally beautiful but personally rather sad technosocial adjustment is adumbrated in this note. In it, Ashby responds to two or possibly three resistances that he felt had arisen in his research. The one that he failed to mention must have been his lack of technical success in developing DAMS as a synthetic brain. The second was the escalating cost and lack of commensurate institutional support for developing DAMS, as just discussed. And the third was what he perceived, at least, to be a developing lack of interest in his research in the European cybernetics community. How far he was correct in this perception is difficult to judge; it is certainly true, however, that youngsters like Stafford Beer and Gordon Pask were bursting onto the scene by the late 1950s—Beer was thirty-four in 1958, Pask thirty-two; Ashby was becoming a grand old man of cybernetics at the age of fifty-four. And all of these resistances were accommodated by Ashby's strategy. Technically, building small demonstration machines presented him with a finite task (unlike the never-ending difficulties with DAMS as a research machine), reduced the cost to a bearable level, and, socially, positioned Ashby as a pedagogue.

In important respects, Ashby went through with this plan. Especially at the University of Illinois in the 1960s, his demonstration machines became legendary, as did his qualities as a pedagogue.[37] It is certainly not the case that he gave up his research after 1958—his "hobby" was always his raison d'être—but his major subsequent contributions to cybernetics and systems theory were all in the realm of theory, as foreshadowed in the first quotation above. As a full professor at a major American university, Ashby's funding problems appear to have been significantly alleviated in the 1960s, and there is one indication that he returned then to some version of DAMS as a research project. In an obituary, Oliver Wells recalled that Ashby's "love of models persuaded von Foerster to have constructed what was called the 'The Grandfather Clock' which was designed as a seven foot noisy model of state-determined complex 'systems' running through trajectories of cycles of stabilisation and 'randomness'" (Wells 1973). One has to assume that nothing significant emerged from this project; like the English DAMS, it was never the subject of anything that Ashby published.

The stars were in a strange alignment for Ashby in the late 1950s. Immediately after the deflationary post-Namur note he added an interstitial, undated note which reads: "Here came the Great Translation, from a person at B. H. to Director at B. N. I. [the Burden] (Appointment, but no more till May '59)" (p. 6060). But now we, too, can take a break and go back to madness.

Madness Revisited

At the beginning of this chapter I noted that Ashby's career in Britain was based in mental institutions and that he was indeed active in research related to his profession, publishing many papers on explicitly psychiatric topics. I want now to discuss the relation between the two branches of Ashby's work, the one addressed to questions of mental illness and the cybernetic work discussed in the preceding sections.

My starting point is Ashby's 1951 assertion, already quoted, that his cybernetics, as developed in his journal, "was to me merely a delightful amusement, a hobby I could retreat to, a world where I could weave complex and delightful patterns of pure thought." This assertion deserves to be taken seriously, and it is tempting to read it as saying that his cybernetic hobby had nothing to do with his professional research on pathological brains and ECT. It is also possible to read his major works in cybernetics, above all his two books, as exemplifications of this: there is remarkably little of direct psychiatric interest in them. The preceding discussions of the homeostat and DAMS should likewise make clear that this aspect of Ashby's work had its own dynamic. I nevertheless want to suggest that this reading is untenable, and that there were in fact interesting and constitutive relationships between the two branches of Ashby's oeuvre—that psychiatry was a surface of emergence and return for Ashby's cybernetics, as it was for Walter's.

We can start by noting that in the 1920s Englishmen took up many hobbies, and theorizing the adaptive brain is hardly the first that comes to mind. If in 1928 Ashby had taken up stamp collecting, there would be nothing more to say. But it is evident that his professional interests structured his choice of hobby. If his cybernetics, as discussed so far, was an attempt to understand the go of the normal brain, then this related to his professional concerns with mental illness, at minimum, as a direct negation rather than a random escape route. More positively, Ashby's materialism in psychiatry, shared with Golla and Walter, carried over without negation into his hobby. The hobby and the professional work were in exactly the same space in this respect. And we should also remember that in medicine the normal and the pathological are two sides of the same coin. The pathological is the normal somehow gone out of whack, and thus, one way to theorize the pathological is first to theorize the normal. The correlate of Ashby's interest in adaptation, in this respect, is the idea going back at least to the early twentieth century, that mental illnesses can be a sign of *maladaptation* (Pressman 1998). Simply by virtue of this reciprocal implication of the normal and the pathological, adaption and maladaptation, it would have been hard for Ashby to keep the two branches of his research separate, and he did not.

The most obvious link between the two branches of Ashby's research is that most of Ashby's early cybernetic publications indeed appeared in psychiatric journals, often the leading British journal, the *Journal of Mental Science*. And, as one should expect, all of these papers gestured in one way or another to the problems of mental illness. Sometimes these gestures were largely rhetorical. Ashby would begin a paper by noting that mental problems were problems of maladaptation, from which it followed that we needed to understand adaptation, which would lead straight into a discussion of tilted cubes, chicken incubators, beads and elastic, or whatever. But sometimes the connections to psychiatry were substantial. Even Ashby's first cybernetic publication, the 1940 essay on dynamic equilibrium, moves in that direction. Ashby there discusses the "capsule" which controls the fuel flow in a chicken incubator and then asks what would happen if we added another feedback circuit to control the diameter of the capsule. Clearly, the capsule would not be able to do its job as well as before, and the temperature swings would be wilder. Although

Ashby does not explicitly make the point, this argument about "stabilizing the stabilizer" is of a piece with the conventional psychiatric idea that some mental fixity lies behind the odd behavior of the mentally ill—mood swings, for example. What Ashby adds to this is a mechanical model of the go of it. This simple model of the adaptive brain can thus be seen as *at once* a model for thinking about pathology, too. Likewise, it is hard not to relate Ashby's later thoughts on the density of connections between homeostat units, and their time to reach equilibrium, with lobotomy. Perhaps the density of neural interconnections can somehow grow so large that individuals can never come into equilibrium with their surroundings, so severing a few connections surgically might enable them to function better. Again, Ashby's understanding of the normal brain immediately suggests an interpretation of mental pathology and, in this case, a therapeutic response.

Ashby often failed to drive home these points explicitly in print, but that proves very little. He contributed, for example, the entry "Cybernetics" to the first *Recent Progress in Psychiatry* to appear in Britain after World War II (Fleming 1950).[38] There he focused on pathological positive feedback in complex machines—"runaway"—as a model for mental illness, leading up to a lengthy discussion of the stock ways of curing such machine conditions: "to switch the whole machine off and start again," "to switch out some abnormal part," and "to put into the machine a brief but maximal electric impulse" (Ashby 1950b, 107). We saw this list before in the previous chapter, and when Walter produced it he was not shy of spelling out the equivalences to sleep therapy, lobotomy, and ECT, respectively. Given a pulpit to preach to the psychiatric profession, Ashby could bring himself to say only, "These methods of treatment [of machines] have analogies with psychiatric methods too obvious to need description" (1950b, 107).

To find more specific and explicit connections between Ashby's cybernetics and his professional science, it is interesting to begin with a paper I mentioned before, his 1953 essay "The Mode of Action of Electro-convulsive Therapy" (Ashby 1953a). As I said, the body of this paper is devoted to reporting biochemical observations on rats that had beeen subjected to electroshock, and the theoretical introduction accordingly lays out a framework for thinking about ECT and brain chemistry. But Ashby also throws in a second possible interpretation of the action of ECT:

> There is a possibility that E. C. T. may have a direct effect on the cortical machinery, not in its biochemical but in its cybernetic components. . . . It has been shown [in *Design for a Brain*] that one property such systems [of many interacting

> elements] will tend to show is that their responses . . . will tend to diminish. When the stimulus is repeated monotonously, the phenomenon is well known under the name of "habituation." We can also recognise, in everyday experience, a tendency for what is at first interesting and evocative to become later boring and uninspiring. Whether the extreme unresponsiveness of melancholia is really an exaggeration of this process is unknown, but the possibility deserves consideration. What makes the possibility specially interesting is that the theory of such statistical systems makes it quite clear that any complex network that has progressed to a non-responding state can, in general, be made responsive again by administering to it any large and random disturbance. The theory also makes clear that such a disturbance will necessarily disturb severely the system's memory: the parallel with E. C. T.'s effect on memory is obvious. Whether, however, E. C. T. acts in essentially this way is a question for the future.

This passage is remarkable in at least two ways. First, it does not belong in Ashby's essay at all. If taken seriously, it undercuts the entire rationale for the biochemical investigations reported there. Second, and more important in the present context, it makes an explicit connection between Ashby's cybernetics and his work on DAMS on the one hand, and his interest in ECT and its functioning on the other, and we can return to DAMS here.[39] A journal entry of 25 August 1951 records that "while working with DAMS I found I was unconsciously expecting it to 'run down,' then I realised what was happening, & that my expectation was not unreasonable, was a new idea in fact." Then follows the first discussion of "habituation" in DAMS (though Ashby does not use the word here): "there is therefore a tendency for the neons to change their average 'readiness' from 'more' to 'less.'" And Ashby immediately moves from this observation to a consideration of the antidotes to habituation: "After this initial reserve of changeable neons has been used up the system's possibilities are more restricted. The only way to restore the possibilities is to switch the set off, or perhaps to put in some other change quite different from those used during the routine. This fact can obviously be generalised to a principle." As just mentioned, there was a stock equation in the cybernetics of this period between switching off a machine and sleep therapy for mental illness, though Ashby does not comment on this in his note. However, there then follows a quick sketch of the argument that in its response to a new and different input, DAMS will regain its prehabituation sensitivity to the old one, after which Ashby concludes: "Summary: A multistable system tends to lose reactivity, which will often be restored by applying some strong, but unrelated stimulus, at the cost of some forgetting. ? Action of E. C. T. (Corollary p. 3464)" (pp. 3434–3437).

This is the argument Ashby relied upon above but did not provide in his 1953 essay on the functioning of ECT, but here we find it right in the heartland of his hobby, engaging directly with his major cybernetic project of the early 1950s, DAMS. And it is revealing to follow this story a little further in his journal. The reference forward from the last note takes us to a journal entry dated 12 September 1951, which begins, "From p. 3464, it is now obvious how we make DAMS neurotic: we simply arrange the envt. so that it affects two (or more) essl. variables so that it is impossible that both should be satisfied." Page 3464 in fact takes us to a discussion of Clausewitz, which I will come back to in the next section. In this entry, though, Ashby draws a simple circuit diagram for DAMS as subject to the conflicting demands of adapting to two different voltages at once (fig. 4.10) and comments that "both E.V.'s will now become very noisy," seeking first to adapt to one voltage and then the other, "and the system will be seriously upset. It is now very like a Masserman cat that must either starve or get a blast in the face. The theme should be easily developed in many ways" (pp. 3462–63). We thus find ourselves explicitly back in the psychiatric territory I associated in the previous chapter with Grey Walter and the CORA-equipped tortoise, now with DAMS as a model of neurosis as well as normality and of the functioning of ECT.[40]

Ashby's journal entry refers forward to another dated 22 September 1951, where Ashby remarks that DAMS will simply hunt around forever when posed an insoluble problem, but that "the animal, however, . . . will obviously have some inborn reflex, or perhaps several, for adding to its resources. . . . A snail or tortoise may withdraw into its shell. . . . The dog may perhaps simply bite savagely. . . . A mere total muscular effort—an epileptic fit—may be the last resort of some species. . . . My chief point is that the symptoms of the unsolvable problem, whether of aggression, of apathy, of catatonia, of epilepsy, etc are likely to be of little interest in their details, their chief importance clinically being simply as indicators that an unsolvable problem has been set" (pp. 3479–81). Here Ashby covers all the bases, at once addressing a whole range of pathological clinical conditions, while dismissing the importance of symptoms in favor of his cybernetic analysis of the underlying cause of all of them—and, in the process, perhaps putting down Grey Walter, for whom epilepsy—"a mere total muscular effort"—was a major research field in its own right.

Habituation and dehabituation, then, were one link between Ashby's cybernetics and his psychiatry, and, indeed, it is tempting to think that the

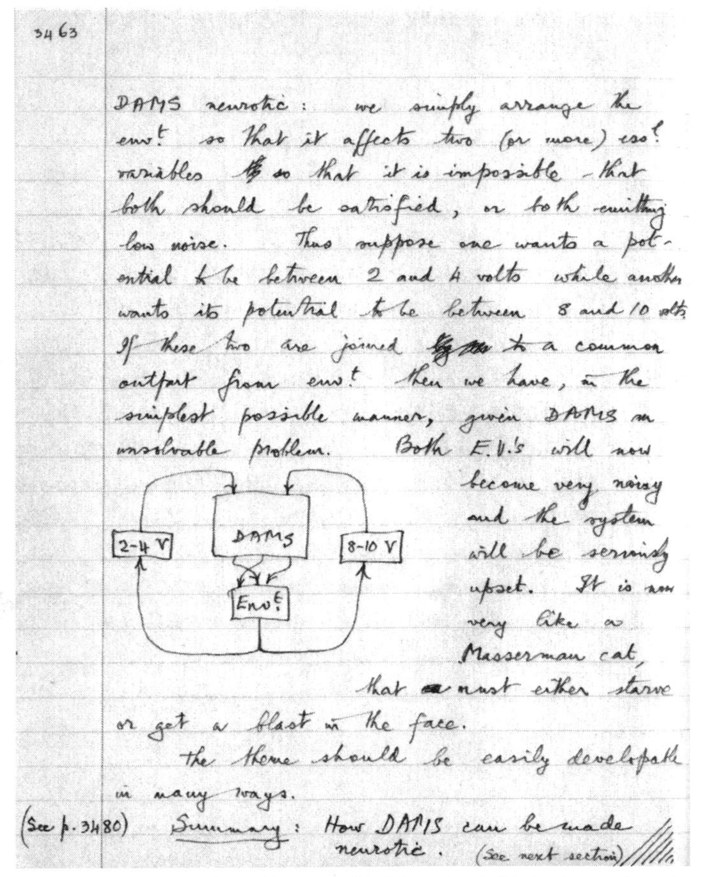

3463

DAMS neurotic: we simply arrange the envt so that it affects two (or more) essl variables so that it is impossible that both should be satisfied, or both emitting low noise. Thus suppose one wants a potential to be between 2 and 4 volts while another wants its potential to be between 8 and 10 volts. If these two are joined to a common output from envt then we have, in the simplest possible manner, given DAMS an unsolvable problem. Both E.V.'s will now become very noisy and the system will be seriously upset. It is now very like a Masserman cat, that must either starve or get a blast in the face.

The theme should be easily developable in many ways.

(See p. 3480) Summary: How DAMS can be made neurotic. (See next section)

Figure 4.10. "How DAMS can be made neurotic." Source: Ashby's journal, entry dated 12 September 1951 (p. 3463). (By permission of Jill Ashby, Sally Bannister, and Ruth Pettit.)

possibility of this link explains some of the energy Ashby invested during the 1950s in this otherwise hardly exciting topic. But it is worth emphasizing that it was by no means the only possible link that Ashby discerned. To get at the range of his thinking it is enough to look at his published record, and here we can focus on a 1954 paper, "The Application of Cybernetics to Psychiatry" (Ashby 1954).[41] This tentatively outlines several different ways of thinking cybernetically about mental illness. I will just discuss a couple.[42]

One carried further Ashby's theorizing of the chemistry of electroshock. As mentioned at the beginning of this chapter, Ashby's own measurements had shown, he believed, that electroshock was followed by "a brisk outpouring of steroids." Here the question addressed was this: The level of steroids in the brain is presumably a quantity which varies continuously, up or down. Insanity, in contrast, appears to be dichotomous—one is either mad or not. How then can a continuous cause give rise to a discontinuous effect? "What is not always appreciated is that the conditions under which instability appears are often sharply bounded and critical even in a system in which every part varies continuously. . . *Every dynamic system is potentially explosive*. . . . These facts are true universally. . . . They are necessarily true of the brain" (1954, 115–16). And Ashby had, in fact, addressed this topic mathematically in a 1947 paper (Ashby 1947). There he considered a complex system consisting of interlinked autocatalytic chemical reactions of three substances, with rates assumed to be controlled by the presence of some enzyme, and he showed by numerical computation that there was an important threshold in enzyme concentration. Below that threshold, the concentration of one of the reacting chemicals would inevitably fall to zero; above the threshold, the concentration would rise to unity. This mathematical result, then, showed in general how discontinuous effects can emerge from continuous causes, and, more specifically, it shed more light on the possible go of ECT—how the outpouring of steroids might conceivably flip the patient's brain into a nonpathological state.[43]

The other suggestion was more directly cybernetic. Ashby supposed that when the essential variables exceed their limits in the brain they open a channel to signals from a random source, which in turn pass into the cortex and initiate homeostat-like reconfigurations there (fig. 4.11). Both the source and the channel were supposed to be real anatomical structures (1954, 120): "*V* [the random source] could be small, perhaps even of molecular size. It won't be found until specially looked for. The channel *U* [carrying the random signal to the cortex], however, must be quite large. . . . One thinks naturally of a tract like the mammillo-thalamic . . . [and] of the peri-ventricular fibres . . . but these matters are not yet settled; they offer an exceptional opportunity to any worker who likes relating the functional and the anatomical." And, having hypothesized this cybernetic channel *U*, Ashby was in a position to describe the pathologies that might be associated with it. If it was unable to carry sufficient information, the brain would be unable to change and learn from its mistakes, while if it carried too much, the brain would be continually experimenting and would never reach equilibrium—conditions which Ashby associated with melancholia and mania, respectively. Here then, he came back to the idea that he had unsuccessfully explored in the 1930s—that there exists an identifiable organic basis for the various forms of mental pathology—but now at a much greater level of specificity. Instead of examining gross features of brains in

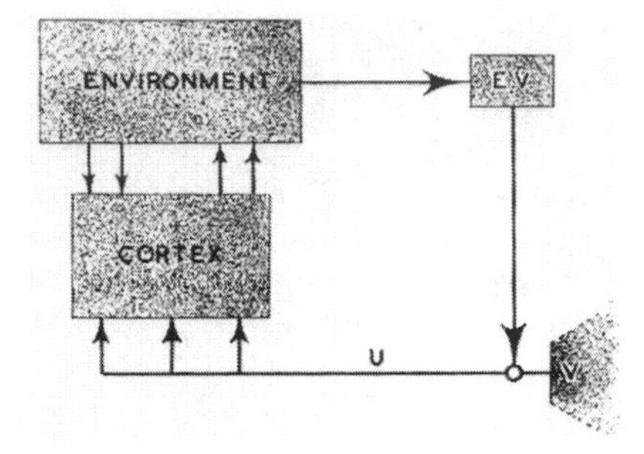

Figure 4.11. The brain as homeostat. Signals from the essential variables (*E.V., top right*) open the channel *U* to the random source (*V, bottom right*). Reproduced with permission from W. R. Ashby, "The Application of Cybernetics to Psychiatry," *Journal of Mental Science, 100* (1954), 120. (© 1954 The Royal College of Psychiatrists.)

pursuit of differences, one should above all look for this channel *U* and its possible impairments. This idea that the brain contains a special organ to accomplish its homeostatic adaptations—a whole new kind of bodily structure lying outside the classifications of contemporary medical and biological science—is a striking one. As far as I know, however, no one took this suggestion up in anatomical research.

There is more to be said about Ashby's cybernetic psychiatry, but that will take us in different directions, too, so I should briefly sum up the relation between his cybernetics and psychiatry as we have reviewed it thus far. First, as I said of Grey Walter in the previous chapter, psychiatry was a surface of emergence for Ashby's cybernetics: his cybernetics grew out of psychiatry, partly by a reversal (the normal instead of the pathological brain as the focus of his hobby) but still remaining in the same space (the normal and the pathological as two sides of the same coin). There is no doubt that Ashby's hobby represented a significant detour away from the mental hospital in his thinking; as I said, his cybernetic research had its own dynamics, which cannot be reduced to a concern with mental illness. But still, psychiatry remained very much present in Ashby's cybernetics as a potential surface of return. Especially during his years at Barnwood House, 1947–59, the key years in the flowering of his cybernetics, Ashby was more than ready to see how his cybernetics could grow back into psychiatry. And we should not see this as some cynical maneuver, simply pandering to the profession that paid him. The appearance of psychiatric concerns in his journal—where, for example, his wife and children never get a look in, and where his own appointment to the directorship of the Burden only warranted an interstitial remark—testifies to his own continuing interest in psychiatry. This, I believe, is how we should think of the relation between cybernetics and psychiatry in Ashby's work: psychiatry as both a surface of emergence and return for a cybernetics that was, nevertheless, a scientific detour away from it.[44]

Adaptation, War, and Society

SUPPOSE WE CONSIDERED WAR AS A LABORATORY?

THOMAS PYNCHON, *GRAVITY'S RAINBOW*

We have been following the development of Ashby's cybernetics as a science of the brain, but I mentioned at the start the instability of the referent of his work, and now we can pick up this thread. In the next section I will discuss Ashby's transformation of cybernetics into a theory of everything, but first I want to follow some passages in Ashby's journal that constitute more focused extensions of his cybernetics into the field of *the social*—specifically, questions of war and planning. These interest me for two reasons. First, they are further manifestations of the protean character of cybernetics, spilling over beyond the brain. Second, Ashby's thoughts on war and planning manifest diametrically opposed ways—asymmetric and symmetric, respectively—of imagining adaptation in multiagent systems. This is an important contrast we need to keep in mind for the rest of the book. Ashby assimilated psychiatry to the asymmetric adaptation he associated with warfare, while we will see that Bateson and Laing took the other route, emphasizing a symmetry of patient and therapist (and Beer and Pask also elaborated the symmetric stance). This difference in stance goes to the heart of the difference between the psychiatry of Ashby and Walter and the "antipsychiatry" of Bateson and Laing.

Ashby started making notes on DAMS on 11 August 1950, and one of his lines of thought immediately took on a rather military slant. In the long second note he wrote that day he began to struggle with the central and enduring problem of how DAMS could associate specific patterns of its inner connections with specific environmental stimuli—something he took to be essential if DAMS was to accumulate adaptations. Clearly, DAMS would have to explore its environment and find out about it in order to adapt, and "[when]one is uncomfortable [there] is nothing other than to get restless. (3) Do not suffer in silence: start knocking the env[ironmen]t about, & watch what happens to the discomfort. (4) This is nothing other than 'experimenting': forcing the environment to reveal itself. (5) Only by starting a war can one force the revelation of which are friends & which foes. (6) Such a machine does not solve its problems by thinking, just the opposite: it solves them by forcing action. . . . So, in war, does one patrol to force the enemy to reveal himself and his characteristics" (p. 2971).

A year later, we find similar imagery. "A somewhat fearsome idea!" begins the entry for 7 September 1951 (pp. 3451–52):

> In evolution, the fact that survival rules everything means that organisms will not only develop those features that help them to survive against their environment but will also force them to develop those features that help them to survive against each other. The "killer" Paramecium, or the aggressive male stag, is favoured as compared with its more neutral neighbours. . . . If the cerebral cortex evolves similarly, by "survival" ruling everything in that world of behaviour & subsystems, then those subsystems should inevitably become competitive under the same drive. . . . In a really large cortex I would expect to find, eventually, whole armies of subsystems struggling, by the use of higher strategy, against the onslaught of other armies.

Ashby was a great reader, and his next note on the following day begins thus (pp. 3452–7):[45]

> I have always held that war, scientific research, and similar activities, being part of the organism's attempt to deal with its environment, must show, when efficient & successful, the same principles that are used by the organism in its simpler & more direct interactions with an environment. I have hunted through the Public Library for some book on the essentials of military method, but could find nothing sufficiently abstract to be usable. So I borrowed "Clausewitz." Here is my attempt to translate his principles into the psychological. He starts 'What is war? War is an art of violence, and its object is to compel our opponent to comply with our will.' Comment: Clearly he means that step-functions must change, and those are not to be ours.

War among the homeostats! It is worth continuing this passage. Ashby remarks that the approximate symmetry between opponents in war (he is thinking of old-fashioned wars like World War II) "is quite different from the gross asymmetry usually seen in the organism-environment relation," and continues:

> Where, then, do we find such a struggle between equals? Obviously in a multistable system between adapted sub-systems, each of which, being stable, "tries" to force the other to change in step-functions. . . . If two systems interact, how much information should each admit? . . . If I am wrestling, there is a great practical difference between (1) getting information by looking at my opponent with

> open eyes and (2) setting his hands around my throat & feeling what he is going to do. Obviously the difference is due to the fact that effects from the throat-gripping hands go rapidly & almost directly to the essential variables, whereas the effects from the retina go through much neural network & past many effectors before they reach the E.V.'s. In war, then, as discussed by Clausewitz, we must assume that the systems have essential variables. Is this true of the cortical sub-systems? Probably not if we are talking about purely cortical sub-systems. . . . It would, however, be true of subsystems that have each some of the body's essential variables and that are interacting: [see fig. 4.12]. Now we have something like two armies struggling. . . . Summary: The art of war—in the cortex.

What should we make of these ruminations? The first point to note is the extension of Ashby's ontological vision: here warfare and brain processes are understood on the same basic plan, as the interaction of adaptive entities. But second, an *asymmetry* has entered the picture. Warfare, on Ashby's reading of Clausewitz, is *not* a process of *reciprocal* adaptation: in war each party seeks to remain constant and to oblige the other to adapt.[46] Third, it is evident that in the early 1950s Ashby's cybernetics evolved in a complex interplay between his thinking on war and brain science and his struggles with DAMS. And, furthermore, we can get back to the topic of the previous section by throwing psychiatry back into this heady mix. Figure 4.12, for example, is almost identical to a circuit diagram that Ashby drew four days later, except that there the central box was labeled "DAMS." This latter figure was reproduced above as figure 4.10, which I labeled with a quotation from Ashby, "how DAMS can be made neurotic." We thus return very directly to the topics of psychiatry, once more in the heartland of Ashby's journal. In this phase of his research, then, it is fair to say that DAMS, adaptation, war, and neurosis were bound up together. Ashby's thinking on each was productively engaged with his thoughts on the other.

This line of thought on Clausewitz and war never made it explicitly into Ashby's published writings, and I have not tracked its evolution systematically through his journal, but it makes a striking reappearance seven years later, in the entry immediately following the note that he had just been appointed director of the Burden. On 3 November 1958 he remarked (pp. 6061–2) that

> treating a patient is an imposition of the therapist's will on the patient's; it is therefore a form of war. The basic principles of war are therefore applicable. They may actually be very useful, for an opposing army is like a patient in that both are [very complex, inherently stable, etc.]. A basic method much used in war is to use a maximal concentration of all possible forces on to a small part,

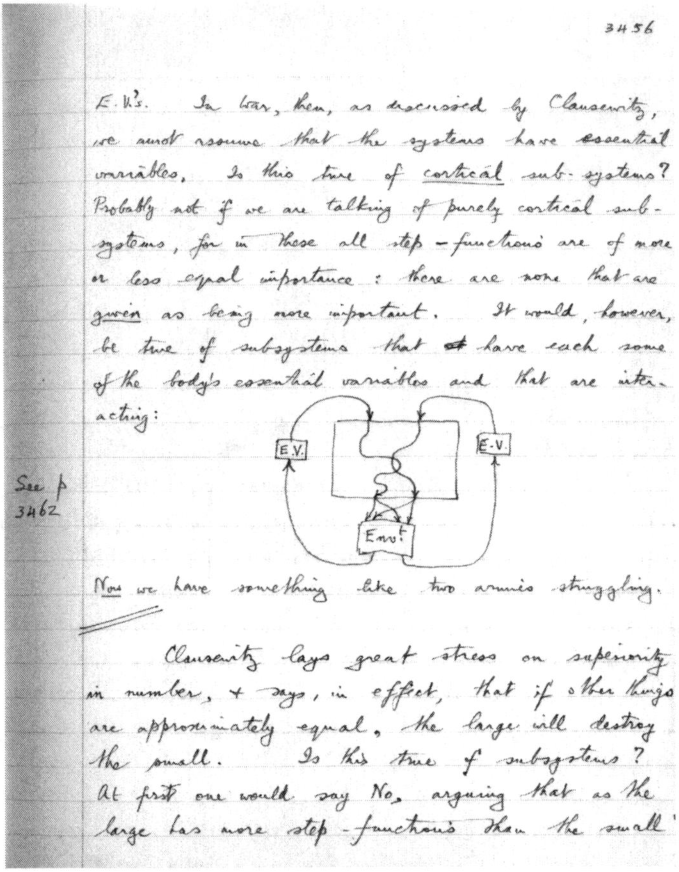

3456

E.V.'s. In war, then, as discussed by Clausewitz, we must assume that the systems have essential variables. Is this true of cortical sub-systems? Probably not if we are talking of purely cortical sub-systems, for in these all step-functions are of more or less equal importance: there are none that are given as being more important. It would, however, be true of subsystems that have each some of the body's essential variables and that are interacting:

See p 3462

Now we have something like two armies struggling.

Clausewitz lays great stress on superiority in number, & says, in effect, that if other things are approximately equal, the large will destroy the small. Is this true of subsystems? At first one would say No, arguing that as the large has more step-functions than the small'

Figure 4.12. War among subsystems in the cortex. Source: Ashby's journal, entry dated 8 September 1951 (p. 3456). (By permission of Jill Ashby, Sally Bannister, and Ruth Pettit.)

to try to get it unstabilised. The gain here may be semi-permanent, so that, with this holding, the forces can then attack another point. With this in mind, a Blitz-therapy would be characterised by:- (1) Use of techniques in combination, simultaneously. E.g. LSD, then hypnosis while under it, & ECT while under the hypnosis. (2) Not waiting to "understand" the patient's pathology (psycho-, somato-, neuro-) but hitting hard & seeing what happens. (3) Get a change anyhow, then exploit it; when it comes to a stop, take violent action to get another change somehow. (4) Get normal every point you possibly can. (5) Apply pressure everywhere & notice whether any part of the psychosis shows signs of cracking. (6) Let the psychiatric team focus on one patient, others being ignored meanwhile. Summary: Blitz-therapy.

LSD, hypnosis *and* electroshock. . . . As I said of Grey Walter in the previous chapter, Ashby was hardly one of Deleuze and Guattari's disruptive nomads within the world of professional psychiatry, and we can no doubt understand that along similar lines. But this horrendous image of "Blitz-therapy"—what a combination of words!—does help to bring to the fore a characteristic feature of British psychiatry in the 1950s which is worth emphasizing for future reference, namely its utter *social asymmetry*. In Ashby's world, it went without saying that the only genuine agents in the mental hospital were the doctors. The patients were literally that, subject to the will of the psychiatrist, whose role was to apply whatever shocks might jolt the mentally ill into a homeostat-like change of state. In this world, Blitz-therapy and the association between psychiatry and war made perfect sense, psychiatrically and cybernetically. In the next chapter we can explore the form of psychiatry that took the other fork in the road, on the model of symmetric and reciprocal adaptation between patient and psychiatrist.[47]

— — — — —

One can see Ashby's military musings as a drift toward a more general social elaboration of his cybernetics. War, as Ashby thought of it, following Clausewitz, was an extreme form that the relations between adaptive systems might take on, but it was not the only form. I have been quoting from Ashby's notes on DAMS, psychiatry, and warfare from early September 1951, and right in the middle of them is an entry dated 12 September, which begins, "On arranging a society" (pp. 3460–62): "Here is an objection raised by Mrs Bassett, which will probably be raised by others. May it not happen for instance that the planner will assume that full mobility of labour is available, when in fact people don't always like moving: they may have friends in the district, they may like the countryside, they may have been born and bred there, or they may dislike change. What is to stop the planner riding rough-shod over these 'uneconomic' but very important feelings?" Mrs. Bassett was, I believe, a researcher at the Burden Neurological Institute with whom Ashby later published a paper on drug treatment for schizophrenia (Ashby, Collins, and Bassett 1960). She was evidently also an early spokeswoman for the Big Brother critique of cybernetics, and her argument drove Ashby to think about real everyday social relations:

> The answer, of course, is that one sees to it that feedback loops pass through the people so that they are fully able to feel their conditions and to express opinions and take actions on them. One of the most important class of "essential

> variables" in such a society would be those that measure the "comfort" of the individual. . . . It is obvious that the original objection was largely due to a belief that the planner must understand every detail of what he plans, & that therefore the Plan must be as finite as the intelligence of the Planner. This of course is not so. Using the principles of the multistable system it should be possible to develop, though not to understand, a Plan that is far superior to anything that any individual can devise. Coupled with this is the new possibility that it can be self-correcting. Summary: Society.

Here we see the usual emphasis on performativity as prior to representation, even in planning—"though not to understand"—and temporal emergence, but expressed now in a much more socially symmetric idiom than Ashby's remarks on warfare and psychiatry. Now planners do not dictate to the planned how their lives will develop; instead planners and planned are envisaged as more or less equivalent parts of a single multistable system, entangled with one another in feedback loops from which transformations of the plan continually emerge. The image is the same as the vision of evolutionary design that Ashby articulated in relation to DAMS, transferred from the world of machines to that of people—now social designs and plans are to be understood not as given from the start and imposed on their object but as growing in the thick of things.

This is just one entry in Ashby's journal. He never systematically developed a cybernetic sociology. I mention it now because these remarks can serve as an antidote to the idea that Ashby's only vision of society was warfare, and, more important, because here he crudely sketches out a symmetric cybernetic vision of society that we shall see elaborated in all sorts of ways in the following chapters.

In conclusion, however, we can note that all traces of hierarchy were hardly purged from Ashby's thinking. The sentences that I skipped above contain his reflections on just how "the people" should make themselves felt in the feedback loops that pass through them. "The 'comfort' of the individual . . . can easily be measured. One simply makes a rule that every protest or appeal must be accompanied by a sum of money, & the rule is that the more you pay the more effective will your appeal be. You can have a sixpenny appeal which will adjust trivialities up to a hundred-pound appeal that will move mountains." This from a medical professional with a weakness for fast sports cars in a class-ridden society recovering from the devastations of war. It would be nice to think he was joking.

Cybernetics as a Theory of Everything

From the late 1920s until well into the 1950s Ashby's research aimed to understand the go of the brain. But this project faltered as the fifties went on. As we have just seen, Ashby's ambition to build a synthetic brain came to grief over his failure to get DAMS to accumulate adaptations. And, at the same time, as we saw in the previous chapter, the psychiatric milieu in which Ashby's cybernetics had grown started to shrink—as psychoactive drugs began to replace ECT and whatever, and as the antipsychiatric reaction to materialist psychiatry began to gain force. Where did those developments leave Ashby? Did he just give up? Evidently not. His mature cybernetics—that for which he is best remembered among cyberneticians today—in fact grew out of this smash-up, in ways that I can sketch out.

We can begin with what I called the "instability of the referent" of Ashby's cybernetics. Even when his concern was directly with the brain, he very often found himself thinking and writing about something else. His 1945 publication that included the bead-and-elastic device, for example, was framed as a discussion of a "dynamic system" or "machine" defined as "a collection of parts which (a) alter in time, and (b) interact on one another in some determinate and known manner. Given its state at any one moment it is assumed we know or can calculate what its state will be an instant later." Ashby then asserted that "consideration seems to show that this is the most general possible description of a 'machine' . . . not in any way restricted to mechanical systems with Newtonian dynamics" (1945, 14). Ashby's conception of a "machine" was, then, from early on exceptionally broad, and correspondingly contentless, by no means tied to the brain. And the generality of this conception was itself underwritten by a mathematical formalism he first introduced in his original 1940 protocybernetic publication, the set of equations describing the temporal behavior of what he later called a *state-determined system*, namely,

$$dx_i/dt = f_i(x_1, x_2, \ldots, x_n) \quad \text{for } i = 1, 2, \ldots, n,$$

where t stands for time, x_i are the variables characterizing the system, and f_i is some mathematical function of the x_i.

Since Ashby subsequently argued that almost all the systems described by science are state-determined systems, one can begin to see what I mean by the instability of the referent of his cybernetics: though he was trying to understand the brain as a machine, from the outset his concept of a machine was more or less coextensive with all of the contents of the universe. And this accounts for some of the rhetorical incongruity of Ashby's early cybernetic writings. For example, although it was published in the *Journal of General Psychology*, Ashby's 1945 bead-and-elastic essay contains remarkably little psychological content in comparison with its discussion of machines. It opens with the remark that "it is the purpose of this paper to suggest that [adaptive] behavior is in no way special to living things, that it is an elementary and fundamental property of all matter," it defines its topic as "all dynamic systems, whether living or dead" (13), and it closes with the assertion that "this type of adaptation (by trial and error) is therefore an essential property of matter, and no 'vital' or 'selective' hypothesis is required" (24). One wonders where the brain has gone in this story—to which Ashby's answer is that "the sole special hypothesis required is that the animal is provided with a sufficiency of breaks" (19), that is, plenty of elastic bands. "The only other point to mention at present is that the development of a nervous system will provide vastly greater opportunities both for the number of breaks available and also for complexity and variety of organization. Here I would emphasize that the difference . . . is solely one of degree and not of principle" (20).

So we see that in parallel to his inquiries into the brain, and indeed constitutive of those inquiries, went Ashby's technical development of an entire worldview—a view of the cosmos, animate and inanimate, as built out of state-determined machines. And my general suggestion then is that, as the lines of Ashby's research specifically directed toward the brain ran out of steam in the 1950s, so the cybernetic worldview in general came to the fore. And this shift in emphasis in his research was only reinforced by the *range* of disparate systems that Ashby described and analyzed in enriching his intuition about the properties of state-determined machines. I have already mentioned his discussions of chicken incubators and bead-and-elastic contrivances (the latter described as a "typical and clear-cut example of a dynamic system" [Ashby 1945a, 15]). The homeostat itself was first conceived as a material incarnation of Ashby's basic set of equations; his analysis of discontinuities in autocatalytic chemical reactions, discussed above, likewise concerned a special case of those equations. In *Design for a Brain* Ashby outlined the capabilities of a homeostatic autopilot—even if you wire it up backward so that its initial tendency is to destabilize a plane's flight, it will adapt and learn to keep the plane level anyway. And later in the book he spelled out the moral for evolutionary biology—namely, that complex systems will tend over time to arrive at complicated and interesting equilibriums with their environment. Such equilibriums, he argued are definitional of *life*, and therefore, "the development of life on earth must thus *not* be seen as something remarkable. On the contrary, it was inevitable" (233)—foreshadowing the sentiments of Stuart Kauffman's book *At Home in the Universe* (1995) four decades in advance. Ashby's single venture into the field of economics is also relevant. In 1945, the third of his early cybernetic publications was a short letter to the journal *Nature*, entitled "Effect of Controls on Stability" (Ashby 1945b). There he recycled his chicken-incubator argument about "stabilizing the stabilizer" as a mathematical analysis of the price controls which the new Labour government was widely expected to impose, showing that they might lead to the opposite result from that intended, namely a destabilization rather than stabilization of the British economy.[48] This reminds us that, as we have just seen, in his journal he was also happy to extend his analysis of the multistable system to both social planning and warfare.

Almost without intending it, then, in the course of his research into normal and pathological brains, Ashby spun off a version of cybernetics as a supremely general and protean science, with exemplifications that cut right across the disciplinary map—in a certain kind of mathematics, engineering, chemistry, evolutionary biology, economics, planning, and military science (if one calls it that), as well as brain science and psychiatry. And as obstacles were encountered in his specifically brain-oriented work, the brain lost its leading position on Ashby's agenda and he turned more and more toward the development of cybernetics as a freestanding general science. This was the conception that he laid out in his second book, *An Introduction to Cybernetics*, in 1956, and which he and his students continued to elaborate in his Illinois years.[49] I am not going to go in any detail into the contents of *Introduction* or of the work that grew out of it. The thrust of this work was formal (in contrast to the materiality of the homeostat and DAMS), and to follow it would take us away from the concerns of this book. I will mention some specific aspects of Ashby's later work in the following sections, but here I need to say a few words specifically about *An Introduction to Cybernetics*, partly out of respect for its author and partly because it leads into matters discussed in later chapters.[50]

An Introduction to Cybernetics presents itself as a textbook, probably the first and perhaps the last introductory textbook on cybernetics to be written. It aims to present the "basic ideas of cybernetics," up to and including "feedback, stability, regulation, ultrastability, information, coding, [and] noise" (Ashby 1956, v). Some of the strangeness of Ashby's rhetoric remains in it. Repeatedly and from the very start, he insists that he is writing for "workers in the biological sciences—physiologists, psychologists, sociologists" (1960, v) with ecologists and economists elsewhere included in the set. But just as real brains make few appearances in *Design for a Brain*, the appearances of real physiology and so on are notable by their infrequency in *An Introduction to Cybernetics*. The truly revealing definition of cybernetics that Ashby gives is on page 2: cybernetics offers "the framework on which all individual machines may be ordered, related and understood."[51]

An Introduction to Cybernetics is distinguished from *Design for a Brain* by one major stylistic innovation, the introduction of a matrix notation for the transformation of machine states in discrete time steps (in contrast to the continuous time of the equations for a state-determined system). Ontologically, this highlights for the reader that Ashby's concern is with change in time, and, indeed, the title of the first substantive chapter, chapter 2, is "Change" (with subheadings "Transformation" and "Repeated Change"). The new notation is primarily put to work in an analysis of the regulatory capacity of machines. "Regulation" is one of the new terms that appeared in Ashby's list of the basic ideas of cybernetics above, though its meaning is obvious enough. All of the machines we have discussed thus far—thermostats, servomechanisms, the homeostat, DAMS—are regulators of various degrees of sophistication, acting to keep some variables within limits (the temperature in a room, the essential variables of the body). What Ashby adds to the general discussion of regulation in *An Introduction to Cybernetics*, and his claim to undying eponymous fame, is the law of requisite variety, which forms the centerpiece of the book and is known to his admirers as Ashby's law. This connects to the other novel terms in *An Introduction to Cybernetics*'s list of basic ideas of cybernetics—information, coding, and noise—and thence to Claude Shannon's foundational work in information theory (Shannon and Weaver 1963 [1949]). One could, in fact, take this interest in "information" as definitive of Ashby's mature work. I have no wish to enter into information theory here; it is a field in its own right. But I will briefly explain the law of requisite variety.[52]

Shannon was concerned with questions of efficiency in sending messages down communication channels such as telephone lines, and he defined the quantity of information transmitted in terms of a selection between the total number of possible messages. This total can be characterized as the *variety* of the set of messages. If the set comprised just two possible messages—say, "yes" or "no" in answer to some question—then getting an answer one way or the other would count as the transmission of one bit (in the technical sense) of information in selecting between the two options. In effect, Ashby transposed information theory from a representational idiom, having to do with messages and communication, to a performative one, having to do with machines and their configurations. On Ashby's definition, the variety of a machine was defined precisely as the number of distinguishable states that it could take on. This put Ashby in a position to make quantitative statements and even prove theorems about the regulation of one machine or system by another, and preeminent among these statements was Ashby's law, which says, very simply, that "only variety can destroy variety" (Ashby 1956, 207).

To translate, as Ashby did in *An Introduction to Cybernetics*, a regulator is a blocker—it stops some environmental disturbance from having its full impact on some essential variable, say, as in the case of the homeostat. And then it stands to reason that to be an effective blocker one must have at least as much flexibility as that which is to be blocked. If the environment can take on twenty-five states, the regulator had better be able to take on at least twenty-five as well—otherwise, one of the environment's dodges and feints will get straight past the regulator and upset the essential variable. I have stated this in words; Ashby, of course, used his new machine notation as a means to a formal proof and elaboration; but thus Ashby's law.

To be able to make quantitative calculations and produce formal proofs was a major step forward from the qualitative arguments of *Design for a Brain*, in making cybernetics more recognizably a science like the modern sciences, and it is not surprising that much of the later work of Ashby and his students and followers capitalized on this bridgehead in all sorts of ways. It put Ashby in a position, for example, to dwell repeatedly on what he called Bremermann's limit. This was a quantum-mechanical and relativistic estimate of the upper limit on the rate of information processing by matter, which sufficed to make some otherwise plausible accounts of information processing look ridiculous—they could not be implemented in a finite time even if the entire universe were harnessed just to that purpose.[53] But there I am going to leave this general topic; Ashby's law will return with Stafford Beer in chapter 6.[54]

Cybernetics and Epistemology

I have been exploring Ashby's cybernetics as ontology, because that is where his real originality and certainly his importance for me lies. He showed how a nonmodern ontology could be brought down to earth as engineering which was also brain science, wth ramifications extending in endless directions. That is what I wanted to focus on. But Ashby did epistemology, too. If the Ur-referent of his cybernetics was preconscious, precognitive adaptation at deep levels of the brain, he was also willing to climb the brain stem to discuss cognition, articulated knowledge, science, and even painting and music, and

I want just to sketch out his approach to these topics. I begin with what I take to be right about his epistemology and then turn to critique.

How can we characterize Ashby's vision of knowledge? First, it was a deflationary and pragmatic one. Ashby insisted that "knowledge is finite" (Ashby 1963, 56). It never exceeds the amount of information on which it rests, which is itself finite, the product of a finite amount of work. It is therefore a mistake to imagine that our knowledge ever attains the status of a truth that transcends its origins—that it achieves an unshakeable correspondence to its object, as I would put it. According to Ashby, this observation ruled out of court most of the contemporary philosophical discourse on topics like induction that has come down to us from the Greeks. And, having discarded truth as the key topic for epistemological reflection, he came to focus on "the practical usefulness of models" (Ashby 1970, 95) in helping us get on with mundane, worldly projects.[55] The great thing about a model, according to Ashby, is that it enables us to *lose* information, and to arrive at something more tractable, handle-able, manipulable, than the object itself in its infinite complexity. As he put it, "No electronic model of a cat's brain can possibly be as true as that provided by the brain of another cat, yet of what *use* is the latter as a model?" (1970, 96). Models are thus our best hope of evading Bremermann's limit in getting to grips with the awful diversity of the world (1970, 98–100).

For Ashby, then, knowledge was to be thought of as engaged in practical projects and worldly performances, and one late essay, written with his student Roger Conant, can serve to bring this home. "Every Good Regulator of a System Must Be a Model of That System" (Conant and Ashby 1970) concerned the optimal method of feedback control. The authors discussed two different feedback arrangements: error- and cause-controlled. The former is typified by a household thermostat and is intrinsically imperfect. The thermostat has to wait until the environment drives the living-room temperature away from its desired setting before it can go to work to correct the deviation. Error control thus never quite gets it right: some errors always remain—deviations from the optimum—even though they might be much reduced by the feedback mechanism. A cause-controlled regulator, in contrast, does not need to wait for something to go wrong before it acts. A cause-controlled thermostat, for example, would monitor the conditions outside a building, predict what those conditions would do to the interior temperature, and take steps in advance to counter that—turning down the heating as soon as the sun came out or whatever. Unlike error control, cause control might approach perfection: all traces of environmental fluctuations might be blocked from affecting the controlled system; room temperature might never fluctuate at all. And the result that Conant and Ashby formally proved in this essay (subject to formal conditions and qualifications) was that the minimal condition for optimal cause control was that the regulator should contain a model of the regulated system.

Intuitively, of course, this seems obvious: the regulator has to "know" how changes in the environment will affect the system it regulates if it is to predict and cancel the effects of those changes, and the model is precisely that "knowledge." Nevertheless, something interesting is going here. In fact, one can see the cause-controlled regulator as an important elaboration of Ashby's ontological theater. The servomechanism, the homeostat, and DAMS staged, with increasing sophistication, an image of the brain as an adaptive organ performatively engaged with a lively world at the level of doing rather than knowing. This is undoubtedly the place to start if one wants to get the hang of the ontology of cybernetics. But, like CORA and *M. docilis*, the cause-controlled regulator invites us to think about the insertion of *knowledge* into this performative picture in a specific way. The virtue of knowledge lies not in its transcendental truth but in its usefulness in our performative engagements with the world. Knowledge is engaged with performance; epistemology with ontology. This performative epistemology, as I called it before, is the message of the cause-controlled regulator as ontological or epistemological theater; this is how we should think about knowledge cybernetically. Conversely, the cause-controlled regulator is a concrete example of how one might include the epistemic dimension in bringing ontology down to earth in engineering practice. That is what interests me most about this example.[56]

BASIC RESEARCH IS LIKE SHOOTING AN ARROW INTO THE AIR, AND, WHERE IT LANDS, PAINTING A TARGET.

HOMER ADKINS, CHEMIST, QUOTED IN BUCHANAN (2007, 213)

Now we can return to the critique I began earlier. In discussing the homeostat I noted that it had a fixed and pregiven goal—to keep its essential variables within limits, and I suggested that this is a bad image to have in general. At that stage, however, the referent of the essential variables was still some inner parameter analogous to the temperature of the blood—a slippery concept to criticize. But in his more epistemological writings, Ashby moved easily to a discussion of goals which clearly pertain to states of the outer, rather than the inner, world. An essay on "Genius," written with another of his students,

Crayton Walker, can serve to illustrate some consistent strands of Ashby's thinking on this (Ashby and Walker 1968).

The topic of "Genius" is more or less self-explanatory. In line with the above discussion, Ashby and Walker aim at a deflationary and naturalistic account of the phenomena we associate with word "genius." But to do so, they sketch out an account of knowledge production in which the importance of *predefined goals* is constantly repeated. "On an IQ test, *appropriate* [selection of answers in a multiple choice test] means correct, but not so much in an objective sense as in the sense that it satisfies a decision made *in advance* (by the test makers) about which answers show high and which low intelligence. In evaluating genius, it makes an enormous difference whether the criterion for appropriateness [i.e., the goal] was decided *before* or *after* the critical performance has taken place. . . . Has he succeeded or failed? The question has no meaning in the absence of a declared goal. The latter is like the marksman's saying he really meant to miss the target all along" (Ashby and Walker 1968, 209–10). And, indeed, Ashby and Walker are clear that they understand these goals as explicit targets in the outer world (and not, for example, keeping one's blood temperature constant): "In 1650, during Newton's time, many mathematicians were trying to explain Galileo's experimental findings. . . . In Michelangelo's day, the technical problems of perspective . . . were being widely discussed" (210). The great scientist and the great artist thus both knew what they were aiming for, and their "genius" lay in hitting their specified targets (before anyone else did).

I can find nothing good to say about this aspect of Ashby's work. My own historical research has confronted me with many examples in which great scientific accomplishments were in fact bound up with shifts in goals, and without making a statistical analysis I would be willing to bet that most of the accomplishments we routinely attribute to "genius" have precisely that quality. I therefore think that while it is reasonable to regard the fixity of the homeostat's goals as possibly a good model for some biological processes and a possibly unavoidable electromechanical limitation, it would be a mistake to follow Ashby's normative insistence that fixed goals necessarily characterize epistemological practice. This is one point at which we should draw the line in looking to his cybernetics for inspiration.

Beyond that, there is the question of how cognitive goals are to be achieved. Once Ashby and Walker have insisted that the goals of knowledge production have to be fixed in advance, they can remark that "the theorems of information theory are directly applicable to problems of this kind" (Ashby and Walker 1968, 210). They thus work themselves into the heartland of Ashby's mature cybernetics, where, it turns out, the key question is that of *selection*.[57] Just as the homeostat might be said to select the right settings of its uniselectors to achieve its goal of homeostasis, so, indeed, should all forms of human cultural production be considered likewise (210):

> To illustrate, suppose that Michelangelo made one million brush strokes in painting the Sistine Chapel. Suppose also that, being highly skilled, at each brush stroke he selected one of the two best, so that where the average painter would have ranged over ten, Michelangelo would have regarded eight as inferior. At each brush stroke he would have been selecting appropriately in the intensity of one in five. Over the million brush strokes the intensity would have been one in $5^{1,000,000}$. The intensity of Michelangelo's selection can be likened to his picking out one painting from five-raised-to-the-one-millionth-power, which is a large number of paintings (roughly 1 followed by 699,000 zeroes). Since this number is approximately the same as $2^{3,320,000}$, the theorem says that Michelangelo must have processed at least 3,320,000 "bits" of information, in the units of information theory, to achieve the results he did. He *must* have done so, according to the axiom, because appropriate selections can only be achieved if enough information is received and processed to make them happen.

Ashby and Walker go on to deduce from this that Michelangelo must have worked really hard over a long period of time to process the required amount of information, and they produce a few historical quotations to back this up. They also extend the same form of analysis to Newton, Gauss, and Einstein (selecting the right scientific theories or mathematical axioms from an enormous range of possibilities), Picasso (back to painting), Johann Sebastian Bach (picking just the right notes in a musical composition), and even Adolf Hitler, who "had many extraordinary successes before 1942 and was often acclaimed a genius, especially by the Germans" (207).

What can one say about all this? There is again something profoundly wrong about the image of "selection" that runs through Ashby's epistemology and even, before that, his ontology. There is something entirely implausible in the idea of Michelangelo's picking the right painting from a preexisting set or Einstein's doing the same in science. My own studies of scientific practice have never thrown up a single instance that could be adequately described in those terms (even if there is a branch of mainstream philosophy of science that does conceive "theory choice" along those lines). What I have found instead are many instances of open-ended, trial-and-error *extensions* of scientific culture. Rather than selecting between existing possibilities, scientists (and artists, and everyone else, I think) continually construct new ones and see how they play out. This is also a cybernetic image of epistemology—but one that emphasizes creativity and the appearance of genuine novelty in the world (both human and nonhuman) that the homeostat cannot model. The homeostat can only offer us selection and combinatorics. I have already discussed the homeostat's virtues as ontological theater at length; here my suggestion is that we should not follow it into the details of Ashby's epistemology.[58]

— — — — —

I want to end this chapter by moving beyond Ashby's work, so here I should offer a summary of what has been a long discussion. What was this chapter about?

One concern was historical. Continuing the discussion of Walter's work, I have tried to show that psychiatry, understood as the overall problematic of understanding and treating mental illness, was both a surface of emergence and a surface of return for Ashby's cybernetics. In important ways, his cybernetics can be seen to have grown out of his professional concerns with mental illness, and though the development of Ashby's hobby had its own dynamics and grew in other directions, too, he was interested, at least until the late 1950s, in seeing how it might feed back into psychiatry. At the same time, we have explored some of the axes along which Ashby's cybernetics went beyond the brain and invaded other fields: from a certain style of adaptive engineering (the homeostat, DAMS) to a general analysis of machines and a theory of everything, exemplified in Ashby's discussions of autopilots, economics, chemistry, evolutionary biology, war, planning, and epistemology. Ashby even articulated a form of spirituality appropriate to his cybernetics: "I am now . . . a Time-worshipper." In this way, the chapter continues the task of mapping out the multiplicity of cybernetics.

Another concern of the chapter has been ontological. I have argued that we can see the homeostat, and especially the multihomeostat setups that Ashby worked with, as ontological theater—as a model for a more general state of affairs: a world of dynamic entities evolving in performative (rather than representational) interaction with one another. Like the tortoise, the homeostat searched its world and reacted to what it found there. Unlike the tortoise's, the homeostat's world was as lively as the machine itself, simulated in a symmetric fashion by more homeostats. This symmetry, and the vision of a lively and dynamic world that goes with it, was Ashby's great contribution to the early development of cybernetics, and we will see it further elaborated as we go on. Conversely, once we have grasped the ontological import of Ashby's cybernetics, we can also see it from the opposite angle: as ontology in action, as playing out for us and exemplifying the sorts of project in many fields that might go with an ontology of performance and unknowability.

We have also examined the sort of performative epistemology that Ashby developed in relation to his brain research, and I emphasized the gearing of knowledge into performance that defined this. Here I also ventured into critique, arguing that we need not, and should not, accept all of the ontological and epistemological visions that Ashby staged for us. Especially, I argued against his insistence on the fixity of goals and his idea that performance and representation inhabit a given space of possibilities from which selections are made.

At the level of substance, we have seen that Ashby, like Walter, aimed at a modern science of the brain—at opening up the Black Box. And we have seen that he succeeded in this: the homeostat can indeed be counted as a model of the sort of adaptive processes that might happen in the brain. But the hybridity of Ashby's cybernetics, like Walter's, is again evident. In their mode of adaptation, Ashby's electromechanical assemblages themselves had, as their necessary counterpart, an unknowable world to which they adapted performatively. As ontological theater, his brain models inescapably return us to a picture of engagement with the unknown.

Furthermore, we have seen that that Ashby's cybernetics never quite achieved the form of a classically modern science. His scientific models were revealing from one angle, but opaque from another. To know how they were built did not carry with it a predictive understanding of what they would do. The only way to find out was to run them and see (finding out whether multihomeostat arrays with fixed internal settings would be stable or not, finding out what DAMS would do). This was the cybernetic discovery of complexity within a different set of projects from Walter's: the discovery that beyond some level of complexity, machines (and mathematical models) can themselves become mini–Black Boxes, which we can take as ontological icons, themselves models of the stuff from which the world is built. It was in this context that Ashby articulated a distinctively cybernetic philosophy of evolutionary design—design in medias res—very different from the blueprint attitude of modern engineering design, the stance of a detached observer who commands matter via a detour through knowledge.

Finally, the chapter thus far also explored the social basis of Ashby's cybernetics. Like Walter's, Ashby's distinctively cybernetic work was nomadic, finding a home in transitory institutions like the Ratio Club, the Macy and Namur conferences, and the Biological Computer Laboratory, where Ashby ended his career. I noted, though, that Ashby was hardly a disruptive nomad in his professional home, the mental hospital. There, like Walter, he took for granted established views of mental illness and therapy and existing social relations, even while developing novel theoretical accounts of the origins of mental illness in the biological brain and of the mechanisms of the great and desperate cures. This was a respect in which Ashby's cybernetics reinforced, rather than challenged, the status quo.

The last feature of Ashby's cybernetics that I want to stress is its seriousness. His journal records forty-four years' worth of hard, technical work, 7,189 pages of it, trying to think clearly and precisely about the brain and machines and about all the ancillary topics that that threw up. I want to stress this now because this seriousness of cybernetics is important to bear in mind throughout this book. My other cyberneticians were also serious, and they also did an enormous amount of hard technical work, but their cybernetics was not as unremittingly serious as Ashby's. Often it is hard to doubt that they were having fun, too. I consider this undoing of the boundary between serious science and fun yet another attractive feature of cybernetics as a model for practice. But there is a danger that it is the image of Allen Ginsberg taking LSD coupled to a flicker machine by a Grey Walter–style biofeedback mechanism, or of Stafford Beer invoking the Yogic chakras or the mystical geometry of the enneagram, that might stick in the reader's mind. I simply repeat here, therefore, that what fascinates me about cybernetics is that its projects could run the distance from the intensely technical to the far out. Putting this somewhat more strongly, my argument would have to be that the technical development of cybernetics encourages us to reflect that its more outré aspects were perhaps not as far out as we might think. The nonmodern is bound to look more or less strange.

A New Kind of Science: Alexander, Kauffman, and Wolfram

In the previous chapter, I explored some of the lines of work that grew out of Grey Walter's cybernetics, from robotics to the Beats and biofeedback, and I want to do something similar here, looking briefly at other work up to the present that resonates with Ashby's. My examples are taken from the work of Christopher Alexander, Stuart Kauffman, and Stephen Wolfram. One concern is again with the protean quality of cybernetics: here we can follow the development of distinctively Ashby-ite approaches into the fields of architecture, theoretical biology, mathematics, and beyond. The other concern is to explore further developments in the Ashby-ite problematic of complexity.

The three examples carry us progressively further away from real historical connections to Ashby, but, as I said in the opening chapters, it is the overall cybernetic stance in the world that I am trying to get clear on here, rather than lines of historical filiation.

— — — — —

IN ALEXANDER'S VIEW, MODERNITY IS A SORT OF TEMPORARY ABERRATION.

HILDE HEYNEN, *ARCHITECTURE AND MODERNITY* (1999, 20)

Christopher Alexander was born in Vienna in 1936 but grew up in England, graduated from Cambridge having studied mathematics and architecture, and then went to the other Cambridge, where he did a PhD in architecture at Harvard. In 1963 he became a professor of architecture at the University of California, Berkeley, retiring as an emeritus professor in 1998. British readers will be impressed, one way or the other, by the fact that from 1990 to 1995 he was a trustee of Prince Charles's Institute of Architecture. Alexander is best known for his later notion of "pattern languages," but I want to focus here on his first book, *Notes on the Synthesis of Form* (1964), the published version of his prize-winning PhD dissertation.[59]

The book takes us back to questions of design and is a critique of contemporary design methods, in general but especially in architecture. At its heart are two ideal types of design: "unselfconscious" methods (primitive, traditional, simple) and "selfconscious" ones (contemporary, professional, modern), and Alexander draws explicitly on *Design for a Brain* (the second edition, of 1960) to make this contrast.[60] The key concept that he takes there from Ashby is precisely the notion of adaptation, and his argument is that unselfconscious buildings, exemplified by the Mousgoum hut built by African tribes in French Cameroon, are well-adapted buildings in several senses: in the relation of their internal parts to one another, to their material environment, and to the social being of their inhabitants (Alexander 1964, 30). Contemporary Western buildings, in contrast, do not possess these features, is the claim, and the distinction lies for Alexander in the way that architecture responds to problems and misfits arising in construction and use. His idea is that in traditional design such misfits are localized, finite problems that are readily fixed in a piecemeal fashion, while in the field of self-conscious design, attempts to fix misfits ramify endlessly: "If there is not enough light in a house, for instance, and more windows are added to correct this failure, the change may improve the light but allow too little privacy; another change for more light makes the windows bigger, perhaps, but thereby makes the house more likely to collapse" (1964, 42).

The details here are not important, but I want to note the distinctly Ashbyite way in which Alexander frames the problem in order to set up his own solution of it, a solution which is arguably at the heart of Alexander's subsequent career. As discussed earlier, in a key passage of *Design for a Brain* Ashby gave estimates of the time for multihomeostat systems to achieve equilibrium, ranging from short to impossibly long, depending upon the density of interconnections between the homeostats. In the second edition of *Design*, he illustrated these estimates by thinking about a set of rotors, each with two positions labeled A and B, and asking how long it would take various spinning strategies to achieve a distribution of, say, all As showing and no Bs (Ashby 1960, 151). In *Notes on the Synthesis of Form*, Alexander simply translates this illustration into his own terms, with ample acknowledgment to Ashby but with an interesting twist.

Alexander invites the reader to consider an array of one hundred lightbulbs that can be either on, standing for a misfit in the design process, or off, for no misfit. This array evolves in time steps according to certain rules. Any light that is on has a 50-50 chance of going off at the next step. Any light that is off has a 50-50 chance of coming back on if at least one light to which it is connected is on, but no chance if the connected lights are all off. And then one can see how the argument goes. The destiny of any such system is eventually to become dark: once all the lights are off—all the misfits have been dealt with—none of them can ever, according to the rules, come back on again. So, following Ashby exactly, Alexander remarks, "The only question that remains is, how long will it take for this to happen? It is not hard to see that apart from chance this depends only on the pattern of interconnection between the lights" (1964, 40).[61]

Alexander then follows Ashby again in providing three estimates for the time to darkness. The first is the situation of independent adaptation. If the lights have no meaningful connections to one another, then this time is basically the time required for any single light to go dark: 2 seconds, if each time step is 1 second. At the other extreme, if each light is connected to all the others, then the only way in which the lights that remain on can be prevented from reexciting the lights that have gone off is by all of the lights happening to go off in the same time step, which one can estimate will take of the order of 2^{100} seconds, or 10^{22} years—one of those hyperastronomical times that were crucial to the development of Ashby's project. Alexander then considers a third possibility which differs in an important way from Ashby's third possibility. In *Design for a Brain*, Ashby gets his third estimate by thinking about the situation in which any rotor that comes up A is left alone and the other rotors are spun again, and so on until there are no Bs left. Alexander, in contrast, considers the situation in which the one hundred lights fall into *subsystems* of ten lights each. These subsystems are assumed to be largely independent of one another but densely connected internally. In this case, the time to darkness of the whole system will be of the order of the time for any one subsystem to go dark, namely 2^{10} seconds, or about a quarter of an hour—quite a reasonable number.

We recognize this line of thought from *Design*, but the advantage of putting it this way is that it sets up Alexander's own solution to the problem of design. Our contemporary problems in architecture stem from the fact that the variables we tinker with are not sufficiently independent of one another, so that tinkering with any one of them sets up problems elsewhere, like the lit lightbulbs turning on the others. And what we should do, therefore, is to "diagonalize" (my word) the variables—we should find some new design variables such that design problems only bear upon subsets of them that are loosely coupled to others, like the subsystems of ten lights in the example. That way, we can get to grips with our problems in a finite time and our buildings will reach an adapted state: just as in unselfconscious buildings, the internal components will fit together in all sorts of ways, and whole buildings will mesh with their environments and inhabitants. And this is indeed the path that Alexander follows in the later chapters of *Notes on the Synthesis of Form*, where he proposes empirical methods and mathematical techniques for finding appropriate sets of design variables. One can also, though I will not go into this, see this reasoning as the key to his later work on pattern languages: the enduring patterns that Alexander came to focus on there refer to recurring design problems and solutions that can be considered in relative isolation from others and thus suggest a realistically piecemeal approach to designing adapted buildings, neighborhoods, cities, conurbations, or whatever (Alexander et al. 1977).

What can we take from this discussion? First, evidently, it is a nice example of the consequentiality of Ashby's work beyond the immediate community of cyberneticians. Second, it is another example of the undisciplined quality of the transmission of cybernetics through semipopular books like *Design for a Brain*. I know of no evidence of contact between Alexander and Ashby or other cyberneticians; it is reasonable to assume that Alexander simply read *Design* and saw what he could do with it, in much the same way as both Rodney Brooks and William Burroughs read Grey Walter. Along with this, we have another illustration of the protean quality of cybernetics. Ashby thought

Figure 4.13. The Linz Café. Source: Alexander 1983, 48.

he was writing about the brain, but Alexander immediately extended Ashby's discussion of connectedness to a continuing program in architecture and design, a field that Ashby never systematically thought about. We can thus take both Alexander's distinctive approach to architectural design and the actual buildings he has designed as further exemplars of the cybernetic ontology in action.[62] Finally, we can note that Alexander's architecture is by no means uncontroversial. Alexander's "Linz Café" (1983) is an extended account of one of his projects (fig. 4.13) that includes the text of a debate at Harvard with Peter Eisenman. Alexander explains how the cafe was constructed around his "patterns" (58–59) but also emphasizes that the design elements needed to be individually "tuned" by building mock-ups and seeing what they felt like. The goal was to construct spaces that were truly "comfortable" for human beings. This tuning harks back to and exemplifies Alexander's earlier discussion of how problems can be and are solved on a piecemeal basis in traditional architecture, and the last section of his article discusses resonances between the Linz Café and historical buildings (59). In debate Eisenman tries to problematize Alexander's comfort principle and suggests a different, less harmonious idea of architecture (theoretically inspired). Egged on by a sympathetic audience, Alexander remarks that "people who believe as you do are really fucking up the whole profession of architecture right now by propagating these beliefs" (67)—another marker of the fact that ontology makes a difference. We can return to this theme in a different and less "comfortable" guise when we come to Gordon Pask's version of adaptive architecture.

IT IS A FUNDAMENTAL QUESTION WHETHER METABOLIC STABILITY AND EPIGENESIS REQUIRE THE GENETIC REGULATORY CIRCUITS TO BE PRECISELY CONSTRUCTED. HAS A FORTUNATE EVOLUTIONARY HISTORY SELECTED ONLY NETS OF HIGHLY ORDERED CIRCUITS WHICH ALONE CAN INSURE METABOLIC STABILITY; OR ARE STABILITY AND EPIGENESIS, EVEN IN NETS OF RANDOMLY CONNECTED INTERCONNECTED REGULATORY CIRCUITS, TO BE EXPECTED AS THE PROBABLE CONSEQUENCE OF AS YET UNKNOWN MATHEMATICAL LAWS? ARE LIVING THINGS MORE AKIN TO PRECISELY PROGRAMMED AUTOMATA SELECTED BY EVOLUTION, OR TO RANDOMLY ASSEMBLED AUTOMATA WHOSE CHARACTERISTIC BEHAVIOR REFLECTS THEIR UNORDERLY CONSTRUCTION, NO MATTER HOW EVOLUTION SELECTED THE SURVIVING FORMS?

STUART KAUFFMAN, "METABOLIC STABILITY AND EPIGENESIS IN RANDOMLY CONSTRUCTED GENETIC NETS" (1969B, 438)

Now for Stuart Kauffman, one of the founders of contemporary theoretical biology, perhaps best known in the wider world for two books on a complex systems approach to the topics of biology and evolution, *At Home in the Universe* (1995) and *Investigations* (2002). I mentioned his important and explicitly cybernetic notion of "explanation by articulation of parts" in chapter 2, but now we can look at his biological research.[63]

The pattern for Kauffman's subsequent work was set in a group of his earliest scientific publications in the late 1960s and early 1970s, which concerned just the same problem that Alexander inherited from Ashby, the question of a large array of interacting elements achieving equilibrium. In *Design for a Brain*, Ashby considered two limits—situations in which interconnections between the elements were either minimal or maximal—and argued that the time to equilibrium would be small in one case and longer than the age of the universe in the other. The question that then arose was what happened in between these limits. Ashby had originally been thinking about an array of interacting homeostats, but one can simplify the situation by considering an array of binary elements that switch each other on and off according to some rule—as did Alexander with his imaginary lightbulbs. The important point to stress, however, is that even such simple models are impossible to solve analytically. One cannot calculate in advance how they will behave; one simply has to run through a series of time steps, updating the binary variables at each step according to the chosen transformation rules, and see what the system will in fact do. This is the cybernetic discovery of complexity transcribed from the field of mechanisms to that of mathematical formalisms. Idealized binary arrays can remain Black Boxes as far as their aggregate behavior is concerned, even when the atomic rules that give rise to their behavior are known.

The only way to proceed in such a situation (apart from Alexander's trick of simply assuming that the array breaks up into almost disconnected pieces) is brute force. Hand calculation for a network of any size would be immensely tedious and time consuming, but at the University of Illinois Crayton Walker's 1965 PhD dissertation in psychology reported on his exploration of the time evolution of one-hundred-element binary arrays under a variety of simple transformation rules using the university's IBM 7094–1401 computer. Walker and Ashby (1966) wrote these findings up for publication, discussing how many steps different rule systems took to come to equilibrium, whether the equilibrium state was a fixed point or a cycle, how big the limit cycles were, and so on.[64] But it was Kauffman, rather than Walker and Ashby, who obtained the most important early results in this area, and at the same time Kauffman switched the focus from the brain to another very complex biological system, the cell.

Beginning in 1967, Kauffman published a series of papers grounded in computer simulations of randomly connected networks of binary elements, which he took to model the action of idealized genes, switching one another on and off (like lightbulbs, which indeed feature in *At Home in the Universe*). We could call what he had found a *discovery of simplicity* within complexity. A network of N binary elements has 2^N possible states, so that a one-thousand-element network can be in 2^{1000} distinct states, which is about 10^{300}—another one of those hyperastronomical numbers. But Kauffman established two fundamental findings, one concerning the inner, endogenous, dynamics of such nets, the other concerning exogenous perturbations.[65]

On the first, Kauffman's simulations suggested that if each gene has exactly two inputs from other genes, then a randomly assembled network of one thousand genes would typically cycle among just twelve states—an astonishingly small number compared with 10^{300} (Kauffman 1969b, 444). Furthermore the lengths of these cycles—the number of states a network would pass through before returning to a state it had visited before—were surprisingly short. He estimated, for example, that a network having a million elements would "possess behavior cycles of about one thousand states in length—an extreme localization of behavior among $2^{1,000,000}$ possible states" (446). And beyond that, Kauffman's computer simulations revealed that the number of distinct cycles exhibited by any net was "as surprisingly small as the cycles are short" (448). He estimated that a net of one thousand elements, for example, would possess around just sixteen distinct cycles.

On the second, Kauffman had investigated what happened to established cycles when he introduced "noise" into his simulations—flipping single elements from one state to another during a cycle. The cycles proved largely resistant to such exogenous interference, returning to their original trajectories around 90% of the time. Sometimes, however, flipping a single element would jog the system from one cyclic pattern to one of a few others (452).

What did Kauffman make of these findings? At the most straightforward level, his argument was that a randomly connected network of idealized genes could serve as the model for a set of cell types (identified with the different cycles the network displayed), that the short cycle lengths of these cells were consistent with biological time scales, that the cells exhibited the biological requirement of stability against perturbations and chemical noise, and that the occasional transformations of cell types induced by noise corresponded to the puzzling fact of cellular differentiation in embryogenesis.[66] So his idealized gene networks could be held to be models of otherwise unexplained biological phenomena—and this was the sense in which his work counted as "theoretical biology." At a grander level, the fact that these networks were randomly constructed was important, as indicated in the opening quotation from Kauffman. One might imagine that the stability of cells and their pathways of differentiation are determined by a detailed "circuit diagram" of control loops between genes, a circuit diagram laid down in a tortuous evolutionary history of mutation and selection. Kauffman had shown that one does not have to think that way. He had shown that complex systems can display self-organizing properties, properties arising from within the systems themselves, the emergence of a sort of "order out of chaos" (to borrow the title of Prigogine and Stengers 1984). This was the line of thought that led him eventually to the conclusion that we are "at home in the universe"—that life is what one should expect to find in any reasonably complex world, not something we should be surprised at and requiring any special explanation.[67]

This is not the place to go into any more detail about Kauffman's work, but I want to comment on what we have seen from several angles. First, I want to return to the protean quality of cybernetics. Kauffman was clearly working in the same space as Ashby and Alexander—his basic problematic was much the same as theirs. But while their topic was the brain (as specified by Ashby) or architecture (as specified by Alexander), it was genes and cells and theoretical biology when specified by Kauffman.

Second, I want to comment on Kauffman's random networks, not as models of cells, but as ontological theater more generally. I argued before that tortoises, homeostats, and DAMS can, within certain limitations, be seen as electromechanical models that summon up for us the cybernetic ontology more broadly—machines whose aggregate performance is impenetrable. As discussed, Kauffman's idealized gene networks displayed the same character, but as emerging within a formal mathematical system rather than a material one. Now I want to note that as world models Kauffman's networks can also further enrich our ontological imaginations in important ways. On the one hand, these networks were livelier than, especially, Ashby's machines. Walter sometimes referred to the homeostat as *Machina sopora*—the sleeping machine. Its goal was to become quiescent; it changed state only when disturbed from outside. Kauffman's nets, in contrast, had their own endogenous dynamics, continually running through their cycles whether perturbed from the outside or not. On the other hand, these nets stage for us an image of systems with which we can genuinely interact, but not in the mode of command and control. The perturbations that Kauffman injected into their cycling disturbed the systems but did not serve to direct them into any other particular cycles.

This idea of systems that are not just performative and inscrutable but also dynamic and resistant to direction helps, I think, to give more substance to Beer's notion of "exceedingly complex systems" as the referent of cybernetics. The elaborations of cybernetics discussed in the following chapters circle around the problematic of getting along with systems fitting that general description, and Kauffman's nets can serve as an example of the kinds of things they are.[68]

My last thought on Kauffman returns to the social basis of cybernetics. To emphasize the odd and improvised character of this, in the previous chapter (note 31) I listed the range of diverse academic and nonacademic affiliations of the participants at the first Namur conference. Kauffman's CV compresses the whole range and more into a single career. With BAs from Dartmouth College and Oxford University, he qualified as a doctor at the University of California, San Francisco, in 1968, while first writing up the findings discussed above as a visitor at MIT's Research Laboratory of Electronics in 1967. He was then briefly an intern at Cincinnati General Hospital before becoming an assistant professor of biophysics and theoretical biology at the University of Chicago from 1969 to 1975. Overlapping with that, he was a surgeon at the National Cancer Institute in Bethesda from 1973 to 1975, before taking a tenured position in biochemistry and biophysics at the University of Pennsylvania in 1975. He formally retired from that position in 1995, but from 1986 to 1997 his primary affiliation was as a professor at the newly established Santa Fe Institute (SFI) in New Mexico. In 1996, he was the founding general partner of Bios Group, again in Santa Fe, and in 2004 he moved to the University of Calgary as director of the Institute for Biocomplexity and Informatics and professor in the departments of Biological Sciences and Physics and Astronomy.[69]

It is not unreasonable to read this pattern as a familiar search for a congenial environment for a research career that sorts ill with conventional disciplinary and professional concerns and elicits more connections across disciplines and fields than within any one of them. The sociological novelty that appears here concerns two of Kauffman's later affiliations. The Santa Fe Institute was established in 1984 to foster a research agenda devoted to "simplicity, complexity, complex systems, and particularly complex adaptive systems" and is, in effect, an attempt to provide a relatively enduring social basis for the transient interdisciplinary communities—the Macy and Namur conferences, the Ratio Club—that were "home" to Walter, Ashby, and the rest of the first generation of cyberneticians. Notably, the SFI is a freestanding institution and not, for example, part of any university. The sociologically improvised character of cybernetics reappears here, but now at the level of institutions rather than individual careers.[70] And two other remarks on the SFI are relevant to our themes. One is that while the SFI serves the purpose of stabilizing a community of interdisciplinary researchers, it does not solve the problem of cultural transmission: as a private, nonprofit research institute it does not teach students and grant degrees.[71] The other is that the price of institutionalization is, in this instance, a certain narrowing. The focus of research at the SFI is resolutely technical and mathematical. Ross Ashby might have been happy there, but not, I think, any of our other principals. Their work was too rich and diverse to be contained by such an agenda.

Besides the SFI, I should comment on Kauffman's affiliation with the Bios Group (which merged with NuTech Solutions in 2003). "BiosGroup was founded by Dr. Stuart Kauffman with a mission to tackle industry's toughest problems through the application of an emerging technology, Complexity Science."[72] Here we have an attempt is establish a stable social basis for the science of complexity on a business rather than a scholarly model—a pattern we have glimpsed before (with Rodney Brooks's business connections) and which will reappear immediately below. And once more we are confronted with the protean quality of cybernetics, with Kauffman's theoretical biology morphing into the world of capital.

— — — — —

WE HAVE SUCCEEDED IN REDUCING ALL OF ORDINARY PHYSICAL BEHAVIOR TO A SIMPLE, CORRECT THEORY OF EVERYTHING ONLY TO DISCOVER THAT IT HAS REVEALED EXACTLY NOTHING ABOUT MANY THINGS OF GREAT IMPORTANCE.

R. B. LAUGHLIN AND DAVID PINES,
"THE THEORY OF EVERYTHING" (2000, 28)

IT'S INTERESTING WHAT THE PRINCIPLE OF COMPUTATIONAL EQUIVALENCE ENDS UP SAYING. IT KIND OF ENCAPSULATES BOTH THE GREAT STRENGTH AND THE GREAT WEAKNESS OF SCIENCE. BECAUSE ON THE ONE HAND IT SAYS THAT ALL THE WONDERS OF THE UNIVERSE CAN BE CAPTURED BY SIMPLE RULES. YET IT ALSO SAYS THAT THERE'S ULTIMATELY NO WAY TO KNOW THE CONSEQUENCES OF THESE RULES—EXCEPT IN EFFECT JUST TO WATCH AND SEE HOW THEY UNFOLD.

STEPHEN WOLFRAM, "THE GENERATION OF FORM
IN A *NEW KIND OF SCIENCE*" (2005, 36)

If the significance of Kauffman's work lay in his discovery of simplicity within complexity, Wolfram's achievement was to rediscover complexity within simplicity. Born in London in 1959, Stephen Wolfram was a child prodigy, like Wiener: Eton, Oxford, and a PhD from Caltech in 1979 at age twenty; he received a MacArthur "genius" award two years later. Wolfram's early work was in theoretical elementary-particle physics and cosmology, but two interests that defined his subsequent career emerged in the early 1980s: in cellular automata, on which more below, and in the development of computer software for doing mathematics. From 1983 to 1986 he held a permanent position at the Institute for Advanced Study in Princeton; from 1986 to 1988 he was professor of physics, mathematics and computer science at the University of Illinois at Urbana-Champaign, where he founded the Center for Complex Systems Research (sixteen years after Ashby had left—"shockingly, I don't think anyone at Illinois ever mentioned Ashby to me"; email to the author, 6 April 2007). In 1987 he founded Wolfram Research, a private company that develops and markets what has proved to be a highly successful product: Mathematica software for mathematical computation. Besides running his company, Wolfram then spent the 1990s developing his work on cellular automata and related systems, in his spare time and without publishing any of it (echoes of Ashby's hobby). His silence ended in 2002 with a blaze of publicity for his massive, 1,280-page book, *A New Kind of Science*, published by his own company.[73]

The key insight of the new kind of science, which Wolfram abbreviates to NKS, is that "incredibly simple rules can give rise to incredibly complicated behavior" (Wolfram 2005, 13), an idea grounded in Wolfram's explorations of simple, one-dimensional cellular automata. "Cellular automaton" is a forbidding name for a straightforward mathematical system. A one-dimensional CA is just a set of points on a line, with a binary variable, zero or one, assigned to each point. One imagines this system evolving in discrete time steps according to definite rules: a variable might change or stay the same according to its own present value and those of its two nearest neighbors, for example. How do such systems behave? The relationship of this problematic to Ashby's, Alexander's, and Kauffman's is clear: all three of them were looking at the properties of CAs, but much more complicated ones (effectively, in higher dimensions) than Wolfram's. And what Wolfram found—"playing with the animals," as he once put it to me—was that even these almost childishly simple systems can generate enormously complex patterns.[74] Some do not: the pattern dies out after a few time steps; all the variables become zero, and nothing happens thereafter. But Wolfram's favorite example is the behavior of the rule 30 cellular automaton shown in figure 4.14 (one can list and number all possible transformation rules for linear CAs, and Wolfram simply ran them all on a computer).

If Kauffman was surprised that his networks displayed simple behavior, one can be even more surprised at the complexities that are generated by Wolfram's elementary rules. He argues that rule 30 (and other rules, too) turn out to be "computationally irreducible" in the sense that "there's essentially no way to work out what the system will do by any procedure that takes less computational effort than just running the system and seeing what happens." There are no "shortcuts" to be found (Wolfram 2005, 30). And this observation is the starting point for the new kind of science (31):

> In traditional theoretical science, there's sort of been an idealization made that the observer is infinitely computationally powerful relative to the system they're observing. But the point is that when there's complex behavior, the Principle of Computational Equivalence says that instead the system is just as computationally sophisticated as the observer. And that's what leads to

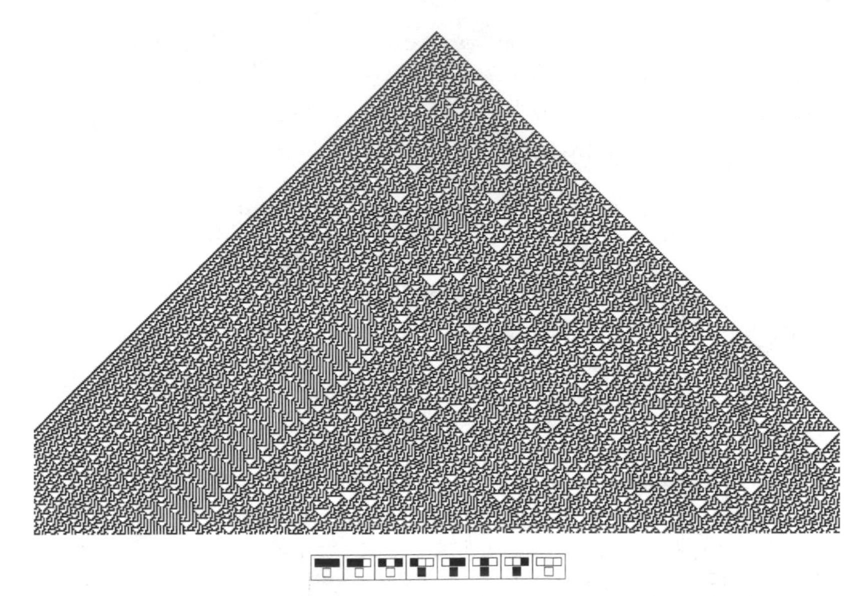

Figure 4.14. Rule 30 cellular automaton. Time steps move from the top downward; 1s are denoted by black cells, starting from a single 1. The transformation rule is shown at the bottom. Source: Wolfram 2005, 4. (Image courtesy of Wolfram Research, Inc. [] and Stephen Wolfram LLC, as used in Stephen Wolfram's *New Kind of Science* © 2002.)

> computational irreducibility. And that's why traditional theoretical science hasn't been able to make more progress when one sees complexity. There are always pockets of reducibility where one can make progress, but there's always a core of computational irreducibility.

The classical sciences thus address just those "pockets" of the world where the traditional shortcuts can be made to work, while the reference of NKS is to all of the other aspects of the world where brute complexity is the rule, and much of Wolfram's work has been devoted to bringing this ontological perspective down to earth in all sorts of fields: mathematics; a sort of crystallography (e.g., snowflake structures); studies of turbulence; biology, where Wolfram's discussion echoes Kauffman's.[75] Having compared the patterns on mollusc shells to those generated by various CAs, Wolfram notes that (22)

> it's very much as if the molluscs of the Earth are little computers—sampling the space of possible simple programs, and then displaying the results on their shells. You know, with all the emphasis on natural selection, one's gotten used to the idea that there can't be much of a fundamental theory in biology—and

> that practically everything we see must just reflect detailed accidents in the history of biological evolution. But what the mollusc shell example suggests is that that may not be so. And that somehow one can think of organisms as uniformly sampling a space of possible programs. So that just knowing abstractly about the space of programs will tell one about biology

And, of course, reflecting his disciplinary origins, Wolfram also sees the NKS as offering a "truly fundamental theory of physics." Space, time and causality are merely appearances, themselves emerging from a discrete network of points—and the ultimate task of physics is then to find out what rule the system is running. "It's going to be fascinating—and perhaps humbling—to see just where our universe is. The hundredth rule? Or the millionth? Or the quintillionth? But I'm increasingly optimistic that this is all really going to work. And that eventually out there in the computational universe we'll find our universe. With all of our physics. And that will certainly be an exciting moment for science" (27).

We can thus see Wolfram's work as a further variant on the theme that Ashby set out in 1952 in his considerations of the time to reach equilibrium of multihomeostat assemblages, but differing from the other variants in interesting and important ways. Unlike Alexander and Kauffman, Wolfram has generalized and ontologized the problematic, turning it into an account of how the world is, as well as respecifying it in the domains mentioned above and more. Beyond that, from our point of view, Wolfram's distinctive contribution has been to focus on systems that do not settle down into equilibrium, that perform in unpredictable ways, and to suggest that that is the world's ontological condition. His NKS thus offers us a further enrichment of our ontological imaginations. Systems like the rule 30 CA genuinely *become*; the only way to find out what they will do next is run the rule on their present configuration and find out. As ontological theater, they help us to imagine the world that way; they add becoming to our models of what Beer's "exceedingly complex systems" might be like. If we think of the world as built from CA-like entities, we have a richer grasp of the cybernetic ontology.

It remains only to comment on the social basis of Wolfram's work. We have seen already that after a meteoric but otherwise conventional career in academic research Wolfram (like Kauffman) veered off into business, and that this business enabled him to sustain his unusual hobby (like Ashby)—providing both a living and research tools. There is the usual improvised oddity here, evident in the biographies of all our cyberneticians. What I should add is that having launched NKS with his 2002 book, Wolfram has since sought to foster the growth of the field with an annual series of conferences and summer schools. Organized by Wolfram's group, these parallel the Santa Fe Institute in existing outside the usual academic circuits, and one can again see them as an attempt to stabilize a novel social base for a novel kind of science. Nine of the eleven people listed as faculty for the 2005 NKS summer school worked for, or had worked for, Wolfram Research, including Wolfram himself, and the website for the school mentions that, in the past, "some of our most talented attendees have been offered positions at Wolfram Research."[76] Wolfram also imagines a permanent NKS research institute, supported, perhaps, by software companies, including his own (personal communication). Bios, the SFI, NKS: a nascent social formation for the latter-day counterparts of cybernetics begins to appear here beyond the frame of the usual instititutions of learning—a parallel world, a social as well as ontological—a socio-ontological—sketch of another future.

5

GREGORY BATESON AND R. D. LAING

SYMMETRY, PSYCHIATRY, AND THE SIXTIES

I THINK THAT THE FUNCTIONING OF SUCH HIERARCHIES MAY BE COMPARED WITH THE BUSINESS OF TRYING TO BACK A TRUCK TO WHICH ONE OR MORE TRAILERS ARE ATTACHED. EACH SEGMENTATION OF SUCH A SYSTEM DENOTES A REVERSAL OF SIGN, AND EACH ADDED SEGMENT DENOTES A DRASTIC DECREASE IN THE AMOUNT OF CONTROL. . . . WHEN WE CONSIDER THE PROBLEM OF CONTROLLING A SECOND TRAILER, THE THRESHOLD FOR JACKKNIFING IS DRASTICALLY REDUCED, AND CONTROL BECOMES, THEREFORE, ALMOST NEGLIGIBLE. AS I SEE IT, THE WORLD IS MADE UP OF A VERY COMPLEX NETWORK (RATHER THAN A CHAIN) OF SUCH ENTITIES WHICH HAVE THIS SORT OF RELATION TO EACH OTHER, BUT WITH THIS DIFFERENCE, THAT MANY OF THE ENTITIES HAVE THEIR OWN SUPPLIES OF ENERGY AND PERHAPS EVEN THEIR OWN IDEAS OF WHERE THEY WOULD LIKE TO GO.

GREGORY BATESON, "MINIMAL REQUIREMENTS FOR A THEORY OF SCHIZOPHRENIA" (1959, 268)

The two previous chapters covered the emergence of cybernetics in Britain from the 1940s onward. At their heart were Walter and Ashby's electromechanical brain models, the tortoise, the homeostat, and DAMS, and the discovery of complexity that went with them—the realization that even simple models can display inscrutably complex behavior. I emphasized that this first-generation cybernetics was born in the world of psychiatry, and that, despite its ramifications outside that field, it left clinical psychiatry itself largely untouched. Walter and Ashby's cybernetics in effect endorsed existing psychiatric practice by modelling and conceptualizing the action of electroshock, lobotomy, and so on. In this chapter, I want to look at a very different approach to psychiatry that grew up in the fifties and sixties that was also identifiably cybernetic, and that I associate primarily with the work of Gregory Bateson and R. D. Laing.

The pivot here can be Ashby's contrasting analyses of war and planning. Ashby understood both on the model of interacting homeostats searching for a shared equilibrium, but he thought of war and psychiatry ("blitz therapy") in an asymmetric fashion. The general and the psychiatrist try to stay the same and force the other—the enemy, the patient—to adapt to them: the defeated enemy accedes to the terms of the victor; the patient returns to the world of sanity and normality embodied in the psychiatrist. This asymmetric vision was the key to the reconciliation between early cybernetics and its psychiatric matrix. On the other hand, Ashby envisaged the possibility of a more symmetric relation between planner and planned: each party, and the plan that links them, can adapt homeostatically to the other. In this chapter, we will be exploring what psychiatry looked like when it took the other fork in the road and understood social relations in general on the symmetric rather than the asymmetric model. As we will see, big transformations in practice accompanied this. This chapter can also serve as a transition to the following chapters on Beer and Pask, who also took the symmetric fork in thinking about reciprocal adaptations of people, animals, machines, and nature. It is this symmetric version of the cybernetic ontology of performative adaptation that interests me most in this book.

Four more introductory points are worth making. First, the object of Walter and Ashby's cybernetics was the biological brain: they wanted to understand the material go of it, and the go of existing psychiatric therapies. This was not the case with Bateson and Laing. Neither of them was concerned with the biological brain; the referent of their work was something less well defined, which I will refer to as *the self*. Their work remained cybernetic inasmuch as their conception of the self was again performative and adaptive, just like the cybernetic brain more narrowly conceived. Second, we will see below how this concern with the performative self provided further openings to the East and accompanied an interest in strange performances and altered states more generally. One can, in fact, specify the connection between madness and spirituality more tightly in Laing and Bateson's work than was possible in the previous chapters. Third, I can mention in advance that while Walter and

Ashby's psychiatric interests were not tied to any specific form of mental pathology, Bateson and Laing's work focused in particular on schizophrenia, and the "visionary" quality of schizophrenia was central to their extension of psychiatry in a spiritual direction. And, fourth, we will also have a chance here to examine in more detail connections between cybernetics and the sixties.

Unlike the four principals of this book, Laing and Bateson have been much written about, so this chapter does not explore their work in depth comparable that of the chapters 3 and 4. Bateson was interested in many topics during the course of his life, but I will only cover his psychiatric phase. Laing was a professional psychiatrist throughout his working life, but I focus only on the period of his greatest fame and notoriety, the sixties—partly because I am interested in the sixties, but also because the therapeutic communities established in the sixties by Laing's Philadelphia Association offer us a stark example of what the symmetric version of cybernetics can look like in practice. Neither Bateson nor Laing worked alone, so their names often feature here as a convenient shorthand for groups of collaborators.

Gregory Bateson

THE TRUE CHALLENGE IS HOW NOT TO PLAY THE GAME BY THE RULES OF NATURAL SCIENCE . . . HOW TO ESTABLISH AN AUTHORITY THAT ENABLES THE PURSUIT OF THE POSSIBILITIES OF AN ALTERED SCIENCE, ONE THAT IS FAR LESS DESTRUCTIVE.

PETER HARRIES-JONES, "UNDERSTANDING ECOLOGICAL AESTHETICS" (2005, 67)

MY PERSONAL INSPIRATION HAS OWED MUCH TO THE MEN WHO OVER THE LAST TWO HUNDRED YEARS HAVE KEPT ALIVE THE IDEA OF A UNITY BETWEEN MIND AND BODY: *LAMARCK* . . . *WILLIAM BLAKE* . . . *SAMUEL BUTLER* . . . *R. G. COLLINGWOOD* . . . AND *WILLIAM BATESON*, MY FATHER, WHO WAS CERTAINLY READY IN 1894 TO RECEIVE THE CYBERNETIC IDEAS.

GREGORY BATESON, *STEPS TO AN ECOLOGY OF MIND* (2000, XXI–XXII)

Gregory Bateson (fig. 5.1) was born in Grantchester, near Cambridge, in 1904, the son of the eminent geneticist William Bateson, and died in San Francisco in 1980. He studied at Cambridge, completing the natural science tripos in 1924 and the anthropological tripos in 1926, and was a research fellow at St. John's College from 1931 to 1937. He made his prewar reputation as an

Figure 5.1. Gregory Bateson in the mid-1950s. (Used courtesy of Lois Bateson.)

anthropologist in Bali and New Guinea, and in 1940 he moved from Britain to the United States, where he worked for the Office of Strategic Services, the forerunner of the CIA, from 1943 until 1945. Bateson was married to the American anthropologist Margaret Mead from 1936 until 1950, and together they were among the founding members of the Macy cybernetics conferences held between 1946 and 1953. In the same period Bateson's interests took a psychiatric turn, as he lectured at the Langley Porter Clinic in San Francisco and then worked as an ethnologist at the Veterans Administration Hospital in Palo Alto, California (1949–63).[1] What follows seeks to trace out some of the main features of Bateson's psychiatric work as it developed in a ten-year project which formally began in 1952 with a two-year grant from the Rockefeller Foundation. Bateson was joined in this project by Jay Haley, John Weakland, and William Fry in 1953 and by Don Jackson in 1954.[2]

In 1956 the Bateson group published the first of a series of important papers, "Towards a Theory of Schizophrenia" (Bateson et al. 1956). There Bateson advanced his famous concept of the *double bind*, and we should note that this is entirely cybernetic. Much like the contradictory conditioning of Pavlov's dogs and Walter's tortoises, the double bind was envisaged as a repeated situation to which the sufferer could find no satisfactory response.[3] The first schizophrenia paper gave as an example a mother who encouraged her son to display conventionally loving behavior but froze and repelled him whenever he did, and then asked him what was wrong with him when he moved away. Like Pavlov and Walter, then, Bateson understood schizophrenia as a possible response to this sort of contradictory situation. If there is no normal way to go on, one has to find some abnormal response—total withdrawal from communication, paranoid suspicion, an inability to take anything literally, mistaking inner voices for the outside world, and so on.[4]

Thus the basic plot, and two points need clarification here. One is the thought that Bateson's interest in communication patterns might seem to move us away from the cybernetic concern with performance and toward the more familiar representational brain and self. He is, however, better seen as again elaborating a performative understanding of communication, both verbal and nonverbal—a notion of speech as a representational detour leading out of and back to performance. The mother and son in the example are not exchanging information so much as eliciting and responding to the behavior of the other.[5]

This leads to the second point. Bateson did not think that the mother in the example *caused* her son's schizophrenia in any linear fashion. Instead, as I mentioned earlier, on the model of the homeostat, he thought of all the parties as adapting to one another in a trial-and-error search through the space of performance, and of schizophrenia as an instance of the whole system reaching a state of equilibrium having bizarre properties.[6]

SCHIZOPHRENIA AND ENLIGHTENMENT

IN THE EASTERN RELIGION, ZEN BUDDHISM, THE GOAL IS TO ACHIEVE ENLIGHTENMENT. THE ZEN MASTER ATTEMPTS TO BRING ABOUT ENLIGHTENMENT IN HIS PUPIL IN VARIOUS WAYS. ONE OF THE THINGS HE DOES IS TO HOLD A STICK OVER THE PUPIL'S HEAD AND SAY FIERCELY, "IF YOU SAY THIS STICK IS REAL, I WILL STRIKE YOU WITH IT. IF YOU SAY THIS STICK IS NOT REAL, I WILL STRIKE YOU WITH IT. IF YOU DON'T SAY ANYTHING, I WILL STRIKE YOU WITH IT." WE FEEL THAT THE SCHIZOPHRENIC FINDS HIMSELF CONTINUALLY

IN THE SAME SITUATION AS THE PUPIL, BUT HE ACHIEVES SOMETHING LIKE DISORIENTATION RATHER THAN ENLIGHTENMENT. THE ZEN PUPIL MIGHT REACH UP AND TAKE THE STICK AWAY FROM THE MASTER—WHO MIGHT ACCEPT THIS RESPONSE, BUT THE SCHIZOPHRENIC HAS NO CHOICE SINCE WITH HIM THERE IS NO NOT CARING ABOUT THE RELATIONSHIP, AND HIS MOTHER'S AIMS AND AWARENESS ARE NOT LIKE THE MASTER'S.

GREGORY BATESON ET AL., "TOWARDS A THEORY OF SCHIZOPHRENIA" (1956, 208)

In the same 1956 publication, Bateson made another important move which again echoes the general concerns of cybernetics, this time with strange performances and the East, but making a much tighter connection to psychiatry than Walter. Bateson noted a formal similarity between the double bind and the contradictory instructions given to a disciple by a Zen master—Zen koans.[7] In the terms I laid out before, the koan is a technology of the nonmodern self that, when it works, produces the dissolution of the modern self which is the state of Buddhist enlightenment. And Bateson's idea was that double binds work in much the same way, also corroding the modern, autonomous, dualist self. The difference between the two situations is, of course, that the Zen master and disciple both know what is going on and where it might be going, while no one in the schizophrenic family has the faintest idea. The symptoms of schizophrenia, on this account, are the upshot of the sufferer's struggling to retain the modern form while losing it—schizophrenia as the dark side of modernity.

This, then, is where Eastern spirituality entered Bateson's approach to psychiatry, as a means of expanding the discursive field beyond the modern self.[8] And here it is interesting to bring in two more English exiles to California, Alan Watts and Aldous Huxley. Watts was a very influential commentator on and popularizer of Zen Buddhism in the United States in the 1950s, and he was also a consultant on Bateson's schizophrenia project (Haley 1976, 70). Two of the project's principals, Haley and Weakland, "took a course from Watts on the parallels between Eastern philosophy and Western psychiatry, back in the days when he was Director of the American Academy of Asian Studies. I think the focus on Zen offered us an alternative to the ideas about change offered in psychiatry in the 1950s" (Haley 1976, 107). It makes sense, then, to see Zen as a constitutive element of the Batesonian approach to schizophrenia. And, interestingly, Bateson's cybernetics also fed back into Watts's expositions of

Buddhism. In *The Way of Zen* (1957, 57–58), Watts drew on cybernetics as "the science of control" to explain the concept of *karma*. His models were an oversensitive feedback mechanism that continually elicits further corrections to its own performance, and the types of logical paradox that Bateson took to illuminate the double bind. Watts also discussed the circular causality involved in the "round of birth-and-death," commenting that in this respect, "Buddhist philosophy should have a special interest for students of communication theory, cybernetics, logical philosophy, and similar matters." This discussion leads Watts directly to the topic of nirvana, which reminds us of the connection that Walter and Ashby made between nirvana and homeostasis. Watts later returns to a discussion of cybernetics (135ff.), now exemplified by pathologies of the domestic thermostat, to get at a peculiar splitting of the modern mind—its tendency to try to observe and correct its own thought patterns while in process—and he also mentions the double bind (142), though not in connection with madness, citing Jurgen Ruesch and Bateson (1951). Here, then, we have a very interesting instance of a two-way flow between cybernetics and Buddhist philosophy, with the nonmodern self as the site of interchange.

Next, to understand Laing's extension of Bateson it helps to know that Aldous Huxley had also evoked a connection between schizophrenia and enlightenment two years prior to Bateson (neither Bateson nor Laing ever mentioned this in print, as far as I know; Huxley cited D. T. Suzuki as his authority on Zen, rather than Watts). In what became a countercultural classic of the sixties, *The Doors of Perception* (1954), Huxley offered a lyrical description of his perceptions of the world on taking mescaline for the first time and tried to convey the intensity of the experience via the language of Zen philosophy—he speaks of seeing the dharma body of the Buddha in the hedge at the bottom of the garden, for example. But he also linked this experience to schizophrenia. Having described his experience of garden furniture as a "succession of azure furnace-doors separated by gulfs of unfathomable gentian," he went on (45–47):

> And suddenly I had an inkling of what it must feel like to be mad. . . . Confronted by a chair which looked like the Last Judgement. . . . I found myself all at once on the brink of panic. This, I suddenly felt, was going too far. The fear, as I analyse it in retrospect, was of being overwhelmed, of disintegrating under a pressure of reality greater than a mind accustomed to living most of the time in a cosy world of symbols could possibly bear. . . . The schizophrenic is like a man permanently under the influence of mescalin, and therefore unable to shut off

> the experience of a reality which he is not holy enough to live with, which he cannot explain away . . . [and which] scares him into interpreting its unremitting strangeness, its burning intensity of significance, as the manifestations of human or even cosmic malevolence, calling for the most desperate of countermeasures, from murderous violence at one end of the scale to catatonia, or psychological suicide, at the other.

Huxley's first-person account of his mescaline experience served to fill in a phenomenology of enlightenment and madness that Bateson had left undeveloped, and it was this specific phenomenology that informed the sixties imagination of both—and that, conversely, made schizophrenia a key referent (among the much wider field of mental conditions that concerned Walter, Ashby, and orthodox psychiatry).[9]

We can return to Bateson. In 1961 he took the development of his thinking on schizophrenia one step further in his new edition of *Perceval's Narrative*, a first-person account of madness and spontaneous remission dating from the early nineteenth century. In his introduction to the book, Bateson described madness as a temporally extended *process* with a distinctive structure, which constituted a *higher level of adaptation* than those modelled by Walter's tortoises or Ashby's homeostats.[10] Here is the key passage (Bateson 1961, xiv):

> Perceval's narrative and some of the other autobiographical accounts of schizophrenia propose a rather different view of the psychotic process [from that of conventional psychiatry]. It would appear that once precipitated into psychosis the patient has a course to run. He is, as it were, embarked upon a voyage of discovery which is only completed by his return to the normal world, to which he comes back with insights different from those of the inhabitants who never embarked on such a voyage. Once begun, a schizophrenic episode would appear to have as definite a course as an initiation ceremony—a death and rebirth—into which the novice may have been precipitated by his family life or by adventitious circumstance, but which in its course is largely steered by endogenous process.
>
> In terms of this picture, spontaneous remission is no problem. This is only the final and natural outcome of the total process. What needs to be explained is the failure of many who embark on this voyage to return from it. Do these encounter circumstances either in family life or institutional care so grossly maladaptive that even the richest and best organised hallucinatory experience cannot save them?

There is more to be said about Bateson, but this is as far as we need to go in exploring his cybernetic understanding of schizophrenia. In this passage he arrives at an image of the schizophrenic as an exceedingly complex system, in Stafford Beer's terms—a system with its own dynamics, with which one can possibly interfere but which one cannot control, "a voyage of discovery . . . largely steered by endogenous process." From one perspective, the model for the voyage could be the homeostat or DAMS or one of Stuart Kauffman's simulations of gene networks, but Bateson alluded instead to richer and more substantive referents: initiation ceremonies and alchemy (the motif of death and rebirth). Schizophrenia and recovery appear here as a sort of *gymnastics of the soul*, as Foucault might have said—a plunge beyond the modern self, precipitated by adaptation to double binds, with psychosis as a higher level of adaptation that returns to a transformed self.

THERAPY

WE DO NOT LIVE IN THE SORT OF UNIVERSE IN WHICH SIMPLE LINEAL CONTROL IS POSSIBLE. LIFE IS NOT LIKE THAT.

GREGORY BATESON, "CONSCIOUS PURPOSE VERSUS NATURE" (1968, 47)

At this point it would be appropriate to move from psychiatric theory to practice, but since Laing and his colleagues went further than Bateson in that direction, much of this discussion can be postponed for a while. What I should emphasize here is that Bateson's understanding of schizophrenia hung together with a principled critique of orthodox psychiatry, and this gets us back to the fork in the road where Bateson and Laing split off from Ashby and Walter. Just as one can think of relations within the family on the model of interacting homeostats all searching for some sort of joint equilibrium, one can also think of relations between sufferers and psychiatrists on that model. Ashby, of course, thought of the psychiatric encounter asymmetrically, as a site where the psychiatrist used electric shocks or surgery to try to jolt the patient back into normality. Bateson, instead, thought such an approach was worse than useless. Implicit in the notion of the self as an exceedingly complex system is the idea that it is not subject to any sort of determinate, linear control. One can impinge on the dynamics of the self as one can impinge on the dynamics of a homeostat, but not with any determinate outcome. From this perspective, the chance that blasting someone's brain with current would simply straighten that person out psychiatrically is very small. And if one adds in Bateson's later idea of psychosis as a voyage with an adaptive course to run, then such interventions appear entirely counterproductive—"hindering and even exacerbating circumstances during the progress of the psychosis" (Bateson 1961, xvi)—which simply leave sufferers stuck in their double binds without any possibility of escape.

Already in the early 1950s Bateson echoed Harry Stack Sullivan's critique of "mechanistic thinking which saw [man] so heavily determined by his internal psychological structure that he could be easily manipulated by pressing the appropriate buttons." In contrast, Bateson favored

> the Sullivanian doctrine [which] places the therapeutic interview on a human level, defining it as a significant meeting between two human beings. . . . If . . . we look at the same Sullivanian doctrine of interaction with the eyes of a mathematician or circuit engineer, we find it to be precisely the theory which emerges as appropriate when we proceed from the fact that the two-person system has circularity. From the formal, circularistic point of view no such interactive system can be totally determined by any of its parts: neither person can effectively manipulate the other. In fact, not only humanism but also rigorous communications theory leads to the same conclusion. (Ruesch and Bateson 1951, quoted by Heims 1991, 150)

Adumbrated here, then, is a symmetric version of cybernetic psychiatry, in which the therapist as well as the patient appears within the frame, more or less on the same plane as each other, as part of a continuing process that neither can control.[11] But still, just what should this process look like? The most enduring legacy of Batesonian psychiatry is family therapy, in which the therapist enters into the communication patterns of families and tries to help them unravel double binds (Lipset 1980; Harries-Jones 1995). Bateson's own favored approach seems to have been simply an unstructured and open-ended engagement with sufferers—chatting, eating together, playing golf (Lipset 1980, chap. 12).[12] More on this when we get to Laing.

AS NOMAD

UNTIL THE PUBLICATION OF *STEPS* [BATESON 1972], GREGORY MUST HAVE GIVEN THE IMPRESSION, EVEN TO HIS STRONGEST ADMIRERS, OF TAKING UP AND THEN ABANDONING A SERIES OF DIFFERENT DISCIPLINES; SOMETIMES HE MUST HAVE FELT HE HAD FAILED IN DISCIPLINE AFTER DISCIPLINE. LACKING

A CLEAR PROFESSIONAL IDENTITY, HE LACKED A COMFORTABLE PROFESSIONAL BASE AND A SECURE INCOME.

MARY CATHERINE BATESON (2000, VIII)

We can leave Bateson by examining the social basis of his cybernetics, and the point to dwell on is his nomadism. Even more than Walter and Ashby, Bateson was a wanderer. He never held a permanent position in his life; his work always lacked a secure institutional base. Instead, apart from temporary teaching positions, he took advantage of the ample funding opportunities available in the postwar United States, although this sometimes left him with no support at all. The schizophrenia project was funded in its first two years by the Rockefeller Foundation, but the grant was not renewed after that and "my team stayed loyally with me without pay." Eventually, "the Macy Foundation saved us," followed by grants from the Foundations Fund for Psychiatry and the National Institute of Mental Health. "Gradually it appeared that . . . I should work with animal material, and I started to work with octopuses. My wife, Lois, worked with me, and for over a year we kept a dozen octopuses in our living room. This preliminary work was promising but needed to be repeated and extended under better conditions. For this no grants were available. At this point, John Lilly came forward and invited me to be the director of his dolphin laboratory in the Virgin Islands. I worked there for a year and became interested in the problems of cetacean communications, but I think I am not cut out to administer a laboratory dubiously funded in a place where the logistics are intolerably difficult" (M. C. Bateson 2000, xx–xxi). And so on.

Bateson's octopuses in the living room remind me of the robot-tortoises in Walter's kitchen.[13] Again we are in the presence of a life lived at odds with and transversely to the usual institutional career paths. What should we make of this? Bateson was a scholar with no scholarly place to be, and we could think of this in terms of both repulsion and attraction. On the former, Bateson tended to be critical of the fields whose terrain he crossed, and none more so than psychiatry. Unlike Walter and Ashby, Bateson was intensely critical of orthodox psychiatry, and his analysis of the double bind implied a drastic departure from orthodox modes of therapy, as we can explore further below. Here we approach Deleuze and Guattari's sense of the nomad as a threat to the state and the established social order. From the side of attraction, Bateson was always searching for like-minded people to interact with, but never with great success. Lipset (1980, 232) records that in 1959 Bateson applied for a three-year fellowship at the Institute for Advanced Study in Princeton. The director,

Robert Oppenheimer, put him off, on the grounds that the institute was not interdisciplinary enough, to which Bateson replied, "I sometimes think that the ideal sparring partners died off like the dinosaurs at the end of the eighteenth century." In the absence of a stable group of sparring partners, Bateson tried several times to assemble temporary groups in the form of intense conferences (echoing the Macy conferences), the best known of which were two long meetings in 1968 and 1969 sponsored by the Wenner-Grenn Foundation, organized around Bateson's ecological concerns (Gordon Pask was one of the invitees).[14] Again we return to the improvised basis of cybernetics.

Having said all that, we can now note that there was one social location that offered Bateson a home. Though Bateson's biographers show little interest in this, one can easily make a case for a close association between him and the West Coast counterculture, especially in the later years of his life. The connection Bateson made between madness and enlightenment became a standard trope of the sixties, of course, but Bateson's interests in strange performances and altered states ranged far beyond that. In 1974, for example, he returned to a topic of his prewar anthropological research: the states of trance he had studied in Bali (Lipset 1980, 282–84). Lipset (281) records a conversation about LSD at around the same between Bateson and Carter Wilson: Wilson "finally asked him if he thought there was something truly different about the *kind* of experience LSD provides. Long pause. Then Gregory said slowly that yes, he did think you could say that the experience under LSD was different in kind from other experiences. And that once you had had it then you knew—a very long pause—that it was an experience you could have again for a dollar." One can deduce that Bateson was no stranger to the acid culture of the time, even if this entailed a degree of distance.[15] Along much the same lines, on 28 October 1973 Bateson wrote a brief account of his experience of floating for an hour in John Lilly's sensory deprivation tank: "Mostly away—no—just no words? Briefly a dream—switching to and fro between the others (?Lilly. J.) is a boy; I am a man. And vice versa he is a man, I a boy—mostly just floating. . . . Relaxing from all that—very definite process, interrupted by Lilly calling to me 'Are you all right?' Opened lid which for two of us sort of joke" (Lilly 1977, 189).[16]

Bateson scholars refer to Lilly as a man who did research on dolphins and gave Bateson a job. But he was also a leading figure in the U.S. counterculture and the New Age movement, one of the great explorers of consciousness, finding his spiritual guides while spending long hours in his tanks under the influence of LSD (Lilly 1972). Another friend of Bateson's was Stewart Brand, the founder of *The Whole Earth Catalog*, another key figure in the psychedelic sixties, closely associated with the Grateful Dead and Ken Kesey's acid tests, for example. Brand read Bateson's *Steps to an Ecology of Mind* when it first appeared in 1972, "to find that it spoke to the 'clear conceptual bonding of cybernetic whole-systems thinking with religious whole-systems thinking.' . . . Relations between the two men expanded in the spring of 1974, when Brand founded the *CoEvolution Quarterly*. . . . Part of his focus was now in homage of Bateson." In 1975 Brand introduced Bateson to Jerry Brown, the governor of California, who appointed him to the board of regents of the University of California in late 1976 (Lipset 1980, 286, 290).[17]

At an institutional level, Bateson had various temporary teaching appointments over the course of his life. The last of these was from 1972 to 1978, part-time in the Department of Anthropology at the new Santa Cruz campus of the University of California. He affiliated himself there with Kresge College, "the most radical of the Santa Cruz experiments in undergraduate education [which] tended towards crafts, meditation, utopias, gardening, and poetry writing" (Lipset 1980, 280). His last appointment, from 1978 to 1980 was as scholar in residence at the Esalen Institute at Big Sur, the epicenter of the nascent New Age movement.[18] Bateson died on 4 July 1980 at the Zen Center in San Francisco.

What should we make of this? Evidently, northern California was a key site at which cybernetics crossed over into the broader field of the counterculture, as it hung on there into the 1970s and mutated into New Age.[19] Gregory Bateson was at the heart of this, and the medium of exchange was a shared interest in what I called earlier the performative brain but which is better described in Bateson's case as the performative self—a nonmodern self capable of strange performances and the achievement of altered states, including a pathological disintegration into madness in one direction, and dissolution into nirvana in the other. There is one other link between Bateson and the counterculture that will get us back across the Atlantic. He was friends with R. D. Laing.

R. D. Laing

I WAS TRYING TO DESCRIBE THE FAMILY "MANGLE," THE WAY FAMILIES MANUFACTURE PAIN FOR THEIR MEMBERS.

R. D. LAING, INTERVIEW, QUOTED IN BURSTON (1996, 101)

R. D. Laing (as he was named on his book jackets and known to popular culture), Ronald David Laing, was born in Glasgow on 7 October 1927 and

Figure 5.2. R. D. Laing. Used courtesy of University of Glasgow Library.

died playing tennis in the south of France in 1989 (fig. 5.2).[20] He studied medicine at Glasgow University from 1945 to 1951, when, after six months of neurosurgical internship, he was called up by the army and "summarily informed" that he was now a psychiatrist. He left the army in 1953 and found a position at Glasgow Royal Mental Hospital. In 1957 he headed south, joining the Tavistock Institute in London in 1957 and remaining there until 1967 (Howarth-Williams 1977, 4–5).[21] During the 1960s he published seven books and many articles while developing and implementing an increasingly radical psychiatric stance, becoming a key figure in what is often referred to as the "antipsychiatry movement."[22] He also became a central figure in the British "underground" scene, and the publication in 1967 of his popular book *The Politics of Experience* brought him national and international attention and even notoriety.

Laing's writings portray him as a scholar and an intellectual, drawing upon works in Continental philosophy and current sociology to construct a "social phenomenology" that might inform our understanding of mental illness and psychiatric practice (see Howarth-Williams 1977, on which I have drawn extensively). Laing did not describe himself as a cybernetician, but his work was certainly cybernetic, inasmuch as from 1958 onward he was strongly influenced by Gregory Bateson, which is why he bears attention here.[23] Laing's second book, *Self and Others*, for example, includes a long discussion of the double bind, including the statement that "the work of the Palo Alto group [Bateson et al.], along with Bethesda, Harvard, and other studies, has . . . *revolutionized* the concept of what is meant by 'environment' and has already rendered obsolete most earlier discussions on the relevance of 'environment' to the origins of schizophrenia" (Laing 1961, 129).[24]

Laing was not uncritical of Bateson, however. Bateson had a fondness for formal logic and wanted to understand the double bind on the model of a logical paradox—the Cretan who says "all Cretans are liars," and so on. Laing put something more empirically and phenomenologically satisfying in its place. His 1966 book *Interpersonal Perception*, for example (written with H. Phillipson and A. Robin Lee), explores the levels and structures of interpretation that go into the formation of "what he thinks she thinks he thinks," and so on. The end result is much the same as the double bind, though: pathological reflections back and forth in communication, "whirling fantasy circles," Laing calls them, from which escape is difficult or impossible, and that are "as destructive to relationships, individual (or international), as are hurricanes to material reality" (Laing, Phillipson, and Lee 1966, 22). As it happens, this is the context in which Laing came closest to explicitly cybernetic language. When he remarked of these spirals that "the control is reciprocal . . . the causality is circular" (118), he was echoing the subtitle of the Macy conferences: "Circular Causal and Feedback Mechanisms in Biological and Social Sciences." It is also worth noting that this strand of Laing's work fed straight back into the mainstream of cybernetics: Gordon Pask made it the basis of his formal theory of "conversation," meaning any kind of performative interactions between men and machines.[25]

But it was Bateson's notion of madness as an inner voyage that Laing really seized upon, no doubt because it spoke to his own psychiatric experience, and that contributed greatly to Laing's reputation and impact in the sixties. Two talks that Laing gave in 1964, which were revised in 1967 as the key chapters (5 and 6) of *The Politics of Experience*, take this idea very seriously (reproducing my Bateson quote in its entirety) and offer vivid elaborations, with strong echoes of Huxley's *The Doors of Perception*: the space of the nonmodern self as experientially a place of wonder and terror, "the living fount of all religions" (131), schizophrenia as an unexpected and unguided plunge into "the infinite reaches of inner space" (126–27), the need for a "guide" to inner space—Laing's colleague David Cooper was the first to invoke the figure of the shaman in print, I think (Cooper 1967)—modernity as a denial of the nonmodern self and nonmodern experience which leaves the voyager at a loss: modernity as, in this sense, itself a form of madness.

ON THERAPY

DR. LAING, I AM TOLD THAT YOU ALLOW YOUR SCHIZOPHRENIC PATIENTS TO TALK TO YOU.

A CHIEF PSYCHIATRIC SOCIAL WORKER, QUOTED IN LAING, *WISDOM, MADNESS AND FOLLY* (1985, 142)

This quick sketch of Laing's thought and writing is enough to establish that his understanding of madnesss and therapy was in very much the same cybernetic space as Bateson's. Now I want to see what this approach looked like in practice. I can first sketch the basic problematic in general terms and then we can look at a series of implementations.

My quotation from Bateson included the idea that psychotic inner voyages have their own endogenous dynamics. This is integral to the idea of psychosis as an adaptive mechanism. But the example of the Zen master and Laing's idea of "the guide" both entail the notion that one can somehow participate in that dynamics from the outside, even if one cannot control it (which is, again, the sense of the word "steersman," from which Wiener derived the word "cybernetics"). The question for Laing and his fellows was, then, how to latch on, as it were, to schizophrenics—how to get in touch with them, how to adapt to them—when schizophrenia was more or less defined by the disruption of conventional patterns of communication. The only answer to that question that I can see is trial-and-error experimentation with behavior patterns to see what works. One thus arrives at the symmetric image of sufferers and psychiatrists as assemblages of homeostats running through sequences of configurations in pursuit of a joint equilibrium, with this difference: Ashby's homeostats were hard wired to be sensitive to specific variables, whereas the psychiatric experiment necessarily included a search for the relevant variables. In Ashby's terms (though he himself did not think of psychiatry in this way), the psychiatrists had to expand the variety of their performances in their attempts to latch onto schizophrenia. Now we can look at some examples of what this meant in practice.

1. An enduring thread in Laing's psychiatry was that it might help to treat the mentally disturbed "simply as human beings" (Howarth-Williams 1977, 8). This seems quite an obvious thing to do and hardly radical until one remembers that that it was just what orthodox psychiatry did not do. Laing *talked* to his patients, an activity strongly discouraged in orthodox psychiatric circles as stimulating the "schizophrenic biochemical processes" that drugs were intended to inhibit.[26] And it is worth noting that Laing undertook such interactions, both verbal and nonverbal, in a performative spirit, as a way of getting along with the patients, rather than a representational and diagnostic one (Laing 1985, 143):

> In a recent seminar that I gave to a group of psychoanalysts, my audience became progressively aghast when I said that I might accept a cigarette from a patient without making an interpretation. I might even offer a patient a cigarette. I might even give him or her a light.
>
> "And what if a patient asked you for a glass of water?" one of them asked, almost breathlessly.
>
> "I would give him or her a glass of water and sit down in my chair again."
>
> "Would you not make an interpretation?"
>
> "Very probably not."
>
> A lady exclaimed, "I'm totally lost."

In this instance, then, latching onto schizophrenics involved just the same tactics as one might deploy with the girl or boy next door. Hardly radical in themselves, as I said, but utterly divergent from the mainstream psychiatry of Laing's day. The expansion of the therapist's variety in performance, relative to standard practice, is evident. As usual, we see that ontology (the symmetric rather than the asymmetric version of cybernetics) makes a difference.

2. It once surprised to me to discover the enormous amount of serious scientific, clinical, and philosophical attention that was generated by LSD in the 1950s and 1960s (see, for example, Solomon 1964; and Geiger 2003). In psychiatry, LSD figured in at least three ways. One was as a psychotomimetic, capable of inducing psychotic symptoms in the subjects of laboratory experiments. Another was as yet another weapon in the arsenal of psychic shocks—as in Ashby's inclusion of LSD in his blitz therapy. But third, from the other side, LSD also featured as a technology of the nonmodern self, a means

of destabilizing the everyday self of the *therapist* and thus helping him or her to gain some sort of access to the experiential space of his or her patients.[27] Laing never wrote about this, as far as I can ascertain, but it is clear from various sources that he was indeed heavily involved with LSD in the early 1960s. Martin Howarth-Williams (1977, 5) records that in 1961 Laing was "experimenting with the (legal) use of hallucinogens such as LSD" and that by 1965 he was "reputedly taking (still legal) LSD very frequently." In what appears to be a thinly disguised account of the period, Clancy Sigal's novel *Zone of the Interior* (1976) has the narrator doing fabulous amounts of acid supplied by a therapist who sounds remarkably like Laing, and often wrestling naked with him while tripping.[28]

Here, then, we find the cybernetic concern with altered states in a new and performative guise, with LSD as a means to put the therapist into a new position from which possibly to latch onto the patient. We are back to what I just called the gymnastics of the soul, and the contrast with orthodox psychiatric therapy is stark. Likewise, LSD exemplifies nicely the idea of expanding the variety of the therapist as a way of coming alongside the sufferer—entry into an altered state. We could also notice that while I described verbal communication earlier as a detour away from and back to performance, here LSD features as a dramatic contraction of the detour—a nonverbal tactic for getting alongside the sufferer as a base for not necessarily verbal interaction (wrestling!). This theme of curtailing the detour will reappear below.[29]

3. The two examples so far have been about microsocial interactions between therapist and patient. Now we can move toward more macrosocial and institutional instantiations. Laing's first publication (Cameron, Laing, and McGhie 1955) reported a yearlong experiment at Glasgow Royal Mental Hospital in which eleven of the most socially isolated chronic schizophrenics spent part of each day in a room with two nurses. The nurses had no direct instructions on how to perform, and Laing and his coauthors regarded this project simply as an experiment in which the patients and nurses had a chance "to develop more or less enduring relations with one another" (Cameron, Laing, and McGhie 1955, 1384). This "rumpus room" experiment, as it has been called, was Laing's tactic of relating to patients as human beings writ large and carried through by nurses instead of Laing himself. Over a year, the nurses and patients were left to adjust and adapt to one another, without any prescription how that should be accomplished. And, first, we can note that this tactic worked. The patients changed for the better in many ways (Cameron, Laing, and McGhie 1955, 1386): "They were no longer isolates. Their conduct became more social, and they undertook tasks which were of value in their small

community. Their appearance and interest in themselves improved as they took a greater interest in those around them. . . . The patients lost many of the features of chronic psychoses; they were less violent to each other and to the staff, they were less dishevelled, and their language ceased to be obscene. The nurses came to know the patients well, and spoke warmly of them." Second, we can return to the image of interacting homeostats searching for some joint equilibrium and note that the nurses as well as the patients changed in the course of their interactions. In the first few months, the nurses tried giving the patients orders, sedated them before they walked over to the allotted space, and so on. But after some time (1385), "the [two] nurses [became] less worried and on edge. They both felt that the patients were becoming more 'sensible.' From a survey of what the patients were saying at this period, however, it is clear that the change lay with the nurses, in that they were beginning to understand the patients better. They ceased always to lock the stair door, and to feel it necessary for the patients to be sedated in the mornings. They began to report more phenomenological material. They became more sensitive to the patients' feelings and more aware of their own anxieties." Third, this change in nursing practice again points to the fact that ontology makes a difference. Giving orders and sedatives and locking doors were standard ways of handling "chronic deteriorated schizophrenics" (1384) (who were presumed to be beyond the help of ECT, etc.), but in the new experimental setup, nursing grew away from that model as the nurses and patients adapted to one another. Fourth, we should note that this experiment provoked institutional frictions, inasmuch as it departed from the usual practices of the hospital (1386): "The problems which arise when a patient is felt to be 'special' caused difficulty. Comments that all the 'rumpus room patients' were being specially treated were common among the staff. The group nurses, who did not work on a shift system but were free in the evenings and the weekends, were considered to have a cushy job. The group nurses reacted in a defensive way to this, adopting a protective attitude when their patients were criticized during staff discussions." This quotation points both to the fact that symmetric psychiatry differed in institutional practice from orthodox psychiatry and to the frictions that arise when one form of practice is embedded within the other. Laing and his colleagues remarked, "Tensions of this kind are lessening" (1386), but they will resurface below.

4. We can contextualize the rumpus room. British psychiatry in and after World War II included a variety of developments in "social" or "communal psychiatry." These entailed a variety of elements, but we should focus on a leveling of the social hierarchy and a transformation of power relations within

the mental hospital.[30] As discussed already, the traditional mental hospital had a top-down power structure in which doctors gave orders to nurses who gave orders to patients. Social psychiatry, in contrast, favored some measure of bottom-up control. Patients and staff might meet as a group to discuss conditions in the hospital or individual mental problems.[31] The Glasgow rumpus room experiment can be seen as a radical early variant of this approach, and one later development from the early sixties is particularly relevant here—David Cooper's experimental psychiatric ward, Villa 21, at Shenley Hospital, where patients were encouraged to take care of their surroundings and each other. "The venture, while it lasted, was a modest success and many interesting lessons were learnt," but it ran into many of the same problems as had the rumpus room before it: embedding this kind of a bottom-up structure within a top-down institution created all sorts of problems and tensions, nicely evoked in *Zone of the Interior*. Villa 21 and its inmates made the whole hospital look untidy; the patients disrupted the orderly routines of the institution; nurses feared for their jobs if the patients were going to look after themselves. "Cooper concluded that any future work of this kind had to be done outside the great institutions" (Kotowicz 1997, 78).[32]

This gets us back to Laing. In 1965, Laing, Cooper, Aaron Esterson, Sidney Briskin, and Clancy Sigal decided to found a therapeutic community entirely outside the existing institutions of mental health care in Britain.[33] In April 1965, they established the Philadelphia Association as a registered charity with Laing as chairman, and with the object of taking over, two months later, a large building in the East End of London, Kingsley Hall, as the site for a new kind of institution (Howarth-Williams 1977, 52).[34] Kingsley Hall itself closed down in 1970, but the Philadelphia Association continued the project into the 1970s with a series of therapeutic communities that moved between condemned houses in Archway, north London. Life in these communities is the best exemplification I know of what a symmetric cybernetic psychiatry might look like in practice, and I therefore want to examine it in some detail. The proviso is, alas, that documentary information is thin on the ground. The only book-length account of life at Kingsley Hall is Mary Barnes and Joe Berke's *Two Accounts of a Journey through Madness* (1971), though *Zone of the Interior* is illuminating reading. On Archway, the only written source is *The David Burns Manuscript* (Burns 2002), written by one of the residents, unpublished but available online. There is also a documentary film made at one of the Archway houses over a period of seven weeks in 1971, *Asylum*, by Peter Robinson.[35] This lack of information is itself an interesting datum, given the impressive amounts of time and emotional energy expended at Kingsley Hall especially, though I am not sure what to make of it. The obvious interpretation is that even people who are committed to transforming practice remain happier writing about ideas.[36]

KINGSLEY HALL

ALL LIFE IS MOVEMENT. FOR INSTANCE, ONE MAY BE HIGH OR LOW, BE BESIDE ONESELF . . . GO BACK OR STAND STILL. OF THESE MOVEMENTS, THE LAST TWO IN PARTICULAR TEND TO EARN THE ATTRIBUTION OF SCHIZOPHRENIA. PERHAPS THE MOST TABOOED MOVEMENT OF ALL IS TO GO BACK. . . . AT COOPER'S VILLA 21 AND IN OUR HOUSEHOLDS, THIS MOVEMENT HAS NOT BEEN STOPPED. IF ALLOWED TO GO ON, A PROCESS UNFOLDS THAT APPEARS TO HAVE A NATURAL SEQUENCE, A BEGINNING, MIDDLE AND END. INSTEAD OF THE PATHOLOGICAL CONNOTATIONS AROUND SUCH TERMS AS "ACUTE SCHIZOPHRENIC BREAKDOWN," I SUGGEST AS A TERM FOR THIS WHOLE SEQUENCE, METANOIA.

R. D. LAING, 1967 (QUOTED BY HOWARTH-WILLIAMS 1977, 80–81)

PERHAPS THE MOST CENTRAL CHARACTERISTIC OF AUTHENTIC LEADERSHIP IS THE RELINQUISHING OF THE IMPULSE TO DOMINATE OTHERS. . . . THE MYTHICAL PROTOTYPE OF THE INAUTHENTIC LEADER IS WILLIAM BLAKE'S URIZEN, THE MAN OF HORIZON, OF LIMITS, CONTROL, ORDER. . . . THE NAZI DEATH CAMPS WERE ONE PRODUCT OF THIS DREAM OF PERFECTION. THE MENTAL HOSPITAL, ALONG WITH MANY OTHER INSTITUTIONS IN OUR SOCIETY, IS ANOTHER.

DAVID COOPER, *PSYCHIATRY AND ANTI-PSYCHIATRY* (1967, 96–97)

THERE WAS A SPECIAL PSYCHIC ATMOSPHERE WITHIN THE COMMUNITIES; THERE WAS A HOPE AND A PROMISE; THERE WAS A FEELING OF THE GROWTH OF CONSCIOUSNESS, OF EVOLUTION. . . . IT WAS A SPIRITUAL REFUGE, A PLACE WHERE ONE COULD GROW AND CHANGE AND LEARN IN A WAY THAT WAS IMPOSSIBLE OUTSIDE, LIKE A MONASTERY OR A CAVE IN THE MOUNTAINS.

DAVID BURNS, *THE DAVID BURNS MANUSCRIPT* (2002, 20)

Kingsley Hall (fig. 5.3) could accommodate up to fourteen people; it included shared spaces, such as a dining room, a kitchen, a game room, and a library that became a meditation room; there was also a meeting room that could accommodate about a hundred people.[37] The group that lived there was

Figure 5.3. Kingsley Hall, London. (Photograph by Gordon Joly, used under Creative Commons Share Alike 2.5 Generic License.)

mixed. It included schizophrenics and psychiatrists—hence "therapeutic community"—but, according to an American resident therapist, Joe Berke, "the majority of the community, and visitors, were not medical or paramedical men and women. Many were artists, writers, actors or dancers" (Barnes and Berke 1971, 260). Kingsley Hall thus offered a kind of support community for the mentally ill; put very crudely, it was a place designed to help people through the sorts of inner voyages that Bateson had first conjured up in *Perceval's Narrative*, free from the interruptions of mainstream psychiatry.

Kingsley Hall was run as a commune—an association of adults who chose to live together, all paying rent and free to come and go as they pleased (including to work, if they had jobs). And the key point to grasp is thus that at

Kingsley Hall the mentally troubled and psychotherapists (and others) came together symmetrically within a very different frame of power relations from those of the conventional mental hospital. Except for the usual mundane considerations of communal living, the therapists and the mad were on the same plane.[38] The therapists were not in charge, they did not make the rules, and they did not deploy the standard psychotherapeutic techniques—they did not prescribe drugs or ECT for the other residents, for example.

It is also worth noting right away that life at Kingsley Hall asked a lot of its residents, including the psychiatrists. Conditions there, and later at Archway, were often, by conventional standards of domesticity, hellish. The behavior of schizophrenics is, almost by definition, often bizarre. It can take the form of catatonic withdrawal, which is disturbing enough, but David Burns (2002) mentions residents at Archway who would shout continually for days on end, frequent trashings of the kitchen, a tendency to disrobe and stroll off to the shops naked (with accompanying hassles with neighbors and police); the ubiquity of potential violence; and a resident who stabbed a cat to death.[39] At Kingsley Hall, psychotic behavior also included urinating and smearing excrement all over the place (Barnes and Berke 1971). No picnic, and not surprisingly therapists and others in residence tended to burn out from stress in a period of weeks or months, typically moving out but continuing to visit the community. Laing, in fact, lasted longer than most, staying at Kingsley Hall for its first year, before establishing a smaller community in his own home.[40] Staging a place where madness could be acted out carried a significant price; conventional psychiatry looks like an easy way out in comparison with antipsychiatry.

What did life at Kingsley Hall look like? There are accounts of Laing's role as dinner-time raconteur and guru, dancing through the night and annoying the neighbors. Clancy Sigal (1976) portrays Laing as an evil genius and claims to have gone mad there just to please him. To get much further, we have to turn to Barnes and Berke's *Two Accounts of a Journey through Madness* (1971). Barnes was a mentally disturbed woman who found her way to Laing and moved to Kingsley Hall when it opened, determined finally to live out an inner voyage; Joe Berke was an American therapist at Kingsley Hall who took much of the responsibility for looking after Mary. The book interweaves descriptions of Mary's journey written by both, and on these accounts Berke's strategy in latching onto Barnes was a double one.

One prong was performative. "As soon as I got to Kingsley Hall, I realized the best way to *learn* about psychosis would be for me to help Mary 'do her things.' And so I did. . . . Getting to know Mary was simple. I imagined where she was at and then met her on that level. Our first encounter consisted of my growling at her and she growling back at me" (Barnes and Berke 1971, 221). Mary's voyage consisted in going back to her early childhood, and much of Berke's engagement with her consisted in setting up such childlike physical games (Mary was forty-two when she went to Kingsley Hall). Besides bears, they also played together at being sharks and alligators. Mary would often hit Joe, and on a couple of occasions Joe hit Mary and made her nose bleed. He fed Mary, when necessary, with milk from a baby's bottle, and bathed her, including one occasion when she had smeared herself with feces. He provided her with drawing and painting materials and Mary responded avidly, producing a series of large paintings and later becoming a successful artist. And the trial-and-error aspect of these experiments in engagement is evident in the fact that not all of them worked (Barnes and Berke 1971, 224):

> It became obvious that it wasn't words that mattered so much as deeds, and even when the words and deeds coincided and were seemingly accepted by her, the ensuing state of relaxation could revert to one of agony for the barest of reasons. All I had to do was turn my head, or look inattentive, or blink an eye while feeding her, and Mary began to pinch her skin, twist her hair, contort her face, and moan and groan. Worse shrieks followed if I had to leave the room and get involved in another matter at about the time she was due for a feed. Suffice to say that if my acts and/or interpretations had been sufficient, such agonies could have been averted. So I said to myself, "Berke, you had better stop trying to tell Mary what you think she is wanting, and pay more attention to that with which she is struggling."

Berke's interactions with Barnes thus put more flesh on the earlier idea that latching onto schizophrenics as exceedingly complex systems necessarily entailed trial-and-error performative experimentation, and also the idea that such experimentation might well entail an expansion of the therapist's variety—Berke was probably not in the habit of playing bears, sharks, and alligators with other adults. Here, then, we have another instance of ontological theater: Berke's interactions with Barnes stage for us a more general image of homeostat-like systems performatively interfering with each other's dynamics without controlling them. And, from the other angle, those interactions again exemplify how one might go in practice—here, in psychiatry—if one thinks of the other on the model of the cybernetic ontology.

Having said that, I should turn to Berke's other mode of engagement with Barnes. Berke continually constructed *interpretations* of the nature of Barnes's problems and fed them back to her. He concluded, for example, that much of her strange behavior derived from anger. This anger was related to guilt: because of her inability to distinguish between inner and outer states, she tended to blame herself for anything that went wrong at Kingsley Hall, often including things which had nothing at all to do with her. Barnes also tended to interpret any situation on models derived from her childhood: Berke concluded that sometimes she was treating him as her mother, or her father, or her brother, and so on. Barnes at first rejected much of this but eventually came to share many, though not all, of Berke's interpretations, and this acceptance seems to have been an integral part of her recovery.

What should we make of this? The first thing to say is that there is nothing especially cybernetic about Berke's interpretive interventions into Barnes's life. They take us back instead to the field of representation rather than performance; they belong to the other ontology, that of knowable systems. But we can also note that the epistemological aspects of the interaction—Berke's interpretations of Barnes's performances—were parasitic upon their performative engagement. They did not flow from a priori understandings that determined those interactions from beginning to end: "It became obvious that it wasn't words that mattered so much as deeds." I have also emphasized that this performative engagement had an experimental quality; Berke had to *find out* how to relate to Barnes, and his psychotherapeutic interpretations grew out of that relation as reflections upon its emerging substance. And, further, the interpretations were themselves threaded though a performative feedback loop running between Berke and Barnes, and the value of specific interpretations depended upon their contributions to Mary's behavior: "If my acts and/or interpretations had been sufficient, such agonies could have been averted." This aspect of the Barnes-Berke interaction thus again stages for us a performative epistemology, in which articulated knowledge functions as *part of* performance—as arising from performance and returning to it—rather than as an externality that structures performance from without.

ARCHWAY

When we turn to the Archway communities that succeeded Kingsley Hall, we find a similar pattern, though at Archway the interpretive aspect of therapy receded still further.[41] Burns (2002, 23) mentions that during his time in

Archway he had three years of formal therapy with the psychiatrist Leon Redler but says nothing about the interpretive aspect of their interactions and instead emphasizes the performative techniques that evolved there (Burns 2002, 14–15):

> We obviously had to find ways of coping with . . . extreme and distressing behaviors that did not contradict our philosophy of not interfering violently with what might be valuable inner experience. We learned the hard way, perhaps the only way. At Kingsley Hall, when a resident had screamed for forty-eight hours continually and we were trying to have dinner, someone briefly sat on him with his hand over his mouth. For a moment we had calm and silence but of course it could not last. He soon started screaming and running about again. This did not work.
>
> Compassion, understanding, acceptance, all these were important and necessary. But they were not sufficient. Eventually we found a way to contain and lovingly control the behavior of a person under stress. We needed to do this for the sake of our own peace of mind and also because of the problems that occurred when a person took their screaming or nakedness into the outside world. . . . One resident at Archway . . . behaved in such distressing ways that we had to give her total attention. She would fight, kick, scream, pick up a knife, urinate in the kitchen or walk out the door, down our street and into the street of shops completely naked. She was nevertheless beloved by many of us. She was the first person to receive twenty-four-hour attention. To control her violence and keep her from going outside naked we had to keep her in the common space and make sure someone was always with her. We found this painful at first, but over months the twenty-four-hour attention became an institution of its own, and a major way of restoring order to community life.

"Twenty-four-hour attention" was a technique sui generis at Archway. In this passage it appears to have a purely negative function, but it quickly developed a positive valence, too (15–16): "Usually a group will gather and there will be something of a party or learning atmosphere. Change will occur not only in the person in crisis but in others who are there. . . . A resident may wish to attempt some project, exploring his inner world, overcoming his loneliness, his fear or his sadness, or coming off medications, drugs and alcohol. If the support group is large and strong enough a resident may request similar twenty-four-hour attention; or he may be encouraged to accept twenty-four-hour care, for example to come off phenothiazines or other substances." Other techniques were introduced to Archway from the outside, largely, it seems, by

Leon Redler, and largely of Eastern origin (Redler invited Zen masters to stay at his apartment). Burns (2002, 28) mentions hatha yoga and "sitting"—Zen meditation—and the list continues:

> Other techniques include Aikido and tae-kwon-do, oriental martial arts without the aggressive factor. Zen walking, moving through hatha yoga postures and Aikido are all forms of dance. Massage became an important part of community life at different times; one of our residents set up as a practicing giver of massage. . . . Various herbalists and acupuncturists applied their techniques. We realized the importance of the body, of the body-mind continuum. To think of mental illness outside of its physical context seems absurd. Thus much of the cooking at the community was vegetarian; there I received my introduction to the virtues of rice, beans, and vegetables. We had become aware of dance, of the movement of the body; we also became aware of music. . . . Music was always important to us, whether listening to records, playing the flute or chanting the Heart Sutra. Laing is an accomplished pianist and clavichordist. He would come visit us and play the piano, or organize a group beating of drums.

Various aspects of these developments are worth noting. "Twenty-four-hour attention" clearly continues the Barnes-Berke story of experimental performative engagement, culminating here in a relatively stable set of arrangements to protect members of the community from each other and the outside world while supporting their endogenous dynamics. One has the image of a set of homeostats finally coming more or less into equilibrium through the operation of this technique.[42]

Burns's list that starts with yoga requires more thought. We could start by noting that here we have another instance of the connection between the cybernetic ontology and the East, though now at the level of performance rather than representation. Bateson appealed to Zen as a way of expanding the discursive field beyond the modern self in order to conceptualize the inner experience of schizophrenics; at Archway Zen techniques appeared as material practices, ways of dealing with inner experiences.

Next, the items on Burns's list are technologies of the self in very much Foucault's original sense—ways of caring for, in this instance, the nonmodern, schizophrenic self. They were not understood as curing schizophrenia, but as ways of making it bearable, as "specific techniques for relieving stress or exploring one's inner world" (Burns 2002, 27). And this returns us to the question of power. As discussed so far, cybernetic psychiatry appears as a leveling of traditional hierarchies consistent with the symmetric version of the multihomeostat image, putting psychiatrists and sufferers on a level playing field. The Archway experiment went beyond that: psychiatrists became hangovers from the old days, and sufferers were treated as able, literally, to care for themselves. We find here the possibility, at least, of a full reclamation of the agency that traditional psychiatry had stripped from the schizophrenic.

Last, we could note that the Archway experiment converged on a position where communication was itself marginalized—the detour through words and other forms of interpersonal interaction receded into the background (at least on Burns's account) in relation to what one might think of as performative adaptation *within a multiple self*—the body and mind as two poles of an interactive "continuum." The Archway residents thus themselves arrived at a nonmodern ontology of the nondualist self: "In fact we [the Archway residents] gradually realized that much of what is called 'mental illness' is actually physical suffering, whether it be skin rashes, insomnia, vomiting, constipation, or general anxiety-tension. The schizophrenic process is endurable and can be meaningful in a context of minimal physical stress. . . . Zen and yoga have traditionally been means toward physical health and inner illumination" (Burns 2002, 28).

COUPLED BECOMINGS, INNER VOYAGES, AFTERMATH

A CHANGE IN ONE PERSON CHANGES THE RELATION BETWEEN THAT PERSON AND OTHERS, AND HENCE THE OTHERS, UNLESS THEY RESIST CHANGE BY INSTITUTIONALISING THEMSELVES IN A CONGEALED PROFESSIONAL POSTURE. . . . ANY TRANSFORMATION OF ONE PERSON INVITES ACCOMMODATING TRANSFORMATIONS IN OTHERS.

R. D. LAING, "METANOIA" (1972, 16)

NO AGE IN THE HISTORY OF HUMANITY HAS PERHAPS SO LOST TOUCH WITH THIS NATURAL *HEALING* PROCESS THAT IMPLICATES *SOME* OF THE PEOPLE WHOM WE LABEL SCHIZOPHRENIC. NO AGE HAS SO DEVALUED IT, NO AGE HAS IMPOSED SUCH PROHIBITIONS AND DETERRENCES AGAINST IT, AS OUR OWN. INSTEAD OF THE MENTAL HOSPITAL, A SORT OF RESERVICING FACTORY FOR HUMAN BREAKDOWNS, WE NEED A PLACE WHERE PEOPLE WHO HAVE TRAVELLED FURTHER AND, CONSEQUENTLY, MAY BE MORE LOST THAN PSYCHIATRISTS AND OTHER SANE PEOPLE, CAN FIND THEIR WAY *FURTHER* INTO INNER SPACE AND TIME, AND BACK AGAIN. INSTEAD OF THE *DEGRADATION* CEREMONIAL OF PSYCHIATRIC EXAMINATION . . . WE NEED . . . AN *INITIATION* CEREMONIAL, THROUGH

WHICH THE PERSON WILL BE GUIDED WITH FULL SOCIAL ENCOURAGEMENT AND SANCTION INTO INNER SPACE AND TIME, BY PEOPLE WHO HAVE BEEN THERE AND BACK AGAIN. PSYCHIATRICALLY, THIS WOULD APPEAR AS EX-PATIENTS HELPING FUTURE PATIENTS TO GO MAD.

R. D. LAING, *THE POLITICS OF EXPERIENCE* (1967, 127–28)

An asymmetry remains in my account of Kingsley Hall and Archway. I have written as if they existed solely for the benefit of the mad and with the object of returning them to a predefined normality. But to leave it at that would be to miss an important point. Laing's idea was that in modernity, the apparently sane are themselves mad, precisely in the sense of having lost any access to the realms of the nonmodern self that go unrecognized in modernity. Hence the sense in the above quotation of having dramatically lost touch with a natural healing process. And hence the idea that the Philadelphia communities might be a place of *reciprocal* transformation for the mad and sane alike: "This would appear as ex-patients helping future patients to go mad." Clearly, the sort of variety expansion I talked about above was, to some degree, a transformative experience for the nurses in the rumpus room, for example, or Laing on LSD, or Berke playing biting games with a middle-aged woman—and all of these can stand as examples of what is at stake here. But to dramatize the point, I can mention the one example recorded by Burns of a transformative inner voyage undertaken by one of the "normal" residents at Archway.

Burns (2002, 56) talks about John, who joined the community as one of the sane, a student. He had not been in a mental hospital before or diagnosed as a schizophrenic. But at Archway he, too, took advantage of twenty-four-hour attention and embarked on an inner voyage:

> He had moved into his emotional center and he had moved into the space in the common room and accepted the attention and care of guardians who sat with him day and night. He had taken off his clothes. He had shaved his head. He had listened into himself. He had become silent and private, undergoing the inner journey as had the others. . . . Under the attention of those who gathered John experienced a change. To be paranoid means that one feels hostile or malicious feelings directed at one. . . . But it is a different matter to be in a room with a group who are gathered with the expressed purpose of letting one be at the center and to accept their mindfulness. . . . The trembling and insecurity of one's consciousness need not be so intense. One need not fear the unknown other . . . . John found that he need not fear them, that he could trust them,

> that he could use them for his own purposes of growth. . . . Perhaps it was here that he learned that if people were laughing at him it was a laughter in which he could join. John remained a student but applied his happy new awareness to the way he went about it. He remained in London studying Japanese with the intention of translating ancient Zen manuscripts.

The sane could have transformative experiences in these communities, then, just like the mad, and, in fact, very similar experiences. And this, in the end, is how one should think about the Philadelphia communities. Not as a place where the mentally ill could be restored to some already given definition of normality, but as a place where all the residents could open-endedly explore their own possibilities in an interactive and emergent process with no predefined end point—a collective exploration of nonmodern selfhood. This is the final sense in which these communities staged a symmetric, rather than asymmetric, ontological vision. They aimed to make it possible for *new kinds of people* to emerge, beyond the modern self. And here we could note one final aspect under which Laingian psychiatry went beyond the basic image of the homeostat. As discussed in the previous chapter, Ashby's homeostats as real electromechanical devices had fixed goals: to keep certain electrical currents within preset bounds. Clearly, as ontological theater, Kingsley Hall and the Archway communities acted out a more adventurous plot, in which the very goals of performative accommodation were themselves emergent in practice rather than a given telos. Who knows what a body and mind can do? This is the version of the cybernetic ontology that interests me especially and that will remain at the center of attention in the chapters that follow.

— — — — —

For the sake of completeness, I want to comment on two further aspects of these Philadelphia Association communities. First, one might wonder whether they worked. This is not such an easy question to answer. I know of no quantitative comparative data which would enable one to say that it was more or less beneficial to spend time at Archway or a conventional mental hospital, and since the mainstream aspiration is to control rather than cure schizophrenia, now with drugs, it seems likely that such data will never be forthcoming.[43] It is also clear, I hope, that these communities problematized the very idea of success. If the ambition of conventional psychiatry is to produce people like, shall we say, Ross Ashby, then counting successes is not so much a problem. If the aim is to let a new kind of person in touch with his or her inner self emerge, the counting becomes difficult in the extreme. It is safe to say, I think, that those who spent time at Kingsley Hall and Archway were significantly marked by the experience. David Burns's manuscript is a remarkably lucid and insightful piece of writing by someone who went to Kingsley Hall in search of therapy, and he clearly felt that he benefited it from it enormously. And this is perhaps the key point to grasp. No one made anyone live in these communities (in contrast to involuntary confinement within the British state system at the time). The residents chose to go there and remain there, not necessarily in the expectation of dramatic cures, but in preference to other places such as their family homes or mental hospitals.[44] The Philadelphia communities offered them another kind of place to be, to live their lives, however odd those lives might appear to others, and these communities needed no other justification than this choice of its residents to be there.

Second, I should say something more about inner voyages. The visionary aspect of these, at least in literary portrayals, had much to do with the grip of Laing's psychiatry on the sixties imagination. One should not overemphasize their importance in the life of the Philadelphia Association communities. Mary Barnes's transformation at Kingsley Hall seems to have been entirely free from visions or interesting hallucinations, and the *Asylum* documentary depicts life at Archway at its most mundane. There is nothing specifically cybernetic at stake here, except as it concerns altered states and strange performances more generally, but still one wonders whether such visionary experience was ever forthcoming. According to Burns (2002, 64–67, 70),

> A number of residents [at Archway] seemed to go through a similar experience in its outward form and I learned that they shared to some degree an inner experience. I came to know Carl best of those who found their way to the center through the tension, violence and turmoil they expressed and the terrible pain and fear they felt. Carl told me his story. Sitting in his room Carl began to feel that he had become transparent, that the barrier between his self and the outside world had faded. He felt that his thoughts were being perceived by others and he heard voices responding to what he was thinking . . . . Carl felt that he was the center of the universe, that he was the focus of loving energy. But it was necessary to move into this other and alien dimension without reservation. . . . This was the difference between heaven and hell. . . . One day . . . Carl felt that a wise old man from China had decided many years ago to be reborn as himself and to live his life with all its ignorance. . . . Everyone in the world except him [Carl] had decided to take on animal form while retaining human consciousness. This was an eminently sensible decision as they would have beauty and resilience and freedom from the oppressions of human culture. . . . [Carl] knew

> it was really a fantasy but he was fascinated by the truth of this inner experience. He had had glimpses of peace and glory. But now he found that he could inhabit these spaces that he passed through. Especially during the nights when he could let go his defenses and drop his everyday persona he was swimming in a sea of meaning and breathing an air that was more than a mixture of gases. . . . He had been able to inhabit a land of ecstasy and intense feeling where effort had value, where questions of the meaning of life became irrelevant. This was a land to be explored, an adventure that held joys and terrors.

In Carl's case, then, which Burns thought was typical, a resident did go through a voyage with the otherworldly quality that fascinated Laing. That said, one should also recognize that after inhabiting a land of ecstasy and intense feeling, Carl eventually found himself "cast out. Perhaps he was only living in the everyday world once more but it seemed worse than he remembered it. It seemed a place of filth and degradation and trivia. A place of confusion and obsession. . . . If he had been 'schizophrenic' before and had been able to learn from it and glory in it, then he was 'obsessive-compulsive-neurotic' now" (Burns 2002, 70). The moral of his story for psychiatry, then, is hardly a clear one, though Carl himself found some virtue in it. Carl "survived and he told me that although he still did not understand why it had been necessary he had learned some invaluable lessons from the experience. . . . He began to learn to forgive. This came about because he had realized that he could never know what someone else might be going through. He knew a depth of suffering he had not known before. More important, he told me, he began to learn to forgive himself" (71–72).

— — — — —

There are two other aspects of the story that I should not leave hanging: what happened to the Archway communities, and what happened to R. D. Laing?

The Archway communities closed down in 1980 (A. Laing 1994, 207). The Philadelphia Association still exists and continues to run community houses.[45] I have no detailed information on life there, but it is clear that the character of the communities changed after Archway. "Along the years what we were offering changed from a model, which valued the free expression of emotions in whatever extreme form it might unleash, to one which prioritised containment and worked with respected concepts psychoanalysis was using widely, such as transference, repression and repetition. . . . In the end we were funded by local authorities, something Laing had always opposed. . . . One change we also faced was that psychotic episodes were not welcomed by the community but feared instead; the prospect of broken windows in the middle of winter, sleepless nights and keeping constant vigil were no longer an activity the community or the PA at large regarded as desirable. These days florid psychosis ends up controlled by psychiatric care" (Davenport 2005). As I said earlier, life at places like Kingsley Hall and Archway was not a picnic. In the sixties that was no deterrent; in the Thatcherite era it became insufferable.

As far as Laing himself is concerned, his life and work changed direction after Kingsley Hall shut down in 1970. In July 1971 he left England for the East, spending six months in a Buddhist monastery in Sri Lanka, where he was reported to spend seventeen hours a day in meditation, followed by seven months in India and a brief visit to Japan to study yoga and Zen (Howarth-Williams 1977, 5, 97). Returning to Britain, he was less actively involved with the Philadelphia Association, his writings lost their overtly social and political edge, and he became less of a public figure: "With hindsight it is easy to realize that by the late seventies Ronnie was becoming more and more marginalized" (A. Laing 1994, 202).[46] He remained as radical as ever in psychiatry, but in the early 1970s came to focus on "the politics of the birth process and the importance of intrauterine life. Inspired by the work of American psychotherapist Elizabeth Fehr, [he began] to develop a team of 'rebirthing workshops' in which one designated person chooses to re-experience the struggle of trying to break out of the birth canal represented by the remaining members of the group who surround him/her" (Ticktin n.d., 5). Laing's *The Facts of Life* (1976) elaborates on this new perspective, including the story of his first meeting with Fehr, in New York, on 11 November 1972, and emphasizing, for example, the effects of cutting of the umbilical cord: "I have seen a global organismic reaction occur the instant the cord is cut. It would appear to be neurologically impossible. There are no nerves in the umbilical cord. But it *does* happen. I've *seen* it happen" (1976, 73). Characteristically, however, the book gives no details of the practice of rebirthing.

What interests me about these shifts is that they parallel the moves at Archway, away from language and toward performance, revealing, and technologies of the nonmodern self.[47] Meditation as a technology for exploring the inner space of the therapist (echoing LSD ten years earlier), prelinguistic experiences in the womb and at birth as the site of psychiatric problems in later life (rather than communicative double binds), physical performance as therapy—staging rebirths. As I said about Archway, the detour through interpersonal communication also shrank to almost nothing in Laing's post–Kingsley Hall psychiatry. Like the residents at Archway, Laing's life and work came to stage a relatively pure form of a performative ontology.

PSYCHIATRY AND THE SIXTIES

We can turn now to the social basis of Laingian psychiatry up to and including Kingsley Hall and look at it from two angles—as psychiatry and as a part of the sixties. The first is important but can be covered briefly, since we have already gone over much of the ground. We have seen before in the cases of Walter, Ashby, and Bateson that ontology and sociology hang together, with practice as the linking term. All three men encountered mismatches with both the substance and the institutional forms of the fields they crossed, and in all three instances the response was to improvise some novel but transitory social base, picking up support wherever it could be found and setting up temporary communities transverse to the usual institutions, in dining clubs, conference series, and societies. Laingian psychiatry illustrates this socio-ontological mismatch starkly in the contrasting social forms of the Philadelphia Association communities and established mental hospitals, and in the problems experienced in embedding one form within another (especially at Cooper's Villa 21). Beyond that, though, these communities displayed a rather different kind of accommodation to the social mismatch between cybernetics and modernity. Instead of an opportunistic and ad hoc response, the Philadelphia Association paid serious attention to the social side of its work, and attempted, at least, to create at Kingsley Hall and Archway an entirely new and stable social basis where its form of psychiatry could sustain and reproduce itself outside the established system. Like the Santa Fe Institute, Bios, and Wolfram's NKS initiative in later years (chap. 4), Kingsley Hall, we could say, sketched out another future at the social and institutional level as well as that of substantive practice. Kingsley Hall was an instance of, and helped one to imagine, a wider field of institutions existing as a sort of parallel world relative to the mainstream of Western psychiatry and so on.[48] We can continue this thought below.

— — — — —

Now for the sixties. I have concentrated so far on the specifically psychiatric aspects of Laing's (and Bateson's) work. But, of course, I might just as well have been talking about the psychedelic sixties and the counterculture. Altered states; technologies of the self; an idea of the self, social relations, and the world as indefinitely explorable; a notion of symmetric and reciprocal adaptation rather than hierarchical power relations; experimentation with novel forms of social organization; a mystical and Eastern spirituality as a counterpoint to madness—all of these were just as much the hallmarks of the sixties as they were of cybernetic psychiatry. And the basic point I want to stress is that these resonances are again markers of the ontological affinity that we have met before between other strands of cybernetics and the sixties. In different ways, Laingian psychiatry and the counterculture (and situated robotics, complexity theory, and Ashby on planning) staged much the same ontological vision: of the world as a multiplicity of exceedingly complex systems, performatively interfering with and open-endedly adapting to one another.

In this instance, though, we can go beyond ideas of resonance and affinity. One can argue that Laing and his colleagues had a constitutive role in shaping the counterculture itself; they helped to make it what it was. I want to examine this role briefly now as the furthest I can go in this book in tying cybernetics and the sixties together. It is ironic that this example concerns technical practice that did not explicitly describe itself as cybernetic, but I hope I have said enough about Laing and his colleagues and followers to make the case for seeing their work as a continuation of cybernetics, as playing out the basic cybernetic ontology.

We can begin with Laing himself. It is said that in the late 1960s Laing was the best-known psychiatrist in the world. I know of no hard evidence to support that, but it certainly points toward his prominence (and to the fact that the sixties were perhaps the last time the world was much interested in psychiatry—as distinct from pharmaceuticals). And his public fame and notoriety derived from his writings, rather than his day-to-day practice, and especially from his 1967 book *The Politics of Experience*.[49] *Politics* therefore bears examination. The first few chapters are cogent, but hardly best-seller material. They run through social-constructivist and labeling theories of madness, drawing on Laing's earlier work as well as the work of now-eminent scholars such as the sociologists Howard Garfinkel and Erving Goffman. None of this seems remarkable in retrospect. The book explodes in its concluding three chapters, however, and these are the chapters in which Laing quotes Bateson on the "inner voyage" and then embroiders on the theme, focusing in particular on the inner experience of the voyage and "transcendental experience" (the title of chap. 6).

I mentioned chapters 5 and 6 before, and all that needs to be added concerns the book's last chapter—chapter 7, "A Ten-Day Voyage." Originally published in 1964 (before the establishment of Kingsley Hall), the chapter consists of extracts from Laing's tape recordings of his friend Jesse Watkins's recollections of a strange episode that he had lived through twenty-seven years earlier.[50] Watkins's experiences were certainly strange (Laing 1967, 148, 149, 153–54, 156, 158):

> I suddenly felt as if time was going back. . . . I had the—had the feeling that . . . I had died. . . . I actually seemed to be wandering in a kind of landscape with—um—desert landscape. . . . I felt as if I were a kind of rhinoceros . . . and emitting sounds like a rhinoceros. . . . I felt very compassionate about him [another patient in the hospital to which Watkins was taken for observation], and I used to sit on my bed and make him lie down by looking at him and thinking about it, and he used to lie down. . . . I was also aware of a—um—a higher sphere, as it were . . . another layer of existence lying above the—not only the antechamber but the present. . . . I had feelings of—er—of gods, not only God but gods as it were, of beings which are far above us capable of—er—dealing with the situation that I was incapable of dealing with, that were in charge and were running things and—um—at the end of it, everybody had to take on the job at the top. . . . At the time I felt that . . . God himself was a madman.

It is clear that for Laing, Watkins's voyage was a paradigm for the uninterrupted psychotic experience, a trip to another world from that of mundane reality, both wonderful and horrifying, even encompassing the acquisition of new and strange powers in the everyday world—Watkins's new-found ability to control his fellows by just looking and thinking. And if we want to understand the appeal of such writing to the sixties, we have only to think of the sixties' determined interest in "explorations of consciousness," and of *The Politics of Experience* as an extended meditation on that theme, with chapter 7 as an empirical example of where they might lead. Perhaps the best way to appreciate the wider significance of the book beyond psychiatry proper is to situate it as a major contribution to the countercultural canon, in which Aldous Huxley's glowing description of the mescaline experience in *The Doors of Perception* (1954) was a key landmark from the 1950s, shortly to be followed by Carlos Castaneda's otherworldly explorations in *The Teachings of Don Juan* (1968) and John Lilly's descriptions of his transcendental experiences in sensory deprivation tanks in *The Center of the Cyclone: An Autobiography of Inner Space* (1972).

At another and entirely nonliterary extreme, Laing's interest in LSD coupled with his psychiatric expertise gave him an important place in the London drug scene of the sixties. Laing's home, for example, figured as a place to take people who were having a bad trip. "I felt great all of a sudden and I didn't give a shit about Sarah any more. Ronnie was looking after her. The Man. I'd taken her to the Man. I went and lay on the bed and in the end it was the greatest trip I ever took." Syd Barrett, the leader of the early Pink Floyd, was progressively wiped out by acid, and his friends took him to see Laing, though without much effect. In 1964, Laing visited the acid guru of the U.S. East Coast, Timothy Leary, who returned the compliment some years later, remarking, "You will not find on this planet a more fascinating man than Ronald Laing."[51]

Beyond Laing's individual connections to the sixties, Kingsley Hall was a key institutional site for the counterculture. Experimental artists and composers would go there and perform, including Cornelius Cardew and Allen Ginsberg, and here some circles begin to close. We have already met Ginsberg taking acid for the first time in a flicker-feedback setup when we examined the connections between Walter's cybernetics and the Beats.[52] Cardew gets a mention in the next chapter, as assimilated to the cybernetic musical canon by Brian Eno, himself much influenced by Stafford Beer. One begins to grasp a significant intertwining of cybernetic and countercultural networks in the sixties, though it would be another project to map this out properly. Kingsley Hall also figures prominently as a countercultural meeting place in *Bomb Culture*, Jeff Nuttall's classic 1968 description and analysis of the British underground. Going in the opposite direction, the Kingsley Hall community would issue forth en masse to countercultural events, including the famous 1965 poetry reading at the Albert Hall, one of the formative events of the British underground scene. "Ronnie Laing decanted Kingsley Hall patients for the night, thought they'd have a good evening out. Real schizophrenics running around the flat bit in the middle. . . . All the nutcases ran jibbering round," was one rather uncharitable description (Sue Miles, quoted in J. Green 1988, 72).

Kingsley Hall was also for a while a blueprint for another institutional future. In 1967 Laing left the Tavistock Institute and he and David Cooper, Joseph Berke, and Leon Redler founded the Institute of Phenomenological Studies (Howarth-Williams 1977, 5), which in turn sponsored the establishment of the Anti-University of London with an interest-free loan.[53] The Anti-University opened its doors on 12 February 1968 (Howarth-Williams 1977, 71): "Again [at the Anti-University] we find a concern to break down internal role structures. Foremost amongst its aims . . . was 'a change in social relations among people.' Primary amongst these relations was, of course, that of staff/student. Although there were lecturers (who were only paid for the first term), the emphasis was on active participation by all. . . . There were, of course, no exams, and fees were minimal (£8 a quarter plus 10s per course attended; goods or services were accepted in lieu of cash . . .)." In our terms, the Anti-University was an attempt to extend the symmetric cybernetic model from psychiatry and Kingsley Hall to the field of higher education. Laing gave lectures there on psychology and religion, "specifically on the accounts of 'inner space' to be found in various mythologies and religions" (Howarth-Williams 1977, 71). "Huxley gave a course on dragons and another on how to stay alive; Laing gave a course; there were courses on modern music from Cornelius Cardew; Yoko Ono did a course; I taught advanced techniques for turning on and all my students had prescriptions for tincture. I'd give them lessons on joint-rolling and so on" (Steve Abrams, quoted in J. Green 1988, 238).[54] Interestingly, like Kingsley Hall, "The Anti-University had a commune associated with it; a significant number of the prominent members lived in it, and the two 'institutions' became synonymous. Indeed, it seems to have been one of the major lessons learned from the Anti-University that such enterprises need the domestic stability plus intimacy yet fluidity of a commune to flourish" (Howarth-Williams 1977, 71).[55] The Anti-University was, however, short lived: "[It] was a wonderful place. It provided a platform for people who didn't have one, to lecture and talk. Either people who didn't have a platform at all or who had perhaps an academic appointment and could only lecture on their own subject. . . . The students were almost anybody: it was £10 to register for a course and it went on for a year or two. But in the second year or the second term it started getting out of hand. The idea became to charge the teachers and pay the students" (Abrams, quoted in J. Green 1988, 238).

In 1967 the Institute of Phenomenological Studies also sponsored a "Dialectics of Liberation" conference, held at the Roundhouse in London. This was an important countercultural gathering—"the numero uno seminal event of '67"—which brought together over a period of two weeks in July many of the era's luminaries, including Allen Ginsberg, Gregory Bateson (the only card-carrying cybernetician), Emmett Grogan, Simon Vinkenoog, Julian Beck, Michael X, Alexander Trocchi, Herbert Marcuse, and Timothy Leary, "in order to figure out what the hell is going on" (Howarth-Williams 1977, 69, quoting Joe Berke).[56] The meeting itself was not a great success. Laing famously fell out with Stokely Carmichael of the Black Panthers over the political role of hippies. Laing felt that the hippies were already acting out a nonviolent revolution at the level of communal lifestyles; Carmichael later replied, "You will have to throw down your flowers and fight" (Howarth-Williams 1977, 73).[57]

In Britain, at least, the counterculture had little chance of success when it came to a literal fight with the establishment, and at the dialectics conference

Laing argued instead for the possibility of radical political change at a meso-social level (Laing 1968b, 16): "In our society, at certain times, this interlaced set of systems may lend itself to revolutionary change, not at the extreme micro or macro ends; that is, not through the individual pirouette of solitary repentance on the one hand, or by a seizure of the machinery of the state on the other; but by sudden, structural, radical qualitative changes in the intermediate system levels: changes in a factory, a hospital, a school, a university, a set of schools, or a whole area of industry, medicine, education, etc." Kingsley Hall, of course, was a real example of what Laing was talking about, a new form of midlevel institution that enacted a novel set of social arrangements outside the established institutional framework; the Anti-University followed the next year.

To wrap this discussion up, I can note that the most systematic institutional theorist of the counterculture in Britain was Alexander Trocchi—Laing's friend and fellow Glaswegian. From the early 1960s onward, Trocchi laid out a vision of what he called sigma (for "sum," and favoring the lowercase Greek symbol), an institutional form that would link together existing countercultural institutions and accelerate their propagation. Trocchi was clear that the aim should indeed be a sort of parallel social universe, through which the counterculture could supersede rather than take over older institutional forms: "History will not overthrow national governments; it will outflank them." As exemplified at Kingsley Hall and later Archway, and again instantiating the cybernetic ontology of exceedingly complex systems, Trocchi imagined sigma as a site for the endless emergence of nonmodern selves: "We must reject the fiction of 'unchanging human nature.' There is in fact no such permanence anywhere. There is only *becoming*." Concretely,

> at a chosen moment in a vacant country house (mill, abbey, church or castle) not too far from the City of London we shall foment a kind of cultural "jam session;" out of this will evolve the prototype of our *spontaneous university*. The original building will stand deep within its own grounds, preferably on a river bank. It should be large enough for a pilot group (astronauts of inner space) to situate itself, orgasm and genius, and their tools and dream-machines and amazing apparatus and appurtenances; with outhouses for "workshops" large as could accommodate light industry; the entire site to allow for spontaneous architecture and eventual *town planning*. . . . We envisage the whole as a vital laboratory for the creation (and evaluation) of conscious *situations*; it goes without saying that it is not only the environment which is in question, plastic, subject to change, but men also.

Trocchi even thought about the economic viability of such a project, envisaging sigma as an agent acting on behalf of its associated artists and designers in their capacities as producers and consultants. A nice vision. In *Bomb Culture* (1968, 220–27) Jeff Nuttall records an abortive 1966 "conference" aimed at launching sigma. Attended by several of the principals of Kingsley Hall (Laing, Esterson, Cooper, Berke, and Sigal) as well as Nuttall, Trocchi, and several others, the meeting staged again the usual socio-ontological clash. On the one side, "the community at Brazier's Park, a little colony of quiet, self-sufficient middle-class intellectuals, totally square with heavy overtones of Quakerism and Fabianism, was anxious to extend every kindness and expected, in return, good manners and an observation of the minimal regulations they imposed" (222). On the other, amid much drinking, the artist John Latham built one of his famous skoob towers—a pile of books that he then set alight (226). Later, "in the house everybody was stoned. . . . Latham had taken a book (an irreplaceable book belonging to a pleasant little Chinese friend of Alex's) stuck it to the wall with Polyfilla, and shot black aerosol all over the book and wall in a black explosion of night. . . . At mid-day we fled from one another with colossal relief" (227).

Trocchi himself had noted that sigma "will have much in common with Joan Littlewood's 'leisuredrome' [and] will be operated by a 'college' of teacher-practitioners, with no separate administration."[58] Littlewood's leisure drome, otherwise known as the Fun Palace, was the most sustained attempt in the sixties to create such an experimental institution, and we can return to it with Gordon Pask in chapter 7.

ONTOLOGY, POWER, AND REVEALING

The key ontological thread that has run through this chapter, which we will continue to follow below, is the symmetric vision of a dance of agency between reciprocally and performatively adapting systems—played out here in psychiatry. We began with Bateson's ideas of schizophrenia as the creative achievement of a bad equilibrium and of psychosis as an index to a higher level of adaptation than could be modelled with homeostats and tortoises, and we followed the evolution of Bateson's thought into Laingian psychiatry, Kingsley Hall, and the Archway communities as ontology in action: the playing out of the symmetric cybernetic vision in psychiatric practice. We could say that Bateson and Laing were more cybernetic, in a way, than Walter and Ashby, in a quasi-quantitative fashion: Bateson and Laing understood psychiatric therapy as a two-way process, enmeshing the therapist as well as the patient; while for Walter, Ashby, and the psychiatric establishment, the process was only one-way: therapy as ideally an adaptation of the patient but not of the therapist. But as we have seen, this quantitative difference hung together with a dramatic contrast in practices. If the paradigmatic therapy of conventional psychiatry was ECT in the 1950s moving on to drugs in the sixties, the paradigm of the Bateson-Laing line was that of symmetric, open-ended, and reciprocal interaction.

We can phrase this in terms of *power*. The conventional mental hospital staged a linear model of power in a hierarchical frame of social relations: doctors, then nurses, then patients. The aim of Kingsley Hall was to set everyone on a level plane without any fixed locus of control. Of course, neither of these visions was perfectly instantiated. Formally hierarchical relations are always embedded in informal and transverse social relations; on the other hand, doctors, nurses and patients cannot easily escape from their traditional roles. Nevertheless, I hope to have shown above that these different ontological visions did indeed hang together with distinctly different practices and institutional forms. Ontology made a real difference here.

My last observation is that conceiving these differences in terms of a notion of power is not really adequate. The contrast between Kingsley Hall and a contemporary mental hospital did not lie simply in the fact that the "staff" of the former thought that hierarchy was *bad* in the abstract, or that it would be *nice* in principle not to exercise control over the "patients." Something more substantial was at stake, which can be caught up in the Heideggerian contrast between enframing and revealing. Conventional psychiatry, one could say, already knows what people should be like, and it is the telos of this sort of psychiatry to reengineer—to enframe—mental patients back into that image. That is why a hierarchical system of social relations is appropriate. Power relations and understandings of the self go together. The Bateson-Laing line, of course, was that selves are endlessly complex and endlessly explorable, and the antihierarchical approach of Kingsley Hall was deliberately intended to facilitate such exploration in both the mad and the sane. This is the mode of revealing, of finding out what the world has to offer us. We can, then, name this contrast in social relations in terms of power and hierarchy, but that is not enough. The sociological contrast echoed and elaborated a contrast in ontological stances—enframing versus revealing—which is, I think, very hard to grasp from the standpoint of modernity.[59]

PART TWO

BEYOND THE BRAIN

6

STAFFORD BEER

FROM THE CYBERNETIC FACTORY TO TANTRIC YOGA

Our topic changes character here. Grey Walter and Ross Ashby (and Gregory Bateson) were first-generation cyberneticians, born in the 1900s and active until around 1970. Stafford Beer and Gordon Pask were central figures in the second generation of British cybernetics, twenty years younger and active in cybernetics until their deaths in 2002 and 1996, respectively. What the two generations had in common was the defining interest in the adaptive brain. Where they diverged was in the question of how the brain fitted into their cybernetics. To a degree, Beer and Pask carried forward the attempt to build synthetic brains that they inherited from Walter and Ashby, in their work on biological and chemical computers discussed in this chapter and the next. Even there, however, the emphasis in Beer and Pask's work was not on understanding the brain per se, but in putting these "maverick machines," as Pask called them (Pask and Curran 1982, chap. 8), to work in the world. More generally, psychiatry was not a central concern for either Beer or Pask. Instead, they found inspiration in ideas about the adaptive brain in their extensions of cybernetics into new fields: Beer in his work in management and politics and even in his spiritual life; Pask in his work on training and teaching machines, and in the arts, entertainment, theater, and architecture. This is what interests me so much about the cybernetics of both men: the many projects they engaged in help us extend our range of examples of ontology in action. What also interests me is that, like Bateson and Laing, and unlike Ashby in his understanding of clinical psychiatry, Beer and Pask took the symmetric fork in the road. The referent of their cybernetics was always reciprocally adapting systems.

I should add that Beer and Pask were extraordinary individuals. Beer displayed fabulous energy and creativity. Reading a diary that he kept during his first visit to the United States, from 23 April to 12 June 1960, leaves one limp (Beer 1994 [1960]); his career in management was accompanied by awesome literary productivity (in terms of quality as well as quantity), mainly on cybernetic management and politics, though he was also a published poet (Beer 1977); he painted pictures, and some of his works were displayed in Liverpool Cathedral and elsewhere (Beer 1993b); he also taught tantric yoga, loved women, slept only briefly, and drank continually (white wine mixed with water in his later years).

After an outline biography, I turn to Beer's work in management and politics, focusing in turn on his early work on biological computers, his viable system model of organizations, and the team syntegrity approach to decision making. Then we can examine the spiritual aspects of Beer's cybernetics and the cybernetic aspects of his spirituality. The chapter ends with an examination of the relation between Beer's cybernetics and Brian Eno's music.

— — — — —

Stafford Beer was born in Croydon, near London, on 25 September 1926, nearly five years the elder of two brothers (his younger brother Ian went on to be headmaster of Harrow Public School and on his retirement wrote a book called *But, Headmaster!* [2001]).[1] Like Ashby, Walter, and Pask, Stafford had a first name that he never used—Anthony—though he buried it more deeply than the others. His brother's third name was also Stafford, and when Ian was sixteen, Stafford "asked me to sign a document to promise that I would never use Stafford as part of my name. I could use it as I. D. S. Beer, or, indeed, using the four names together but he wanted the 'copyright' of Stafford Beer and so it was forever more."[2] Early in World War II, their mother, Doris, took Stafford and Ian to Wales to escape the German bombing, and at school there Stafford met Cynthia Hannaway, whom he married after the war. In 1942 the family returned to England, and Stafford completed his education at Whitgift School, where "he was a difficult pupil as he was found to be unsuitable for certain Sixth Form courses or he demanded to leave them for another. He could not stand the specialization and talked all the time of holistic teaching and so on. He wanted to study philosophy but that was not taught at school. He was precocious to a degree. A letter written by him was published in the

Spectator or the Economist, no-one could understand it." He went on to study philosophy and psychology at University College London—which had then been evacuated to Aberystwyth, back in Wales—for one year, 1943–44.[3] At University College he swam for the college team and was English Universities backstroke champion as well as getting a first in his first-year examinations. In 1944 he joined the British Army as a gunner in the Royal Artillery. In 1945 he went to India as a company commander in the Ninth Gurkha Rifles and later became staff captain intelligence in the Punjab. In 1947 he returned to Britain, remaining with the Army as army psychologist with the rank of captain.

Back in England, Beer married Cynthia, and they had six children together, though the first was stillborn. Following a divorce, Beer married Sallie Steadman, a widow and mother of a daughter, Kate, and they had two more children, for a total of eight, but this marriage, too, ended in divorce, in 1996. From 1974 onward Beer lived alone in Wales for much of the year (see below). In 1981 he met and fell in love with another cybernetician, Allenna Leonard (then a mature graduate student and later president of the American Society for Cybernetics), and she was Beer's partner for the remainder of his life.

Leaving the army, Beer hoped to do a PhD in psychology at University College, but when told that he would have to recommence his studies as a first-year undergraduate he turned his back on the academic life, and in 1949 he began work for Samuel Fox in Sheffield, a subsidiary company of United Steel, where he created and ran its Operational Research Group (probably the first such group to exist in Britain outside the armed forces). From 1956 until 1961 he was head of the Operational Research and Cybernetics Group of United Steel, with more than seventy scientific staff based in the appropriately named (by Beer) Cybor House in Sheffield. In 1961 he founded Britain's first operational research consulting firm, SIGMA (Science in General Management). In 1966 he moved on to become development director of the International Publishing Corporation (IPC), then the largest publishing company in the world, where his work largely concerned future initiatives around computing and information systems. In 1970, Beer left IPC "following a boardroom disagreement about development policy." From 1970 until his death in Toronto on 23 August 2002 he operated as an independent consultant in a variety of arenas, some of which are discussed below.

Besides his career in management and consultancy, Beer was a prolific writer of scholarly and popular works, including more than two hundred publications and ten books on cybernetics, which he referred to as "ten pints

Figure 6.1. Beer as businessman. Source: Beer 1994a, facing p. 1. (This and other Beer images in this chapter, where otherwise unattributed, are courtesy of Cwarel Isaf Institute and Malik Management Zentrum St. Gallen [www.management.kybernetik.com, www.malik-mzsg.ch].)

of Beer" (Beer 2000). After 1970, he occupied many institutional roles and gained many honors. At different times he was president of the Operational Research Society, the Society for General Systems Research, and the World Organization of Systems and Cybernetics. He had several footholds in the academic world, though none of them full-time. His most enduring academic base was at the Business School of Manchester University, where he was visiting professor of cybernetics from 1969 to 1993. He was research professor of managerial cybernetics at University College Swansea from 1990 to 1997, visiting professor of management science at the University of Durham from 1990 to 1995, visiting professor of cybernetics at the University of Sunderland and life professor of organizational transformation at Liverpool John Moores University, both from 1997 until his death. And so on, including visiting professorships at many other universities in Britain, Canada, Sweden (Stockholm), and the United States dating from 1970 onward. He was awarded major prizes for his work in operations research and cybernetics by the Operations Research Society of America, the American Society for Cybernetics, the Austrian Society for Cybernetics, and the World Organization of Systems and Cybernetics. A festschrift in Beer's honor was published in 2004 (Espejo 2004), and two volumes of his key papers have also appeared (Beer 1994a; Whittaker 2009).

Figure 6.1 is a photograph of Beer in the early 1960s when he was director of SIGMA—the smartly trimmed hair and beard, the three-piece suit, the cigar: the very model of a successful English businessman. In the early 1970s, however, Beer changed both his lifestyle and appearance. Partly, no doubt, this was in disgust at events in Chile with which he had been deeply involved, culminating in the Pinochet coup in 1973 (as discussed below). But also, as he told me, approaching his fiftieth birthday, he was moved to take stock of his life—"I had had two wives, I had eight children, a big house and a Rolls-Royce"—and the upshot of this stock taking was that in 1974 Beer renounced material possessions and went to live in a small stone cottage in a remote part of Wales.[4] He retained the cottage for the rest of his life, but after the mideighties he divided his time between there and a small house he shared with Allenna Leonard in Toronto. This break in Beer's life was registered by a change in his appearance (fig. 6.2) and also in his writing style. Until this change, Beer's writing took a fairly conventional form. His first book in its wake was *Platform for Change: A Message from Stafford Beer*, printed on paper of four different colors, signaling different modes of argument and presentation. The introduction, printed on yellow paper, begins thus (Beer 1975, 1):

> HELLO
> I would like to talk to you
>   if you have the time
> in a new sort of way
> about a new sort of world.

It ends (6):

> I am fed up with hiding myself
>   an actual human being
> behind the conventional anonymity
> of scholarly authorship.

Figure 6.2. Beer after the move to Wales. Source: Beer 1994a, 315. Photo: Hans-Ludwig Blohm. © Hans-Ludwig Blohm, Canada.)

From Operations Research to Cybernetics

Beer's route into cybernetics began with his work in operations research (OR) which in turn grew out of his work in the British Army in India. We do not need to delve deeply into the history of OR, but some brief remarks are relevant. As its name suggests, OR developed in World War II as a scientific approach to military operations. "Scientific" is to be contrasted here with traditional approaches to tactical and strategic planning based on the accumulated expertise of military commanders, and wartime OR can be broadly characterized in terms of a quantifying spirit aimed at modelling military activities with an eye to optimizing performance. One could try to calculate, for example, the optimal U-boat search pattern to be flown by a specified number of aircraft of given speed and range. OR was first developed in Britain in conjunction with new radar technologies but was also taken to a high art in the United States.[5]

Beer was not himself involved in the wartime development of OR. On his own account, he rather wandered into it while he was in the army, first by attempting to use symbolic logic, which he had studied at University College, to organize large numbers of men into functioning systems.[6] He first heard of OR as a field on his return to England and plunged himself into it as he moved into civilian life. Two early papers, published in 1953 and 1954, for example, outline novel statistical indices for measuring the productivity of manufacturing processes which he developed and implemented at the Samuel Fox steel company. These papers have a very practical bent, including ideas on how the sampling of productivity should be done and how the information could be systematically and routinely collected, assembled, and presented. The aim of the indices in question was the ability to forecast how long it would take to perform any given operation, a topic of interest both to the managers and customers of the mill (Beer 1953, 1954).

Beer's career in OR was very successful, as is evident from the biographical sketch above, and OR continued to play an important part throughout his subsequent work, both as an employee and as a consultant. But at an early stage he began to look beyond it. The second of the OR papers just mentioned is largely devoted to the development and use of performance measures for individual production operations in the factory, but it concludes with a section entitled "The Future Outlook" (also the title of Grey Walter's novel in its English publication two years later) looking forward to the development of "models . . . which would embrace the whole complex manufacturing structure of, say, an integrated steelworks." Beer notes that such models would themselves be very complex to construct and use and mentions some relevant mathematical techniques already deployed by OR practitioners, including game theory and linear programming, before continuing, "Advances in the increasingly discussed subject of cybernetics, allied with the complex models mentioned, might result in a fully mechanized form of control based on the technique described here" (1954, 57).

What did cybernetics mean, in assertions like that, for Beer, and how did it differ from OR? This takes us straight back to questions of ontology and a concept that I have been drawing on all along, that of an exceedingly complex system. Here we need only return briefly to its origin. In his first book, *Cybernetics and Management* (1959), Beer distinguished between three classes of systems (while insisting that they in fact shaded into one another): "simple," "complex," and "exceedingly complex" (fig. 6.3). He gave six examples of the

SYSTEMS	*Simple*	*Complex*	*Exceedingly complex*
Deterministic	Window catch	Electronic digital computer	EMPTY
	Billiards	Planetary system	
	Machine-shop lay-out	Automation	
Probabilistic	Penny tossing	Stockholding	The economy
	Jellyfish movements	Conditioned reflexes	The brain
	Statistical quality control	Industrial profitability	THE COMPANY

Figure 6.3. Beer's classification of systems. Source: S. Beer, *Cybernetics and Management* (London: English Universities Press, 1959), 18.

first two types (subdividing them further into "deterministic" and "probabilistic" systems). Under "simple" came the window catch, billiards, machine shop layout, penny tossing, jellyfish movements, and statistical quality control; under "complex" we find electronic digital computers, planetary systems, automation, stockholding, conditioned reflexes and industrial profitability. What those examples have in common, according to Beer, is that they are in principle knowable and predictable, and thus susceptible to the methods of the traditional sciences. Physics tells us about billiard balls; statistics about penny tossing; OR about stockholding and industrial profitability—this last, of course, being especially relevant to Beer. OR was, then, a classical science of production, a science appropriate to those aspects of the world that are knowable and predictable, in the same space as modern physics. However, under "exceedingly complex" systems (which, according to Beer, can have only probabilistic forms) we find just three examples: the economy, the brain, and the company. And Beer's claim was that these are "very different" (Beer 1959, 17):

> The country's economy, for example, is so complex and so probabilistic that it does not seem reasonable to imagine that it will ever be fully described. The second, living, example—the human brain—is also described in this way. Moreover, it is notoriously inaccessible to examination. . . . Inferential investigations about its mode of working, from studies such as psychiatry and electroencephalography, are slowly progressing.
>
> Probably the best example of an *industrial* system of this kind is the Company itself. This always seems to me very much like a cross between the first

> two examples. The Company is certainly not alive, but it has to *behave* very much like a living organism. It is essential to the Company that it develops techniques for survival in a changing environment: it must adapt itself to its economic, commercial, social and political surroundings, and it must learn from experience.

Beer's exceedingly complex systems, were, then, as discussed already, in a different ontological space from the referents of OR (or physics). They were not fully knowable or adequately predictable, and they were "the province of cybernetics" (18). Beer's enduring goal was precisely to think about management cybernetically—to inquire into how one would run a company, or by extension any social organization, in the recognition that it had to function in and adapt to an endlessly surprising, fluctuating and changing environment.[7]

Toward the Cybernetic Factory

MY GOD, I'M A CYBERNETICIAN!

STAFFORD BEER, ON FIRST READING WIENER'S *CYBERNETICS* (BEER 1994C)

Beer first read Norbert Wiener's *Cybernetics* in 1950 and plunged into the field, establishing an individual presence in it and close personal connections as he went. By 1960, "I had known McCulloch for some years, and he would stay at my house on his Sheffield visits. . . . The British pioneers in cybernetics were all good friends—notably Ross Ashby, Frank George, Gordon Pask, Donald MacKay and Grey Walter" (Beer 1994 [1960], 229). "Norbert Wiener, as founder of cybernetics, was of course my great hero," but Beer did not meet him until his first trip to the United States when, on 25 May 1960, Wiener "almost vaulted over his desk to embrace me," greeting Beer with the words "I have become increasingly conscious that the growing reputation of my work [Wiener's] in Europe derives in large measure from your lectures and writings, and from the fact that you have built Cybor House. For this I should like to thank you" (Beer 1994 [1960], 281, 283).

In what follows, we will be largely concerned with connections between Beer's cybernetics and Ashby and Pask's. Beer and Pask actively collaborated in the work on biological and chemical computers discussed below and in the next chapter, and one can trace many parallels in the development of their work. But the defining features of Beer's cybernetics were Ashby's homeostat as key model for thinking about adaptive systems and Ashby's law of requisite variety, as a tool for thinking realistically about possibilities for adaptive control. Much of what follows can be understood as a very creative extension of Ashby's cybernetics into and beyond the world of organizations and management. During the 1950s, Beer experimented with a whole range of cybernetic approaches to management (e.g., Beer 1956), but two ideas quicky came to dominate his thinking. First, one should think of the factory (or any complex organization) in analogy with a biological organism. Second, and more specifically, to be adaptive within an unknowable environment, the factory as organism should be equipped with an adaptive brain.

Beer laid out an early and striking version of this vision in a paper he presented to a symposium on self-organization held at the University of Illinois's Allerton Park on the 8 and 9 June 1960 (Beer 1962a). He opened the discussion with the notion of the "automatic factory," then attracting great interest, especially in the United States. This was a vision of industrial automation taken, one might think, to the limit. In the automatic factory, not only would individual machines and productive operations be controlled by other machines without human interference, but materials would be automatically routed from one operation to the next. In the "lights out" factory, as it was sometimes called, the entire production process would thus be conducted by machines, and human labor made redundant—literally as well as metaphorically.[8]

Beer was not at this stage a critic of the automatic factory, except that he did not feel it was automatic enough. He compared it to a "spinal dog"—that is, a dog whose nervous system had been surgically disconnected from the higher levels of its brain. The automatic factory (1962a, 164) "has a certain internal cohesion, and reflex faculties at the least. [But] When automation has finished its work, the analogy may be pursued in the pathology of the organism. For machines with over-sensitive feedback begin to 'hunt'—or develop ataxia; and the whole organism may be so specialized towards a particular environment that it ceases to be adaptive: a radical change in the market will lead to its extinction." Beer's argument was that to make it adaptive and to avoid extinction in market fluctuations, the automatic factory would need a brain.

> At present, such an automatic factory must rely on the few men left at the top to supply the functions of a cerebrum. And . . . the whole organism is a strange one—for its brain is connected to the rest of its central nervous system at discrete intervals of time by the most tenuous of connections. The survival-value of such a creature does not appear to be high. . . .

> This will not do. The spinal dog is short of a built-in cerebrum; and the automatic factory is short of a built-in brain. The research discussed in this paper is directed towards the creation of a brain artefact capable of running the company under the evolutionary criterion of survival. If this could be achieved, management would be freed for tasks of eugenics; for hastening or retarding the natural processes of growth and change, and for determining the deliberate creation or extinction of whole species. (Beer 1962a, 165)

The reference to eugenics is provocative to say the least, but the idea is an interesting one. The cybernetic factory, as Beer imagined it, would be *viable*—a key term for Beer: it would react to changing circumstances; it would grow and evolve like an organism or species, all without any human intervention at all. The role of humans in production would thus become that of metamanagement—managers would survey the field of viable production units and decide on which to promote or retard according to metacriteria residing at a level higher than production itself. Figure 6.4 is Beer's schematic vision of what the cybernetic factory should look like. and much of his essay is devoted

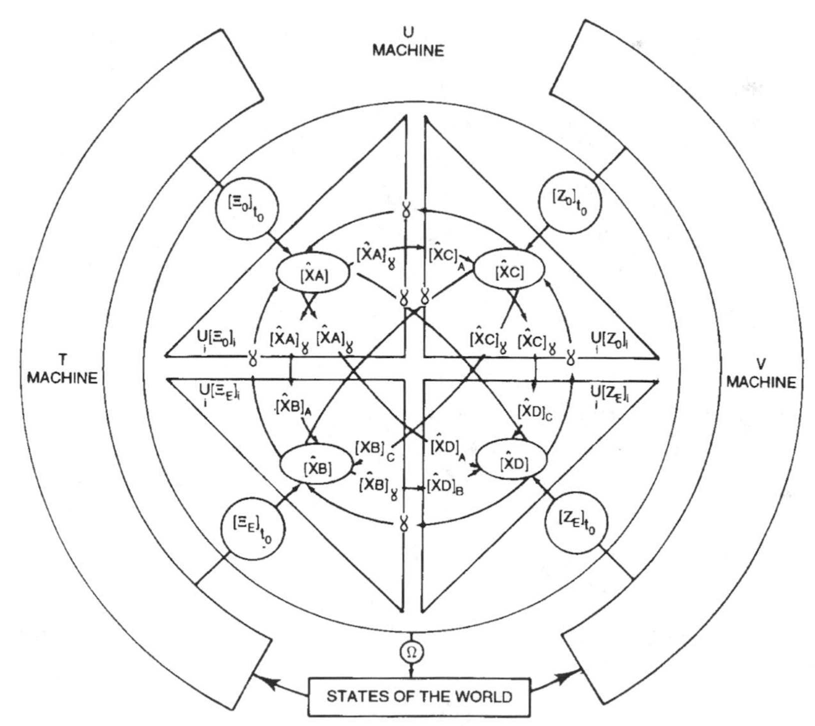

Figure 6.4. Schematic of the cybernetic factory. Source: Beer 1994a, 192.

to a formal, set-theoretic definition of its contents and their relations. This is not the place to go into the details of the formalism; for present purposes, the important components of the diagram are arranged around the circumference: the T- and V-machines at the left and right, bridged by the U-machine and "states of the world" at the bottom. The symbols within the circumference represent processes internal to the U-machine.

Beer envisaged the T-machine as something like Pitts and McCulloch's scanning device (Pitts and McCulloch 1947, discussed in chap. 3) updated in the light of more recent neurophysiological research. The "senses" of the T-machine would be numerical inputs representing the state of the factory's environmment (supplies and orders, finance) and its internal state (stocks, performance measures, etc.). The function of the T-machine was "scansion, grouping and pattern recognition" (Beer 1962a, 173). It would, that is, turn atomistic raw data into a meaningful output, in much the same way as the human brain picks out "universals" from our sensory data. The V-machine was conceived essentially as a T-machine running backward. Its inputs would be framed in the language of T-machine outputs; its outputs would be instructions to the motor organs of the plant—directing production operations and flows, ordering stock, or whatever.

Between the T- and V-machines lay, yes, the U-machine. The U-machine was to be "some form of Ashbean ultrastable machine" (Beer 1962a, 189)—a homeostat, the brain artifact of the firm. The job of the U-machine was continually to reconfigure itself in search of a stable and mutually satisfactory relationship between the firm and its environment. The U-machine was thus the organ that would enable the factory to cope with an always fluctuating and changing, never definitively knowable environment. It was the organ that could take the automatic factory to a level of consciousness beyond that of a spinal dog. Figure 6.5 summed up Beer's abstract presentation, accompanied by the words "The temptation to make the outline look like a coronal section of the living brain was irresistible and I apologize to cerebra everywhere for such insolence" (197).[9]

The second major section of Beer's essay was a progress report on how far he had gone toward realizing a cybernetic factory at the Templeborough Rolling Mills, a division of United Steel engaged in the manufacture of steel rods.[10] This can help us think more concretely about the cybernetic factory, and here we need to refer to figure 6.6. The top level of the diagram represents various material systems relating to the flow of steel within the plant and their interconnections: the "Supplying system" feeds the "Input stocking system" which feeds the "Producing system." and so on. The next level down,

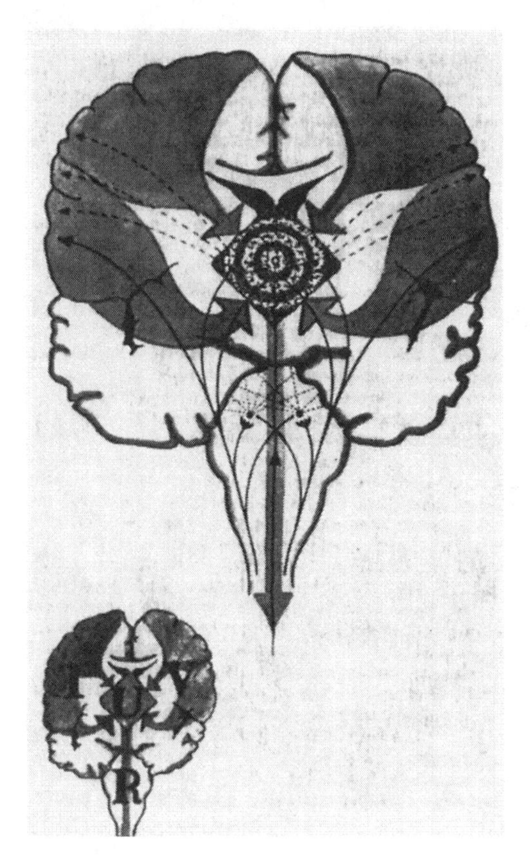

Figure 6.5. The cybernetic factory as brain. Painting by Stafford Beer. The *T*, *U*, and *V* machines are labeled on the smaller painting in the bottom left. Source: Beer 1994a, 198, fig. 3.

"Sensations," is the most important. Nineteen "sensations" are shown in the diagram, running from "a. tons bought" to "s. tons requested." Each of these sensations should be understood as taking the form of numerical data relating to aspects of the plant or its environment—the current state of production, the profit and loss account, the balance sheet, as shown in lower levels of the figures. The "sensation" aspect of this diagram relates to the T-machine of Beer's formal discussion, and his claim was to have sufficiently *simulated* a T-machine to make it clear that an automatic one could be built. The grouping of data into nineteen categories, for example, entailed "a large number of decisions . . . which, ultimately, the brain artefact itself is intended to take by its multiple multiplexing techniques. The research team in the field has, however, taken these decisions on an informed basis, by operational research methods" (Beer 1962a, 202).

The "sensations," then, were to be considered inputs to the T-machine, and further numerical transformations were supposed to correspond to the

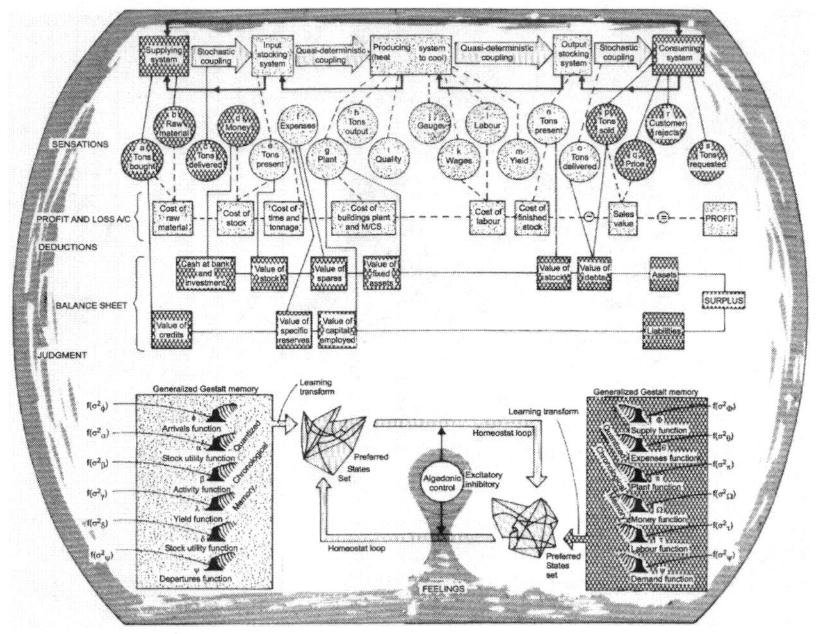

Figure 6.6. The steel mill as cybernetic factory. Source: Beer 1994a, 200–201, fig. 4.

functioning of "the T-Machine proper" (Beer 1962a, 203). These transformations, derived in practice from OR studies, first recombined the nineteen sensations into twelve "functions"—six referring primarily to the company and six to its environment. The functions all depended on ratios of expected behavior to actual behavior of precisely the form of the indices developed in Beer's earlier OR work, discussed above. "This last point," Beer wrote (204–5),

> is important, since it incorporates in this exemplification the essential "black box" treatment of unknowns and imponderables common to all cybernetic machines. For a model of performance in any field may be inadequate: predictions and judgements based upon it will be effectual only insofar as the model *is* adequate. But in exceedingly complex and probabilistic systems *no* analytic model can possibly be adequate. The answer to this paradox, which I have used successfully for 10 years, is to load the raw predictions of any analytic model with a continuous feedback measuring its own efficiency as a predictor. In this way, everything that went unrecognized in the analytic work, everything that proved too subtle to handle, even the errors incurred in making calculations, is "black boxed" into an unanalyseable weighting which is error-correcting.

Here, then, we have an example of one way in which Beer's cybernetics tried to handle the unknown—a predictor that reviewed its own performance in the name of predicting better.[11]

The values of the twelve parameters were measured daily in the steel mill and "were plotted on boards in an Operations Room for the benefit of management, as a by-product of this research" (Beer 1962a, 205). A plot of a year's readings is shown in figure 6.7, which Beer referred to as an encephalogram (205). He was reaching here for a suggestive connection between his work in management and brain science à la Grey Walter, referring to emergent periodicities in the data and noting that the "encephalographer finds this structural component of information (the brain rhythm) of more importance than either its amplitude or voltage" (182). This tempting idea seems to have proved a red herring, alas; I am not aware of any subsequent development of it, by Beer or anyone else. Several other, readily automatable statistical and mathematical transformations of these data then followed, and the work of the T-machine, as simulated at Templeborough, was said to be complete. Given that "the T-Machine was said to be set-theoretically equivalent to a V-Machine," the problem of constructing the latter could be said to have been shown to be soluble, too (208). But figure 6.4 also shows the intervention of the U-machine, the homeostatic brain, into the life of the cybernetic factory: what about that?

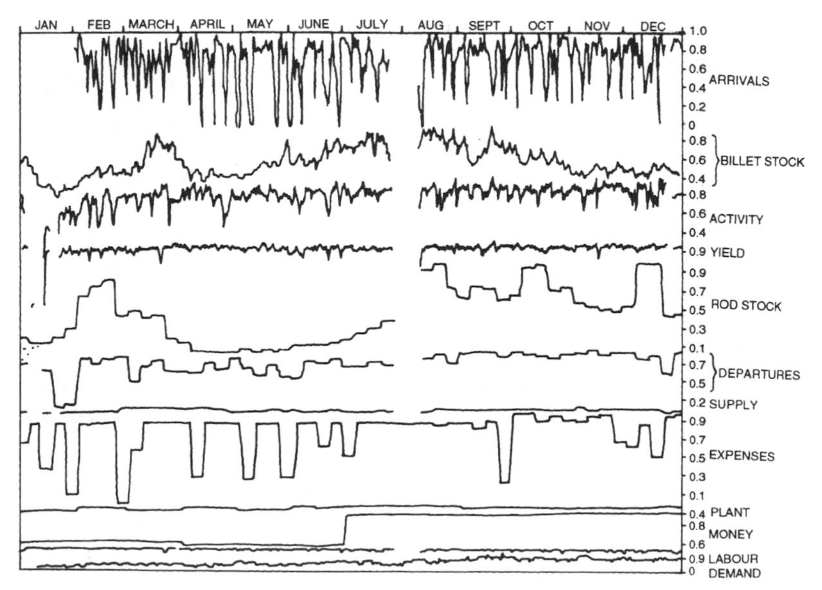

Figure 6.7. An EEG of the firm. Source: Beer 1994a, 206, fig. 5.

The outputs of the simulated T-machine in successive time steps were recorded at Templeborough as a "generalized gestalt memory" indicated in the lower left and right of figure 6.6, the left portion relating to inner states of the factory, the right to its environment. These memories could be thought of defining "two phase spaces in which the company and the environment can respectively operate." And the U-machine was intended to search for a set of "preferred states" within this space via a "mutually vetoing system by which the homeostatic loop in the diagram continues to operate until both company and environmental points in phase-space (representing vectors of functions) lie in the appropriate preferred states set" (Beer 1962a, 208).[12] This notion of mutual or reciprocal vetoing was very important in Beer's work (and Pask's), so I want to digress briefly here to explain it.

The idea of mutual vetoing came directly from Ashby's cybernetics, and here Beer, like Bateson and Pask, took the symmetric fork in the road. Imagine an interconnected setup of just two of Ashby's homeostats, both of which are free to reconfigure themselves. Suppose homeostat 1 finds itself in an unstable situation in which its essential variable goes out of whack. In that case, its relay trips, and its uniselector moves to a new setting, changing the resistance of its circuit. Here one can say that homeostat 2—with its own internal parameters that define the transformation between its input from and output to homeostat 1—has *vetoed* the prior configuration of homeostat 1, kicking it into a new condition. And likewise, of course, when homeostat 2 finds itself out of equilibrium and changes to a new state, we can say that homeostat 1 has vetoed the first configuration of homeostat 2. Eventually, however, this reconfiguration will come to an end, when both homeostats achieve equilibrium at once, in a condition in which the essential variables of both remain within limits in their mutual interactions. And this equilibrium, we can then say, is the upshot of a reciprocal vetoing: it is the condition that obtains when the vetoing stops and each machine finds a state of dynamic equilibrium relative to the other's parameters.

This is enough, I think, to unravel the above quotation from Beer. One can think of the U-machine and the firm's environment as two reciprocally vetoing homeostats, and the U-machine itself attempts to find a relation between its inputs from the T-machine and its outputs to the V-machine that will keep some essential variable standing for the "health" of the company within limits. Beer never reached the stage of defining exactly what that essential variable should be at this stage in his work. For the sake of concreteness, we could imagine it as a measure of profitability, though Beer proposed interestingly different measures in subsequent projects that we can review below.

It was clear enough, then, what the U-machine should do, though in 1960 Beer still had no clear vision of how it should be made, and at Templeborough "management itself," meaning the actual human managers of the plant, "plays the role of the U-Machine" (Beer 1962a, 208). The state of the art was thus that by that date a cybernetic factory had been simulated, though not actually built. Beer was confident that he could construct automated versions of the T-machine, as the factory's sensory organ, and the V-machine, as its motor-organ equivalent. Neither of these had actually been constructed, but their working parts had been simulated by OR studies and data collection and transformation procedures. The U-machine, which figured out the desirable place for the factory to sit in the factory-environment phase space, continued to be purely human, simulated by the managers who would review the "gestalt memory" generated by the T-machine and figure out how to translate that into action via the inputs to the V-machine. The U-machine, then, was the key (209):

> As far as the construction of cybernetic machinery is concerned, it is clear that the first component to transcend the status of mere exemplification must be the U-Machine. For exemplifications of T- and V-input are already available, and can be fed to a U-Machine in parallel with their equivalent reporting to management. . . . Having succeeded in operating the cybernetic U-Machine, the research will turn to constructing cybernetic T- and V-Machines. . . . After this, management would be free for the first time in history to manage, not the company in the language of the organism, but the T-U-V(R) control assembly in a metalanguage.

But what was the U-machine to be? Beer ended his talk at Allerton Park with the words "Before long a decision will be taken as to which fabric to use in the first attempt to build a U-Machine in actual hardware (or colloid, or protein)" (212). Colloid or protein?

Biological Computing

Beer's thinking about the U-machine was informed by some strikingly imaginative work that he and Pask engaged in in the 1950s and early 1960s, both separately and together—work that continued Ashby's goal of a synthetic brain but with an original twist. Ashby had built an adaptive electromagnetic device, the homeostat, which he argued illuminated the go of the adaptive brain. Following his lead, Beer and Pask realized that the world is, in effect, already full of such brains. Any adaptive biological system is precisely an adaptive brain in this sense. This does not get one any further in understanding how the human brain, say, works, but it is an observation one might be able to exploit in practice. Instead of trying to build a superhomeostat to function as the U-machine—and Beer must have known in the mid-1950s that Ashby's DAMS project was not getting far—one could simply try to enroll some naturally occurring adaptive system as the U-machine. And during the second half of the 1950s, Beer had accordingly embarked on "an almost unbounded survey of naturally occurring systems in search of materials for the construction of cybernetic machines" (Beer 1959, 162). The idea was to find some lively system that could be induced to engage in a process of reciprocal vetoing with another lively system such as a factory, so that each would eventually settle down in some agreeable sector of its environment (now including each other).

In 1962 Beer published a brief and, alas, terminal report on the state of the art, which makes fairly mind-boggling reading (Beer 1962b), and we can glance at some of the systems he discussed there to get a flavor of this work. The list begins with quasi-organic electrochemical systems that Beer called "fungoids," which he had worked on both alone and in collaboration with Pask. This was perhaps the aspect of the project that went furthest, but one has to assume Pask took the lead here, since he published several papers in this area in the late 1950s and early 1960s, so I postpone discussion of these systems to the next chapter. Then follows Beer's successful attempt to use positive and negative feedback to train young children (presumably his own) to solve simultaneous equations without teaching them the relevant mathematics—to turn the children into a performative (rather than cognitive) mathematical machine. Beer then moves on to discuss various thought experiments involving animals (1962b, 28–29):

> Some effort was made to devise a "mouse" language which would enable mice to play this game—with cheese as a reward function. . . . In this way I was led to consider various kinds of animal, and various kinds of language (by which I mean intercommunicating boxes, ladders, see-saws, cages connected by pulleys and so forth). Rats and pigeons have both been studied for their learning abilities. . . . The Machina Speculatrix of Grey Walter might also be considered (with apologies to the organic molecule). . . . However no actual machines were built. . . . By the same token, bees, ants, termites, have all been systematically considered as components of self-organizing systems, and various "brainstorming" machines have been designed by both Pask and myself. But again none has been made.

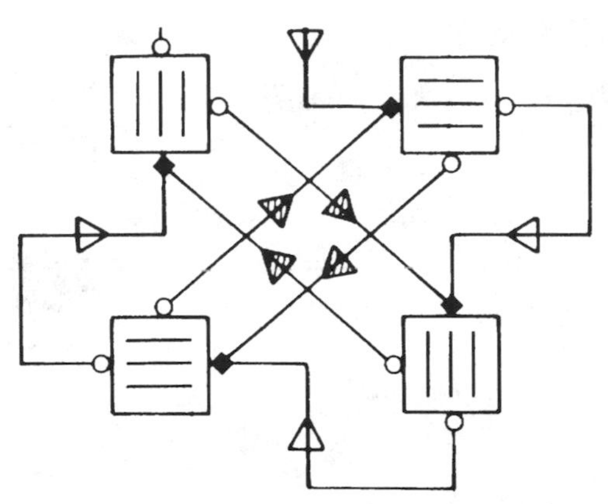

Figure 6.8. The *Euglena* homeostat. *Square, Euglena* culture, with tropism displayed as shown; *solid diamond,* stimulus; *circle,* sensory receptor; *hatched triangle,* inhibiting influence, and, *open triangle,* stimulating influence, of *a*'s sensation on *b*'s stimulus. Source: Beer 1994a, 30, fig. 2.

Beer had, however, devoted most of his own efforts to systems composed from simpler organisms: colonies of *Daphnia*, a freshwater crustacean (Pask had considered *Aedes aegypti*, the larva of the yellow fever mosquito), of *Euglena* protozoa, and an entire pond ecosystem. The key question with all three systems was how to interest these biological entities in us, how to couple them to our concerns, how to make a U-machine that would respond to and care about the state of the cybernetic factory. And this coupling was where Beer's attempts foundered (1962b, 29):

> Many experiments were made with [*Daphnia*]. Iron filings were included with dead leaves in the tank of *Daphnia*, which ingested sufficient of the former to respond to a magnetic field. Attempts were made to feed inputs to the colony of *Daphnia* by transducing environmental variables into electromagnets, while the outputs were the consequential changes in the electrical characteristics of the phase space produced by the adaptive behaviour of the colony. . . . However, there were many experimental problems. The most serious of these was the collapse of any incipient organization—apparently due to the steadily increasing suspension of tiny permanent magnets in the water.

Euglena are sensitive to light (and other disturbances) in interesting ways, and Beer sought to achieve optical couplings to a tank full of them "using a point source of light as the stimulus, and a photocell [to measure the absorption of light by the colony] as the sensory receptor" (fig. 6.8).

> However, the culturing difficulties proved enormous. *Euglena* showed a distressing tendency to lie doggo, and attempts to isolate a more motile strain failed. So pure cultures were difficult to handle. Moreover, they are not, perhaps, ecologically stable systems. Dr. Gilbert, who had been trying to improve the *Euglena* cultures, suggested a potent thought. Why not use an entire ecological system, such as a pond? . . . Accordingly, over the past year, I have been conducting experiments with a large tank or pond. The contents of the tank were randomly sampled from ponds in Derbyshire and Surrey. Currently there are a few of the usual creatures visible to the naked eye (*Hydra*, *Cyclops*, *Daphnia*, and a leech); microscopically there is the expected multitude of micro-organisms. [The coupling is via light sources and photocells, as in the *Euglena* experiments.] . . . The state of this research at the moment is that I tinker with this tank from time to time in the middle of the night. (Beer 1962b, 31–32)

Clearly, however, Beer failed to enroll the pond ecosystem, too, as a U-machine. The cybernetic factory never got beyond the simulation stage; we do not live in a world where production is run by *Daphnia* and leeches, and Beer's 1962 status report proved to be a requiem for this work. I now want to comment on it ontologically and sociologically, before moving on to later phases in Beer's career in management.

Ontology and Design

The sheer oddity of trying to use a pond to manage a factory dramatizes the point that ontology makes a difference. If one imagines the world as populated by a multiplicity of interacting exceedingly complex systems, as modelled by Ashby's homeostats, then one just might come up with this idea. It follows on from what has gone before, though even then some sort of creative leap is required. In contrast, it is hard to see how one would ever come to think this way from a modern technoscientific perspective. One would think instead of trying to program a computer to do the job of management, but that is a very different approach, in ways that are worth pondering.

We could start with issues of representation and performance. In the discussion that followed Beer's presentation at the 1960 Allerton conference, Beer made an interesting contrast between digital and biological computing in just these terms. When the subject of the former came up, he remarked that "this analogy with computers I do not like for two reasons." One had to do with the dynamics of memory and whether memory should be understood like the storage of "a parcel in a cloakroom" or as a "path of facilitation through phase space." The other went like this (1962a, 220–21):

> The big electronic machines . . . are preoccupied with digital access. Now why is this? It is always possible, given an output channel which you can fit on somewhere, to say what is happening just there, and to get an enormous printout. Now we [Beer and Pask] are not concerned with digital access, but with outcomes. Why do we pay so much money to make it [digital output] available? In the sort of machines that Gordon and I have been concerned with, you cannot get at the intermediate answer. If you take out [one?] of Gordon's dishes of colloid, you may be effectively inverting a matrix of the order 20,000. The cost of the computer is perhaps 10 cents. The only trouble is you do not know what the answer is. Now this sounds absurdly naïve, but it is not, you know, because you do not *want* the answer. What you do want is to *use* this answer. So why ever digitise it?

We are back to the notion of representation as a detour away from performance. Digital computing, in this sense, is an enormous detour away from its object—the functioning of a factory for example—into and through a world of symbols. In the previous chapter we discussed the discovery at Kingsley Hall and Archway that this detour could be drastically shortened or even done away with in therapeutic practice. But Beer *started* from this realization: in a world of exceedingly complex systems, for which any representation can only be provisional, performance is what we need to care about. The important thing is that the firm adapts to its ever-changing environment, not that we find the right representation of either entity. As ontological theater, then, Beer and Pask's biological computers stage this performative ontology vividly for us, dispensing entirely with representation, both exemplifying an ontology of sheer performance and indicating how one might go on in computing if one took it seriously. I could note here that this concern for performance and a suspicion of representation per se is a theme that ran through all of Beer's work.[13]

There is second and related sense of a detour that also deserves attention here. As Beer put it (1962a, 209, 215), "As a constructor of machines man has become accustomed to regard his materials as inert lumps of matter which have to be fashioned and assembled to make a useful system. He does not normally think first of materials as having an intrinsically high variety which has to be constrained. . . . [But] we do not want a lot of bits and pieces which we have got to put together. Because once we settle for [that], we have got to have a blueprint. We have got to design the damn thing; and that is just what we do not want to do." The echoes of Ashby on DAMS and the blueprint attitude are clear. We are back to the contrasting conceptions of design that go with the modern ontology of knowable systems and the cybernetic ontology of unknowable ones. Within the frame of modern science and engineering, design entails figuring out what needs to be done to achieve some result and then arranging "inert lumps of matter" to achieve those specifications. Digital computers depend on this sort of design, specifying material configurations right down to the molecular level of chemical elements on silicon chips. Beer's idea instead was, as we have seen, to find lively (not inert) chunks of matter and to try to enroll their agency directly into his projects. This gets us back to the discussion of the hylozoist quality of biofeedback music (chap. 3) and the idea that it's all there already in nature (as in the extraction of music from the material brain). We could say that the modern stance on design has no faith in matter and relies upon human representations and agency to achieve its effects. The cybernetic ontology, as Beer staged it, entailed a faith in the agency of matter: whatever ends we aim at, some chunk of nature probably already exists that can help us along the way. We don't need these long detours through modern design. We can explore Beer's hylozoism further later in the chapter in a broader discussion of his spirituality.

There is, of course, yet a third sense of detour that comes to mind here. The mastery of matter, from the molecular level upward, required to build a digital computer has been painstakingly acquired over centuries of technscientific effort. Beer's argument was, in effect, that perhaps we didn't need to make the trek. Just to be able to suggest that is another striking manifestation of the difference that ontology makes.

Now, Heidegger. It makes sense to see modern computer engineering as operating in the mode of enframing. It is not that semiconductor engineers, for example, have actually achieved some magical mastery over matter. For all their representational knowledge, they remain, like the rest of us, in medias res, obliged to struggle with the performance of obstinate stuff (Lécuyer and Brock 2006). Nevertheless, a successful chip is one that fits in with our preconceived plans: the chip either manipulates binary variables in a regular fashion, or it does not—in which case it is junk. Bending matter to our will like that is just what Heidegger meant by enframing. And then we can begin, at least, to see that the cybernetic ontology in this instance has more in common with a stance of revealing. Beer wanted to find out what the world—assemblages of mice, *Daphnia*, his local pond—could offer us. Against this, one might argue that Beer had some definite end in view: a replacement for the human manager of the factory. But the important point to note is that the pond was not envisaged as an identical substitute for the human. We will see in the next chapter that Pask, who thought this through in print further than Beer, was clear that biological computers would have their own management style, not identical to any human manager—and that we would, indeed, have to find out what that style was, and whether we could adapt to and live with it. This is the sense in which this form of cybernetic design in the thick of things is a stance of revealing rather than enframing.

One last thought in this connection. Somewhere along the line when one tries to get grips with Beer on biological computing, an apparent paradox surfaces. Beer's goal, all along, was to improve management. The cybernetic factory was supposed to be an improvement on existing factories with their human managers. And yet the cybernetic brain of the factory was supposed to be a colony of insects, some dead leaves for them to feed on, the odd leech. Did Beer really think that his local pond was cleverer than he was? In a way, the answer has to be that he did, but we should be clear what way that was. Recall that Beer thought that the economic environment of the factory was itself an exceedingly complex system, ultimately unknowable and always becoming something new. He therefore felt that this environment would always be setting managers problems that our usual modes of cognition are simply unable to solve. This connects straight back to the above remarks on Beer's scepticism toward representational knowledge. On the other hand, according to Beer, biological systems *can* solve these problems that are beyond our cognitive capacity. They *can* adapt to unforeseeable fluctuations and changes. The pond survives. Our bodies maintain our temperatures close to constant whatever we eat, whatever we do, in all sorts of physical environments. It seems more than likely that if we were given conscious control over all the parameters that bear on our internal milieu, our cognitive abilities would not prove equal to the task of maintaining our essential variables within bounds and we would quickly die. This, then, is the sense in which Beer thought that ecosystems are smarter than we are—not in their representational cognitive abilities, which one might think are nonexistent, but in their performative ability to solve problems that exceed our cognitive ones. In biological computers, the hope was that "solutions to problems simply *grow*" (1962a, 211).

The Social Basis of Beer's Cybernetics

At United Steel, Beer was the director of a large operations research group, members of which he involved in the simulation of the cybernetic factory at the Templeborough Rolling Mills. This was a serious OR exercise, supported by his company. The key ingredient, however, in moving from the simulation to the cybernetic reality, was the U-machine, and, as Beer remarked in opening his 1962 status report on biological computing, "everything that follows is very much a spare time activity for me, although I am doing my best to keep the work alive—for I have a conviction that it will ultimately pay off. Ideally, an endowed project is required to finance my company's Cybernetic Research Unit in this fundamental work" (1962b, 25). I quoted Beer above on tinkering with tanks in the middle of the night, evidently at home, and Beer's daughter Vanilla has, in fact, fond childhood memories of weekend walks with her father to collect water from local ponds (conversation with the author, 22 June 2002). We are back once more on the terrain of amateurism, ten years after Walter had worked at home on his tortoises and Ashby on his homeostat.

Again, then, a distinctive cybernetic initiative sprang up and flourished for some years in a private space, outside any established social institution. And, as usual, one can see why that was. Beer's work looked wrong. Tinkering with tanks full of pond water looked neither like OR nor like any plausible extension of OR. It was the kind of thing an academic biologist might do, but biologists are not concerned with managing factories. The other side of the protean quality of cybernetics meant that, in this instance, too, it had no obvious home, and the ontological mismatch found its parallel in the social world. I do not know whether Beer ever proposed to the higher management of United Steel or to the sponsors of his consulting company, SIGMA, that they should support his research on biological computing, but it is not surprising that he should be thinking wistfully of an endowed project in 1962, or that such was not forthcoming. We should, indeed, note that Beer failed to construct a working U-machine, or even a convincing prototype. This is, no doubt, part of the explanation for the collapse of Beer's (and Pask's) research in this area after 1962. But it is only part of the explanation. The electronic computer would not have got very far, either, if its development had been left solely to a handful of hobbyists.

Of course, Beer did not carry on his cybernetics in total isolation. As mentioned above, having read Wiener's *Cybernetics* in 1950, he sought out and got to know many of the leading cyberneticians in the United States as well as Britain. In the process, he quickly became a highly respected member of the cybernetics community which existed transversely to the conventional institutions to which its members also belonged. It was Beer who first brought Ashby and Pask together, by inviting both of them to a lecture he gave in the city hall in Sheffield in 1956, and his recollection of the meeting sheds some light on the characters of both (S. Beer 2001, 553): "Gordon was speaking in his familiar style—evocative, mercurial, allusory. He would wave his arms about and try to capture some fleeting insight or to give expression to some half-formed thought. I was used to this—as I was to Ross's rather punctilious manner. So Ashby would constantly interrupt Gordon's stream of consciousness to say, 'Excuse me, what exactly do you mean by that?' or 'Would you define that term?' Both were somewhat frustrated, and the evening was close to disaster." Beyond his personal involvement in the cybernetics community, Beer appreciated the importance of establishing a reliable social basis for the transmission and elaboration of cybernetics more than the other British cyberneticians. Ross Ashby also presented his work at the 1960 conference at which Beer presented "Towards the Cybernetic Factory," and while there Beer conspired with Heinz von Foerster to offer Ashby the position that took him to the University of Illinois (Beer 1994 [1960], 299–301). In the second half of the 1960s, when Beer was development director of the International Publishing Corporation, he conceived the idea of establishing a National Institute of Cybernetics at the new Brunel University in Uxbridge, London, aiming to create academic positions for both Gordon Pask and Frank George. Beer persuaded the chairman of IPC, Cecil King, to fund part of the endowment for the institute and a fund-raising dinner for the great and good of the British establishment was planned (with Lord Mountbatten, the queen's uncle, and Angus Ogilvy, the husband of Princess Alexandra, among the guests). Unfortunately, before the dinner could take place there was a palace coup at IPC—"in which, ironically, I [Beer] was involved"—which resulted in the replacement of King by Hugh Cudlipp as chairman.

> I had never managed to explain even the rudiments of cybernetics to him [Cudlipp]. Moreover, it is probably fair to say that he was not one of my greatest fans. . . . At any rate the dinner broke up in some disorder, without a single donation forthcoming. Dr Topping [the vice-chancellor at Brunel] went ahead with the plan insofar as he was able, based on the solitary commitment that Cecil King had made which the new Chairman was too late to withdraw. Gordon was greatly disappointed, and he could not bring his own operation (as he had intended) [System Research, discussed in the next chapter] into the ambit of the diminished Institute which soon became a simple department at Brunel. The funding was just not there. However, both he and Frank George used their Chairs on the diminished scale. (S. Beer 2001, 557)

Though Beer had not fully achieved his ambition, the establishment of the Department of Cybernetics at Brunel was the zenith of the institutional career of cybernetics in Britain, and we shall see in the next chapter that Pask made good use of his position there in training a third generation of cyberneticians. Characteristically, the trajectory of cybernetics in Britain was further refracted at Brunel, with Pask's PhD students focusing on such topics as teaching machines and architecture. The Brunel department closed down in the early 1980s, and, given the lack of other institutional initiatives, these students were once more left to improvise a basis for their careers.[14]

In the 1960s, then, Beer helped find academic positions for three of Britain's leading cyberneticians and played a major role in establishing an academic department of cybernetics. Conversely, as remarked already, in 1974 Beer effectively deinstitutionalized himself in moving to a cottage in Wales. Partly, as I said, this was an aspect of an overall shift in lifestyle; partly it was a response to events in Chile. Partly, too, I think, it was a reflection of his failure in the later 1960s to persuade Britain's Labour government of the importance of cybernetics. He wrote of his "disappointment in the performance of Harold Wilson's 'white heat of technology' government. This was operating at a barely perceptible glow, and the ministers with whom I had been trying to design a whole new strategy for national computing failed to exert any real clout. There were five ministers involved—the Postmaster General himself (John Stonehouse) 'did a runner' and was discovered much later in Australia" (S. Beer 2001, 556). Beer was an exceptionally well connected spokesman for cybernetics in the 1960s, but the fruits of his efforts were relatively few. As he once put it to me, speaking of the sixties, "the Establishment beat us" (phone conversation, 3 June 1999).[15]

The Afterlife of Biological Computing

Neither Beer nor Pask ever repudiated his biological computer work; both continued to mention it favorably after the 1960s. In his 1982 popular book, *Micro Man*, Pask discusses a variety of "maverick machines," including his electrochemical systems, which he describes as "dendritic." He mentions that improved versions of them have been built by R. M. Stewart in California and comments that "there is now a demand for such devices, which are appropriate to non-logical forms of computation, but dendrites . . . are physically too cumbersome for such demand to be met practically. It now seems that biological media may perform in similar fashion but on a more manageable scale" (Pask and Curran 1982, 135). A few pages later he actually reproduces a picture of a pond, with the caption "A real-life modular processor?" Likewise, Beer in the text he wrote for a popular book on the history of computing, *Pebbles to Computers*: "Some thirty years ago, some scientists began to think that biological computers might be constructed to outpace even electronic achievement. At that time it was not clear that transistors themselves would become reliable! Attempts were made to implicate living cells—microorganisms—in computations. In England in the 'fifties, one such computer solved an equation in four hours that a bright school girl or boy could solve in (maximum) four minutes. Its time had not yet come!" (Blohm, Beer, and Suzuki 1986, 13).

Biological computing enjoyed a happier fate in science fiction, making its way into the popular imagination. With Beer's experiments on mice with cheese as a "reward function" we are surely in the presence of the mouse-computer that turns up in both Douglas Adams's *The Hitchhiker's Guide to the Galaxy* (1979) and Terry Pratchett's *Discworld* series of fantasy novels.[16] The most convincing representations of biological computing that I have come across include the obviously organic control systems of alien space ships that featured in various episodes of *Doctor Who* and, more recently, in Greg Bear's novel *Slant* (1997), which includes a biological computer called Roddy (recombinant optimized DNA device) that is an entire ecosystem of bees, wasps, ants, peas, and bacteria (and which succeeds in subverting the world's most sophisticated conventional AI, Jill).

And back in the material world biological computing has, in fact, recently been experiencing a resurgence. Figure 6.9 shows a cockroach-controlled robot, recently built by Garnet Hertz in the Arts, Computing, Engineering Masters Program at the University of California, Irvine. A giant Madagascan cockroach stands on the white trackball at the top of the assembly, attached by Velcro on its back to the arm which loops above the other components. Motions of the cockroach's legs rotate the trackball, which in turn controls the motions of the cart (much as a trackball can be used to control the motion of the cursor on a computer screen). Infrared sensors detect when the cart is approaching an obstacle and trigger the appropriate light from an array that surrounds the roach. Since roaches tend to avoid light, this causes the roach to head off in another direction. The entire assemblage thus explores its environment without hitting anything or getting stuck—ideally, at least. The cybernetic filiations of this robot are obvious. From one angle, it is a version of Grey

Walter's tortoise, five decades on. From the other, a lively biological agent replaces the precisely designed electronic circuitry of the tortoise's brain, exemplifying nicely the sense of "biological computing."[17] Figure 6.10 shows another biorobot, this one built by Eduardo Kac as part of his installation *The Eighth Day*. This time, the robot is controlled by a slime mold. These machines have no functional purpose. They are artworks, staging for the viewer a cybernetic ontology of entrained lively nonhuman agency. We can return to the topic of cybernetic art at the end of this chapter. For now, we might note that back in the 1950s and early 1960s Beer and Pask were aiming at something much more ambitious than Hertz and Kac, to latch onto the adaptive properties of biological systems, rather than their basic tropic tendencies.[18]

Figure 6.9. Cockroach-controlled robot. (Photograph by Garnet Hertz. Used by permission.)

Figure 6.10. Eduardo Kac, *The Eighth Day*, 2001. Transgenic artwork with biological robot (biobot), GFP plants, GFP amoebae, GFP fish, GFP mice, audio, video, and Internet (dimensions variable). The photograph shows the biobot in the studio, with its internal amoebae already in place, before it was introduced into the transgenic ecology that constitutes *The Eighth Day*. Source: www.ekac.org/8thday.html. Used courtesy of Eduardo Kac.

The Viable System Model

When Beer's dreams of biological computing came to an end in the early 1960s, this implied not an abandonent of his vision of the cybernetic factory but a transformation of it. Beginning in 1972, a trilogy of books developed his account of what he called the viable system model—the VSM for short: *Brain of the Firm* (1972; 2nd ed., 1981), *The Heart of the Enterprise* (1979), and

Diagnosing the System for Organizations (1985). The VSM was at the forefront of Beer's thinking and consulting work from the 1960s to the 1990s and attracted a considerable following. A two-day workshop on the VSM held at the Manchester Business School in January 1986 led to the production of an edited volume describing further interpretations and applications of the VSM by a range of academics, consultants, and people in industry and the military (Espejo and Harnden 1989), and variants of the VSM are still practiced and taught today.

The VSM transformed Beer's earlier vision of the cybernetic factory along two axes. First, the simulation of the cybernetic factory discussed above, where human management filled in for the not-yet-built U-machine, became in effect the thing itself. Beer continued to look forward to as much computerization of information gathering, transmission, and transformation as possible (as in the T- and V-machines). But the ambition to dispense with the human entirely was abandoned. Instead, human managers were to be positioned within purposefully designed information flows at just those points that would have been occupied by adaptive ponds or whatever (e.g., the position they in fact occupied in the earlier simulations).

Second, Beer extended and elaborated his conception of information flows considerably. In *Brain of the Firm*, the first of the VSM trilogy, he argued thus: The aim of the firm had, as usual, to be to survive in an environment that was not just fluctuating but also changing—as new technologies appeared in the field of production and consumption for example. How was this to be accomplished? What would a viable firm look like? The place to look for inspiration, according to Beer, was again nature, but now nature as the source of inspiration in the design of viable organizations, rather than nature as the immediate source of adaptive materials. Beer's idea was to read biological organisms as exemplary of the structure of viable systems in general, and to transplant the key features of their organization to the structure of the firm. In particular, he chose the human nervous system as his model. In the VSM, then, Beer's strategy was to transplant the organic into the social, but not as literally as before. The firm would no longer contain trained mice or *Daphnia* at its heart; instead, information flows and processing would be laid out as a diagram of human bodily flows and transformations.

The spirit of the VSM is strikingly expressed in the juxtaposition of two figures from *Brain of the Firm*. Figure 6.11A is a schematic of the body; figure 6.11B is a schematic of the firm. *Brain* goes into considerable detail in rehearsing the then-current understanding of human neurophysiology—the pathways both

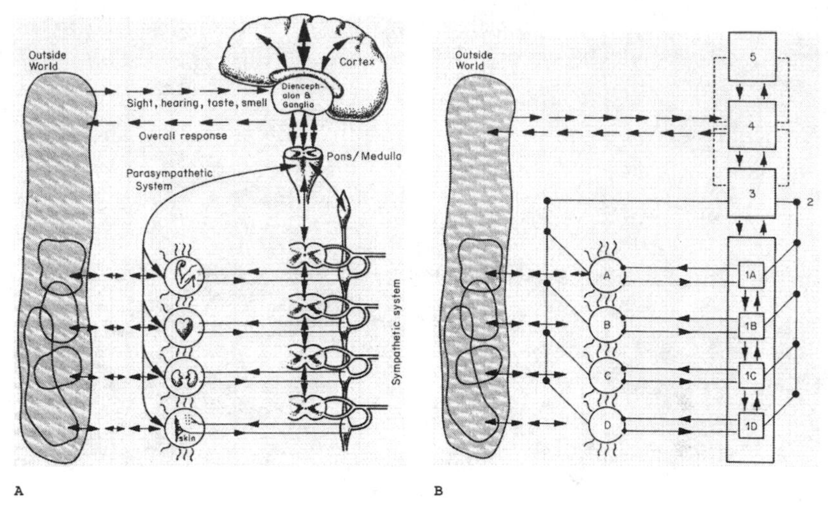

A B

Figure 6.11. Control systems: *A*, in the human body; *B*, in the firm. Source: S. Beer, *Brain of the Firm*, 2nd ed. (New York: Wiley, 1981), 131, figs. 23, 22. Permission: John Wiley & Sons.

nervous and biochemical along which information flows, the operations performed upon it at different points—and how it might be transcribed into the organization of the firm. I am not going to attempt an extensive review. But some key features need to be singled out. The VSM divided the "nervous system" of the firm into five subsystems, numbered 1–5 in figure6.11B. Although the VSM was supposed to be applicable to any organization (or any viable system whatsoever), for illustrative purposes figure 6.11B is a diagram of a company having four subsidiaries, labeled *1A–1D*, and as indicated in figure6.11A, one can think of these in analogy to systems in the body, controlling the limbs, the heart, the kidneys, and so on. A notion of autonomy arises here, because such systems in the body largely control themselves without reference to the higher levels of the brain. The heart just speeds up or slows down without our ever having to think about it. It adapts to the conditions it finds itself in by reflex action largely mediated somewhere down the spinal column. Beer's contention was that subsidiaries of the firm should be like that. They should act in the world and on one another (supplying materials to one another, say) as indicated by the circles with wavy lines and arrows moving off to the left from them, and their performance would be monitored at appropriate points on the "spinal column"—the square boxes labeled *1A* and so on. This monitoring would consist of a comparison of their performance in relation to a plan already given by the higher management of the firm, and deviations could be compensated for by appropriate adjustments to their behavior. The model here would be a simple servo-controlled negative feedback mechanism.

But even at this level of the body, autonomy is not complete. Figure 6.11B shows direct connections between the control systems, *1A*, *1B*, and so on. The idea here is that if something unusual is happening in subsidiary 1A, say, which supplies parts to *1B*, then *1B* should know about it so that it can take steps to allow for that. There must, that is, be some information channel linking these subsidiaries, as there is between the heart and the lungs. And further, Beer observed, in the human body there are usually several different channels linking levels of the nervous system. Figure 6.11A thus distinguishes two further channels—the sympathetic and the parasympathetic systems—and figure 6.11B shows their equivalents—lines of information flow upward, from the controllers on the spinal cord (the squares) and from the operational sites (the circles). The equivalent of the sympathetic system is system 2 of the VSM. Beer understood this as attempting to damp conflicts that could arise at the system 1 level—the various subsidiaries trying to hoard some material in short supply to each other's detriment, for example. This damping, which Beer knew enough not to expect to be necessarily successful, would be accomplished by reference to system 3. Corresponding to the pons and the medulla at the base of the brain, system 3 would be basically an operations research group, running models of the performance of the entire ensemble of subsidiaries, and thus capable, in principle, of resolving conflicts between subsidiaries in the light of a vision available to none of the subsidiaries alone.[19]

At this stage, no information has traveled upward beyond system 3 into higher layers of management. The parasympathetic system, however, was envisaged to act somewhat differently. This traveled straight up to system 3 and was intended to transmit an "algedonic" "cry of pain." Less metaphorically, production data would be monitored in terms of a set of dimensionless ratios of potential to actual performance of the kind that Beer had introduced in his 1953, paper discussed earlier. If one of those ratios departed from a predecided range, this information would be automatically passed onward to system 3, which, in the light of its models, would act as a filter, deciding whether to pass it on to levels 4 and possibly 5.[20]

I am inclined to think that system 4 was Beer's favorite bit of the VSM. The equivalent of the diencephalon and ganglia of the human brain, this had access to all the information on the performance of the firm that was not filtered out by system 3; it was also the level that looked directly outward on the state of the world. If the level 1 systems had access to information directly relating to their own operations, such as rising or falling stockpiles or order books,

Figure 6.12. World War II operations room, near London, during the Battle of Britain. Source: Beer 1968a, 23.

level 4 had much wider access, to national economic policies and changes therein, say, to the price of money, the results of market research, and what have you. System 4 was, then, the T-U-V system of Beer's earlier model, with the humans left in.

Beer envisaged system 4 as a very definite place. It was, in fact, modelled on a World War II operations room, of the kind shown in figure 6.12 (taken from Beer's 1968 book *Management Science*), as developed further by NASA at "Mission Control in the Space Centre at Houston, Texas, where the real-time command of space operations is conducted" (Beer 1981, 193–94), and updated with all of the decision aids Beer could think of (194–97). All of the information on the state of the firm and of the world was to be presented visually rather than numerically—graphically, as we would now say. Dynamic computer models would enable projections into the future of decisions made by management. Simply by turning knobs (197), managers could explore the effects of, say, investing more money in new plant or of trends in consumption. Feedbacks that had passed the level 3 filters would also arrive at system 4 from the lower levels, "signalled appropriately—that is, if necessary, with flashing red lights and the ringing of bells" (194), alerting management to emerging production problems, perhaps to be passed on again to level 5. In terms of social organization, "I propose a control centre for the corporation which is in continuous activity. This will be the physical embodiment of any System 4. All senior formal meetings would be held there; and the rest of the time, all senior executives would treat it as a kind of club room. PAPER WOULD BE BANNED FROM THIS PLACE. It is what the Greeks called a *phrontisterion*—a thinking shop" (194).

System 4, then, differed from system 3 in the range of its vision, a vision which now encompassed the future as well as the present, and Beer imagined system 4 as a primary locus for decision making on change. If the levels below strove to implement given production plans for the firm, level 4 was the level at which such plans were drawn up and modified.

Finally we arrive at system 5, the equivalent of the human cortex. This was the level where policies were deliberated upon and the most consequential decisions were made (Beer 1981, 201). This was where the human directors of the firm had been imagined to continue to exist in the original blueprint for the cybernetic factory. The firm's viability and continued existence, and even growth and evolution, were maintained by systems 1–4. The job of system 5 was, therefore, to think big at a metalevel superior to questions of mere viability.[21]

This outline of the VSM is now almost complete, but two points need to be added. First, the various levels of the viable system were intended to be coupled *adaptively* to one another. The 3 and 4 systems, for example, would engage in the process of reciprocal vetoing discussed earlier. Level 4 might propose some change in the overall operating plan for the firm; this would be run through the OR models at level 3 and might be rejected there—perhaps it would place excessive strain on one of the subsidiaries. Level 3 could then propose some modified plan back to level 4, which could run it through its models. Perhaps the plan would be vetoed again, once more transformed, and returned to level 3. And so on, back and forth, until some operating plan agreeable to both systems 3 and 4 was discovered.

Second, we should note a *recursive* aspect of the VSM. Beer argued that firms were themselves parts of bigger systems—national economies, say. The entire 1–5 structure of the firm would thus appear as a single system 1 on a diagram of the national economy. This in turn should be a viable system with its own levels 2–5 overseeing the ensemble of firms. Proceeding down the scale instead of up it, each subsidiary of the firm should also be a viable system in its own right, meaning that the level 1 systems of figure 6.11 should actually have their own levels 1–5 within them. Figure 6.13 shows what became Beer's standard diagram of the VSM, depicting two levels of such recursion. The two level 1 subsidiaries in square boxes at the lower end of the spinal column (running up the right-hand side) are shown as having their own 1–5 structure projecting downward at an angle of 45 degrees (and each has two subsidiary

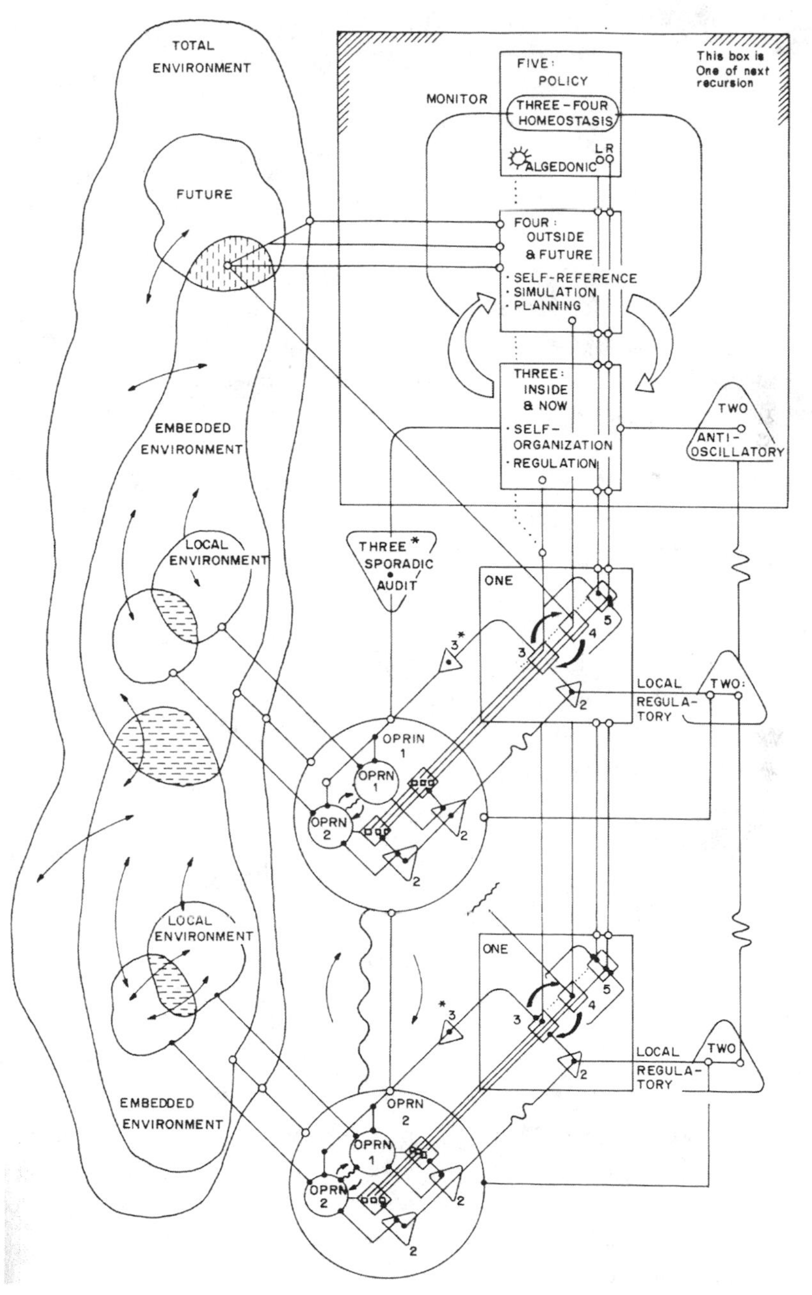

Figure 6.13. The viable system model showing recursive embeddings. Source: S. Beer, *Diagnosing the System for Organizations* (New York: Wiley, 1985), 136, fig. 37.

"operations" within its large circle). Likewise, the 3-4-5 brain at the top of the spinal column is itself enclosed in a square box, indicating that it is part of a level 1 system of some bigger system. Beer felt that such recursivity was a necessary property of viable systems—they had to be nested inside one another "like so many Russian dolls or Chinese boxes" in a chain of embeddings which "descends to cells and molecules and ascends to the planet and its universe" (Beer 1989a, 22, 25).

The VSM was thus a vision of the firm in the image of man. Especially, functions of management and control were envisaged on the lines of the human brain and nervous system. The brain and nervous system were simulated by a combination of information technologies and real human beings appropriately arranged. One could say that the VSM is one of the most elaborated images of the cyborg in postwar history, though the word "cyborg" is tautologous here, standing as it does for "cybernetic organism." Any viable system was exactly that, according to Beer. We should also note that the VSM was the "circuit diagram" (Beer 1981, 123) of a time machine, an adaptive system accommodating itself to the exigencies of the unknown in real time, ranging from mundane disturbances at the level of production to world-historical changes.

The VSM as Ontology and Epistemology

The basic ontological vision that the VSM conjures up is the same as that of the cybernetic factory before it: the world as an ungraspable and unmasterable space of becoming; the organization as open-endedly and performatively adaptable. The VSM, however, also suggests some refinements to that picture. First, my portrayal of the cybernetic factory was centered on the brain of the firm as a unitary entity, the U-machine, in dialogic conversation with the firm's environment. Beer's conception of the VSM, in contrast, was one in which the overall behavior of the firm was the upshot of an interplay of many active but quasi-autonomous elements, the VSM's systems 1–5, themselves also interacting with different aspects of the firm's environment. The recursive aspect of the model adds an indefinite sequence layers of elements to this picture. The VSM thus moves us toward a vision of *ontological multiplicity*, a multiplicity which is, furthermore, *irreducible*: the system 3 of a given organization is not reducible to the organization's system 4, say, or to the system 3 of another organization.[22]

Second, we can return to the question of goals. Walter's and Ashby's devices had fixed goals that organized their adaptation: the homeostat reconfigured itself so as to keep its essential variables within preset limits. Beer's conception of the VSM, in contrast, specified no goals whatsoever, except adaptation itself. And we could think of Heidegger: adaptation in the VSM was a process of revealing rather than enframing. The process that Beer called reciprocal vetoing between levels of the system, for example, was by no means as negative as the phrase suggests. A veto from one level to another was at the same time an invitation for a novel counterproposal, a way of finding out what the other had to offer.

The VSM was explicitly about information flows and transformations, so we can return now to a consideration of cybernetic epistemology as well as ontology. In Beer's vision, viable systems do contain knowledge—representations of their own inner workings and of their environment—principally enshrined in the OR models at level 3 of the VSM and the projective models at system 4. What should we make of this? First, we could recall that in his work on truly biological controllers, Beer had sought to avoid this detour through representation. His biological computers did not contain any representational or symbolic elements; they were intended simply to do their adaptive thing. The VSM, then, one might say, was a concession to representation as a response to the failure of biological computation. And it is appropriate to recall that, as I remarked before, Beer did not much trust representational models. He did not think, for example, that one could arrive at a uniquely correct model of the firm and its environment that could function unproblematically at level 4 of the VSM. This is a direct corollary of the idea that both the firm and its environment are exceedingly complex.

Beer did not, however, take this to imply that the construction of representational models was a useless endeavor. His idea, instead, was that the models in question should be continually examined and updated in relation to performance—"*continuously adapted*" (Beer 1981, 185) or even always "aborting" (Beer 1969 and 1994b, 151). The company should act in the light of the future projections of the model at level 4, but then actual developments in time should be compared with expectations from the model's simulations. These would not, in all probability, match, and the model should be adjusted accordingly.[23] The VSM thus stages for us an image of a performative epistemology—a more elaborated version of what we have seen in the preceding chapters. The "knowledge components" of the VSM were not an end in themselves; they were geared directly into performance as part of the mechanism of adaptation, and they were revisable in performance, just like the other components of the VSM; they were not the controlling center of the action.

Here I need to enter a caveat. What might adaptation of these models in practice mean? I just described adaptation in the VSM as open ended, but

Beer imagined and was prepared to implement something less than this in his models. He understood them as sets of mathematical equations linking long lists of variables such as demand, revenue, technological and economic change, dividends, share prices, and the money market. And the basic form of these sets of equations was not, in itself, revisable, at least as part of Beer's description of the regular functioning of a viable system. What could be revised in practice were the parameters figuring in these equations which specified the intensity of the couplings between variables. Beer's models were thus adaptive, but only to a degree, within a fixed overall form.[24]

One further point. The symbolic models of the VSM were envisaged as conventional simulations programmed on digital computers. In this respect, there was no distinctively cybernetic aspect to the VSM. But it is still instructive to review Beer's thoughts on the computerization of industry. It is important to note that Beer was himself an enthusiast for computers. As early as 1956 at United Steel he had had installed one of the first computers in the world to be dedicated to management science, a Ferranti Pegasus (Harnden and Leonard 1994, 4). He was nevertheless a consistent critic of the way in which computers were being introduced more generally into industry and business. His argument was that "the first and great mistake" was that "people set out to automate the procedures and therefore the organisations they already knew. These themselves were frozen out of history and fixed by professionalism." Computers were, in other words, being used to automate existing clerical tasks while leaving the overall structure of the traditional organization untouched: "Companies have exchanged new lamps for old, and set them in the window as marks of progress. . . . We are using a powerful control instrument competent to reorganise the firm, its departments and functions, and encapsulating it in a received system geared to the quill pen." Instead, Beer argued, we should ask, "What should my enterprise be like, now that computers exist?" (Beer 1967, 214–17).

Beer was especially critical of the use of computers in business to automate and augment record keeping, and this gets us back to the ontological question. If the world is beyond our capacity to know it, and if, even worse, it continually changes, knowing the past is of limited utility. Our information processing should therefore be forward looking, as in the system 4 model of the VSM. "It is worth making a tremendous effort to burst through the barrier marked 'now,' and to make managers concern themselves with what *can* be managed—namely the future, however near—rather than peruse a record of what can be managed no longer—namely the past, however recent. We may learn from that past record of course, but we cannot influence it in retrospect. . . .

Look straight ahead down the motorway when you are driving flat out. Most enterprises are directed with the driver's eyes fixed on the rear-view mirror" (1981, 127, 199). Beer's idea in the VSM was thus that most of the information that one can collect on an organization is useless and can be discarded. This was what the filtering operations at the various levels did, keeping only anomalous signals for transmission to higher levels.

Seen from this angle, the object of the VSM was to reorganize the firm around the computer—to effect a transformation that was social as well as technological, to rearrange the human components as part of an adaptive technosocial system of information flows and transformations. Here, too, then, social relations and ontology hung together. And this contrast between the VSM and the traditional structure of the organization is another nice example of how ontology can make a difference in practice. Further aspects of this are apparent below.

The VSM in Practice

The VSM was a normative vision of the organization. Organizations had to look like the VSM if they were to survive and grow in time. The obvious implication of that would seem to be that they needed to be remade from the ground up to exemplify the VSM. Beer had one serious chance at that, which is reviewed in the next section. But Beer could hardly claim that all existing organizations were nonviable—some of them had been around for a long time, the Catholic Church, for example. He therefore made a more nuanced argument. Just like organisms, organizations could be more or less viable—some struggling to survive, others actually dying, others springing happily into the future: "The amoeba succeeded, the dinosaur failed, the coelacanth muddles along" (Beer 1981, 239). And the problem was that organizations had no way to discuss this temporal viability; they lacked any language or conceptual apparatus for it.

What organizations had instead was organization charts of hierarchical power relationships running downward from the board of directors through vertical chains of command devoted to production, accounting, marketing, and so on. Beer's claim was that such charts did not, and could not, represent how firms actually worked. They functioned, at most, as devices for apportioning blame when things went wrong.[25] Already, then, whether anyone recognized it or not, the VSM was a better description of how the firm really worked, and Beer's pitch was that the formal VSM could therefore function as a diagnostic tool (1981, 155). One could examine the firm, or any other organization, and see just which bits of it corresponded to the five levels of the VSM, and one could examine the ways in which they were connected together. Certain aspects of the firm might thus be identified as especially deficient as compared to the VSM diagram and made the targets of therapeutic intervention. Beer claimed that an experienced VSM practitioner could often walk into a factory and identify the major problems within a day or two and that, once pointed out, management would recognize the veracity of the judgment—such problems having already been subconsciously recognized and papered over (Beer 1989a, 27). Of course, addressing the problems thus identified might take much longer—conceivably a period of years. "My guess would be that organizations cannot face up to more than a quarter of the reshaping that their long-term viability demands. This is of course the reason why so many enterprises are in a state of *continuous* . . . reorganisation" (Beer 1981, 239).

One simple example of what might be at stake here, of continuing practical and scholarly interest, concerns automation and the interlinkages between the systems 1 of figure 6.13. Beer noted that such linkages between different subsidiaries of a single company found no formal representation on the typical organization chart. But (Beer 1981, 107) "I have collected scores of examples of this. Sometimes, very often perhaps, the foremen in the related departments make it their business to keep in intimate touch. Maybe they walk across the road and drink tea together; maybe they telephone: 'You'd better know, Charlie, that . . .' In a few extreme case, it was not possible to discover how the messages were transmitted—but transmitted they certainly were." Beer was quite happy with such informal channels of communication; his only concern was that Ashby's law should be respected—that there should be enough variety at each end to cope with that at the other, and that there be enough bandwidth between them to mobilize and organize those varieties appropriately. Instead, Beer argued, the introduction of computers as information-processing devices often acted to sever such channels completely. Because the channels did not appear on the organization chart, they did not become automated; at the same time, their human conduits—the foremen, in this example—might be forbidden to step outside their own domains, or their positions eliminated entirely. "In the limiting case where the departmental outstation is fully automated, there is no possible way in which the social link can be maintained. Computers do not just happen to develop the trick of shouting to each other across the void, as human beings always do" (108). A technological transformation which appeared progressive on the surface might thus be regressive as seen from the perspective of the VSM.[26]

Beer also claimed that many organizations were entirely lacking a system 2 (1981, 175), and in the absence of the "sympathetic" damping generated by the 1-2-3 system would thus always be prone to pathological competition and "oscillations" between their subsidiaries. More generally, Beer worried about the higher levels of the brain of the firm. Pieces of the organization which he felt should lie directly on the "command" axis were often found to be misplaced. This was true especially of parts of the organization that had grown up since World War II, including management accounting, production control (Beer's first job in the steel industry), and operations research (his first love in management). These had no place on prewar organization charts and thus found themselves a position almost at random (Beer 1981, 82–83). OR groups, for example, might be found buried in subsidiaries and thus serving the overall organization asymmetrically—to the benefit of some subsidiary rather than the whole firm. The moral of the VSM was that there should be an OR group on the command axis itself, at level 3. Beer also argued that "in most firms System 4 is a fiasco" (153–54). Elements of system 4—the monitoring and planning organ at the base of the conscious brain—were usually to be found in any large organization, but they tended to be dispersed across the organization instead of grouped coherently together on the command axis. Certainly very few clubby operations rooms were to be found in industry in this period.

We need to remember that from 1970 onward Beer made his living primarily as an independent management consultant, and his writings on the VSM were integral to that. In 1989, he produced a list of consultancies he had been engaged in (Beer 1989a, 35):

> Small industrial businesses in both production and retailing, such as an engineering concern and a bakery, come to mind; large industrial organizations such as the steel industry, textile manufacturers, ship-builders, the makers of consumer durables, paper manufacturers are also represented. Then there are the businesses that deal in information: publishing in general, insurance, banking. Transportation has figured: railways, ports and harbours, shipping lines. Education, and health (in several countries), the operations of cities, belong to studies of services. Finally comes government at all levels—from the city, to the province, to the state and the nation itself—and the international agencies: the VSM has been applied to several.
>
> Obviously . . . these were not all major undertakings, nor is "success" claimed for massive change. On the other hand, none of these applications was an academic exercise. In every case we are talking about remunerated consultancy, and that is not a light matter. The activities did not necessarily last for very long

> either, since speedy diagnosis is a major contribution of the whole approach. On the other hand, some of them have lasted for years. Undoubtedly the major use of this work to date was in Chile from 1971–1973.

Chile is next. Here I can just emphasize what is obvious from this list: Beer operated not only at the level of commercial companies; many other kinds of social organizations were likewise open to his interventions. We should also remember what was noted earlier—that by the 1980s the VSM had gained a significant following among management consultants and their academic counterparts, leading to the publication of at least one multiauthor book on the VSM (Espejo and Harnden 1989). The interested reader can look there for case studies written up by Beer and his followers, including Beer's sixty-page account of his association over nine years with a mutual life assurance company (Beer 1989a), as well as for various methodological and substantive reflections on and extensions of the VSM. The VSM was never one of those great fads that seem to have periodically overtaken the world of management since the Second World War. Given its subtlety and complexity, to which I have done scant justice here, this does not seem surprising. But it has been at the heart of a significant movement.

Chile: Project Cybersyn

In Chile in the autumn of 1970 Salvador Allende became the world's first democratically elected socialist president. The new government started nationalizing the banks and major companies operating within Chile, operating through an existing organization known as CORFO (Corporacíon de Fomento de la Produccíon). On 13 July 1971, the technical general manager of CORFO, one Fernando Flores, wrote to Beer (Beer 1981, 247): "This letter spoke of 'the complete reorganization of the public sector of the economy,' for which it appeared its author [Flores] would be primarily responsible. He had read my books, and had even worked with a SIGMA team ten years before. He went on to say that he was now 'in a position from which it is possible to implement, on a national scale—at which cybernetic thinking becomes a necessity—scientific views on management and organization.' He hoped that I would be interested. I was." Beer's commitment to the project became "total" (245), and he subsequently published a long account of the project's evolution and termination, in five chapters added to the second edition of *Brain of the Firm* (Beer 1981, 241–399). Beer's chapters are, as usual, very dense, and I can only attempt an overview of his account as a way of sketching in the main features of what was undoubtedly the world's most striking cybernetic project.[27]

Taking up Flores's invitation, Beer flew into the capital of Chile, Santiago, on 4 November 1971, remaining for eight days and returning to London on 13 November. In Santiago he met Flores and his collaborators, and together they made plans to implement the VSM at the level of the national economy. Beer had just completed the manuscript of *Brain of the Firm*; the Chileans studied it while he was there, and it became the basis for their vision of Chile's future. On 12 November Beer met President Allende himself and explained the VSM to him. When Beer drew the box for system 5 of the VSM diagram, he was thinking of it as representing the president, but Allende "threw himself back in his chair: 'at last,' he said, 'el pueblo'" (Beer 1981, 258)—the people. Beer was so impressed by this that he told the story often. Allende was apparently similarly impressed with Beer and the VSM: "'The President says: Go ahead—fast'" (257).

What did the plan sketched out on Beer's first visit look like—Project Cybersyn, for "cybernetic synergy," as it became known? Beer felt that speed was of the essence—"within a year . . . the foreign reserves would run out" (251)—so he aimed to begin by installing a cut-down version of the VSM by, astonishingly, 1 March 1972. This was less than four months after his first visit, and he promised to return on 13 March 1972. The initial plan aimed to achieve real-time (meaning daily) communications between system 1 productive activities at the level of individual factories, and a system 4 control room to be constructed in Santiago.

OR teams were charged "to construct a quantitative flow chart of activities within each factory that would highlight all important activities" (253). OR models would then be used in consultation with management—typically workers' committees, foreign managers having fled the country—to construct indices of performance analogous to those Beer had devised in the steel industry and reported upon in the 1953 OR paper discussed above (163).[28] "In practice, it turned out that some ten or a dozen indices were adequate to monitor the performance of every plant" (253). Among these was to be an index to measure morale as a ratio depending inversely on absenteeism (253).

The question of what to do with all the data thus generated, how to handle it, then arose. Ideally, every plant should have its own computer to "process whatever information turned out to be vital for that factory's management" (252)—this, thinking of each plant as a viable system in its own right. "But such computers did not exist in Chile, nor could the country afford to buy them. . . . Therefore it was necessary to use the computer power available in

Santiago: it consisted of an IBM 360/50 machine and a Burroughs 3500 machine" (252). The remaining technical problem was to connect plants all over the country up to Santiago. This was to be accomplished by requisitioning telex machines, augmented by microwave and radio links whenever necessary. "The plan allowed just four months for this to be accomplished (and it was)" (252). This national information system was known as Cybernet; the data it brought to Santiago were processed there "and examined for any kind of important signal. . . . If there were any sort of warning implied by the data, then an alerting signal would be sent back to the managers of the plant concerned" (253). Beer himself took two tasks back to England with him (256): "I had to originate a computer program capable of studying tens of thousands of indices a day, and of evaluating them for the importance of any crucial information which their movements implied. . . . I had done this kind of system building many times before. . . . Secondly, I should need to investigate prospects for a simulation system in the operations room that could accept the input of real-time data. This would be a completely novel development in operational research technique." The basic blueprint and timetable for Cybersyn were thus set. Beer's own account covers subsequent developments in some detail; we can review some of the main features.

As indicated above, the Cybernet national information system was indeed established by the deadline of March 1972. The first computer program mentioned in the above quotation took longer than hoped to construct, partly because of the incorporation of very new OR techniques in forecasting. A temporary version was indeed implemented in March 1972, but the permanent version only became operational in November that year. By that time "something like seventy percent of the socio-industrial economy was operating within this system, involving about four hundred enterprises" (Beer 1981, 262, 264).[29]

These "Cyberstride" programs sat at the system 3 level, contributing to the homeostasis of the 1-2-3 economic assemblage while at the same time filtering data upward into the 3-4-5 system. A key element of the latter was a computer model of the Chilean economy and its national and global environment. This was to be the centerpiece of system 4 planning, intended to enable future projections according to different inputs and assumptions. This program was also Beer's responsibility. Lacking time to design such a model afresh, Beer announced in a January 1972 report that he had decided "to make use of the immediately available DYNAMO compiler extensively developed by J. W. Forrester of MIT. I have directed three projects in the past using this compiler, and have found it a powerful and flexible tool" (266). Forrester's

Figure 6.14. Operations room of Project Cybersyn. Source: Beer 1974a, 330, fig. 12.1.

work had grown by devious routes out of his World War II work at the Servomechanisms Laboratory at MIT and was just about to become famous, or notorious, with the publication of the Club of Rome's *Limits to Growth* report, which, on the basis of DYNAMO simulations, predicted an imminent collapse of the global economy and ecosystems.[30] Work in London and Chile under Chilean direction had developed a tentative version of the Checo (for Chilean economy) program by June 1972, and by September a better model was running. "I wanted to inject information *in real time* into the Checo program *via* Cyberstride. Thus any model of the economy, whether macro or micro, would find its base, and make its basic predictions, in terms of aggregations of low-level data—as has often been done. But Checo would be updated every day by the output from Systems 1-2-3, and would promptly rerun a ten-year simulation; and this has never been done. This was one of my fundamental solutions to the creation of an effective Three-Four homeostat; it remains so, but it remains a dream unfulfilled" (268). This continual updating was the way in which Checo simulations were foreseen as evolving in time, responsively to real-time input, thus exemplifying the performative epistemology of the VSM discussed in general terms in the previous section.

The system 4 operations room loomed ever larger as potentially the visible symbol, the icon, of Project Cybersyn (fig. 6.14). Detailed design was turned over to Gui Bonsiepe in Chile, from which emerged a plan for an octagonal room ten meters wide that would serve as an "information environment." Information on any aspect of the functioning of the economy at the desired level of recursion would be displayed visually on panels on the walls, including flashing warning signals that registered the algedonic "cries of pain" from lower levels, mentioned above, and an animated Checo simulation of the

Chilean economy that could be used to predict the effects over the next decade of decisions taken today. These days, computer graphics could handle what was envisaged with ease, but in the early 1970s in Chile the displays included hand-posted notes (of algedonic warnings), banks of projectors, and slides prepared in advance of meetings (showing quantified flow charts of production). The Checo display "certainly worked visually; but the computer drive behind it was experimental and fragmentary" (Beer 1974a, 329–32). The target date for completion of the control room was set as 9 October 1972; in fact, it was in "experimental working order" by 10 January 1973 (Beer 1981, 270).

Project Cybernsyn evolved very quickly, but so did other developments (Beer 1981, 307):

> As time wore on throughout 1972, Chile developed into a siege economy. How ironic it was that so many eyes were focussed with goodwill on the Chilean experiment in all parts of the world, while governments and other agencies, supposedly representing those liberal-minded observers, resisted its maturation with implacable hostility. The nation's life support system was in a stranglehold, from financial credit to vital supplies; its metabolism was frustrated, from the witholding of spare parts to software and expertise; literally and metaphorically, the well-to-do were eating rather than investing their seed-corn—with encouragement from outside. Even more ironic, looking back, is the fact that every advance Allende made, every success in the eyes of the mass of the people (which brought with it more electoral support) made it less likely that the Chilean experiment would be allowed to continue—because it became more threatening to Western ideology.

Before Allende came to power, copper had been Chile's major source of foreign exchange, and "we were to see the spectacle of the 'phantom ship' full of copper that traipsed around European ports looking for permission to unload" (307). Economic collapse was imminent, and Beer's thought was to "search for novel and evolutionary activity whereby the Chilean economy might very rapidly enhance its foreign earnings" (308). His answer was indigenous crafts, wine, and fish, and in 1972 and 1973 he sought to mobilize his contacts in Europe to expand those markets—without success. There was nothing especially cybernetic about those efforts, but they do indicate Beer's commitment to Allende's Chile.

In 1973 the situation in Chile continued to worsen. In September 1973, the Cybersyn team received its last instruction from the president, which was to move the control room into the presidential palace, La Moneda. "By the 11 September 1973, the plans were nearly ready. Instead La Moneda itself was reduced to a smoking ruin" (Beer 1974a, 332). Salvador Allende was dead, too, in the ruin: the Pinochet coup—Chile's 9/11—brought a definitive end to the Chilean experiment with socialism, and with it went Cybersyn. Beer was in London at the time but had prepared for the end by devising three different codes in which to communicate with his collaborators and friends in Chile, who were, through their association with the Allende government, in very serious trouble. Beer did what he could to help them. On 8 November 1973, he wrote to von Foerster at the University of Illinois: "My dear Heinz, I think you know that I am doing everything possible to rescue my scientific colleagues (at the level of Team Heads) from Chile. It is going well—10 families. There is another problem. My main collaborator is held in a concentration camp, and is coming up for trial. There is a real risk that he will be shot, or sent down for life."[31] The collaborator in question was Fernando Flores, who had risen to become Chile's minister of finance before the coup. Beer enclosed the draft of his personal statement to be read at Flores's trial and urged von Foerster to send his own. In the event, Flores was imprisoned for three years, until Amnesty International helped to negotiate his release, when he moved to the United States, completed a PhD in Heideggerian philosophy, and became a highly successful management consultant.[32]

The Politics of the VSM

THE PROBLEM IS FOR CYBERNETICS TO DISCOVER, AND TO MAKE ABUNDANTLY CLEAR TO THE WORLD, WHAT METASYSTEMS TRULY ARE, AND WHY THEY SHOULD NOT BE EQUATED WITH THE SUPRA-AUTHORITIES TO WHICH OUR ORGANIZATIONAL PARADIGMS DIRECT THEM. IT IS AN APPALLING [*SIC*] DIFFICULT JOB, BECAUSE IT IS SO VERY EASY TO CONDEMN THE WHOLE IDEA AS TOTALITARIAN. HENCE MY USE OF THE TERM: THE LIBERTY MACHINE. WE WANT ONE THAT ACTUALLY WORKS.

STAFFORD BEER, "THE LIBERTY MACHINE" (1975 [1970], 318)

Beer's daughter Vanilla recalls that "Stafford and I generally ran Jesus and Marx together in an attempt to produce metanoyic possibilities," so I turn now to Beer's politics and its relation to his cybernetics; later sections will focus on his spiritual beliefs and practices.[33]

As a schoolboy, Beer shared a bedroom with his brother, Ian, who recalled that Stafford "painted the whole wall . . . with extraordinary apparitions. In the centre of the wall was the original 'Towering Inferno'—a huge skyscraper with flames all around the bottom licking their way up the tower." Vanilla Beer adds that the picture was called *The Collapse of Capitalism*. In the late forties, Stafford fell out with his father, who pressured him into admitting that he had voted for the Labour Party in the recent election (Ian Beer, letter to Stafford's family, 25 August 2002). Later in life, Beer sometimes described himself as "an old-fashioned Leftist" (Medina 2006) or even as "somewhat to the left of Marx," though it would be a mistake to think of him within the conventional frame of British Marxism: "Stafford was fond of telling the story about Marx that had him saying 'Thank God I'm not a Marxist.' He didn't usually describe himself in this context but Stafford had a great deal of admiration for Marx, especially his early writings on alienation. He wasn't much of a fan of Das Capital mostly on the grounds of dull and repetitive."[34]

Little of this found its way into Beer's early writings. Until 1970, his books, essays, and talks were largely couched in a technical idiom and addressed to a management readership. But in 1969 (Beer 1975, 3)

> I had come to the end of the road in my latest job . . . and re-appraised the situation. What was the use of seeking another such job all safe and sound pensions all that from which haven to speak and write as I had done for years about the desperate need for drastic change and how to do it in a sick world? Not even ethical. How to begin? It was almost 1970. A decade opened its doors for business. There were speeches to be made already committed throughout that first year and I must see them through. What's more these platforms gave me the opportunity if I could only seize it to collect my thoughts for a new life and to propound ARGUMENTS OF CHANGE.

This series of talks, with assorted explanatory material, was published in 1975 as *Platform for Change: A Message from Stafford Beer*. In 1973, just before the Pinochet coup, Beer continued to develop his thinking in public, this time in the Canadian Massey Lectures on CBC radio, which were published the next year as *Designing Freedom* (Beer 1974b). The focus of these works, and many to follow, was on liberty, freedom, and democracy. Marx is not mentioned in them, nor any of the classic Marxist concerns such as class struggle. Instead, Beer attempted a distinctly cybernetic analysis, which is what interests me most. Here we can explore another dimension of ontology in action: cybernetics as politics.

The distinctly cybernetic aspect of Beer's politics connected immediately to the ontology of unknowability. Other people, at any scale of social aggregation, are exceedingly complex systems that are neither ultimately graspable nor controllable through knowledge. And along with that observation goes, as I noted in chapter 2, a notion of respect for the other—as someone with whom we have to get along but whom we can never possibly know fully or control. And this was Beer's normative political principle: we should seek as little as practically possible to circumscribe the other's variety, and vice versa—this was the condition of freedom at which Beer thought politics should aim. This, in turn, translated into an explicit view of social relations. If the ontology of knowability sits easily with an image of hierarchical command and control, in which orders are transmitted unchanged from top to bottom, then Beer's notion of freedom entailed a symmetric notion of adaptive coupling between individuals or groups. In a process of reciprocal vetoing—also describable as mutual accommodation—the parties explore each other's variety and seek to find states of being acceptable to all. The ontological and practical resonances here among Beer and Bateson and Laing are obvious, though Beer was operating in the space of organizations rather than psychiatry.

Beer recognized, of course, that any form of social organization entailed some reduction in the freedom of its members, but he argued that one should seek to minimize that reduction. In reference to viable systems, his thought was that freedom was a condition of maximal "horizontal" variety at each of the quasi-autonomous levels, coupled with the minimum of "vertical" variety reduction between levels consistent with maintaining the integrity of the system itself. Hence the notion of "designing freedom": as Beer explained it, the VSM was a diagram of social relations and information flows and transformations that could serve to guarantee the most freedom possible within organized forms of life. As we need to discuss, that view did not go uncontested, but let me emphasize now two features of Beer's vision.

First, there are many absorbing books of political theory which go through immensely subtle arguments to arrive at the conclusion that we need more freedom, fuller democracy, or whatever—conclusions which many of us would accept without ever reading those books. Beer was not in that business. He took it for granted that freedom and democracy are good things. The characteristic of his work was that he was prepared to think through in some detail just how one might arrange people and information systems to make the world freer and more democratic than it is now. Beer's specific solutions to this problem might not have been beyond criticism, but at least he was prepared to think at that level and make suggestions. This is an unusual enterprise, and I find it one of the most interesting and suggestive aspects of Beer's cybernetics. Second, we should note that, as already remarked, Beer's talks and writings did not foreground the usual substantive political variables of left-wing politics: class, gender, race. They foregrounded, instead, a generic or abstract topology in which the exercise of politics, substantively conceived, would be promoted in a way conducive to future adaptations. We should perhaps, then, think of Beer as engaging in a particular form of subpolitics rather than of politics as traditionally understood.

That said, Cybersyn was the only cybernetic project discussed in this book to be subjected to the political critique I mentioned in the opening chapters. I therefore want to examine the critique at some length, which will also help us get Beer's subpolitics into clearer focus and serve to introduce some more features of Cybersyn.

The Political Critique of Cybernetics

The early phases of Project Cybersyn were conducted without publicity, but public announcements were planned for early 1973. Beer's contribution to this was "Fanfare for Effective Freedom," delivered as the Richard Goodman Memorial Lecture at Brighton Polytechnic on 14 February 1973 (Beer1975b [1973]). The previous month, however, reports of Cybersyn had appeared in the British underground press and then in national newspapers and magazines (Beer 1981, 335), and the media response had proved hostile. The day after Beer's "Fanfare" speech, Joseph Hanlon wrote in the *New Scientist* that Beer "believes people must be managed from the top down—that real community control is too permissive. . . . The result is a tool that vastly increases the power at the top," and concluded with the remark that "many people . . . will think Beer the supertechnocrat of them all" (Hanlon 1973a, 347; and see also Hanlon 1973b). Hanlon's article thus sketched out the critique of cybernetics discussed in chapter 2: cybernetics as the worst sort of science, devoted to making hierarchical control more effective.

Beer replied in a letter to the editor, describing Hanlon's report as a "hysterical verbal onslaught" and resenting "the implied charge of liar" (Beer 1973a). One H. R. J. Grosch (1973) from the U.S. National Bureau of Standards then joined in the exchange, explicitly calling Beer a liar: "It is absolutely not possible for Stafford Beer, Minister Flores or the Chilean government or industrial computer users to have since implemented what is described." Grosch further remarked that this was a good thing, since Cybersyn "well merits the horror expressed by Dr Joseph Hanlon. . . . I call the whole concept beastly. It is a good thing for humanity, and for Chile in particular, that it is as yet only a bad dream." Beer's reply (1973b) stated that the Cybersyn project had indeed achieved what was claimed for it, that "perhaps it is intolerable to sit in Washington DC and to realise that someone else got there first—in a Marxist country, on a shoestring," and that "as to the 'horror' of putting computers to work in the service of the people, I would sooner do it than calculate over-kill, spy on a citizen's credit-worthiness, or teach children some brand of rectitude."

The political critique of Cybersyn and the VSM was further elaborated and dogged Beer over the years, and I want now to review its overall form, rather than the details, and how one might respond to it. The critique is fairly straightforward, so I shall present it largely in my own words.[35]

In 1974, Beer said of Cybersyn that it "aimed to acquire the benefits of cybernetic synergy for the whole of industry, while devolving power to the workers at the same time" (Beer 1974a, 322), and there is no doubt of his good intentions. His critics felt that he was deluding himself, however, and Hanlon's description of Beer as a "supertechnocrat" presaged what was to follow. I find it useful to split the critique into four parts.

1. The VSM undoubtedly *was* a technocratic approach to organization, inasmuch as it was an invention of technical experts which accorded technical experts key positions—on the brain stem of the organization at levels 3 and 4. No one had asked the Chilean workers what sort of a subpolitical arrangement they would like. Nor, I believe, did Beer ever envisage the basic form of the VSM changing and adapting once it had been implemented in Chile. There is not a lot one can say in reply to this, except to note that, on the one hand, the fixity of the overall form of the VSM can be seen as a noncybernetic aspect of Beer's cybernetic management. As ontology in action, the critics seized here on a nonexemplary feature of Beer's work. But we might note, too, that expert solutions are not necessarily bad. Beer's argument always was that cyberneticians were the experts in the difficult and unfamiliar area of adaptation, and that they had a responsibility to put their expertise to use (see, e.g., Beer 1975 [1970], 320–21). To say the least, Cybersyn was a new and imaginative arrangement of socioinformatic relations of production, which might, in principle—if the Pinochet coup had not happened—have proved to have increased the freedom of all concerned. Beyond this, though, the critics found more specific causes for concern within the structure of the VSM itself.

2. Another thread of the critique had to do with the algedonic signals that passed upward unfiltered to higher levels of the VSM. Beer spoke of these as "cries for help" or "cries of pain." They were intended to indicate that

problems had arisen at the system 1 level which could not be addressed there, and which therefore needed assistance from higher levels in their resolution. Beer assumed that the upper levels of the system would adapt a benevolent stance relative to the lower ones and would seek to provide genuine assistance on the receipt of an algedonic signal. Critics pointed out instead that such signals could also constitute a surveillance system that would sooner or later (not necessarily under Allende) be used against the lower levels. A profit-maximizing higher management might readily translate too many algedonic warnings into a rationale not for assistance with problems but for plant closures. Again, it is hard to spring to Beer's defense. He might have replied that to think this way is to denature and degrade the biological model behind the VSM. Brains do not jettison arms and legs every time we get pins and needles, but the obvious reply would be that this simply brings into question Beer's biological model for social organizations. For Beer, this was a normative aspect of the model, but no one could guarantee that higher management would accede to this.

A more detailed version of this same critique acknowledged that there must be some vertical communication within organizations but questioned the automaticity of "cries for help." In the VSM, this was simply a matter of statistical filtration of data. If production indices remained anomalous after an agreed period of time, the algedonic signal automatically passed on to the next level. Werner Ulrich (1981, 51–52) pointed out that in a less automated system there would be a place for management learning—managers come to recognize patterns in the signals arriving at their level and thus to discriminate between which needed to be passed on and which did not—thus protecting the lower levels to some extent from vindictiveness above. I do not know whether Beer ever addressed this point, but, again, the VSM was not exemplary of the cybernetic ontology in action to just the degree to which this automaticity was a fixed part of the VSM.

3. Following the lines set down by Hanlon, the VSM's critics asserted that the VSM prescribed a "top-down" mode of organizational control: management or government gave orders that the workers were then expected simply to implement. Cybersyn "has some kind of built-in executive power. . . . Its strongly hierarchical organisation and its concept of 'autonomy' one-sidedly serve the top decision maker, the government" (Ulrich 1981, 52, 54). As before, there is something to this critique, but it is worth taking it slowly. Though the critics seem to have read Cybersyn as implementing a classic "command and control" form of organization, with a unilinear flow of orders descending

from on high, in this they were wrong. Beer did not think of viable systems in that way. This was precisely the significance of the adaptive couplings that pervaded the VSM, especially the couplings between the various levels. As discussed earlier, these were modelled on the reciprocal vetoing in Ashby's multihomeostat setups and implied that the parties at different levels had to cast around for mutually agreeable initiatives and plans, precisely *not* the traditional command-and-control mode. These adaptive couplings were the most definitively cybernetic component of the VSM, and it is significant that the critics failed to get to grips with them or even to recognize their distinctive character. Beer often complained that outsiders erred in a similar way concerning all sorts of cybernetic machines and contrivances, utterly failing to grasp their adaptive aspects, and this seems to have been the case here. If ontology makes a difference, then that difference eluded the VSM's critics. But more needs to be said.

Cybersyn was, on one occasion, operated in both a surveillance and a command-and-control mode. This was the time of the *gremio* strike in October 1972, a "CIA-instigated trucker's strike" in Chile (Ulrich 1981, 54n; Beer 2004 [2001], 860) which threatened to halt flows of goods around the country.[36] The Cybernet information system was then switched temporarily to monitoring shortages around the country and figuring out how to use the transportation available to overcome them. Beer was very pleased that this approach worked and that the strike was defeated (Beer 1981, 312–15), but there was no homeostatic give-and-take involved in this episode in negotiating plans between different levels, and it serves to show just how readily the organic quality of the VSM could be conjured away, and, indeed, this possibility seems to have appealed to Allende's enemies.[37] "At the end of July [1973] . . . several strange messages reached me. . . . They were coming from the political opposition. It seemed that this [Cybersyn] was the best project undertaken under Allende's aegis, and that his (self-assumed) successor would continue it in his own way. This would not, of course, involve any 'nonsense' about worker participation. . . . I found these overtures obnoxious; but our strategies were well prepared" (Beer 1981, 345). The strategies, I believe, were intended to render Cybersyn useless in the event of a coup, but three comments are called for. First, in its genuinely cybernetic aspect—the adaptive couplings between levels—the VSM did serve to undo hierarchies of command and control. Second, these adaptive couplings could easily be "switched off" and replaced by asymmetric ones. It is fair to say, then, that the VSM was hardly a potent bulwark against the institutional arrangements that Beer wanted to obviate. This, too, was much on his critics' minds. But third, as Beer might have advised, we should be concerned here with the future more than the past. Even if key components of the VSM were readily erasable, the VSM remains interesting as a model for a democratic subpolitics.

4. We can return to the question of goals. In chapters 3 and 4 we looked largely at systems with fixed goals. Ashby's homeostats adapted open-endedly, but *so as* to keep their essential variables within given limits. According to Beer, the quasi-organic viable system likewise had goals that patterned its adaptation. But, unlike Ashby, Beer was not attempting to construct models of the adaptive brain, and he therefore did not have to take a sharp position on what the goals of a viable system are. I said earlier that one could think of the profitability of an enterprise as the sort of thing at issue, but actually Beer had something different and more interesting in mind, which we can get at via the critique of the VSM. At the heart of Werner Ulrich's (1981, 35) long critique, for example, is a contrast between "purposive" and "purposeful" systems, which relates to a more familar distinction between means and ends: a "purposive" system is a means to some extrinsically specified end, while a "purposeful" one can deliberate on its own ends. Ulrich criticized the VSM as purposive, and at one level this is correct. Beer was keen not to try to build any substantive goals beyond adaptability into the VSM; this is an aspect of what was entailed in my earlier description of the VSM as a form of subpolitics.

Ulrich, however, went on from this observation to claim that because the VSM had no substantive goals, then whatever goals a system came to manifest would have to be supplied in a top-down fashion, from systems 4 and 5 of the model—we are back to technocracy from a different angle. But here there are some complications worth discussing. One reply would be that Beer was working for a democratically elected government responsive to "the will of the people," but that is an observation about the specific context of Cybersyn rather than an intrinsic feature of the VSM in general. Another reply would go along the lines indicated above: that the adaptive couplings between the VSM's levels are reciprocally adaptive, not one-way. But here, still, some asymmetry remained in the VSM. Beer does not seem to have envisaged the formulation of new plans and goals from below; the higher levels of management and government do seem to have held the advantage here in his thinking (though this assertion will be qualified below when we come to his work on "syntegration," which indeed focused on inclusive processes of goal formation). Nevertheless, Project Cybersyn, as it evolved, did at least try to close the loop between government initiatives and their popular reception in various ways, and I want to examine just one of these.

On Goals

> In March 1972 . . . we addressed the basic issue of the organization of the state that is not economic but societary. . . . I wrote a second paper about a project to examine:
>
> > "the systems dynamics
> > of the interaction
> > between government and people
> > in the light of newly available technology
> > such as TV
> > and discoveries in the realm
> > of psycho-cybernetics"
>
> (Beer 1981, 278)

There were, of course, many channels by which the Chilean government could communicate with the Chilean population at large and vice versa. But the reference to TV immediately suggests an asymmetry. Governments could transmit information over the television in great detail and length—a high-variety channel, in the language of information theory. The people, in contrast, could not reply via the TV at all—an exceedingly low-variety channel. Of course, the people could communicate via other channels, such as forming political parties and voting in elections, but Beer felt that it was necessary to do something to increase the information flow from people to government if a homeostatic equilibrium was to be achieved. He also, as usual, felt that the channel from people to government should be a real-time one, so that the latter could react to how the former felt today rather than last week or last month or last year.[38] The solution Beer proposed, novel and endearing, is shown in figure 6.15. The aim here was to supplement the economic algedonic feedback of the VSM with social feedback. TV viewers, for example, would be provided with very simple "algedonic meters" of the form shown in the lower left of figure 6.15. These would be simple semicircular devices in which a partition could be rotated clockwise (toward "happy") or counterclockwise ("unhappy") in response to whatever was happening before them—a televised political speech, say. Some simple wiring arrangements would aggregate

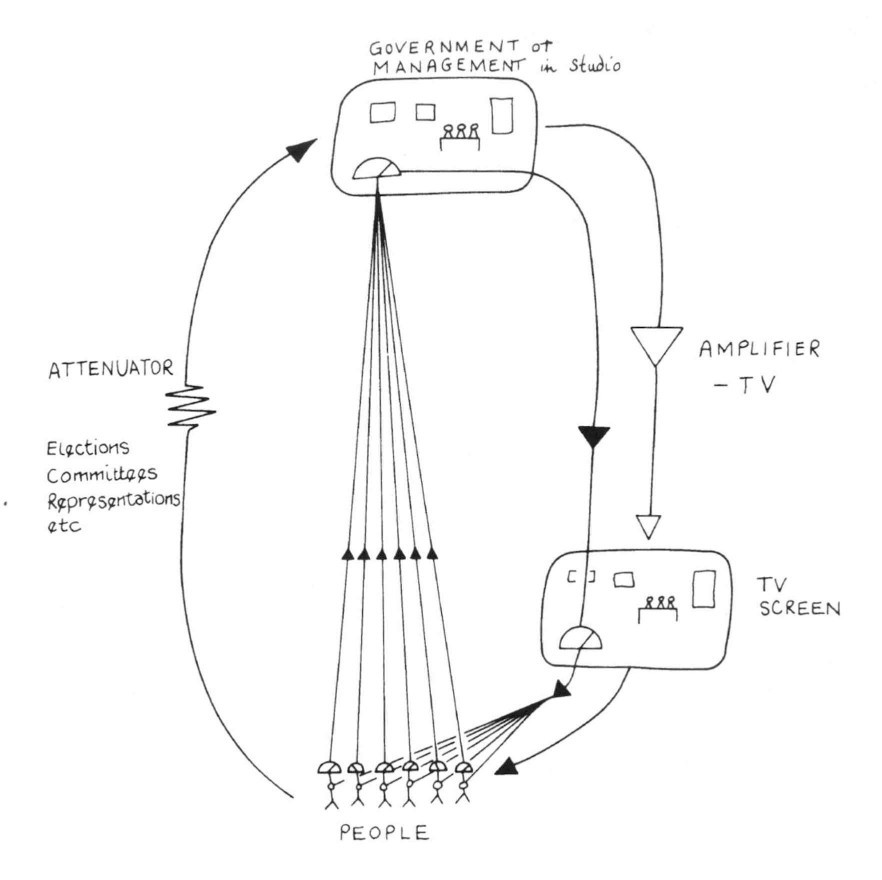

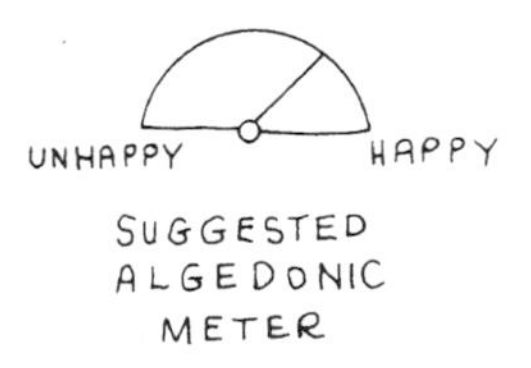

Figure 6.15. Feedback from the people. Source: S. Beer, *Brain of the Firm*, 2nd ed. (New York: Wiley, 1981), 281, fig. 45.

these algedonic signals (the precise arrangements being left open in the initial proposal) and transmit them for display in real time on the TV screen. In this way, the politicians would get instantaneous feedback on their proposals or arguments. And—this is the clever bit—the viewers could also see how the politicians would react to the feedback, and so on in a cascade of feedbacks between the TV studio and its audience (Beer 1981, 285). In effect, some channel, however crude, would thus be opened for mass debate—or, better, a dance of agency—with the government. Again, policy making could thus emerge in real-time interaction.

Like many of the cybernetic devices we have been exploring, these algedonic meters of Beer's were at once serious and amusing, and even startling in spanning the gap between the two. Their origins, I would guess, lay in the clapometers and swingometers of the BBC's popular music TV shows and election reporting.[39] An interesting feature is that they were truly algedonic in being able to register pleasure as well as pain, unlike the algedonic signals in the basic VSM, which were regarded as warnings that something was wrong. Though Beer initially conceived their use in mass communication, they could obviously be deployed in much more limited contexts—in small meetings, say, where some planning group reported to its constituents, or at factory gates as feedback from the workers to management.

Beer's son Simon, an electrical engineer, built a prototype system "of ten algedonic meters, linked by a single wire in a loop through a large summation meter" (Beer 1981, 284), and took it out to Chile, where experiments were done on its use with a group of fifteen friends. These friends, however, "rapidly learned how to rig the system. They joined in plots to 'throw' the lecturer by alternating positive and negative responses, for instance" (286). The algedonic meter was, in this instance, too much fun. And one can easily imagine less amusing forms of rigging—the political party instructing its supporters to slam the indicator to the left whatever an opponent said—or even arguments about whether "unhappy" should be at the left or the right. This form of feedback was thus never introduced in Chile, leaving Beer to reflect that its design was a tricky problem and that more cybernetic research was needed. Nevertheless, it is interesting to stay with them just a little longer.

Beer contrasted his algedometers favorably with another and more familiar form of quasi-real-time feedback from the people to government: questionnaires and opinion polls (Beer 1974a, 334–38). From Beer's perspective, the great thing about the algedonic meters was that they were inarticulate, wordless. They measured "happiness," but the nature of happiness and its causes were left undefined. They simply indicated a positive or negative response on some undefined scale. Beer's enthusiasm for this mode of communication had to do with his intense interest in performance and his associated suspicion of representational knowledge. The trouble with opinion polls, Beer argued, is that the domain of inquiry is circumscribed by the questions asked (themselves framed by politicians, journalists, academics, and so on) and lacks variety. Articulated questions might therefore be able to determine how people feel about specific government policies, but they can never find out whether people's real concerns lie entirely elsewhere. Polls can never contribute, then, to the emergence of real novelty in real-time politics, only to a fine-tuning of the status quo. In contrast, the algedonic meters constituted an open invitation to genuine experiment. If a politician or journalist were to float some wild idea and the integrated meter reading went from lethargically neutral to wildly positive, there would be reason to think that some genuine but hitherto unthought-of social desire had been tapped.

And here we can return to Ulrich's critique of the VSM as purposive rather than purposeful. Though Beer did not try to build into the VSM any substantive goals, he did try to think through the ways in which the system could articulate its own goals, in practice, in a nonhierarchical fashion. We can think of the algedonic meters as expanding the VSM as a subpolitical diagram of social relations and information flows in such a way as to enable any organization to become purposeful, rather than purposive, on its own terms. Ulrich is wrong here about the VSM, at least in principle, though, as above, practical concerns are not hard to find: it would have been less difficult for General Pinochet and his friends to eliminate algedonic meters than, say, rifles in the hands of the workers.

One last thought about the algedonic meters. What did they measure? At the individual level, an unanalyzed variable called "happiness." But for the aggregated, social, level Beer coined a new term—*eudemony*, social well-being (Beer 1974a, 336). Again he had no positive characterization of eudemony, but it is important that he emphasized that it is not any of the usual macrovariables considered by politicians and economists. Eudemony is not, or not necessarily, to be equated with GNP per capita, say, or life expectancy (Beer 1974a, 333). Eudemony is something to be *explored* in the adaptive performance of a viable social system, and, obviously, Beer's algedonic meters were an integral part of that. This thought is perhaps the most radical aspect of Beer's subpolitics: the idea that social systems might continually find out what their collective ends are, rather than, indeed, having those ends prescribed from above (the wonders of the free market, for example). And this remark gets us back to the general question of cybernetics and goals. Beer's cybernetics, unlike that of Walter and Ashby, did not enshrine any idea of fixed goals around which adaptation was structured. Goals, instead, could *become* in Beer's (and Pask's) cybernetics. As ontological theater, then, the VSM staged a vision of open-ended becoming that went an important step beyond that of the first-generation cyberneticians. Beer had not, of course, solved the problem of building a machine that could mimic the human facility of formulating goals; his systems could be adaptive at the level of goal formation precisely because they contained human beings within themselves.

— — — — —

Where does this leave us? After reviewing the critiques of the VSM and Project Cybersyn, I continue to think that we *can* see the VSM as enshrining a very interesting approach to what I have called subpolitics. The VSM offers a considered topology of social locations and relations, information flows and transformations that, to a considerable degree, promises a dispersal of autonomy throughout social organizations. The key elements of the VSM, from this perspective, are the adaptive, homeostat-like couplings between the various levels of the VSM, and the algedonic signals that travel back up the system. Like Beer's earlier experimentation with biological computing, his work on the VSM seems original and singular to me. It is hard to think of any equivalents in more conventional approaches to political theory and practice. And for this reason I am inclined to point to the VSM as another item on my list of striking examples of the cybernetic ontology in action, in politics and management. Here again we can see that the cybernetic ontology of unknowability made a difference.

Turning to the critics, it is significant that they seemed unable ever quite to get the VSM into focus. Beer's overall cybernetic aim, to bolster the adaptability of organizations, was never, as far as I can make out, mentioned by the critics; neither was the key cybernetic idea of adaptive coupling between levels. Instead, the critics focused on a cybernetically denatured version of the VSM, a version from which the distinctively cybernetic elements had been removed, turning it into a nightmare of command and control. The critics mapped the VSM onto a distinctively modern space in which it did not belong, and they found it wanting there. This inability to contemplate the thing in itself I take to be further evidence that ontology makes a difference.[40]

Having said that, I have also recognized that the critics' concerns about the VSM were not empty. It does seem clear that systems like that envisaged in Project Cybersyn could be readily stripped down in practice and turned into rather effective systems of command, control, and surveillance, the very opposite of what both Beer and the critics aimed at. But as I have said before, the object of this book is not to resurrect any specific cybernetic project, including Cybersyn. It is to exhibit and examine a whole range of such projects—as a demonstration of their possibility and their difference from more conventional projects in cognate areas, and as models for the future. A future cybernetic politics that followed Beer's lead into subpolitics might well want to bear in mind the democratic fragility of the VSM—while contemplating algedonic meters as, shall we say, a desperate but entertaining attempt to open up a politically deadening status quo.

Pinochet's coup in Chile was not the end of Beer's involvement in politics at the governmental level, especially in Central and South America. He went on to consult for the governments of Mexico, Venezuala, and Uruguay, as well as, in other directions from the United States, Canada, India, and Israel (Beer 1990a, 318–21), and "bits and pieces of the holistic approach have been adopted in various other countries, but by definition they lack cohesion" (Beer 2004 [2001], 861).[41] I will not pursue that line of development further here; instead, I want to explore Beer's cybernetic politics from another angle.

The Politics of Interacting Systems

LAST MONTH [SEPTEMBER 2001], THE TRAGIC EVENTS IN NEW YORK, CYBERNETICALLY INTERPRETED, LOOK QUITE DIFFERENT FROM THE INTERPRETATION SUPPLIED BY WORLD LEADERS—AND THEREFORE THE STRATEGIES NOW PURSUED ARE QUITE MISTAKEN IN CYBERNETIC EYES. . . . ATTEMPTS TO GUARD AGAINST AN INFINITE NUMBER OF INEXPLICIT THREATS DO NOT HAVE REQUISITE VARIETY.

STAFFORD BEER, "WHAT IS CYBERNETICS?" (2004 [2001], 861–62)

So far we have focused on the internal politics of the VSM—on social arrangements *within* a viable organization. Here, the organization's environment was conceptualized in rather amorphous terms, simply as that to which the organization needed to adapt. As we saw in the previous chapter, in the 1950s Ross Ashby was led to think more specifically about environments that themselves contained adaptive systems and thus about interacting populations of adaptive systems, including the possibility of war between them. Beer's experiences in Chile and of the subversion of the Allende regime by outside states, especially the United States, led him to reflect along similar lines from the 1970s onward. These reflections on the interrelations of distinct systems, usually conceived as nation-states, themselves warrant a short review.

Beer's basic understanding of international relations followed directly from his cybernetic ontology along lines already indicated. Nation-states are obvious examples of exceedingly complex systems, always in flux and never fully knowable. Their interaction should thus take the usual form of reciprocal vetoing or mutual accommodation, exploring, respecting, and taking account of the revealed variety of the other. Beer found little evidence for such symmetric interaction in the contemporary world, and thus, much of his analysis focused on what happens when it is absent. At the other pole from homeostat-like explorations lies the attempt to dominate and control the other, and Beer's argument was that this must fail. According to Ashby's law, only variety (on one side) can control variety (on the other). Any attempt simply to pin down and fix the other—to make it conform to some given political design—is therefore doomed to make things worse. The imposition of fixed structures simply squeezes variety into other channels and manifestations which, more or less by definition, themselves subvert any imposed order.

Beer's general analysis of macropolitics was thus, throughout his career, a pessimistic one: conventional politics is bereft of cybernetic insight and thus continually exacerbates crises at all levels. This rhetoric of crisis is a resounding refrain from his earliest writings to his last. In Beer's first book, the crisis is one of the West in general (the only instance of Cold War rhetoric that I have found in his writing) and of British industry in particular (Beer 1959, ix): "The signs are frankly bad. . . . The index of industrial production has not moved up for four years. We desperately need some radical new advance, something qualitatively different from all our other efforts, something which exploits the maturity and experience of our culture. A candidate is the science of control. Cybernetic research could be driven ahead for little enough expenditure compared with rocketry, for example. And if we do not do it, someone else will." In his later and more political writings, the crisis was often said to be one of the environment and of the conditions of life in the third world, as well as the more usual sense of political crisis: a socialist government in Chile as a crisis for the Americans and British being a prime example.[42]

When I first encountered this language of crisis in Beer's writing, I tended to ignore it. It seemed self-serving and dated. On the one hand, the rhetorical function of "crisis" was so obviously to motivate a need for cybernetics. On the other, we all used to talk like that in the 1960s, but, in fact, the world has not come to an end since then. As it happens, though, while I have been writing about Beer, his stories have started to seem very relevant and, indeed, prescient. Everything that has happened since those planes flew into the World Trade Center and the Pentagon speaks of an American attempt (abetted by

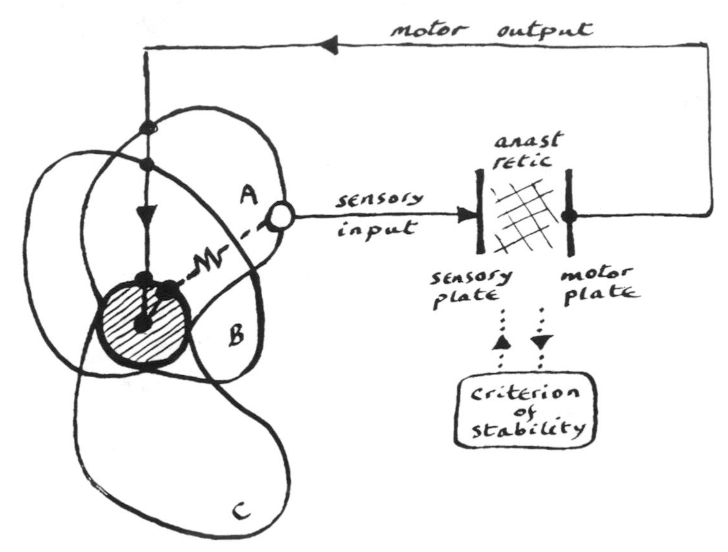

Figure 6.16. The cybernetics of crisis. Source: S. Beer, *Brain of the Firm*, 2nd ed. (New York: Wiley, 1981), 354, fig. 48.S.

the British) at command and control on a global scale, seeking to freeze the world, to stop its displaying any variety at all—running from endless "security" checks and imprisonment without trial to the invasions of Afghanistan and Iraq. And the aftermath of the American invasion of Iraq—what we have been taught to call "the insurgency," the killing, destruction, mayhem, and torture in the name of "democracy"—speaks vividly of the negative consequences of seeking to repress variety.

Little more can be said here—this book is not a treatise on recent world history—but I do want to note that Beer's "cybernetics of crisis" included an analysis of how crises like the present one can arise. Again, Beer's focus was on transformative flows of information. Figure 6.16 is his basic diagram for considering such processes: the hatched area denotes a crisis affecting three different interest groups, which might be nation-states, *A*, *B*, and *C*. The details are less important than Beer's general analysis of the information flow from the crisis region into *A* ("sensory input") and the return action of *A* on the crisis ("motor output"). What Beer emphasized was that such information flows necessarily impoverish variety, and that in a systematic way. His argument was that *representations* of crises are inevitably filtered through low-variety conceptual models, models through which governments interpret crises to themselves and the media interpret them to the public. These models then feed into a low variety of potential actions which return to intensify the variety of the crisis along axes that are unrepresentable in the models, and so on around the loop.

Let me close this section with three comments. First, we can note that this last discussion of the role of models in the production of crises is of a piece with Beer's general suspicion of articulated knowledge and representation. Models might be useful in performance, as in the VSM, but they can also interpose themselves between us and the world of performances, blocking relevant variety (hence the significance of the inarticulacy of Beer's algedonic meters, for example). Second, Beer died before the invasion of Iraq; the above thoughts on that are mine, not his. But, again, I am struck now not by any self-serving quality of his rhetoric, but by the prescience of his analysis. The highly simplifed story of information flows and variety reduction that I just rehearsed illuminates how global politics could have collapsed so quickly into one-bit discriminations (Beer 1993a, 33) between "us" and "them," the goodies and the baddies; how it could have been that a majority of the American population could believe there was some connection between Al Qaeda and Iraq prior to the invasion and in the existence of what we were taught to call "weapons of mass destruction"; how it is that the American public and, perhaps, their government could have expected the invaders to be greeted with flowers and kisses rather than car bombs; and (turning back to the question of controlling variety) why mayhem should have been expected instead. Of course, third, one does not have to be Stafford Beer or a cybernetician to be critical of the war on terror, a "war" in which, "allies are expected to go into battle against an abstract noun, and to assault any nation unwilling to mobilize in such folly" (S. Beer 2001, 862–63). What interests me, though, is the generality of Beer's cybernetic analysis. We all know how to generate simplistic stories of heroes and villains, and much of the political talk of the early twenty-first century takes that form. Take your pick of the goodies and baddies—Saddam Hussein and Osama bin Laden or George W. Bush and the neocons. Such reversible stories will no doubt always be with us. Beer's analysis, instead, did not focus on the particulars of any one crisis. He actually began the most extended exposition of his analysis by mentioning the British abdication crisis of 1936, arguments over Indian independence from Britain in 1946, and the Suez crisis of 1956 (Beer 1981, 352–53). His analysis did not hinge on the question of whether George W. Bush was evil or stupid; his argument was that something was and is wrong at the higher level of large-scale systems and their modes of interaction that persistently produces and intensifies rather than resolves global crises. I take the novelty of this style of analysis to be another example of the ways in which ontology makes a difference.

Team Syntegrity

HOW SHALL WE EVER CONCEIVE

HOWEVER EXPRESS

A NEW IDEA

IF WE ARE BOUND BY THE CATEGORIZATION

THAT DELIVERED OUR PROBLEM TO US

IN THE FIRST PLACE

?

STAFFORD BEER, *BEYOND DISPUTE* (1994B, 8)

From the time of Project Cybersyn onward, the VSM was the centerpiece of Beer's management consultancy. In parallel to the VSM, however, he also developed a rather different approach to organizations that he called "team syntegrity." This grew from the 1950s onward, "flared into considerable activity 20 years ago, and occupied me throughout 1990 in a series of five major experiments" (Beer 1994b, 4). In the 1990s also, the conduct of "syntegrations" became partly a commercial business for Beer and his friends, associates, and followers.[43] Beer only published one book on syntegrity, *Beyond Dispute* (1994b), as distinct from three on the VSM, but he and his collaborators developed and reflected upon syntegration in considerable detail.[44] I am not going to attempt to do justice to that here. My aim is to sketch out the basic form of the approach, to connect it to the cybernetic ontology, and, continuing the above discussion, to examine it as a form of micro-sub-politics.[45]

Put very crudely, the substance of team syntegrity was (and is) an evolving format or protocol for holding a meeting, a rather elaborate meeting called a "syntegration," and we can explore this format in stages. First, there are the connected questions of what the meeting is about and who should come to it. On the latter, Beer offered no prescriptions. The idea was that syntegration was a process focused on some topic of interest to its participants. His model for thinking about this was a group of friends who met regularly in a bar and found themselves returning to some topic, perhaps current politics, but an early example in the development of the technique involved members of the British Operational Research Society seeking to redesign the society's constitution in 1970, and the first experiment in 1990 involved a group of friends and friends of friends thinking about world governance (Beer 1994b, 9, 35). The participants were, then, characterized by their common concern and interest in whatever the syntegration was about. Beer called such a group an "infoset," and, for reasons that will become clear, the basic form of an infoset would comprise thirty people.[46]

But just how should the topic of such a meeting be defined? This was a matter of pressing concern for Beer, a concern that ran along much the same lines as his critique of opinion polls mentioned earlier. The usual way of structuring such a meeting would be to distribute in advance an agenda listing specific topics for discussion and action. Beer's point was that such an agenda prefigures its outcome within lines that can already be foreseen, and "anything truly novel has two minutes as Any Other Business" (Beer 1994b, 9). His idea, therefore, was that the first element of a syntegration should itself be the construction by the infoset in real time of a set of relatively specific topics for discussion. In the mature form of syntegration this entailed a fairly complicated protocol extending over some hours, but, in essence, the procedure was this: Knowing the general topic of the meeting—world governance, say—each participant was asked to write down at least one brief statement of importance (SI) relevant to the topic, aiming to encourage original discussion of some aspect of the overall focus of concern. These statements would then be publically displayed to all of the participants, who would wander around, discussing whichever SIs interested them with others, elaborating them, criticizing them, or whatever (all this, and what follows, with the aid of experienced "facilitators"). Finally, after a prescribed length of time, the participants would vote for the developed SIs they considered of most importance, and the top twelve SIs would be chosen as the focus for the remainder of the meeting (27). In this way, something like a specific agenda would be constructed, *not* as given in advance but as emergent itself in the process of the meeting.

Given a set of thirty participants and twelve SIs, what happens next? In a short but complicated process, participants are each assigned to a pair of SIs, respecting, as much as possible, their preferences. Then the process of syntegration proper begins, and things get complicated to explain. How do you organize the discussion of twelve topics by thirty people? A completely unstructured agora-like situation is imaginable, but experience dictates that it would get nowhere. One might try to structure the meeting by, say, ranking individuals or topics in terms of priority, but this would return to Beer's critique of agendas, one step down the line. Inspired by Buckminster Fuller's geodesic domes (Beer 1994b, 12–14), the solution that Beer arrived at was to structure discussions in the form of a geometric figure, the icosahedron (fig. 6.17).[47]

An icosahedron has thirty edges and twelve vertices, and hence the appearance of these numbers above. Each of the twelve topics is assigned to

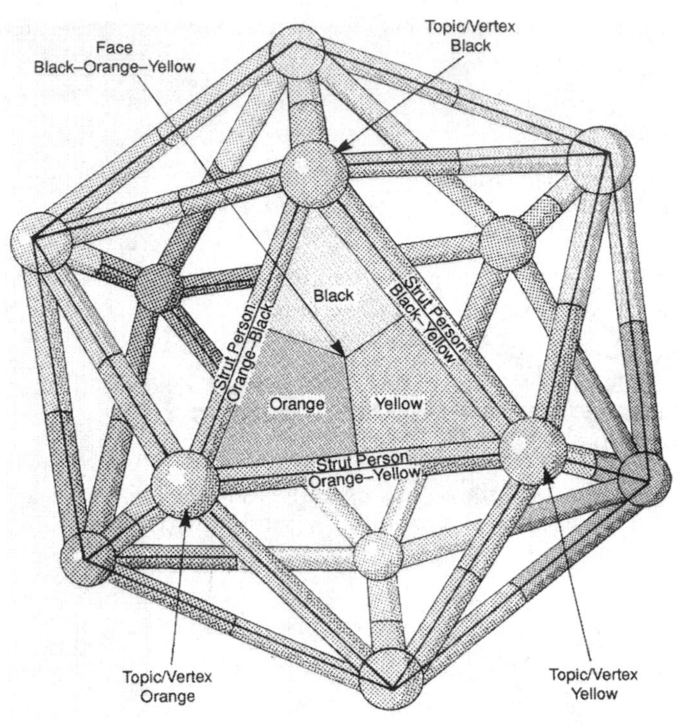

Figure 6.17. The syntegration icosahedron. Source: S. Beer, *Beyond Dispute: The Invention of Team Syntegrity* (New York: Wiley, 1994), 338, fig. S6.2.

one of the twelve vertices of an imaginary icosahedron; each participant is imagined to be placed on an edge and engages in discussions of the two topics assigned to the vertices at the end of his or her edge. In turn, this implies that each participant is a member of two discussion groups of five people, each associated with the five edges that meet at any vertex (plus some additional niceties, including the participation of "critics" from disconnected edges, which I will not go into). These groups then meet repeatedly (three or more times) over a period of days for discussions that take off from SIs at their vertex, adding to, refining, and elaborating these statements in the course of their interactions. These discussions cannot all take place at once—one cannot be a member of two groups discussing two topics simultaneously—so participants alternate in time between their topics. And, according to Beer, the effect of this is that discussions *reverberate* around the icosahedron. On the first occasion, the discussion of any topic has a sui generis quality defined by the interaction of the five people concerned. But by the second iteration, their positions have each been inflected by different discussions at the other end of their edges, themselves inflected by discussions at one further remove. And by the third iteration these inflections are traveling around the geometrically closed figure, and there is the possibility that an earlier contribution returns "to hit its progenitors in the back of the neck" (Beer 1994b, 13). This is what Beer meant by reverberation: ideas travel around the icosahedron in all directions, being transformed and becoming progressively less the property of any individual and more that of the infoset as a whole. At the end of the process, each vertex has arrived at a final statement of importance (FSI), and these FSIs are the collective product of the syntegration (Beer 1994b, 32–33).

Thus, in outline, the form of the syntegration process, and to put a little flesh on what the products of such a process can look like, we can look briefly at a syntegration held in Toronto in late 1990 (Beer 1994b, chap. 6). "The group who came together were recruited mainly by word of mouth. . . . Thus the Infoset was assembled in an unusually arbitrary way: we may call it such a unity only because of its members all being drawn to the heading on the poster: 'What Kind of Future do You Want?'" (87). The first three of the SIs constructed at the start of the syntegration were: 'God is a verb not a noun,' 'Each child spontaneously desires to develop responsibilities commensurate with its abilities,' and 'Censorship is a personal issue.'" In Beer's précis, the first three of the FSIs, the products of the syntegration, were (97–98)

1. *Local Empowerment*: the need to push decision making downwards, especially in the case of abolishing nuclear war.
2. *Law and Government*: the move from ownership to stewardship, control to guardianship, competition to cooperation, winners and losers to winners alone.
3. *How to Make World Peace*: sovereign individuals acknowledge and accept the responsibility of a (human) world social contract, towards environmental protection, security, and evolution of the planet.

What can we say about this example? First, it shows that syntegration can be a genuinely dynamic and open-ended process: the SIs and FSIs were in no sense contained in the original topic; they evidently emerged in the syntegration itself. But what about the statements of importance themselves? They hardly come as singular revelations, at least to scholars interested in such matters, but, as Beer put it, "it could not be claimed that the FSIs . . . embodied major new discoveries, although they may have done for some present. . . . [But] they are hardly banal" (97).

This and similar experiences in other syntegrations led Beer to remark that "amongst many others I have often claimed that in planning it is the process not the product that counts" (Beer 1994b, 97), and *Beyond Dispute* documents in various ways the fact the participants in syntegrations generally found them enjoyable and productive. The phrase "consciousness-raising" comes to mind, and we will see below that such phrases had a very literal meaning for Beer—his idea was that a genuine group consciousness could arise from the reverberations of syntegration. Let me close this section, however, with some general reflections on the syntegrity approach to decision making, with the critique of Beer's VSM in mind—"a topic of which" Beer declared himself in 1990 to be "heartily sick" (Beer 1990b, 124).

Like the VSM, syntegrity can be described as a form of subpolitics, this time at a microscale of small groups. Like the VSM, syntegrity had at its heart a diagram, though now a geometric figure rather than a neurophysiological chart. Again like the VSM, syntegrity, Beer argued, staged an inherently democratic organization, an arrangement of people in which concrete, substantive, political programs could be democratically worked out—indeed, he often referred to syntegration as "complete," idealized," and "perfect democracy" (Beer 1994b, 12; 1990b, 122). And, unlike the VSM, in this case it is hard to dispute Beer's description. Beer's critics were right that the VSM could easily be converted to a system of surveillance, command, and control, but it is hard to contrive such fears about syntegrity. By construction, there are no privileged positions in the syntegration icosahedron, and there is no evident way any individual could control the syntegration process (short of wrecking it beyond recognition).

Once more, too, we can see how ontology and subpolitics are bound up together in syntegrity. As exceedingly complex systems, the participants cannot know in advance what topics will emerge from the syntegration process, and this emergence is orchestrated as a process of multihomeostat-like reciprocal vetoing and creative mutual accommodation between participants and statements of importance. Of course, there is some prestructuring entailed in the assembly of an infoset around a broad topic and in the geometric arrangement of persons and topics, but here we can note two points. First, the syntegration process was even more fully open ended than that of the VSM. If a set of formal if revisable mathematical models were intrinsic to the latter, no such formalisms intervened in syntegration: topics, statements, and goals were all open-endedly revisable in discussion as they reverberated around the icosahedron. Second, the icosahedral structure did undeniably constitute an infringement on individual freedom: individuals could only contribute to the discussion of topics to which they had been assigned. In this sense, and as usual, syntegrity staged a hybrid ontology, partially thematizing and acting out an ontology of becoming, but within a fixed framework. Beer would no doubt have remarked, as he did of the VSM, that any form of organization exacts its price, and that the price here was worth paying for the symmetric openness to becoming that it made possible. One can also note that syntegration was a finite and limited process; participants were not locked into it, in the way that they might be within a business or a nation. So, in the next syntegration participants could take other positions within the diagram, and, of course, the entire general topic could shift.

Throughout this book we have been concerned with the socio-ontological mismatch between cybernetics and modern institutions, with the amateurism of Beer's work on biological computing as our latest example. In the earlier chapters we also ran into examples of a constructive response to the mismatch: Kingsley Hall, for example, as providing a model for a new social basis for cybernetic forms of life, the germ of a parallel social universe as Alexander Trocchi envisaged it. Beer, too, contributed to this constructive project. As we saw, he played a key role in the establishment of the Department of Cybernetics at Brunel University—a partly successful attempt to implant a sustainable cybernetic presence in the established academic order. From another angle, the VSM can be seen as an attempt to reconfigure the world of organizations along cybernetic lines, to make that world an explicitly and self-consciously cybernetic place. And we can understand the team syntegrity approach to decision making similarly—not now as the construction of enduring institutions, but as making available a finite and ephemeral social form lasting for just a few days, that could be mobilized ad hoc by groups at any scale for any purpose, from reorganizing the British OR society up to world governance.[48] One does not have to subscribe to the details of the VSM or team syntegrity; the point here is that Beer's work can further enrich our imaginations with concrete examples of what Trocchi's parallel universe might look like, and that those forms would indeed be importantly different in specific ways from the hegemonic forms of our present social, political, and subpolitical arrangements. Again, ontology makes a difference, here in the domain of subpolitics.

Cybernetics and Spirituality

IN INDIA THERE ARE MANDALAS—PICTURES CONVEYING SACRED INSIGHTS NOT EXPRESSED IN WORDS. OUR MODERN CHIPS MAY NOT BE SACRAMENTALS, BUT

THEY USE NO FORM OF WORDS. COME NOW (SOMEONE MIGHT PROTEST), WE KNOW WHAT THE CHIP DOES, THE FUNCTIONS IT PERFORMS. SO (IT SHOULD BE REPLIED) DID THE YOGIS OF INDIA, THE LAMAS OF TIBET, ALSO UNDERSTAND THEIR OWN MANDALAS.

HANS BLOHM, STAFFORD BEER, AND DAVID SUZUKI, *PEBBLES TO COMPUTERS* (1986, 37)

And now for something completely different. Well, not completely. The previous chapters have looked at some of the connections between cybernetics and Eastern, nonmodern, forms of spirituality, and we can continue the examination here. Beer rigorously excluded all references to spiritual concerns from his writings on management cybernetics, and one can certainly take the latter seriously without committing oneself to the former—many of Beer's associates and followers do just that. But of our cyberneticians it was Beer who lived the fullest and most committed spiritual life, and I want now to explore the relations between his spirituality and his cybernetics, beginning with an outline of his spiritual career.

Beer was born into a High Church family and, according to his brother, before the family moved to Wales to escape the bombing of World War II,

> we all attended the Church of St John the Evangelist, Shirley, where our Father and Stafford were Servers in the choir—indeed both were members of the Guild of Servers and wore their medals. . . . Stafford always sat sideways in his choir stall with one side of his glasses over his ear and the other in his mouth and frowned. The glasses, I believe, had plain glass in them as he wanted to look older than he was. At some moments when the vicar said something (I assume outrageous to Stafford) he took the glasses off and turned to glower at the pulpit. I felt very proud of him. . . . To me they were happy times and prepared us both to take the spiritual dimension of our lives seriously, wherever it took us from that traditional Anglo-Catholic Church in the thirties.[49]

The spiritual dimension of Stafford's life took him in two directions. Sometime after his military service in India, he converted to Catholicism (1965, 301), but he later "gave up Christianity and discovered Christ," and toward the end of his life he described himself as a Buddhist, a tantric yogi. According to Allenna Leonard, he had been fascinated with Eastern philosophy since he was a small child. In his year at University College London he wanted to study Eastern philosophy, but the subject was not taught: "My dear boy, go to

SOAS"—the School of Oriental and African Studies. Instead, as we have seen, he went to India with the British Army in 1944, returning in 1947 "as thin as a rake, a very different person. . . . He was almost totally absorbed in Indian mysticism, had read endless books and had seen death, etc, I recall he told me there was no such thing as pain; it was in the mind and mind over matter and so on. To prove his point he allowed people to press lighted cigarettes onto the inside of his wrist to burn a hole while he felt nothing."[50] So, we have these two sides to Beer's life: the scientific (cybernetics) and the spiritual (Catholicism, Eastern mysticism, and strange performances). There is, of course, nothing especially unusual about that. Many physicists, for example, are deeply religious. But in respect of modern sciences like physics, the scientific and the spiritual are usually held apart, existing, as one might say, in different compartments of life, practiced in different places at different times, in the laboratory during the week and in church on Sunday. Bruno Latour (1993) speaks of the "crossed-out God" of modernity—the Christian God as both almighty and absent from the world of science and human affairs. As usual, cybernetics was not like that. Beer's cybernetics and spirituality were entangled in many ways, and that is what I want to explore here, focusing first on Beer's overall perspective on nature and then on the more esoteric aspects of his spiritual understandings and practices. The earliest of Beer's spiritual writings was an essay published in 1965, "Cybernetics and the Knowledge of God," and this provides a convenient entrée for both topics.

Hylozoism

First, Beer's perspective on nature. "Cybernetics and the Knowledge of God" begins not with nature itself but with a discussion of the finitude of the human mind. "Each of us has about ten thousand million neurons to work with. It is a lot, but it is *the* lot. . . . This means that there is a strict mathematical limit to our capacity to compute cerebrally—and therefore to our understanding. For make no mistake: understanding is mediated by the machinery in the skull" (Beer 1965, 294). As a corollary, beyond our cerebral limits there must exist in the world things which we cannot know.[51] Here we recognize the cybernetic ontology of unknowability—Beer was writing for a readership of nonspecialists; otherwise, he could simply have said that the cosmos was an exceedingly complex system, as he had defined the term in *Cybernetics and Management* in 1959. There is, though, a difference in the way in which Beer develops this thought in this essay. One can think of the economic environment of a firm as being exceedingly complex in a mundane fashion: we can readily comprehend many aspects of the economy; it is just impossible to hold all of them and their interrelations in consciousness at once. In the religious context, in contrast, Beer reaches for a more absolute sense of unknowability, invoking repeatedly "an irreducible mystery: that there is anything" (Beer 1965, 298). And this is where God comes in: "Here is another definition [of God], which I would add to the scholastic list of superlative attributes: *God is what explains the mystery*" (299). This is an odd kind of explanation, since Beer could not offer any independent definition of the *explanans*. One mystery, God, is simply defined here as that which explains another, existence. In ordinary language, at least, there is no "gap" between the two terms, so I am inclined to read Beer as saying here that matter and spirit are one, or that they are two aspects of an underlying unity. This is part of what I want to get at in describing Beer's appreciation of nature as hylozoist—the understanding that nature is infused, one might say, by spirit.

At any rate, we can see here that the ontology of unknowability was a straightforward point of linkage, almost of identity, between Beer's worldly cybernetics and his spirituality: the correlated mysteries of existence and of God are simply the mystery of exceedingly complex mundane systems taken to the *N*th degree, where *N* is infinite. And along with this ontological resonance, we can find an epistemological one. I have remarked several times on Beer's cybernetic suspicion of articulated knowledge and models, as a not necessarily reliable detour away from performance, and he expressed this suspicion, again to the *N*th degree, in relation to the spiritual (Beer 1965, 294–95, 298):

> To people reared in the good liberal tradition, man is in principle infinitely wise; he pursues knowledge to its ultimate. . . . To the cybernetician, man is part of a control system. His input is grossly inadequate to the task of perceiving the universe. . . . There is no question of "ultimate" understanding. . . . It is part of the cultural tradition that man's language expresses his thoughts. To the cybernetician, language is a limiting code in which everything *has* to be expressed—more's the pity, for the code is not nearly rich enough to cope. . . . Will you tell me that science is going to deal with this mystery [of existence] in due course? I reply that it cannot. The scientific reference frame is incompetent to provide an existence theorem for existence. The layman may believe that science will one day "explain everything away"; the scientist himself ought to know better.

Epistemologically as well as ontologically, then, Beer's cybernetics crossed over smoothly into a spiritually charged hylozoism. And we can follow the

Figure 6.18. The Gatineau River, Quebec. Source: Blohm, Beer, and Suzuki 1986, 51. (Photo: Hans-Ludwig Blohm. © Hans-Ludwig Blohm, Canada.)

crossover further by jumping ahead twenty years, to a book published in 1986, *Pebbles to Computers: The Thread*, which combines photographs by Hans Blohm with text by Stafford Beer and an introduction by David Suzuki. It is a coffee-table book with lots of color pictures and traces out a longue durée history of computing, running from simple counting ("pebbles") to digital electronic computers. The history is not, however, told in a linear fashion leading up to the present, but as a topologically complex "thread"—drawn by Beer as a thick red line twisting around photographs and text and linking one page to the next—embracing, for example, Stonehenge as an astronomical computer and Peruvian *quipus*, beautiful knotted threads, as calculational devices. Here Beer develops his ontological vision further. Under the heading "Nature Calculates," he comments on a photograph of the Gatineau River (fig. 6.18) that catches the endless complexity of the water's surface (Blohm, Beer, and Suzuki 1986, 54): "This exquisite photograph of water in movement . . . has a very subtle message for us. It is that nature's computers *are* that which they compute. If one were to take intricate details of wind and tide and so on, and use them . . . as 'input' to some computer simulating water—what computer would one use, and how express the 'output'? Water itself: that answers both those questions." And then he goes on to reproduce one of his own poems, written in 1964, "Computers, the Irish Sea," which reads (Blohm, Beer, and Suzuki 1986, 52; reproduced from Beer 1977):

> That green computer sea
> with all its molecular logic
> to the system's square inch,
> a bigger brain than mine,
> writes out foamy equations from the bow
> across the bland blackboard water.
> Accounting for variables
> which navigators cannot even list,
> a bigger sum than theirs,
> getting the answer continuously right
> without fail and without anguish
> integrals white on green.
> Cursively writes recursively computes
> that green computer sea
> on a scale so shocking
> that all the people sit dumbfounded
> throwing indigestible peel at seagulls
> not uttering an equation between them.
> All this liquid diophantine stuff
> of order umpteen million
> is its own analogue. Take a turn
> around the deck and understand
> the mystery by which what happens
> writes out its explanation as it goes.

In effect, this poem is another reexpression of the cybernetic ontology of unknowability, where the unknowability is conceived to reside in the sheer excess of nature over our representational abilities. The water knows what it is doing and does it faultlessly and effortlessly in real time, a performance we could never emulate representationally. Nature does "a bigger sum than theirs"—exceeding our capacities in way that we can only wonder at, "shocked" and "dumbfounded."[52] But Beer then adds a further point (Blohm, Beer, and Suzuki 1986, 54): "The uneasy feeling that [this poem] may have caused derives, perhaps, from insecurity as to who is supposed to be in charge. Science (surely?) 'knows the score.' Science does the measuring after all. . . .

But if art is said to imitate nature, so does science. . . . Who will realize when the bathroom cistern has been filled—someone with a ruler and a button to press, or the ballcock that floats up to switch the water off? *Nature* is (let it be clear that) *nature* is in charge." There is a clear echo here of Beer's work with biological computers (which, as mentioned earlier, also figure in *Pebbles*): not only can we not hope to equal nature representationally, but we do not need to—nature itself performs, acts, is in charge. This idea of nature as active as well infused with spirit is the definition of hylozoism, which is why I describe Beer's ontology as hylozoist. We could even think of Beer's distinctive approach to biological computing as a form of hylozoist, or *spiritual*, engineering. Aside from the reference to spirit, we can also continue to recognize in this emphasis on the endless performativity of matter the basic ontology of British cybernetics in general.[53] And we can make further connections by looking at Beer's thoughts on mind. In *Pebbles,* he refers to the Buddhist Diamond Sutra: "Think a thought, it says, 'unsupported by sights, sounds, smells, tastes, touchables, or any objects of the mind.' Can you do that?" (Blohm, Beer, and Suzuki 1986, 67). The implicit answer is no. Sensations, feelings, cognition—all emerge from, as part of, the unrepresentable excess of nature, they do not contain or dominate it. And under the heading "The Knower and the Known Are One" Beer's text comes to an end with a quotation from *hsin hsin ming* by Sengstan, the third Zen patriarch (d. 606) (105):

> Things are objects because of the mind;
> The mind is such because of things.
> Understand the relativity of these two
> and the basic reality: the unity of emptiness.
> In this emptiness the two are indistingushable
> and each contains in itself the whole world.

I cannot give a fully cybernetic gloss of this quotation; the notion of "emptiness" presently eludes me. But one can go quite a way in grasping the Zen patriarch's sentiment by thinking about Ashby's multihomeostat setups—one homeostat standing for the brain or mind, the others for its world—or perhaps even better, of the configuration of DAMS in which a subset of its elements could be designated the mind and the others that to which the mind adapts. In the dynamic interplay of mind and world thus instantiated, "objects" and "mind" do reciprocally condition each other.[54]

I can sum this section up by saying that there are two perspectives one might adopt on the relation between cybernetics and Beer's spiritual stance as discussed so far. If one balks at any reference to the spiritual, then one can see Beer's hylozoism as an *extension* of cybernetics proper, adding something to the secular part that we have focused on elsewhere. Then we could say: *This* is how one might extend cybernetics into the realm of the spiritual if one wanted to; *this* is the kind of direction in which it might lead. On the other hand, if one were prepared to recognize that there is a brute mystery of existence, and if one were willing to associate that mystery with the spiritual realm, itself defined by that association, then one could say that Beer's hylozoism just *is* cybernetics—cybernetics taken more seriously than we have taken it before. Beer's spirituality can thus be seen as either continuous with or identical to his worldly cybernetics—a situation very different from the discontinuity between the modern sciences and the crossed-out God of modernity. Once more we see that ontology makes a difference, now in the spiritual realm—the cybernetic ontology aligning itself with Eastern spirituality rather than orthodox Christianity and, at the same time, eroding the boundary between science and spirit.[55]

Tantrism

YOGA MEANS UNION, WHETHER OF SELF AND COSMOS, MAN AND WOMAN, THE DIFFERENT CHAMBERS OF THE MIND. . . . IN THE LIMIT, THEREFORE, OF THE A AND THE NOT-A.

STAFFORD BEER, "I SAID, YOU ARE GODS" (1994 [1980], 385)

The second spiritual topic I need to discuss has to do with esoteric knowledge and practice, and here we can also begin with Beer's 1965 essay "Cybernetics and the Knowledge of God." One might think that having named existence as the ultimate mystery and having defined God as its explanation, Beer would have reduced himself to silence. Instead, the essay opens up a discursive space by thinking along the same lines as Beer did in his management cybernetics. In the latter he insisted that the factory's economic environment was itself ultimately unknowable, but he also insisted that articulated models of the economy were useful, as long as they were treated as revisable in practice and not as fixed and definitive representations of their objects, and the essay follows much the same logic in the spiritual realm.

Just as the factory adapts to its economic environment in a performative fashion without ever fully grasping it, so it might be that, while our finite brains can never rationally grasp the essence of God, nevertheless, the spiritual bears upon us and leaves marks upon the "human condition" (Beer 1965, 294). Beer gives the example of suffering. "The child of loving parents is suddenly seized by them, bound and gagged and locked in a dark cellar. What is the child to make of that? It must be evident to him that (i) his parents have turned against him; but (ii) they have done so without any cause, and therefore (iii) the world is a place where things can happen without causes." In fact, in this story, "what has actually happened is that the home has suddenly been raided by secret police, seeking children as hostages. There was no time to explain; there was too much risk to the child to permit him any freedom" (296). Like the parents in this story, then, Beer had the idea that God moves in a mysterious way *which has effects on us*, though, as child analogues, we cannot grasp God's plan. The marks of God's agency are evident in history.

That means that we *can* accumulate knowledge, though never adequate, of God, just as factory managers learn about their environments. And that, in turn, implies, according to Beer in 1965, that there are two authorities we should consult in the realm of the spiritual. One is the Catholic Church—the "admonitory church" (Beer 1965, 300)—as the repository of our accumulated wisdom in brushing up against and adapting to the spiritual. But since Beer later renounced Catholicism, his second source of authority bears emphasis. It is "the total drift of human knowledge. Though compounded of the work of individual brains . . . the totality of human insight can conceivably be greater than the insight of one brain. For people use their brains in markedly different, and perhaps complementary ways." In cybernetic terms, many brains have more variety than one and thus are better able to latch onto the systems with which they interact. And the reference to "complementary ways" here asserts that there is even more variety if we pay attention to the historical drift of knowledge over a range of spiritual traditions rather than within a single one (301): "Anthropologist friends point out so many alien cultures produce so many similar ideas about God, about the Trinity, about the Incarnation. They expect me to be astonished. They mean that I ought to realise there is something phoney about my specifically Christian beliefs. I am astonished, but for opposite reasons. I am cybernetically impressed . . . by Augustine's precept: 'securus judicat orbis terrarum'—the world at large judges rightly." Beer perhaps verges on heresy in his willingness to find spiritual truths across the range of the world's religions, but he saves himself, if he does, by seizing in this essay on just those truths that the church itself espoused: God, the Trinity, the incarnation of God in Christ. Later, when he had left the church, he seized on other ones, as we will see. For the moment, let me repeat that here Beer had developed a cybernetic rhetoric for smuggling all sorts of positive spiritual knowledge past the ontology of unknowability, and it is worth noting that one example of this figures prominently in the "Knowledge of God" essay (Beer 1965, 297): "In fact, we—that is men—have a whole reference frame, called religion, which distinguishes between orders of creation precisely in terms of their communication capacity. The catalogue starts with inanimate things, works up through the amoeba and jellyfish to the primates, runs through monkeys to men—and then goes gaily on: angels, archangels, virtues, powers, principalities, dominations, thrones, seraphim, cherubim." So here, in the writings of the world's greatest management cybernetician, then director of one of the world's first OR consulting groups, we find the medieval Great Chain of Being, running continuously from rocks and stones to angels and God. There is, of course, no integral connection between this and cybernetics, but, at the same time, it is hard not to read it back into the development of Beer's cybernetics. The recursive structure of the VSM, as discussed so far, is nothing but the Great Chain of Being, sawn off before the angels appear—and, as we shall shortly see, Beer subsequently insisted on recontinuing the series, though in non-Christian terms.

— — — — —

As I said, these maneuvers in "Knowledge of God" open the way for a positive but revisable domain of spiritual knowledge, and we can learn more of where Beer came to stand in this domain from a book that he wrote that was never published, "Chronicles of Wizard Prang" (Beer 1989b).[56] Wizard Prang is the central character in the twenty chapters of the book and clearly stands for Beer himself: he lives in a simple cottage in Wales, has a long beard, wears simple clothes, eats simple food, describes himself as "among other things . . . a cybernetician" (133) and continually sips white wine mixed with water, "a trick he had learned from the ancient Greeks" (12). The thrust of the book is resolutely spiritual and specifically "tantric" (103). Its substance concerns Prang's doings and conversations, the latter offering both cybernetic exegesis of spiritual topics and spiritually informed discussions of topics that Beer also addresses in his secular writings: the failings of an education system that functions to reproduce the world's problems (chap. 2); the sad state of modern economics (chap. 15); the need to beware of becoming trapped within representational systems, including tantric ones (chap. 15).[57] We are entitled, then, to read the book as a presentation of the spiritual system that Beer lived by and taught when he was in Wales, albeit a fictionalized one that remains veiled in certain respects. And with the proviso that I am out of my depth here—I am no expert on the esoteric doctrines and practices to follow—I

Figure 6.19. Beer meditating. (Photo: Hans-Ludwig Blohm. © Hans-Ludwig Blohm, Canada.)

want to explore some of the resonances and connections between Beer's tantrism and his cybernetics.

Tantrism is a hard concept to pin down. In his book *Stafford Beer: A Personal Memoir*, David Whittaker notes that "the word 'tantra' comes from the Sanskrit root tan meaning 'to extend, to expand.' It is a highly ritualistic philosophy of psycho-physical exercises, with a strong emphasis on visualization, including concentration on the yogic art of mandalas and yantras. The aim is a transmutation of consciousness where the 'boundary' or sense of separation of the self from the universe at large dissolves" (Whittaker 2003, 13).[58] And we can begin to bring this description down to earth by noting that meditation was a key spiritual practice for Beer.

Here, then, we can make contact with the discussion from earlier chapters—of meditation as a technology of the nonmodern self, aimed at exploring regions of the self as an exceedingly complex system and achieving "altered states of consciousness" (Beer 1989b, 41).[59] Like the residents in the Archway communities, but in a different register, Beer integrated this technology into his life. Beyond that we can note that, as Whittaker's definition of tantrism suggests, Beer's style of meditation involved visual images. He both meditated upon images—mandalas, otherwise known as yantras (fig. 6.20)—and

Figure 6.20. A yantra. Source: Beer 1989b, chap. 14, 105.

engaged in visualization exercises in meditation. In the tantric tradition this is recognized as a way of accessing a subtle realm of body, energy, and spirit—experiencing the body, for example, as a sequence of chakras ascending from the base of the spine to the top of the head, and eventually aiming at union with the cosmos—"yoga means union," as Beer was wont to put it.[60]

Three points are worth noting here. First, we again find a notion of decentering the self here, relative both to the chakras as lower centers of consciousness and to the higher cosmos. As discussed before, we can understand this sort of decentering on the model of interacting homeostats, though the details, of course, are not integrally cybernetic but derive specifically from the tantric tradition. Second, we should recognize that yantras are, in a certain sense, symbolic and representational. Interestingly, however, Beer has Perny, his apprentice, stress their *performative* rather than symbolic quality when they first appear in Beer's text. Perny remarks on the yantra of figure 6.20 that "I was taught to use this one as a symbol on which to meditate." Another disciple replies, "It's sort of turning cartwheels." "I know what you mean," Perny responds. "This way of communicating, which doesn't use words, seems to work through its physiological effects" (Beer 1989b, 106). We thus return to a performative epistemology, now in the realm of meditation—the symbol as integral to performance, rather than a representation having importance in its own right.

Third, and staying with the theme of performance, we could recall from chapter 3 that Grey Walter offered cybernetic explanations not just for the altered states achieved by Eastern yogis, but also for their strange bodily performances, suspending their metabolism and so on. We have not seen much of these strange performances since, but now we can go back to them. Wizard Prang himself displays displays unusual powers, though typically small ones which are not thematized but are dotted around the stories that make up the book. At one point Prang makes his end of a seesaw ascend and then descend just by intending it: "Making oneself light and making oneself heavy are two of the eight occult powers"; Prang can see the chakras and auras of others and detect their malfunctioning; Perny "change[s] the direction of a swirl [in a stream] by identifying with it rather than by exerting power"; the logs in the fireplace ignite themselves; spilled wine evaporates instantly on hitting the tiles; Prang sends a blessing flying after two of his disciples, "with the result that Toby [slips] and [falls] over with the force of it." More impressively, Perny remarks that "you give me telepathic news and I've seen you do telekinetic acts," and at one point Prang levitates, though even this is described in a humorous and self-deprecating fashion: "The wizard's recumbent form slowly and horizontally rose to the level of where his midriff would be if he were standing up. He stayed in that position for ten seconds, then slowly rotated. His feet described an arc through the air which set them down precisely, smoothly onto the floor. 'My God,' breathed Silica, 'What are you doing?' . . . 'Demonstrating my profession of wizardry, of course.' 'Do you often do things like that?' 'Hardly ever. It's pretty silly, isn't it?' "[61] I paid little attention to these incidents in Beer's text until I discovered that the accrual of nonstandard powers is a recognized feature of spiritual progress by the yogi, and that there is a word for these powers: *siddhis*.[62] Beer's practice was securely within the tantric tradition in this respect, too.

In these various ways, then, Beer's spiritual knowledge and practice resonated with the cybernetic ontology of exceedingly complex performative systems, though, as I said, the detailed articulation of the ontology here derived not from cybernetics but from the accumulated wisdom of the tantric tradition. Having observed this, we can now look at more specific connections that Beer made between his spirituality and his cybernetics.

— — — — —

Beer's worldly cybernetics as I described it earlier is not as worldly as it might seem. This is made apparent in *Beyond Dispute*. For the first ten chapters, 177 pages, this book is entirely secular. It covers the basic ideas and form of team

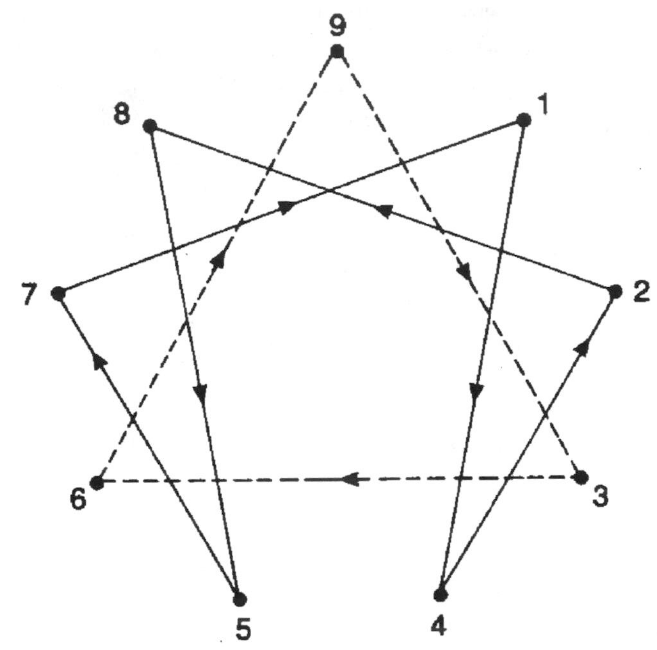

Figure 6.21. The enneagram. Source: S. Beer, *Beyond Dispute: The Invention of Team Syntegrity* (New York: Wiley, 1994), 202, fig. 12.4.

syntegrity, describes various experiments in syntegration and refinements of the protocol, and elaborates many of Beer's ideas that are by now familiar, up to and including his thoughts on how syntegration might play a part in an emergent "world governance." In chapters 11–14, 74 pages in all, the book takes on a different form. As if Beer had done his duty to the worldly aspects of the project in the opening chapters, now he conjures up its spiritual aspects and the esoteric knowledge that informs it. I cannot rehearse the entire content of these latter chapters, but we can examine some of its key aspects.

Chapter 12 is entitled "The Dynamics of Icosahedral Space" and focuses on closed paths around the basic syntegration icosahedron, paths that lead from one vertex to the next and eventually return to their starting points. Beer's interest in such paths derived from the idea mentioned above, that in syntegration, discussions reverberate around the icosahedron, becoming the common property of the infoset. In chapter 12, this discussion quickly condenses onto the geometric figure known as an *enneagram* (fig. 6.21), which comprises a reentrant six-pointed form superimposed on a triangle. Beer offers an elaborate spiritual pedigree for this figure. He remarks that he first heard about it in the 1960s in conversations with the English mystic John Bennett, who had in turn been influenced by the work of Peter D. Ouspensky and George Ivanovich Gurdjieff; that there is also a distinctively Catholic commentary on the properties of the enneagram; and that traces of it can be found in the Vedas, the most ancient Sanskrit scriptures, as well as in Sufism (Beer 1994b, 202–4). Beer also mentions that while working on Project Cybersyn in Chile in the early 1970s he had been given his own personal mandala by a Buddhist monk, that the mandala included an enneagram, and that after that he had used this figure in his meditational practices (205).[63] Once more we can recognize the line of thought Beer set out in "Cybernetics and the Knowledge of God." The enneagram appears in many traditions of mystical thought; it can therefore be assumed to be part the common wisdom of mankind, distilled from varied experience of incomprehensible realms; but its significance is performative, as an aid to meditation, rather than purely representational.

So what? Beer recorded that in the syntegration experiments of the early 1990s he had acquired a new colleague in Toronto, Joe Truss, who had once founded a business based on an enneagrammatic model, and that Truss had then succeeded in finding reentrant enneagrammatic trajectories within the syntegration icosahedron.[64] Truss and Beer were both exceptionally impressed by the fact that these trajectories were three-dimensional, rather than lying in a single plane as in figure 6.21 (Beer 1994b, 206): "Joe came to my house late at night to show me his discovery, and he was very excited. Well, all such moments are exciting. But I was unprepared that he should say, 'Do you see what this means? The icosahedron is the actual *origin* of the enneagram, and the ancients knew it. Could it not be possible that the plane figure was coded esoteric knowledge?' Obviously (now!) it could." From this point on, if not before, syntegration took on for Beer an intense spiritual as well as practical significance, especially as far as its reverberations along closed pathways were concerned.[65] Here, then, we have an example of the sort of very specific and even, one could say, technical continuities that Beer constructed between his worldly cybernetics and his spiritual life, with the enneagram as a pivot between the everyday geometry of the icosahedron and meditative practice. This immediate continuity between the secular and the spiritual contrasts interestingly, as usual, with the separation of these two realms that characterizes modernity. It points to the unusual "earthy" and hylozoist quality of cybernetic spirituality, as a spirituality that does not recognize any sharp separation between the sacred and the profane.

I mentioned earlier the appearance of the Great Chain of Being in Beer's "Knowledge of God" essay, and that this reappeared in a truncated version in his published discussions of the viable system model. We might doubt,

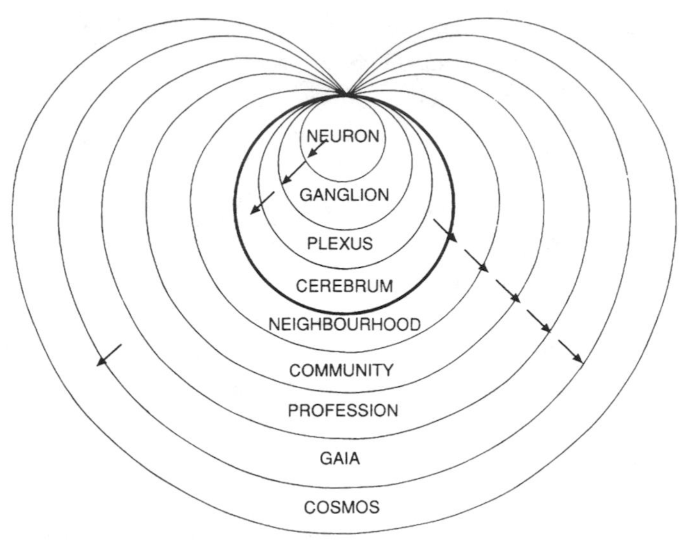

Figure 6.22. "Theory of recursive consciousness." Source: S. Beer, *Beyond Dispute: The Invention of Team Syntegrity* (New York: Wiley, 1994), 253, fig. 14.2.

however, that this truncation had much significance for Beer personally; it seems highly likely that the VSM's recursive structure was always spiritually charged in Beer's imagination—that again in this respect the mundane and the spiritual were continuous for Beer. And in *Beyond Dispute*, the truncation was explicitly undone, as shown in figure 6.22, displaying nine levels of recursion, running from individual neurons in the brain, through individual consciousness ("cerebrum"), up to Gaia and the cosmos. Chapter 14 of *Beyond Dispute* is a fascinating cybernetic-political-mystical commentary on these different levels, and I can follow a few threads as illustrations of Beer's thought and practice.

One point to note is that while the labeling of levels in figure 6.22 is secular, at least until one comes to "Gaia" and "cosmos," Beer's discussion of them is not. It is distinctly hybrid, in two senses. On the one hand, Beer accepts current biological knowledge of the nervous system, as he did in developing the VSM, while, at the same time, conceptualizing it as a cybernetic adaptive system; on the other hand, he synthesizes such biological knowledge with esoteric, mystical, characteristically Eastern accounts of the subtle body accessible to the adept. The connection to the latter goes via a cybernetic analysis of *consciousness* as the peculiar property of reentrant structures.[66] The human brain would be the paradigmatic example of such a structure (containing an astronomical number of reentrant neuronal paths), but Beer's argument was that any reentrant structure might have its own form of consciousness. "Plexus," at the third level of recursion in figure 6.22, refers to various nervous plexuses, concatenations of nerve nets, that can be physiologically identified within the body. Beer regards these as homeostat-like controllers of physiological functions. At the same time, he imputes to them their own form of "infosettic consciousness," remarks that it "is not implausible to identify the six 'spiritual centres' which the yogi calls chakras with plexus activity in the body," and finally asserts that "knowledge *of the existence* of infosettic consciousness within the other five chakras [besides the chakra associated with brain] is possible to the initiate, as I attest from yogic experience myself" (Beer 1994b, 247). With this move Beer deeply intertwines his management cybernetics and his spirituality, at once linking the subtle yogic body with physiological cybernetic structures and endowing the recursive structure of the VSM with spiritual significance, this double move finding its empirical warrant in Beer's meditational practice.

Beer then moves up the levels of consciousness. At the fourth level of figure 6.22 we find "cerebrum," the level of individual consciousness, which Beer identifies in the subtle body with Ajna, the sixth chakra. Then follow four levels having to do with social groupings at increasing scales of aggregation. "Neighbourhood" refers to the small groups of individuals that come together in syntegration, and "it must be evident that the theory of recursive consciousness puts group mind forward as the fifth embedment of consciousness, simply because the neighbourhood infoset displays the usual tokens of consciousness" (Beer 1994b, 248). From this angle, too, therefore, syntegration had a more than mundane significance for Beer. Not simply an apparatus for free and democratic discussion, syntegration produces a novel emergent phenomenon, a group mind which can be distinguished from the individual minds that enter into it, and which continues the spiritually charged sequence of levels of consciousness that runs upward from the neuron to the cerebrum and through the yogic chakras. At higher levels of social aggregation, up to the level of the planet, Beer also glimpses the possibilities for collective consciousness but is of the opinion that practical arrangements for the achievement of such states "work hardly at all . . . there is no cerebrum [and] we are stuck with the woeful inadequacy of the United Nations. The Romans did better than that" (249).

The ninth level of figure 6.22, "cosmos," completes the ascent. Here the mundane world of the individual and the social is left entirely behind in yogic union with "cosmic consciousness," accessible via the seventh chakra, Sahashara, the thousand-petaled lotus. And Beer concludes (255):

> For the yogi, the identification of all the embedments, and particularly his/her own selfhood embodied at the fourth embedment, with the cosmos conceived as universal consciousness, is expressed by the mantra Tat Tvam Asi: "That You Are." These are the last three words of a quotation from one of the Ancient Vedic scriptures, the Chhandogya Upanishad, expressing the cosmic identification:
>
> That subtle essence
> which is the Self of this entire world,
> That is the Real,
> That is the Self,
> That You Are.

— — — — —

Rather than trying to sum up this section, it might be useful to come at the topic from a different angle. Beer, of course, identified the spiritual aspect of his life with the tantric tradition, but it strikes me that his fusion of cybernetics and spirituality also places him in a somewhat more specific lineage which, as far as I know, has no accepted name. In modernity, matter and spirit are assigned to separate realms, though their relations can be contested, as recent arguments about intelligent design show: should we give credit for the biological world to God the Creator, as indicated in the Christian Bible, *or* to the workings of evolution on base matter, as described by modern biology? What interests me about Beer's work is that it refused to fall on either side of this dichotomy—we have seen that his science, specifically his cybernetics, and his understanding of the spiritual were continuous with one another—flowed into, structured, and informed each other in all sorts of ways. This is what I meant by referring to the earthy quality of his spirituality: his tantrism and his mundane cybernetics were one. Once more one could remark that ontology makes a difference, here in the realm of the spiritual.[67] But my concluding point is that Beer was not alone in this nondualist space.

I cannot trace out anything like an adequate history of the lineage of the scientific-spiritual space in which I want to situate Beer, and I know of no scholarly treatments, but, in my own thinking at least, all roads lead to William James—in this instance to his *Varieties of Religious Experience*, as an empirical but nonsceptical inquiry into spiritual phenomena. James's discussion of the "anaesthetic revelation"—transcendental experience brought on by drugs and alcohol—is a canonical exploration of technologies of the nonmodern self, Aldous Huxley *avant la lettre*. Huxley himself lurked in the margins of chapters 3 and 5 above, pursuing a biochemical understanding

Name of Chakra	Position on Surface	Approximate Position of Spinal Chakra	Sympathetic Plexus	Chief Subsidiary Plexuses
Root	Base of spine	4th Sacral	Coccygeal	...
Spleen	Over the spleen	1st Lumbar	Splenic	...
Navel	Over the navel	8th Thoracic	Cœliac or Solar	Hepatic, pyloric, gastric, mesenteric, etc.
Heart	Over the heart	8th Cervical	Cardiac	Pulmonary, coronary, etc.
Throat	At the throat	3rd Cervical	Pharyngeal	...
Brow	On the brow	1st Cervical	Carotid	Cavernous, and cephalic ganglia generally

Figure 6.23. Chakras and plexuses. Reproduced from *The Chakras*, by C. W. Leadbeater (Leadbeater 1990, 41, table 2). (With permission from The Theosophical Publishing House, Adyar, Chennai–600 020, India. © The Theosophical Publishing House, Adyar, Chennai–600 020, India. www.tw-adyar.org.)

of spiritual experience without intending any reduction of the spiritual to base matter, with John Smythies and Humphrey Osmond's research on brain chemistry lurking behind him. Gregory Bateson, Alan Watts, and R. D. Laing likewise ran worldly psychiatry and Buddhism constructively together, with a cybernetic ontology as the common ground. One thinks, too, of the Society for Psychical Research (with which both James and Smythies were associated) as a site for systematic integration of science and the spiritual (chap. 3, n. 62).

From another angle—if it is another angle—a canonical reference on the chakras that loomed so large in Beer's thought and practice is C. W. Leadbeater's *The Chakras*, continuously in print, according to the back cover of my copy (Leadbeater 1990), since 1927, and published by the Theosophical Society. Much of Beer's esoteric writing echoes Leadbeater's text, including, for example, the association between the chakras and nerve plexuses just discussed (fig. 6.23).[68] Theosophy, too, then, helps define the scientific-spiritual space of Beer's cybernetics. And coming up to the present, one also thinks of certain strands of New Age philosophy and practice, already mentioned in chapter 5, as somehow running together science and spirituality, mind, body, and spirit.[69]

We should thus see Beer's cybernetic tantrism as an event within a broader scientific-spiritual history, and I can close with two comments on this. First, to place Beer in this lineage is not to efface his achievement. On the one hand, Beer went much further than anyone else in tying cybernetics—our topic—into the realm of the spirit. On the other hand, from the spiritual side, Beer went much further than anyone else in developing the social aspects of this nonmodern assemblage. Esoteric writings seldom go beyond the realm of the individual, whereas the VSM and team syntegrity were directed at the creation of new social structures and the rearrangement of existing ones in line with cybernetic and, we can now add, tantric sensitivities. Second, placing Beer in relation to this lineage returns us to questions of institutionalization and marginality. The entire lineage could be described as sociologically occult—hidden and suspect. Even now, when New Age has become big business, it remains walled off from established thought and practice. Despite—or, perhaps better, because of—its elision of mind, body, and spirit distinctions, New Age remains invisible in contemporary debates on the relation between science and religion. Like Gysin's Dream Machines, New Age spirituality and Beer's spirituality fail to find a place within modern schemata of classification. And, to change direction again, perhaps we should regret this. The early twenty-first century seems like a time when we should welcome a form of life that fuses science and spirituality rather than setting them at each other's throats. Again, this exploration of the history of cybernetics offers us a sketch of another future, importantly different from the ones that are more readily imagined.

Brian Eno and New Music

[BEER'S WORK] SO FUNDAMENTALLY CHANGED THE WAY THAT I THOUGHT ABOUT MUSIC THAT IT'S VERY DIFFICULT TO TRANSLATE INTO INDIVIDUAL THINGS, IT JUST CHANGED THE WHOLE WAY I WORK. . . . STAFFORD FOR ME WAS THE DOORWAY INTO A WHOLE WAY OF THINKING.

BRIAN ENO, QUOTED IN DAVID WHITTAKER, *STAFFORD BEER* (2003, 57, 63)

We touched on relations between cybernetics and the arts in chapters 3 and 4 as well as briefly here in connection with biological computing, and I want to end this chapter by returning to this topic. The focus now is on music, and the intersection between Beer's cybernetics and the work of the composer and performer Brian Eno. If Beer himself is not widely known, many people have heard of Eno and, with any luck, know his music (which is, like any music, impossible to convey in words, though what follows should help to characterize it). Eno's first claim to fame was as a member of Roxy Music, the greatest rock band to emerge in the early 1970s. Subsequently, he left Roxy and went on to develop his own distinctive form of "ambient" and "generative" music (as well as important collaborations with David Bowie, Talking Heads, and U2), with *Music for Airports* (1978) as an early canonical example.[70] The content of this music is what I need to get at, but first I want to establish Eno's biographical connection to cybernetics.

In an interview with David Whittaker (Whittaker 2003, 53–63), Eno recalled that he first became interested in cybernetics as an art student in Ipswich between 1964 and 1966. The principal of the art school was Roy Ascott, Britain's leading cybernetic artist, who will reappear in the next chapter, and the emphasis at Ipswich was on "process not product. . . . Artists should concentrate on the way they were doing things, not just the little picture that came out at the end. . . . The process was the interesting part of the work" (53). Eno was drawn further into cybernetics in 1974 when his mother-in-law lent him a copy of Beer's *Brain of the Firm*, which she had borrowed from Swiss Cottage Library in London. Eno was "very, very impressed by it" and in 1975 wrote an essay in which he quoted extensively from *Brain*. He sent a copy to Beer, who came to visit him in Maida Vale (Whittaker 2003, 55–56).[71] In 1977 Beer invited Eno for an overnight visit to the cottage in Wales, where Eno recalled that dinner was boiled potatoes and the following conversation took place (55):

> [Beer] said "I carry a torch, a torch that was handed to me along a chain from Ross Ashby, it was handed to him from . . . Warren McCulloch." He was telling me the story of the lineage of this idea . . . and said "I want to hand it to you. I know it's a responsibility and you don't have to accept, I just want you to think about it." It was a strange moment for me, it was a sort of religious initiation . . . and I didn't feel comfortable about it somehow. I said "Well, I'm flattered . . . but I don't see how I can accept it without deciding to give up the work I do now and I would have to think very hard about that." We left it saying the offer is there, but it was very strange, we never referred to it again, I wasn't in touch with him much after that. I'm sure it was meant with the best of intentions and so on but it was slightly weird.

Now we can turn to the substantive connection between Eno's music and Beer's cybernetics: what did Brian get from Stafford? Speaking of this connection and apparently paraphrasing from *Brain of the Firm*, in his interview with David Whittaker Eno said that "the phrase that probably crystallised it [Eno's cybernetic approach to music] . . . says 'instead of specifying it in full detail; you specify it only somewhat, you then ride on the dynamics of the system in the direction you want to go.' That really became my idea of working method" (57).[72] And the easiest way to grasp this idea of riding the dynamics of the system is, in the present context, ontologically. Beer's ontology of exceedingly complex systems conjures up a lively world, continually capable of generating novel performances. Eno, so to speak, picked up the other end of the stick and focused on building musical worlds that would themselves exhibit unpredictable, emergent becomings. And we can get at the substance of this by following a genealogy of this approach that Eno laid out in a 1996 talk titled "Generative Music" (Eno 1996b). This begins with a piece called *In C* by Terry Riley, first performed at the San Francisco Tape Music Center in 1964 (Eno 1996b, 2–3):

> It's a very famous piece of music. It consists of 52 bars of music written in the key of C. And the instructions to the musicians are "proceed through those bars at any speed you choose." So you can begin on bar one, play that for as many times as you want, 20 or 30 times, then move to bar 2, if you don't like that much just play it once, go on to bar three. The important thing is each musician moves through it at his or her own speed. The effect of that of course is to create a very complicated work of quite unpredictable combinations. If this is performed with a lot of musicians you get a very dense and fascinating web of sound as a result. It's actually a beautiful piece.

Here we find key elements of Eno's own work. The composer sets some initial conditions for musical performance but leaves the details to be filled in by the dynamics of the performing system—in this case a group of musicians deciding on the spot which bars to play how often and thus how the overall sound will evolve in time. Eno's second example is a different realization of the same idea: Steve Reich's *It's Gonna Rain*, first performed in 1965, also at the Tape Music Center.[73] In this piece a loop of a preacher saying "It's gonna rain" is played on two tape recorders simultaneously, producing strange aural effects as the playbacks slowly fall out of phase: "Quite soon you start hearing very exotic details of the recording itself. For instance you are aware after several minutes that there are thousands of trumpets in there. . . . You also become aware that there are birds" (Eno 1996b, 3). Again in this piece, the composer specifies the initial conditions for a performance—the selection of the taped phrase, the use of two recorders, and then "rides the dynamics of the system"—in this case the imperfection of the recorders that leads them to drift out of synchronization, rather than the idiosyncratic choices of human musicians—to produce the actual work.

Eno then moves on to one of his own early post-*Brain* pieces composed in this way, from *Music for Airports*. This consists of just three notes, each repeating at a different interval from the others—something like 23 1/2, 25 7/8, and 29 15/16 seconds, according to Eno. The point once more is that the composer defines the initial conditions, leaving the piece to unfold itself in time, as the notes juxtapose themselves in endless combinations (Eno 1996b, 4).

In his talk, Eno then makes a detour though fields like cellular automata and computer graphics, discussing the endlessly variable becomings of the Game of Life (a simple two-dimensional cellular automaton, developed by John Conway: Poundstone 1985), and simple screen savers that continually transform images arising from a simple "seed." In each case, unpredictable and complex patterns are generated by simple algorithms or transformation rules, which connects back to Eno's then-current work on a musical generative system—a computer with a sound card. Eno had contrived this system so as to improvise probabilistically within a set of rules, around 150 of them, which determined parameters such as the instruments and scales to be employed, harmonies that might occur, and steps in pitch between consecutive notes.[74] As usual, one should listen to a sample of the music produced by this system, but at least Eno (1996b, 7) found that it was "very satisfying," and again we can see how it exemplifies the idea of riding the dynamics of what has by now become a sophisticated algorithmic system.

Thus the basic form and a sketchy history of Brian Eno's ambient and generative music, and I want to round off this chapter with some commentary and a little amplification. First, back to Roxy Music. Eno does not include his time with Roxy in any of his genealogies, and one might assume a discontinuity between his Roxy phase and his later work, but the story is more interesting than that. Eno played (if that is the word) the electronic synthesizer for Roxy Music and, as Pinch and Trocco (2002) make clear in their history of the synthesizer, it was not like any other instrument, especially in the early "analog days." In the synthesizer, electronic waveforms are processed via various different modules, and the outputs of these can be fed back to control other modules with unforeseeable effects. As Eno wrote of the EMS synthesizer, for example, "The thing that makes this a great machine is that . . . you can go from the oscillator to the filter, and then use the filter output to control the same oscillator again. . . . You get a kind of squiging effect. It feeds back on itself in interesting ways, because you can make some very complicated circles through the synthesiser" (quoted in Pinch and Trocco 2002, 294). Here we find the familiar cybernetic notion of a feedback loop, not, however, as that which enables control of some variable (as in a thermostat), but as that which makes a system's behavior impenetrable to the user.[75] We can think about such systems further in the next chapter, but for now the point to note is that analog synthesizers were thus inescapably objects of exploration by their users, who had to *find out* what configuration would produce a desirable musical effect.[76] "The resulting music was an exchange . . . between person and machine, both contributing to the final results. This may be why analog synthesists can readily recount feelings of love for their synthesisers" (Pinch and Trocco 2002, 177). In this sense, then, one can see a continuity between Eno's work with Roxy and his later work: even with Roxy Music, Eno was riding the dynamics of a generative system—the synthesizer—which he could not fully control. What he learned from Beer was to make this cybernetic insight explicit and the center of his future musical development.

Second, I want to emphasize that with Eno's interest in cellular automata and complex systems we are back in the territory already covered at the end of chapter 4, on Ashby's cybernetics, with systems that stage open-ended becomings rather than adaptation per se. Indeed, when Eno remarks of *It's Gonna Rain* that "you are getting a huge amount of material and experience from a very, very simple starting point" (Eno 1996b, 3) he is singing the anthem of Stephen Wolfram's "New Kind of Science." In this sense, it would seem more appropriate to associate Eno with a line of cybernetic filiation going back to Ashby than with Beer—though historically he found inspiration in *Brain of the Firm* rather than *Design for a Brain*. We could also recall in this connection that no algorithmic system, in mathematics or in generative music, ever becomes in a fully open-ended fashion: each step in the evolution of such systems is rigidly chained to the one before. Nevertheless, as both Eno and Wolfram have stressed in their own ways, the evolution of these systems can be unpredictable even to one who knows the rules: one just has to set the system in motion and see what it does. For all practical purposes, then, such systems can thematize for us and stage an ontology of becoming, which is what Eno's notion of riding the system's dynamics implies.[77]

Third, we should note that Eno's ambient music sounds very different from the music we are used to in the West—rock, classical, whatever. In terms simply of content or substance it is clear, for instance, that three notes repeating with different delays are never going to generate the richness, cadences, and wild climaxes of Roxy Music. Whatever aesthetic appeal ambient music might have—"this accommodates many levels of listening attention and is as ignorable as it is interesting" (Whittaker 2003, 47)—has to be referred to its own specific properties, not to its place in any conventional canon.[78] And further, such music has a quality of constant novelty and unrepeatability lacking in more traditional music. *In C* varies each time it is performed, according to the musicians who perform it and their changing preferences; *It's Gonna Rain* depends in its specifics on the parameters of the tape players used, which themselves vary in time; the computerized system Eno described above is probabilistic, so any given performance soon differs from all others even if the generative parameters remain unchanged.[79] Perhaps the easiest way to put the point I am after is simply to note that Eno's work, like Alvin Lucier's biofeedback performances (chap. 3), raises the question, Is it music? This, I take it, again, is evidence that ontology makes a difference, now in the field of music. I should add that, evidently, Eno has not been alone in the musical exploitation of partially autonomous dynamic systems, and it is not the case that all of his colleagues were as decisively affected by reading the *Brain of the Firm* as he was. My argument is that all of the works in this tradition, cybernetically inspired and otherwise, can be understood as ontological theater and help us to see where a cybernetic ontology might lead us when staged as music.[80]

Fourth, these remarks lead us, as they did with Beer himself, into questions of power and control. Usually, the composer of a piece of music exercises absolute power over the score, deciding what notes are to be played in what sequence, and thus exercises a great deal of power over musical performers, who have some leeway in interpreting the piece, and who, in turn, have absolute power over the audience as passive consumers. In contrast, "with this generative music . . . am I the composer? Are you if you buy the system the composer? Is Jim Coles and his brother who wrote the software the composer? Who actually composes music like this? Can you describe it as composition exactly when you don't know what it's going to be?" (Eno 1996b, 8). These rhetorical questions point to a leveling of the field of musical production and consumption. No doubt Eno retains a certain primacy in his work; I could not generate music half as appealing as his. On the other hand, the responsibility for such compositions is shared to a considerable extent with elements beyond the artist's control—the material technology of performance (idiosyncratic human performers or tape players, complex probabilistic computer programs)—and with the audience, as in the case of computer-generated music in which the user picks the rules. As in the case of Beer's social geometries, a corollary of the ontology of becoming in music is again, then, a democratization, a lessening of centralized control, a sharing of responsibility, among producers, consumers, and machines.[81]

My fifth and final point is this. It is ironic that Eno came to cybernetics via Beer; he should have read Pask. The musical insights Eno squeezed out of Beer's writings on management are explicit in Pask's writings on aesthetics. As we can see in the next chapter, if Pask had handed him the torch of cybernetics, Eno would not need to have equivocated. Pask, however, was interested in more visual arts, the theater and architecture, so let me end this chapter by emphasizing that we have now added a distinctive approach to music to our list of instances of the cybernetic ontology in action.

7

GORDON PASK

FROM CHEMICAL COMPUTERS TO ADAPTIVE ARCHITECTURE

NOW, WE ARE SELF-ORGANIZING SYSTEMS AND WE WANDER AROUND IN A WORLD WHICH IS FULL OF WONDERFUL BLACK BOXES, DR. ASHBY'S BLACK BOXES. SOME OF THEM ARE TURTLES; SOME ARE TURTLEDOVES; SOME ARE MOCKING BIRDS; SOME OF THEM GO "POOP!" AND SOME GO "POP!"; SOME ARE COMPUTERS; THIS SORT OF THING.

GORDON PASK, "A PROPOSED EVOLUTIONARY MODEL" (1962, 229)

THE DESIGN GOAL IS NEARLY ALWAYS UNDERSPECIFIED AND THE "CONTROLLER" IS NO LONGER THE AUTHORITARIAN APPARATUS WHICH THIS PURELY TECHNICAL NAME COMMONLY BRINGS TO MIND. IN CONTRAST THE CONTROLLER IS AN ODD MIXTURE OF CATALYST, CRUTCH, MEMORY AND ARBITER. THESE, I BELIEVE, ARE THE DISPOSITIONS A DESIGNER SHOULD BRING TO BEAR UPON HIS WORK (WHEN HE PROFESSIONALLY PLAYS THE PART OF A CONTROLLER) AND THESE ARE THE QUALITIES HE SHOULD EMBED IN THE SYSTEMS (CONTROL SYSTEMS) WHICH HE DESIGNS.

GORDON PASK, "THE ARCHITECTURAL RELEVANCE OF CYBERNETICS" (1969A, 496)

Now for the last of our cyberneticians. Andrew Gordon Speedie Pask (fig. 7.1) was born in Derby on 28 June 1928, the son of Percy Pask, a wealthy fruit

Figure 7.1. Gordon Pask in the early 1960s. (Reproduced by permission of Amanda Heitler.)

importer and exporter, and his wife Mary, and died in London on 28 March 1996, at the age of sixty-seven.[1] Gordon, as he was known, was much the youngest of three brothers. His brother Alfred, who trained as an engineer but became a Methodist minister, was twenty years older. The other brother, Edgar, was sixteen years older and was Gordon's "hero and role model" (E. Pask n.d., n.p.), and it is illuminating to note that Gar, as he was known, distinguished himself by bravery in research verging on utter recklessness in World War II. He left his position as an anesthetist at Oxford University to join the Royal Air Force in 1941 and then carried out a series of life-threatening experiments on himself aimed at increasing the survival rate of pilots: being thrown unconscious repeatedly into swimming pools to test the characteristics of life jackets; again being thrown repeatedly, but this time conscious, into the icy waters off the Shetlands to test survival suits; hanging from a parachute breathing less and less oxygen until he became unconscious, to determine at what altitude pilots stood a chance if they bailed out; being anesthetized again and again to the point at which his breathing stopped, to explore the efficacy of different modes of resuscitation. He "won the distinction of being the only man to have carried out all [well, almost all] of his research while asleep," and the Pask Award of the British Association of Anaesthetists for gallantry and distinguished service was instituted in his honor in 1975 (Pain 2002). Gar was a hard act for young Gordon to follow, but he did so, in his own unusual way.[2]

Gordon was educated at Rydal, a private school in Wales, where he also took a course in geology at Bangor University. He was called up for military service in 1945, but "Gordon's career in the RAF was extremely brief. During his first week at camp, he passed out while doing the mandatory session of push-ups, and was returned home in an ambulance" (E. Pask n.d., n.p.). Pask then studied mining engineering at Liverpool Polytechnic, before taking up a place at Downing College, Cambridge, in 1949, where he studied medicine and gained a BA in physiology in the natural science tripos in 1953 (Pask 1959, 878). In 1956, he married Elizabeth Poole (E. Pask [1993] describes their unconventional courtship), and they had two daughters: Amanda (1961) and Hermione (adopted in 1967). In 1964, Pask was awarded a PhD in psychology from University College London and in 1974 a DSc in cybernetics by the Open University. In 1995, the year before his death, Cambridge awarded him an ScD (Scott and Glanville 2001; Glanville and Scott 2001b).

His first book, *An Approach to Cybernetics*, was published in 1961 and was translated into Dutch and Portuguese, and several other books followed (Pask 1975a, 1975b, and 1976a were the major ones; also Pask and Curran 1982; and *Calculator Saturnalia* [Pask, Glanville, and Robinson 1980]—a compendium of games to play on electronic calculators). A list of his publications (journal articles, chapters in books and proceedings, technical reports) runs to more than 250 items (in Glanville 1993, 219–33). At different times he was president of the Cybernetics Society and the International Society for General Systems; he was the first recipient of the *Ehrenmitgleid* of the Austrian Society for Cybernetic Studies and was awarded the Wiener Gold Medal by the American Society for Cybernetics.

From the 1950s onward, Pask enjoyed many university affiliations, including professorial chairs at Brunel University (in the Cybernetics Department, part-time, beginning in 1968) and the University of Amsterdam (in the Centre for Innovation and Co-operative Technology, beginning in 1987; Thomas and Harri-Augstein 1993, 183; de Zeeuw 1993, 202).[3] He also from time to time held visiting positions at several institutions: the University of Illinois, Old Dominion University, Concordia University, the Open University, MIT, the University of Mexico, and the Architecture Association in London. But the principal base for Pask's working life was not an academic one; it was a research organization called System Research that he founded in 1953 together with his wife and Robin McKinnon-Wood.[4] There Pask pursued his many projects and engaged in contract research and consulting work.[5]

So much for the bare bones of Pask's life; now I want to put some flesh on them. Before we get into technical details, I want to say something about Pask the man. The first point to note is that he was the object of an enormous amount of love and affection. Many people cared for him intensely. There are two enormous special issues of cybernetics journals devoted entirely to him, one from 1993 (*Systems Research* [Glanville 1993]), the other from 2001 (*Kybernetes* [Glanville and Scott 2001a]), and both are quite singular in the depth and openness of the feelings expressed. And this was, no doubt, in part because he was not like other men—he was a classic "character" in the traditional British sense (as were Grey Walter and Stafford Beer in their own ways). There are many stories about Pask. His wife recalled that "Gordon always denied that he was born, maintaining that he descended from the sky, fully formed and dressed in a dinner jacket, in a champagne bottle, and that the Mayor and aldermen of Derby were there to welcome him with a brass band and the freedom of the city." It is certainly true that he liked to dress as an Edwardian dandy (bow tie, double-breasted jacket and cape). At school, he built a bomb which caused considerable damage to the chemistry lab (which his father paid a lot of money to put right), and he claimed that "the best thing about his school was that it taught him to be a gangster." At Cambridge, he would cycle between staying awake for forty-eight hours and sleeping for sixteen (E. Pask n.d.). Later in life he became more or less nocturnal. His daughter Amanda told me that she would bring friends home from school to see her father emerge as night fell. Pask's ambition in studying medicine at Cambridge was to follow in his brother Edgar's footsteps, but as one of his contemporaries, the eminent psychologist Richard Gregory, put it, this story "is perhaps best forgotten." Pask apparently tried to learn anatomy by studying only the footnotes of the canonical text, *Gray's Anatomy*, and (Gregory 2001, 685–86) "this saved him for two terms—until disaster struck. He was asked to dissect, I think an arm, which was on a glass dissecting table. Gordon was always very impetuous, moving in sudden jerks. Looking around and seeing that no-one was looking at him, he seized a fire axe, swung it around his head, to sever the arm. He missed, There was an almighty crash, and the arm fell to the floor in a shower of broken glass. Perhaps it is as well that Gordon did not continue in medicine." Pask's partner in that ill-fated anatomy lab was Harry Moore, who later worked with Pask at System Research on many of the projects discussed below (Moore 2001). Among Pask's many behavioral quirks and conceits, one friend "marvelled at his perfect cones of cigarette and pipe ash that he appeared to preserve in every available ash-tray" (Price 2001, 819).[6] Richard Gregory (2001, 685), again, recalls that Pask "was forever taking pills (his brother was an anaesthetist so he had an infinite supply) for real or imagined ailments. These he carried in a vast briefcase wherever he went, and they rattled." Pask apparently felt that he understood medicine better than qualified doctors, which might have had something to do with the decline of his health in the 1990s. Other stories suggest that some of these pills were amphetamines, which might have had something to do with Pask's strange sleeping habits and legendary energy.

Musicolour

Pask's engagement with cybernetics began when he was an undergraduate at Cambridge in the early 1950s. Many people declared themselves cyberneticians after reading Wiener's 1948 *Cybernetics* book, but Pask took his inspiration from the man himself (E. Pask n.d.):

> The epiphany of his Cambridge life came when he was invited by Professor John Braithwaite, Professor of Moral Philosophy, to look after Professor Norbert Wiener, who was visiting Cambridge [and lecturing there on cybernetics]. Gordon who had been struggling for some years to define what he wanted to do, found that Wiener was describing the very science he had longed to work on, but had not known what to call. He had known for some time that what he wanted to do was to simulate how learning took place, using electronics to represent the nervous system . . . [and] in order to study how an adaptive machine could learn. Gordon decided to use his expertise in theatrical lighting to demonstrate the process.

This connection to the theater and the arts is one of the themes that we can pursue in several sections of this chapter. Pask had fallen in love with this world in his schooldays, largely through a friend who ran a traveling cinema in North Wales. At Cambridge, Pask "joined the Footlights club and became a prolific lyric writer for the smoker's concerts where numbers and sketches were tried out. [He also contributed] strange, surreal set design and inventive lighting for shows in Cambridge and in London. Gordon had made friends with Valerie and Feathers Hovenden, who ran a small club theatre in the crypt of a church on Oxford Street." In the same period Pask and McKinnon-Wood, also a Cambridge undergraduate, formed a company called Sirenelle dedicated to staging musical comedies. Both were fascinated with the technology of such performances: "Gordon used to come back [to Cambridge] with bits of Calliope organ, I would come back . . . with bits of bomb sight computer" (McKinnon-Wood 1993, 129). From such pieces, the two men constructed a succession of odd and interesting devices, running from a musical typewriter, through a self-adapting metronome, and up to the so-called Musicolour machine. As we shall see, Pask continued his association with the theater, the arts, and entertainment for the rest of his life.[7]

What, then, of Pask's first sally into cybernetics, the theatrical lighting machine just mentioned? This was the contrivance called Musicolour, for which, as his wife put it, "there were no precedents" (E. Pask n.d.): "Gordon had to design all the circuits used in the machine without any outside assistance.

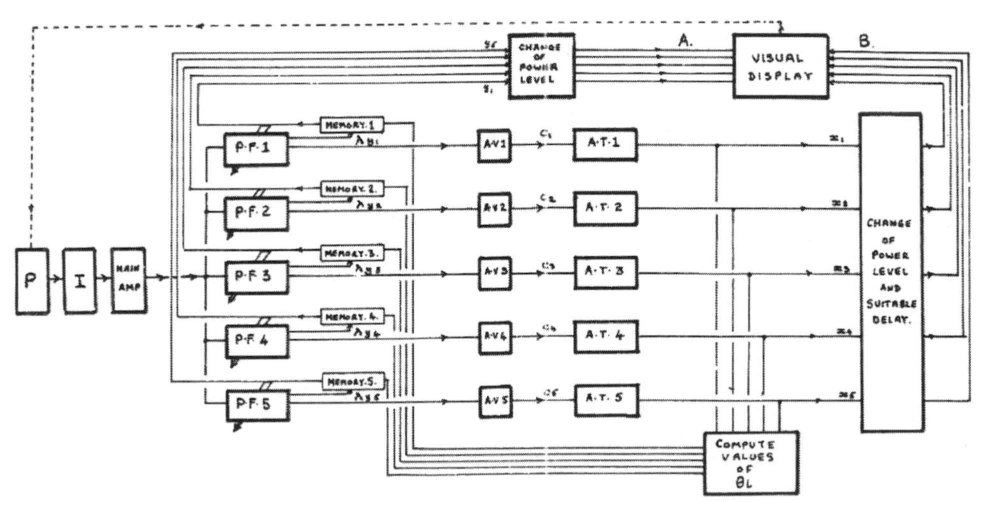

Figure 7.2: Musicolour logic diagram. The original legend reads, "Outline of a typical Musicolour system. *P* = Performer, *I* = Instrument and microphone, *A* = inputs, y_i, to visual display that specify the symbol to be selected, *B* = inputs, x_i, to the visual display that determine the moment of selection, *PF* = property filter, *AV* = averager, *AT* = adaptive threshold device. Memories hold values of (y_i). Control instructions for adjusting the sequence of operation are not shown. Internal feedback loops in the adaptive threshold devices are not shown." Source: G. Pask, "A Comment, a Case History and a Plan," in J. Reichardt (ed.), *Cybernetics, Art, and Ideas* (Greenwich, CT: New York Graphics Society, 1971), 79, fig. 26.

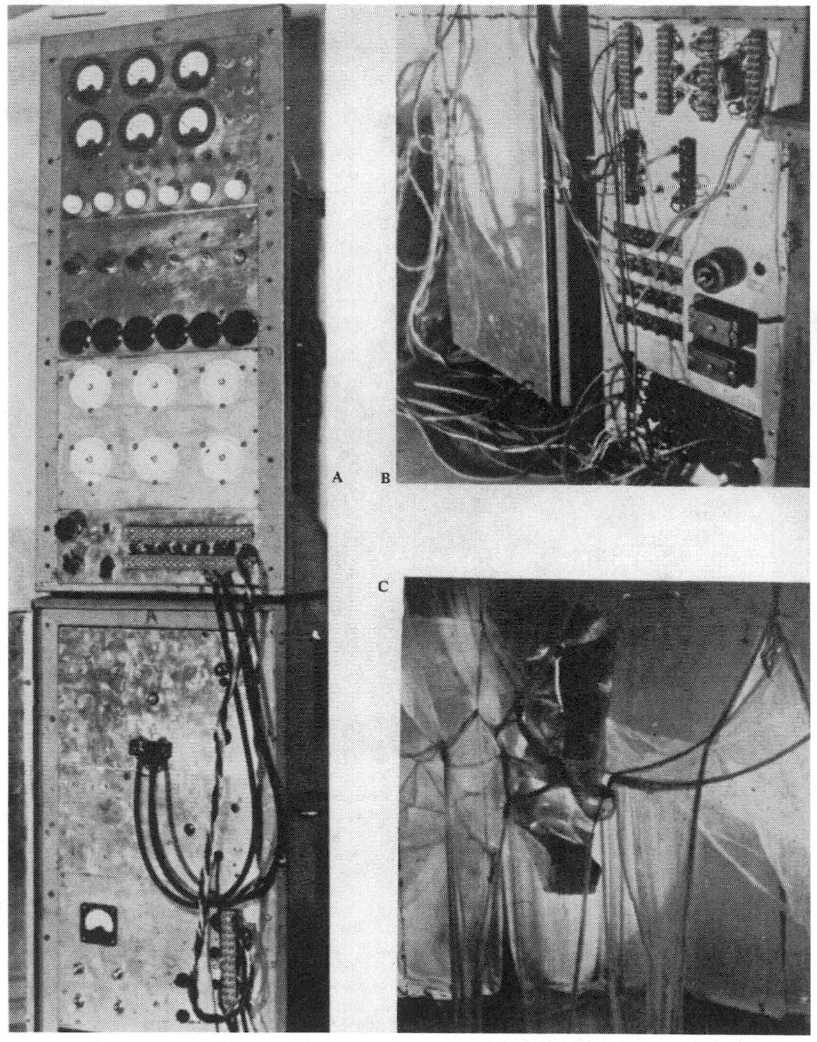

Figure 7.3. *A*, the Musicolour machine; *B*, its power supply; and *C*, a still photograph of a light show. Source: G. Pask, "A Comment, a Case History and a Plan," in J. Reichardt (ed.), *Cybernetics, Art, and Ideas* (Greenwich, CT: New York Graphics Society, 1971), 82, fig. 28.

He built the original machine in his bedroom, on a large, old dining table, which took up most of the room. The process took him several years, during which he took his Tripos examinations and graduated from Cambridge. . . . Gordon had sincerely wanted to be a doctor, like Gar, but once he had begun to work on his Musicolour machine, medicine took second place." Musicolour was a device that used the sound of a musical performance to control a

Figure 7.4. Musicolour display at Churchill's Club, London. Source: G. Pask, "A Comment, a Case History and a Plan," in J. Reichardt (ed.), *Cybernetics, Art, and Ideas* (Greenwich, CT: New York Graphics Society, 1971), 86, fig. 32.

light show, with the aim of achieving a synesthetic combination of sounds and light.[8] Materially, the music was converted into an electrical signal by a microphone, and within Musicolour the signal passed through a set of filters, sensitive to different frequencies, the beat of the music, and so on, and the output of the filters controlled different lights. You could imagine that the highest-frequency filter energized a bank of red lights, the next-highest the blues, and so on. Very simple, except for the fact that the internal parameters of Musicolour's circuitry were not constant. In analogy to biological neurons, banks of lights would only be activated if the output from the relevant filter exceeded a certain threshold value, and these thresholds varied in time as charges built up on capacitors according to the development of the performance and the prior behavior of the machine. In particular, Musicolour was designed to get bored (Pask 1971, 80). If the same musical trope was repeated too often, the thresholds for the corresponding lighting pattern would eventually shift upward and the machine would cease to respond, encouraging the performer to try something new. Eventually some sort of dynamic equilibrium might be reached in which the shifting patterns of the musical performance and the changing parameters of the machine combined to achieve synesthetic effects.[9] Musicolour was central to the subsequent development of much of Pask's cybernetics, so I want to examine it at some length here, historically, ontologically, aesthetically, and sociologically, in that order.

The History of Musicolour

In the early 1950s, first at Cambridge and then at System Research, Gordon and Elizabeth Pask and Robin McKinnon-Wood sought to turn Musicolour into a commercial proposition, beginning in 1953 at the Pomegranate Club in Cambridge—"an eclectically Dadaist organisation"—followed by "a bizarre and eventful tour of the north country" and an eventual move to London (Pask 1971, 78). McKinnon-Wood (1993, 131) later reminisced, "I think my favourite memory of MusiColour was the time we demonstrated the portable version to Mr Billy Butlin [the proprietor of Butlin's holiday camps] in his office. . . . Shortly after his arrival it exploded in a cloud of white smoke. . . . I switched it back on again and it worked perfectly." The first London performance was at Bolton's Theatre and took a traditionally theatrical form. In a piece called "Moon Music," a musician played, Musicolour modulated the lights on a stage set, and, to liven things up, marionettes danced on stage. The marionettes were supposed to move in synchrony with the lights but instead dismembered themselves over the audience (Pask 1971, 82–83). The next show was at Valerie Hovenden's Theatre Club in the crypt of St. Anne's Church on Dean Street. There, in a piece called "Nocturne," attempts were made to link the motions of a human dancer into Musicolour's input—"this proved technically difficult but the aesthetic possibilities are indisputable" (86). Then (86), "since the system was costly to maintain and since the returns were modest, the Musicolour enterprise fell into debt. We secured inexpensive premises above the King's Arms in Tabernacle Street which is a curiously dingy part of the City of London, often engulfed in a sort of beer-sodden mist. There, we set up the system and tried to sell it in any possible way: at one extreme as a pure art form, at the other as an attachment for juke boxes." The story then passed through Churchill's Club, where waiters "dropped cutlery into its entrails [but] the audience reaction was favorable and Musicolour became a permanent feature of the spectacle." After that, Musicolour was used to drive the 120 kilowatt lights at the Mecca Locarno dance hall in Streatham, where, alas, "it became clear that in large scale (and commercially viable) situations, it was difficult or impossible to make genuine use of the system."[10] "Musicolour made its last appearance in 1957, at a ball organized by Michael Gillis. We used a big machine, a small machine and a collection of display media accumulated over the years. But there were other things to do. After the ball, in the crisp, but fragrant air of St. James's Park, the Musicolour idea was formally shelved. I still have a small machine. But it does not work any longer and is of chiefly sentimental value" (Pask 1971, 86–88). We can follow the subsequent mutations of Musicolour in Pask's career below, but one other aspect of its historical development is worth mentioning. As mentioned above by Elizabeth Pask, Gordon had an enduring interest in learning, and we should see how Musicolour fitted into this. The point to note is that in performance the performer learned (performatively rather than cognitively) about the machine (and vice versa), and Pask therefore regarded Musicolour as a machine in which one could learn—scientifically, in a conventional sense—about learning. Thus, in the show at Bolton's Theatre (Pask 1971, 83, 85–86),

> it was possible to investigate the stability of the coupling [between performer and machine]. In this study arbitrary disturbances were introduced into the feedback loop wihout the performer's knowledge. Even though he is ignorant of their occurrence, these disturbances are peculiarly distracting to the performer, who eventually becomes infuriated and opts out of the situation. But there is an inherent stability in the man-machine relation which allows the performer to tolerate a certain level of disturbance. We found that the tolerable level increases as the rapport is established (up to a limit of one hour at any rate). . . . Meanwhile, John Clark, a psychiatrist, had come to the theatre and we jointly observed some phenomena related to the establishment of rapport. First, there is a loss of time sense on the performer's part. One performer, for example, tootled away on his instrument from 10 p.m. to 5 a.m. and seemed unaware that much time had passed; an hour, he thought, at the most. This effect . . . was ubiquitous. Next, there is a group of phenomena bearing on the way in which performers train the machine. As a rule, the performer starts off with simple tricks which are entirely open to description. He says, for example, that he is accenting a chord in a particular passage in order to associate a figure in the display with high notes. . . . Soon . . . the determinate trick gives way to a behaviour pattern which the performer cannot describe but which he adopts to achieve a well-defined goal. Later still, the man-machine interaction takes place at a higher level of abstraction. Goals are no longer tied to properties as sensed by the property filters (though, presumably, they are tied to patterns of properties). From the performer's point of view, training becomes a matter of persuading the machine to adopt a visual style which fits the mood of his perfor-

> mance. At this stage . . . the performer conceives the machine as an extension of himself, rather than as a detached or disassociated entity.

In this sense, Musicolour was, for Pask, an early venture into the experimental psychology of learning and adaptation which led eventually to his 1964 PhD in psychology. I am not going to try to follow this scientific work here, since there was nothing especially cybernetic about it, but we should bear it in mind in the later discussion of Pask's work on training and teaching machines.

Musicolour and Ontology

Musicolour was literally a theatrical object; we can also read it as another piece of ontological theater, in the usual double sense. It staged and dramatized the generic form of the cybernetic ontology; at the same time, it exemplified how one might go on, now in the world of theater and aesthetics, if one subscribed to that ontology. Thus, a Musicolour performance staged the encounter of two exceedingly complex systems—the human performer and the machine (we can come back to the latter)—each having its own endogenous dynamics but nevertheless capable of consequential performative interaction with the other in a dance of agency. The human performance certainly affected the output of the machine, but not in a linear and predictable fashion, so the output of the machine fed back to influence the continuing human performance, and so on around the loop and through the duration of the performance. We are reminded here, as in the case of Beer's cybernetics, of the symmetric version of Ashby's multihomeostat setups, and, like Beer's work and Bateson and Laing's, Pask's cybernetic career was characterized by this symmetric vision.

Beyond this basic observation, we can note that as ontological theater a Musicolour performance undercut any familiar dualist distinction between the human and the nonhuman. The human did not control the performance, nor did the machine. As Pask put it, the performer "trained the machine and it played a game with him. In this sense, the system acted as an extension of the performer with which he could cooperate to achieve effects that he could not achieve on his own" (1971, 78). A Musicolour performance was thus a joint product of a human-machine assemblage. Ontologically, the invitation, as usual, is to think of the world like that—at least the segments that concern us humans, and by analogical extension to the multiplicity of nonhuman elements. This again takes us back to questions of power, which will surface throughout this chapter. In contrast to the traditional impulse to dominate aesthetic media, the Musicolour machine thematized cooperation and revealing in Heidegger's sense. Just as we found Brian Eno "riding the algorithms" in his music in the previous chapter, a Musicolour performer rode the inscrutable dynamics of the machine's circuitry. That is why I said Eno should have read Pask at the end of the previous chapter.

Second, we can note that as ontological theater Musicolour went beyond some of the limitations of the homeostat. If the homeostat only had twenty-five pregiven states of its uniselector, Musicolour's human component had available an endlessly open-ended range of possibilities to explore, and, inasmuch as the machine adapted and reacted to these, so did the machine. (Of course, unlike Ashby, Pask was not trying to build a freestanding electromechanical brain—his task was much easier in this respect: he could rely on the human performer to inject the requisite variety.) At the same time, unlike the homeostat, a Musicolour performance had no fixed goal beyond the very general one of achieving some synesthetic effect, and Pask made no claim to understanding what was required for this. Instead (Pask and McKinnon-Wood 1965, 952),

> other modalities (the best known, perhaps, is Disney's film "Fantasia") have entailed the assumption of a predetermined "synaesthetic" relation. The novelty and scientific interest of this system [Musicolour] emerges from the fact that this assumption is not made. On the contrary, we suppose that the relation which undoubtedly exists between sound (or sound pattern) and light (or light pattern) is entirely personal and that, for a given individual, it is learned throughout a performance. Hence the machine which translates between sound and vision must be a malleable or "learning" device that the performer can "train" (by varying his performance) until it assumes the characteristics of his personally ideal translator.

The Musicolour performer had to find out what constituted a synesthetic relation between sound and light and how to achieve it. We could speak here of a search process and the *temporal emergence of desire*—another Heideggerian revealing—rather than of a preconceived goal that governs a performance. In both of these senses, Musicolour constituted a much richer and more suggestive act of ontological theater than the homeostat, though remaining clearly in the homeostatic lineage.

One subtlety remains to be discussed. I just described Musicolour as one of Beer's exceedingly complex systems. This seems evidently right to me, at least from the perspective of the performer. Even from the simplified description I have given of its functioning, it seems clear that one could not think one's way through Musicolour, anticipating its every adaptation to an evolving sequence of inputs, and this becomes even clearer if one reads Pask's description of all the subtleties in wiring and logic (1971, 78–80). But still, there was a wiring diagram for Musicolour which anyone with a bit of training in electrical engineering could read. So we have, as it were, two descriptions of Musicolour: as an exceedingly complex system (as experienced in practice) and as actually quite simple and comprehensible (as described by its wiring diagram). I am reminded of Arthur Stanley Eddington's two tables: the solid wooden one at which he wrote, and the table as described by physics, made of electrons and nuclei, but mainly empty space. What should we make of this? First, we could think back to the discussion of earlier chapters. In chapter 4 I discussed cellular automata as ontological icons, as exemplifications of the fact that even very simple systems can display enormously complex behavior—as the kinds of objects that might help one imagine the cybernetic ontology more generally. One can think of Musicolour similarly, as a material counterpart to those mathematical systems—thus setting it in the lineage running from the tortoise to DAMS. And second, we could take Musicolour as a reminder that representational understandings of inner workings can often be of little use in our interactions with the world. Though the workings of Musicolour were transparent in the wiring diagram, the best way to get on with it was just to play it. The detour through representation does not rescue us here from the domain of performance.[11]

Ontology and Aesthetics

As one might expect from a cybernetician with roots in the theater, ontology and aesthetics intertwined in Pask's work. I have been quoting at length from an essay Pask wrote in 1968 on Musicolour and its successor, the Colloquy of Mobiles (Pask 1971), which begins with the remark that (76) "man is prone to seek novelty in his environment and, having found a novel situation, to learn how to control it. . . . In slightly different words, man is always aiming to achieve some goal and he is always looking for new goals. . . . My contention is that man enjoys performing these jointly innovative and cohesive operations. Together, they represent an essentially human and inherently pleasurable activity." As already discussed, with this reference to "new goals" Pask explicitly moved beyond the original cybernetic paradigm with its emphasis on mechanisms that seek to achieve predefined goals. This paragraph also continues with a definition of "control" which, like Beer's, differs sharply from the authoritarian image often associated with cybernetics (76): "'Control,' in this symbolic domain, is broadly equivalent to 'problem solving' but it may also be read as 'coming to terms with' or 'explaining' or 'relating to an existing body of experience.'" Needless to say, that Pask was in a position to relax these definitions went along with the fact that he was theorizing and exploring human adaptive behavior, not attempting to build a machine that could mimic it. Musicolour, for example, was a reactive environment; it did not itself formulate new goals for its own performances.

Pask's opening argument was, then, that "man" is essentially adaptive, that adaptation is integral to our being, and to back this up a footnote (76n1) cites the work of "Bartlett . . . Desmond Morris . . . Berlyn . . . Bruner . . . social psychologists, such as Argyll," and, making a connection back to chapter 5, "the psychiatrists. Here, the point is most plainly stated by Bateson, and by Laing, Phillipson and Lee [1966]." Of course, Bateson and Laing and his colleagues were principally concerned with the pathologies of adaptation, while throughout his career Pask was concerned with the pleasures that go with it, but it is interesting to see that he placed himself in the same space as the psychiatrists.[12]

Pask's essay then focused on a discussion of "aesthetically potent environments, that is, . . . environments designed to encourage or foster the type of interaction which is (by hypothesis) pleasurable" (Pask 1971, 76):

> It is clear that an aesthetically potent environment should have the following attributes:
> *a* It must offer sufficient variety to provide the potentially controllable variety [in Ashby's terms] required by a man (however, it must not swamp him with variety—if it did, the environment would be merely unintelligible).
> *b* It must contain forms that a man can learn to interpret at various levels of abstraction.
> *c* It must provide cues or tacitly stated instructions to guide the learning process.
> *d* It may, in addition, respond to a man, engage him in conversation and adapt its characteristics to the prevailing mode of discourse.

Attribute *d* was the one that most interested Pask, and we can notice that it introduces a metaphor of "conversation." An interest in conversation, understood very generally as any form of reciprocally productive and open-ended exchange between two or more parties (which might be humans or machines or humans and machines) was, in fact, the defining topic of all of Pask's work.

Having introduced these general aesthetic considerations, the to the essay then devoted itself to descriptions of the two machines—Musicolour and the Colloquy of Mobiles—that Pask had built that "go *some* way towards explicitly satisfying the requirements of *d*." We have already discussed Musicolour, and we can look at the Colloquy later. Here I want to follow an important detour in Pask's exposition. Pask remarks that (77) "*any* competent work of art is an aesthetically potent environment. . . . Condition *d* is satisfied implicitly and often in a complex fashion that depends upon the sensory modality used by the work. Thus, a painting does not move. But our interaction with it is dynamic for we scan it with our eyes, we attend to it selectively and our perceptual processes build up images of parts of it. . . . Of course, a painting does not respond to us either. But our internal representation of the picture, our active perception of it, does respond and does engage in an internal 'conversation' with the part of our mind responsible for immediate awareness." This observation takes us back to the theme of ontology in action: what difference does ontology make? It seems that Pask has gone through this cybernetic analysis of aesthetics only to conclude that it makes no difference at all. Any "competent" art object, like a conventional painting, can satisfy his cybernetic criterion *d*. So why bother? Fortunately, Pask found what I take to be the right answer. It is not the case that cybernetics *requires* us to do art in a different way. The analysis is not a condemnation of studio painting or whatever. But cybernetics does suggest a new strategy, a novel way of going on, in the creation of art objects. We could try to construct objects which foreground Pask's requirement *d*, which explicitly "engage a man in conversation," which "externalize this discourse" as Pask also put it—rather than effacing or concealing the engagement, as conventional art objects do. Cybernetics thus invites (rather than requires) a certain stance or strategy in the world of the arts that conventional aesthetics does not, and it is, of course, precisely this stance, as taken up across all sorts of forms of life, that interests me.

Beyond the mere possibility of this cybernetic stance, the proof of the pudding is obviously in the eating, though Pask does find a way of recommending it, which has more to do with the "consumption" of art than its production: "The chief merit of externalization . . . seems to be that external discourse correlates with an ambiguity of role. If I look at a picture, I am biased to be a viewer, though in a sense I can and do repaint my internal representation. If I play with a reactive and adaptive environment, I can alternate the roles of painter and viewer at will. Whether there is virtue in this, I do not know. But there might be." So, the cybernetic stance invites both a change in the nature of art objects and, once more, a shift in the power relation between artist and audience, somehow entraining the audience in their production and evolution, as we also saw in the previous chapter in the case of Brian Eno. In the Musicolour performances at Churchill's Club, for example, "we also used the system when people were dancing and discovered that in these circumstances the audience can participate in the performer-machine feedback loop just because they are doing something to music and the band is responding to them" (88), though this turned out not to be the case in the larger setting of the Streatham Locarno.

The Social Basis of Pask's Cybernetics

It is clear that the social dynamics of Pask's formative venture into cybernetics bears much the same marks as the others discussed in earlier chapters. There is, first of all, the undisciplined mode of transmission of cybernetics. Pask did not train to be a cybernetician by enrolling in any disciplinary program; instead, a chance meeting with Norbert Wiener served, as we saw, to crystallize Pask's agenda, an agenda that already existed, though in a relatively formless state. Second, as we have also seen, Pask's first project as a cybernetician was undertaken in an undisciplined space outside any conventional institutional structure—he built the first Musicolour machine in his rooms at Cambridge, out of the detritus of war and a technological society. I mentioned bits of Calliope organs and bomb sight computers earlier; Elizabeth Pask (n.d.) recalled that Gordon and Harry Moore built Musicolour from "old relays and uniselectors junked from post office telephone exchanges"—the same components that Walter and Ashby used in their model brains. One could speak here of a lack of material discipline as well as social discipline. Like our other cyberneticians, then, Pask's cybernetics bubbled up outside the normal channels of society. And along with this undisciplined aspect went the protean quality of Pask's cybernetics: Pask was entirely free to follow his own inclinations in developing his cybernetics in a theatrical direction, a more or less unprecedented development.[13] At the same time, this lack of disciplinary control helps to account for another aspect of the novel form of Pask's cybernetics—his abandonment, already in the early 1950s, of the idea that cybernetic systems seek by definition to pursue fixed goals.

One can think along much the same lines about the fate of Musicolour itself. Pask's recollection, quoted above, that "we . . . tried to sell it in any possible way: at one extreme as a pure art form, at the other as an attachment for juke boxes," goes to the heart of the matter. It was not clear what Musicolour was. It did not fit well into the usual classification of material objects. It had something to do with music, but it wasn't a musical instrument. It drove a light show, but it wasn't just lighting. It was an object, but a pretty ugly one, not an art object in itself. One could say that Musicolour was itself an undisciplined machine, incommensurable with conventional forms of entertainment, and the different modes of presentation and venue that Pask and his friends explored in the 1950s have to be seen as a form of experimentation, trying to find Musicolour a niche in the world. In the end, as we have seen, the world proved recalcitrant, and, like that other odd art object, Gysin's Dreamachine, Musicolour was a commercial failure. Mention of the Dreamachine perhaps reminds us that in the later sixties light shows of all sorts—not all using strobes, and some very reminiscent of Musicolour displays—were de rigueur. But by that time Musicolour had been forgotten and Pask had moved on to other projects. One can only wonder what the Grateful Dead might have got out of one of Pask's devices.

And finally, Pask himself. One should probably understand the boy who built bombs and said that school taught him to be a gangster as someone who enjoyed a lack of discipline—not as someone forced into the margins of society, but who sought them out. No doubt for Pask much of the attraction of Musicolour and cybernetics in general lay in their undisciplined marginality. And this, in turn, helps us to understand his post-Cambridge career, based in a private company, System Research, free from any demands, except that of somehow improvising a living. Now we can pick up the historical thread again.

Training Machines

Pask did not lose interest in adaptive machinery after Musicolour, but he had the idea of putting it to a different and more prosaic use, returning to his formative interest in learning. In the mid-1950s, "there was great demand in the commercial world for keyboard operators, both for punch card machines and typing," and Pask set out to construct an adaptive keyboard trainer. He later recalled that the "first Self Adaptive Keyboard Trainer (SAKI) was constructed in 1956 by myself and prototyped by Robin McKinnon-Wood and me" (Pask 1982, 69; see fig. 7.5). This was displayed at the Inventors and Patentees Exhibition at the Horticultural Hall in London, a meeting also regularly frequented by one Christopher Bailey, technical director of the Solartron Electronic Group, who had "from time to time, made valuable contacts with the

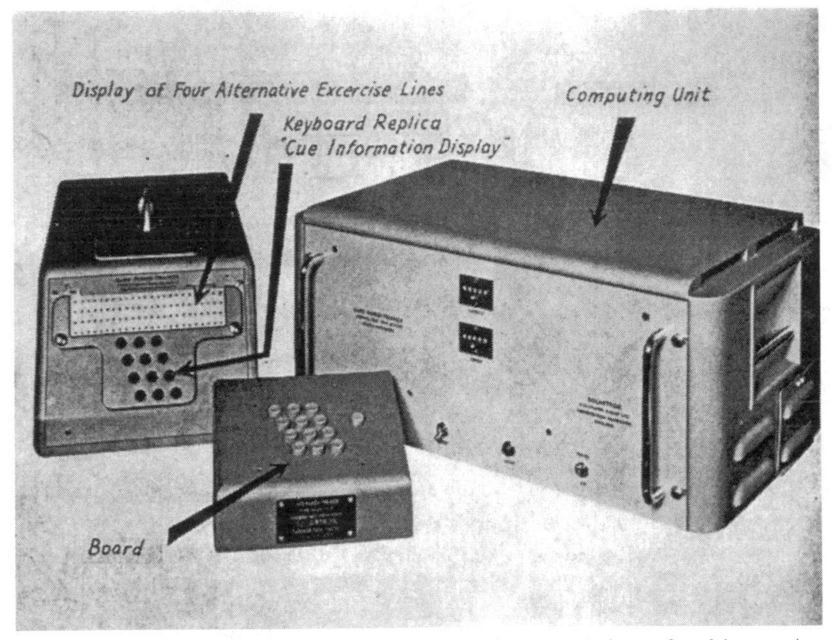

Figure 7.5. SAKI. Source: Pask 1961, pl. II, facing p. 33. (Reproduced by permission of Amanda Heitler.)

less loony exhibitors." Pask got into conversation with Bailey about Grey Walter's robot tortoises, and Bailey in turn proposed that Solartron, which was already expanding into the area of AI, should support the development of adaptive training machines by Pask and System Research (E. Pask n.d.; McKinnon-Wood 1993). Thereafter (Pask 1982, 69, 72),

> Bailey participated in the design and development of this and other systems; notably: EUCRATES [figs. 7.6, 7.7] a hybrid training machine and trainee simulation; a device for training assembly line tasks; a radar simulation training machine and several devices for interpolating adaptively modulated alerting signals into a system, depending upon lapse of attention. The acronym SAKI stood, after that, for *Solartron Adaptive Keyboard Instructor* and a number of these were built and marketed. Details can be found in a U.K. Patent granted in 1961, number 15494/56. The machine described is simply representative of the less complex devices which were, in fact, custom-built in small batches for different kinds of key boards (full scale, special and so on). The patent covered, also, more complex devices like EUCRATES. . . . In 1961 the manufacturing rights for machines covered by these patents were obtained by Cybernetic Developments: about 50 keyboard machines were leased and sold.

A genealogical relationship leading from Musicolour to SAKI and Eucrates is evident.[14] Just as Musicolour got bored with musicians and urged them on to novel endeavors, so these later machines responded to the performance of the trainee, speeding up or slowing down in response to the trainee's emergent performance, identifying weaknesses and harping upon them, while progressively moving to harder exercises when the easier ones had been mastered.[15] Stafford Beer tried Eucrates out in 1958 and recorded, "I began in total ignorance of the punch. Forty-five minutes later I was punching at the rate of eight keys a second: as fast as an experienced punching girl" (Beer 1959, 125).[16] SAKI was an analog machine; like Musicolour, its key adaptive components were the usual uniselectors, relays, and capacitors. Later versions used microprocessors and were marketed by System Research Developments (Sales). SAKI itself formed the basis for the Mavis Beacon typing trainer, widely available as PC software today.

The link to Solartron in the development of adaptive training machines was very consequential for Pask and System Research. Much of Pask's paid work

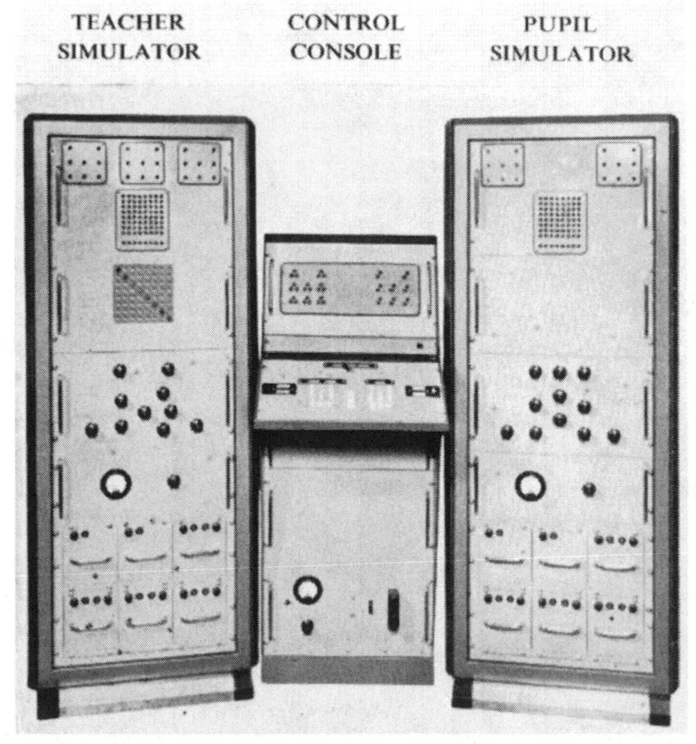

Figure 7.6. Eucrates. Source: Pask 1961, pl. I, facing p. 32. (Reproduced by permission of Amanda Heitler.)

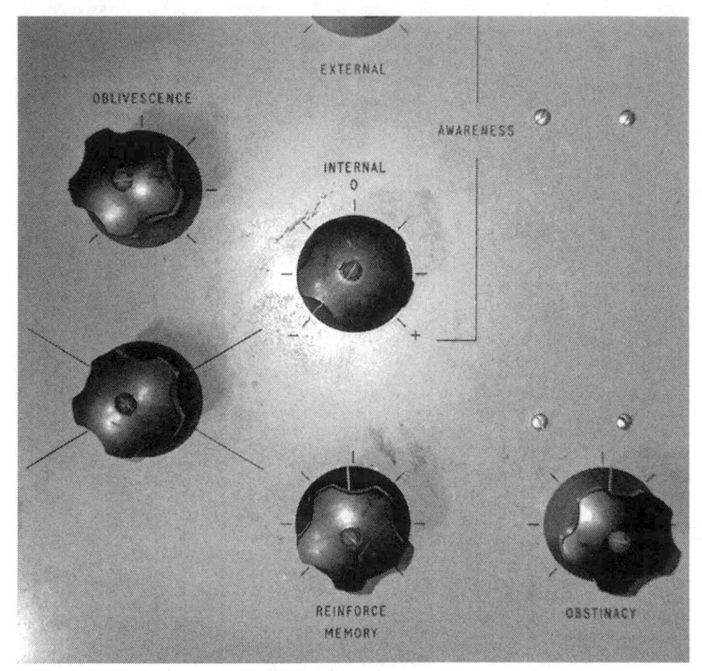

Figure 7.7. Eucrates: detail showing settings. (© 2002 by Paul Pangaro.)

from the late 1950s onward centered on the development of adaptive teaching and training machines and information systems, as discussed further below. And again this episode illustrates some of the social oddity of cybernetics—a chance meeting at an exhibition, rather than any more formalized encounter, and, again, the protean quality of cybernetics, as a peculiar artwork metamorphosed easily into a device for teaching people to type. From an ontological point of view, we can see machines like SAKI as ontological theater much like Musicolour, still featuring a dance of agency between trainee and machine, though now the dance had been domesticated to fit into a market niche—the machine did now have a predetermined goal: to help the human to learn to type efficiently—though the path to the goal remained open ended (like coupled homeostats searching for equilibrium). Interestingly from an ontological angle, Beer recorded witnessing a version of the Turing test carried out with Eucrates. As articulated by Turing, this test relates to machine intelligence: the textual responses of an intelligent machine would be indistinguishable from those of a human being. Pask's demonstrations with Eucrates were a performative rather than representational version of this: "So my very first exposure to Gordon's science was when he sat me in a room with a monitor, in the capacity of metaobserver, and invited me to determine which screen was being driven by the human and which by Eucrates. It was impossible. The behaviour of the two elements was converging, and each was moving towards the other" (S. Beer 2001, 552).

Teaching Machines

The machines we have discussed so far—Musicolour, SAKI, Eucrates—were all directly concerned with performance and nonverbal skills. In the 1970s, however, Pask turned his attention to education more generally and to machines that could support and foster the transmission of representational knowledge. CASTE (for Course Assembly System and Tutorial Environment) was the first example of such a machine, constructed in the early 1970s by Pask and Bernard Scott (who completed a PhD with Pask at Brunel University in 1976; see fig. 7.8). Figure 7.9 shows a more sophisticated version, Thoughtsticker, from around 1977.[17]

There is no doubt that Pask was extremely interested in these machines and the overall project in which they served as markers, or of their worldly

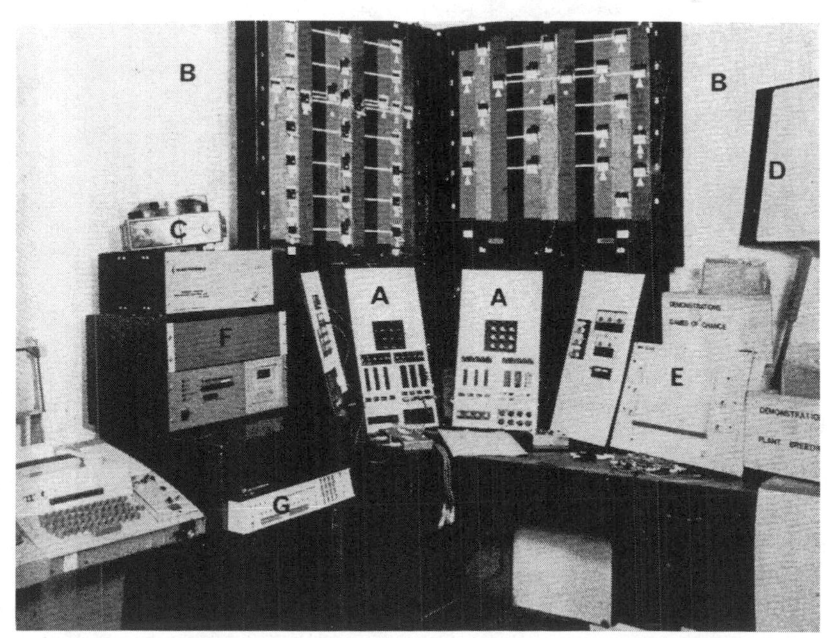

Figure 7.8. Intuition, a portable version of CASTE. Source: G. Pask, "Conversational Techniques in the Study and Practice of Education," *British Journal of Educational Psychology, 46* (1976), 24, fig. 3. (Reproduced with permission from the *British Journal of Educational Psychology*. © The British Psychologcial Society.)

Figure 1. An epistemological laboratory. The figure is placed first to give an idea of how empirical results are obtained but many of the labels will not be intelligible until, at a later stage in the paper, the functioning of the equipment and the types of man machine transactions are spelled out. The key is as follows:

A. Random access slide projector with control keyboard, for displaying slide mounted graphics.
B. Entailment mesh display with overlay multi sheets, containing 60 node positions and 4 × 60 independently addressed coloured signal lamps and touch sensors.
C. Tutorial mode keyboard with special function keys.
D. Course assembly mode keyboard with special function keys.
E. ARDS graphic display tubes with control unit and keyboard used for displaying "pruned" meshes.
F. Video display units with control keyboards used for topic text input–output.
G. Pigeon holes filing system with slots for 60 files and containing 3 × 60 independently addressed signal lamps and 60 sensors.
H. Dual drive floppy disk unit.
I. CAI 32k computer.
J. Digital Cassette unit used as mass storage device.
K. ASR 33 teletype used for "hard copy" output.
L. Electronics rack, containing special electronics and system interface.

Figure 7.9. Thoughtsticker. Source: G. Pask, "Organizational Closure of Potentially Conscious Systems," in M. Zeleny (ed.), *Autopoiesis: A Theory of the Living* (New York: North Holland, 1981), 269, fig. 1.

importance in sustaining him and System Research, but I am going to give them relatively short shrift here. The reason for this is that much of the technical development in this area circled around the problematic of representing articulated knowledge within a machine, and that is not a key problematic of this book. Pask quickly arrived at the position that bodies of knowledge consist of sets of concepts related to one another in what he called an "entailment mesh" (Pask and Scott 1973) an idea which was itself central to what he called "Conversation Theory" and later to his "Interaction of Actors" theory (Pask 1992). I leave it to others to explore the historical evolution of this aspect of Pask's work, set out in three books (Pask 1975a, 1975b, 1976a) and many papers, all of which are, in Bernard Scott's words (2001a, 2), "notoriously difficult" to read.[18]

Here, then, I will confine myself to a few comments. First, CASTE and its descendants remained clearly within Musicolour's lineage. CASTE was another aesthetically potent environment with which the student could interact, exploring different routes around the entailment mesh appropriate to this subject matter or that, being invited to carry out the relevant performative exercises, with apparatus at the laboratory bench if relevant, responding to queries from the machine, and so on. Again we find the performative epistemology that I have associated with cybernetics in the earlier chapters—of articulated knowledge and understanding as part and parcel of a field of performances. Thoughtsticker went so far as to interact with the subject-experts who fed in the contents of entailment meshes—generalizing aspects of the mesh in various ways, for example, and checking back about the acceptability of these generalizations. In this version, the very content of the mesh was, to a degree, a joint product of the human and the machine, as I said earlier about a Musicolour performance.

Second, we can note that, again like Musicolour, Pask's teaching machines also functioned as experimental setups for scientific research, now in the field of educational psychology. Pask and his collaborators toured schools and colleges, including Henley Grammar School and the Architectural Association, with the portable version of CASTE and made observations on how different learners came to grips with them, echoing Pask's observations in the fifties on Musicolour. These observations led him to distinguish two distinct styles of learning—labeled "serial" and "holist" in relation to how people traveled around the mesh—and to argue that different forms of pedagogy were appropriate to each.

Third, we can return to the social basis of Pask's cybernetics. The very ease with which one can skip from a general sense of "adaptation" to a specific sense of "learning" as manifested in schools and universities suggests that education, if anywhere, is a site at which cybernetics stood a good chance of latching onto more mainstream institutions, and so it proved for Pask. In 1982, Pask and Susan Curran (1982, 164) recorded that "over a period of 11 years (five of overlapping projects and a six-year research project) the Social Science Research Council in Britain supported a study by System Research . . . on learning and knowledge." Around the same time, Bernard Scott (1982, 480) noted that "despite the painstaking way in which Pask prepared the ground for the theory's presentation [conversation theory], it is fair to say that it has not won general acceptance in psychology. The dominant attitudes were too strong." But nevertheless, "the theory (through its applications) has had far greater impact in educational circles and is recognised, internationally, as a major contribution to educational praxis" (on which see B. Scott 2001b).

Pask was integrally involved, for example, in pedagogical innovation at Britain's new Open University (OU). David Hawkridge joined the OU as head of their Institute of Educational Technology in 1969. He knew of Pask's work on SAKI and hired Brian Lewis who had worked at System Research as his deputy (Hawkridge 2001, 688–90):

> [Lewis] lost no time in introducing me and other Institute staff to Gordon, his great friend and former boss. In fact, quite a few of us went on expeditions down to System Research Ltd in Richmond in the early 1970s, and came back bemused and sometimes confused. What was Pask up to and could it be turned to advantage in the OU? I suggested to Brian that we should ask Pask to be our Visiting Professor (part-time). That would regularise the Richmond visits, and Brian said he thought Gordon would be delighted to give a few seminars at the Institute. I had little idea what these might involve, but the Institute had just won a large grant from the Ford Foundation, and Gordon's appointment (and, indeed, Bernard Scott's as a consultant) seemed entirely appropriate. He was the pre-eminent British scholar in our field. . . . It was probably Brian who suggested to Gordon that there was a DSc to be had from the OU if only he would submit all his publications. Finding sufficiently knowledgeable referees for his case was not easy, but I had the pleasure of seeing him receive the award to great applause. I think he was delighted with it, and with the robes.[19]

There is much more that could be said on Pask's teaching machines, but I want to close this section by noting that his work on these devices led him to a distinctive general perspective on mind.[20] In earlier chapters we saw how, in different ways, cybernetics shaded into Eastern philosophy and spirituality. None of this figures prominently in Pask's work, nor does religion in general (though he did convert to Catholicism shortly before his death: Amanda Heitler, personal communication). But a 1977 essay, "Minds and Media in Education and Entertainment," is worth examining from this perspective. Here the initial referent of "mind" was the human mind, and the "media" were the usual means of communication between minds: speech, texts, information systems like CASTE. But (Pask 1977, 40)

> there is no *need* to see minds as neatly encapsulated in brains connected by a network of channels called "the media" [fig. 7.10a]. . . . I am inviting the reader

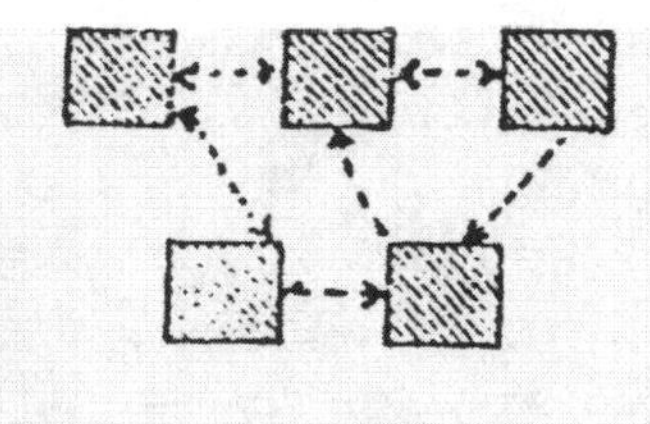

a

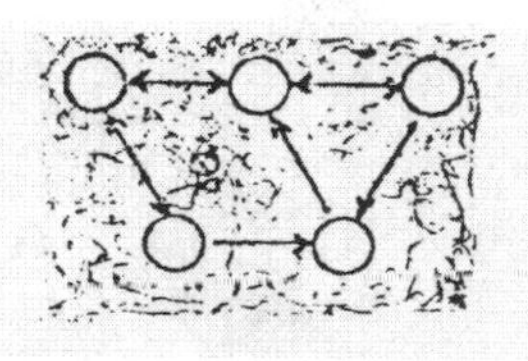

b

Figure 7.10. Two views of minds and media: *a*, linked minds. *Squares*, organisms; *arrows*, media as channels of communication. *b*, embedded minds. *Circles*, individuals; *arrows*, communication as program sharing and linguistic interaction between individuals. Source: Pask 1977, 40, figs. 1, 2.

> to try out a different point of view; namely, the image of a pervasive medium (or media) inhabited by minds in motion. Thus, media are characterized as computing systems, albeit of a peculiar kind. But the statement neither asserts nor denies the homogeneity of a medium. In our present state of knowledge, it seems prudent to regard the medium as heterogeneous, and rendered modular by the existence of specially unrestricted regions (brains, for example), capable of acting as *L* [language] processors (though I have a hankering to imagine that these regions are ultimately determined by programmatic rather than concrete localization). It is surely true that rather powerful computerized systems greatly reduce the differentiation of the medium and coalesce the specially restricted modules, so that "interface barriers" are less obtrusive than they used to be [fig. 7.10b].

Here one might be tempted to think of recent work in cognitive science on "distributed cognition"—the observation that much "mental" activity in fact depends upon external, "non-mental" processing (e.g., Hutchins 1995). But something more is at stake. Even with Musicolour and SAKI, Pask had been impressed by the strength of the coupling established between human and machine, which he argued fused them into a single novel entity: "The teaching machine starts to work, in the sense that it accelerates the learning process and teaches efficiently, just when we, as outsiders, find that it is impossible to say what the trainee is deciding about—in other words, at the stage when interaction between the teaching machine and the trainee has given rise to a dynamic equilibrium which involves parts of both" (Pask 1960a, 975); "although the physical demarcation of the student and the machine is definite, the subsystems representing the student's region of control and the adaptive machine's region of control are arbitrary and (relative to any given criterion) have limits that are continually changing" (Pask and McKinnon-Wood

1965, 962). Pask thus switched gestalt entirely, in favor of an image of mind as an all-pervading medium, with human minds as inflections within the overall flow.

This decentered image of mind as all pervasive, and of individual brains as finite localized nodes, is common to much Eastern philosophy and spirituality, though Pask did not quite put it like that: "There is little originality in the view put forward. The McCluhans [*sic*] (both Marshall and Eric in different style) say that media are extensions of the brain; poets, mystics, and sorcerers have expressed similar sentiments for ages" (Pask 1977, 40). It bears emphasis, however, that like the other cyberneticians, Pask did not simply equate cybernetics with Buddhist philosophy or whatever. We could say that he *added* to Buddhist philosophy an engineering aspect. If Eastern philosophy has been presented for millenia in the form of a reflection on the mind, this lineage of Paskian machines running from Musicolour to Thoughtsticker staged much the same vision in the mundane material world of entertainment and teaching machines. In this sense, we could think of Pask's work, like Beer's, as a sort of spiritual engineering.[21]

Chemical Computers

SELF-ORGANIZING SYSTEMS LIE ALL AROUND US. THERE ARE QUAGMIRES, THE FISH IN THE SEA, OR INTRACTABLE SYSTEMS LIKE CLOUDS. SURELY WE CAN MAKE THESE WORK THINGS OUT FOR US, ACT AS OUR CONTROL MECHANISMS, OR PERHAPS MOST IMPORTANT OF ALL, WE CAN COUPLE THESE SEEMINGLY UNCONTROLLABLE ENTITIES TOGETHER SO THAT THEY CAN CONTROL EACH OTHER. WHY NOT, FOR EXAMPLE, COUPLE THE TRAFFIC CHAOS IN CHICAGO TO THE TRAFFIC CHAOS OF NEW YORK IN ORDER TO OBTAIN AN ACCEPTABLY SELF-ORGANIZING WHOLE? WHY NOT ASSOCIATE INDIVIDUAL BRAINS TO ACHIEVE A GROUP INTELLIGENCE?

GORDON PASK, "THE NATURAL HISTORY OF NETWORKS" (1960B, 258)

Much of Pask's cybernetics grew straight out of Musicolour: the trajectory that ran through the trainers and educational machines just discussed and the work in the arts, theater, and architecture discussed later in this chapter. But in the 1950s and early 1960s there was another aspect to his cybernetics that was not so closely tied to Musicolour and that I want to examine now.[22] This was the work on "biological computers" already mentioned in the previous chapter. Some of this work was done in collaboration with Stafford Beer, but here we can focus on the best-documented work in this area, on what I will now call "chemical computers"—though Pask often referred to them as "organic computers," in reference to their quasi-organic properties rather than the materials from which they were constructed.[23] Beer again figures in this story, though it is clear that the initiative and most of the work was Pask's.

As discussed in the previous chapter, at the center of Beer's vision of the cybernetic factory was the numinous U-machine, the homeostatic controller which not only kept the factory on course in normal conditions but also adapted to changing conditions. Beer's experiments with biological systems aimed at constructing such a machine. In his publications on chemical computers, which first appeared in 1958, Pask set his work in a similar frame, and a review of this work might help us understand the overall problematic more clearly. The opening paragraph of Pask's essay "Organic Control and the Cybernetic Method" (1958, 155) is this: "A manager, being anxious to retire from his position in an industry, wished to nominate his successor. No candidate entirely satisfied his requirements, and after a prolonged but fruitless search, this manager decided that a control mechanism should take his place. Consequently he engaged four separate cyberneticians. Each of them had been recommended in good faith as able to design a control mechanism which would emulate and improve upon the methods of industrial decision making the manager had built up throughout the years." Among other things, this paragraph is evidently a setup for distinguishing between four versions of what cybernetics might be and recommending one of them, namely, Pask's (and Beer's). There is no need to go into the details of all four, but a key contrast among them is brought out in the following hypothetical conversation. One of the cyberneticians is trying to find out how the manager manages (158):

> *Manager.*—I keep telling you my immediate object was to maximise production of piston rings.
> *Cybernetician.*—Right, I see you did this on a budget of £10,000.
> *Manager.*—I bought the new machine and installed it for £8,000.
> *Cybernetician.*—Well, how about the remaining £2,000?
> *Manager.*—We started to make ornamental plaques.
> *Cybernetician.*—Keep to the subject. That has nothing to do with piston rings.
> *Manager.*—Certainly it has. I didn't want to upset Bill Smith. I told you he was sensitive about being a craftsman. So we tried our hand at ornamental plaques, that was my daughter's idea.

Cybernetician.—Which costs you £2,000.
Manager.—Nonsense, Bill Smith enjoys the job. He is a responsible chap, and helps to sober up the hot heads, no it's worth every penny.
Cybernetician.—Very well, as you please. Just one other enquiry, though. What is an appropriate model for this process? What does it seem like to manage a piston ring plant?
Manager.—It's like sailing a boat.
Cybernetician.—Yes.

In case the reader might miss the significance of that final "yes," Pask comments that "they might continue to infuriate each other indefinitely." This "cybernetician" is infuriated because he wants to extract some rules from the manager that can be run on a computer, or perhaps find some statistical regularity between the firm's inputs and outputs that can be likewise encoded. The manager, in contrast, insists that running the factory is not like that; that genuinely novel solutions to problems are sometimes necessary, solutions not given in prior practice and thus not capturable in algorithms, like spending £2,000 just to keep Bill Smith happy for the overall good of the firm. Hence his very cybernetic final reply, that managing a firm is like sailing a boat—a performative participation in the dynamics of a system that is never fully under control (taking us straight back to Wiener's derivation of "cybernetics," and reminding us, for example, of Brian Eno's approach to musical composition).

In this essay, Pask makes it clear that he does not take the search for algorithms to be the defining aspect of cybernetics. People who take that approach are "rightly electronic engineers examining their particular kinds of hypotheses about managers" (Pask 1958, 171). In effect, Pask makes here much the same contrast I made in the opening chapter between symbolic AI and the branch of cybernetics that interests me and to which Pask and our other principals devoted themselves. Pask was interested in machines that could sail boats, to which we can now turn. We can look at how Pask's chemical computers functioned, and then how they might substitute for human managers.

Threads

Figure 7.11 is a schematic of a chemical computer. A set of electrodes dips down vertically into a dish of ferrous sulphate solution. As current is passed through the electrodes, filaments of iron—"threads" as Pask called them—

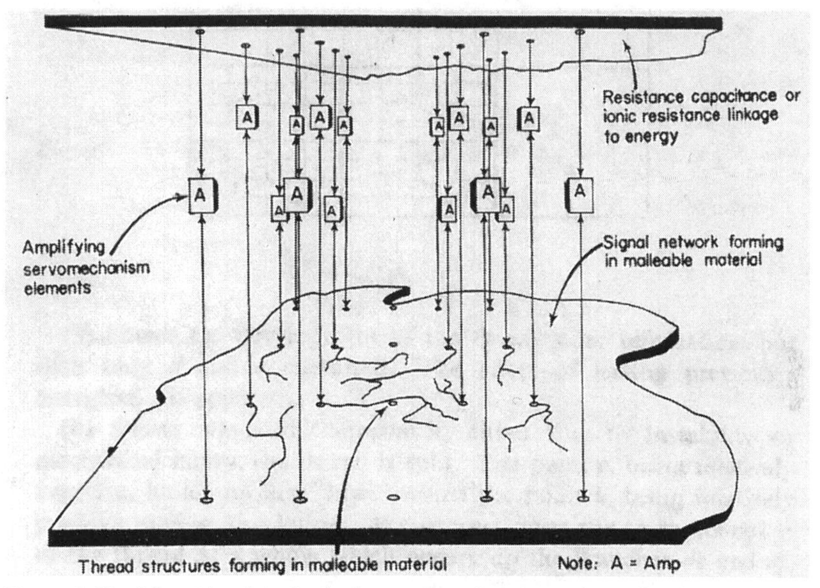

Figure 7.11. Schematic of a chemical computer. Source: Pask 1960b, 247, fig. 4.

grow outward from their tips into the liquid: figure 7.12 is a photograph of a stage in this process. Very simple, but so what? Three points about such devices need to be understood to appreciate Pask's vision. First, the threads are *unstable*: they grow in regions of high current density but dissolve back into solution otherwise. Second, the threads grow *unpredictably*, sprouting new dendritic branches (which might extend further or dissolve)—"The moment to moment development of a thread proceeds via a trial process. Slender branches develop as extensions of the thread in different directions, and most of these, usually all except the one which points along the path of maximum current, are abortive" (Pask 1958, 165). Such a system can be seen as conducting a search through an open-ended space of possibilities, and we can also see that in Ashby's terms its has the high variety required of a controller: it can run through an endless list of material configurations (compare the space of thread geometries with the twenty-five states of the homeostat). Third, as extensions of the electrodes, the threads themselves influence current densities in the dish. Thus, the present thread structure helps determine how the structure will evolve in relation to currents flowing through the electrodes, and hence the growth of the thread structure exhibits a path dependence in time: it depends in detail on both the history of inputs through the electrodes and on the emerging responses of the system to those. The system thus has a

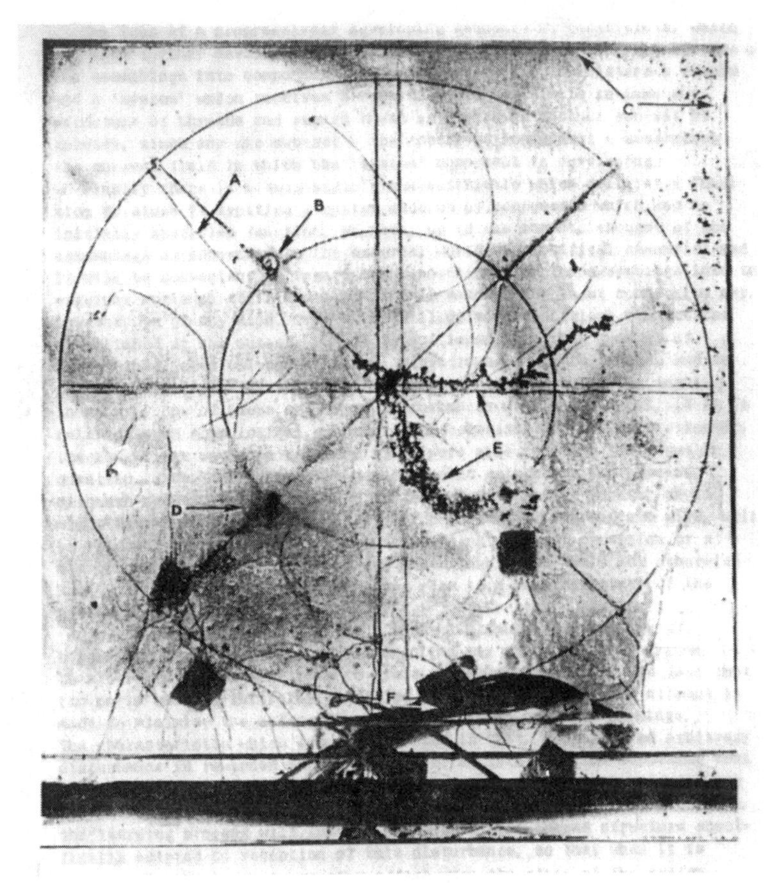

Figure 7.12. Threads growing in a chemical computer. *A*, connecting wires for electrodes; *B*, platinum pillar electrodes; *C*, edges of glass tank containing ferrous sulfate; *D*, chemical reaction in progress; *E*, "tree" threads being formed; *F*, connecting cables. Source: Pask 1959, 919, fig. 12.

memory, so it can learn. This was Pask's idea: the chemical computer could function as an adaptive controller, in the lineage of the homeostat. In this, of course, it was not so far removed from Musicolour and SAKI, though realized in a much more flexible and lively medium than that supplied by uniselectors, relays, and capacitors.

The question now becomes one of how such a system might be interested in *us*: how can a chemical computer be induced to substitute for the human manager of a factory? As with Beer's biological computers, the answer is simple enough, at least in principle. Imagine there are two different sets of electrodes dipping into the dish of ferrous sulphate with its thread structure.

One set is inputs: the currents flowing through them reflect the parameters of the factory (orders, stocks, cash-flow, etc.). The other set is outputs: the voltages they detect represent instructions to the factory (buy more raw materials, redirect production flows). There will be some determinate relationship between these inputs and outputs, fixed by the current thread structure, but this structure will itself evolve in practice in a process of reciprocal vetoing, as Beer callled it, and, as Ashby would have said, the combined system of factory plus controller will inevitably "run to equilibrium." Like a set of interacting homeostats, the chemical computer and the factory will eventually find some operating condition in which both remain stable: the factory settles down as a viable system, in Beer's terms, and the chemical computer, too, settles down into a state of dynamic equilibrium (at least until some uncontrollable perturbation arrives and disturbs the equilibrium, when the search process starts again).

The magic is done—well, almost. Pask thought through at least two further complications. First, there is the question of how to get the process of coupling the computer to the factory going. One answer was to envisage a "catalyst," a system that would send current through the "least visited" electrodes, thus fostering a variety of interactions with the factory and enabling the computer to interrogate the factory's performance on a broad front. Of course, second, the procedure of simply letting the computer and the factory search open-endedly for a mutual equilibrium would almost certainly be disastrous. Who knows what terminally idiotic instructions the computer would issue before stability was approached? Pask therefore imagined that the manager would be allowed to *train* the controller before he retired, monitoring the state of the factory and the machine's responses to that and approving or disapproving those responses by injecting pulses of current as appropriate to reinforce positive tendencies in the machine's evolution, as indicated in figure 7.13. Pask noted that this kind of training would not take the form of the manager dominating the controller and dictating its performance; there was no way that could be done. In fact, and as usual, the interaction would have to take the form of a "partly competitive and partly collaborative game" or conversation (Pask 1958, 170): "After an interval, the structured regions [in the controller] will produce a pattern of behaviour which the manager accepts, not necessarily one he would have approved of initially, but one he accepts as a compromise." Thus the manager and the controller come into homeostatic equilibrium at the same time, in the same way, and in the same process as the controller comes into equilibrium with the factory. "At this point the structured region will replicate indefinitely so that its replica produces the same

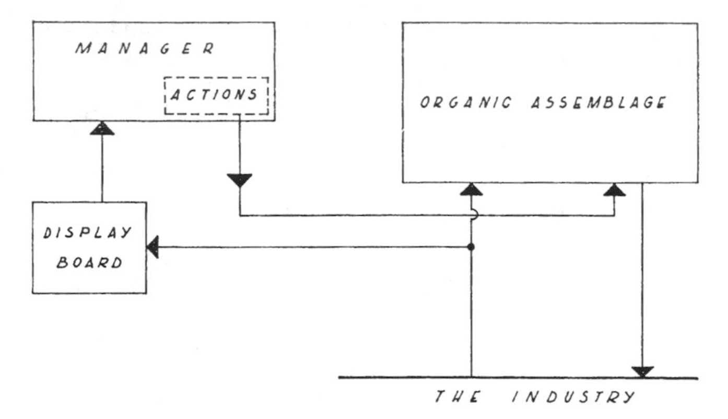

Figure 7.13. Training a chemical computer. Source: Pask 1958, 169, diagram 2.

pattern of behaviour. The manager may thus be removed and the assemblage will act as an organic control mechanism in the industry" (169).

Not much new commentary is needed here. As ontological theater, Pask's chemical computers were in much the same space as Beer's biological ones, staging a direct performative coupling between exceedingly complex dynamic systems (the threads, the factory, the manager) free from any representational detour—a coupling originally enacted by Musicolour in Pask's career and that could take us all the way back to Walter's tortoises, except that the threads displayed more variety than the tortoise and, of course, they grew without any painstaking design, exploiting the liveliness of matter instead (and taking us back to Ashby's thoughts on evolutionary design in chapter 4, as well as Beer in chapter 6). As I said in the previous chapter, I am struck by the imagination required to even begin contemplating the use of an electrochemical device such as this as an adaptive controller for any sort of system. It is hard to imagine arriving at such a vision within the symbolic AI tradition, for example.

But there is another striking feature of Pask's chemical computers that remains to be discussed. We have forgotten about Bill Smith. His function in the hypothetical conversation with the cybernetician is to introduce a consideration of what Pask called the "relevance conditions" for control systems, the question of what variables the system needs to pay attention to, the ones that figure as its inputs and outputs. Bill Smith's contentedness was not something the manager needed to think about under the old regime of production—Bill was happy enough—but suddenly becomes a key variable when the new machine is installed and his work is deskilled. Now, it is one thing to design a control system when these relevance conditions are fixed and known in advance, but quite another to control a system where the relevance conditions change and have continually to be found out. This brings us to the most magical aspect of Pask's chemical computers—an aspect that went forgotten until an important and very insightful essay published by Peter Cariani in a 1993 festschrift for Pask, to which Stafford Beer added historical detail in a 2001 tribute in a similar volume.

New Senses

Beer recalled that in 1956 or 1957 he was visiting London from Sheffield and spent most of the night with Pask at the latter's flat in Baker Street, as he often did. They first had the idea of exploring the robustness of Pask's chemical computers by chiseling out sections of established threads and seeing what happened. It turned out that the systems were very robust and that the gaps healed themselves, though in an unexpected way—instead of joining up from either end, they traveled along the thread until they disappeared. "And yet these demonstrations, though exciting at the time, were somehow recognized to be trivial" (S. Beer 2001, 554–55):

> "Adaptation to the unexpected" should mean more than this, and yet there must be limits. I was already developing my theory of viable systems, and often used myself as an example. But what if someone pulled out a gun and shot me. Would that be proof that I am not after all a viable system? Surely not: the system itself would have been annihilated. We fell to discussing the limiting framework of ultrastability. Suddenly Gordon said something like, "Suppose that it were a survival requirement that this thing should learn to respond to sound? If there were no way in which this [sound] 'meant' anything [to the device], it would be equivalent to your being shot. It's like your being able to accommodate a slap rather than a bullet. We need to see whether the cell can learn to reinforce successfully by responding to the volume of the sound."
>
> It sounded like an ideal critical experiment. I cannot remember what exactly the reinforcement arrangements were, but the cell already had them in place in order to study the rate of adaptation to input changes, and we had created various gaps in the filigree by now.[24] And so it was that two very tired young men trailed a microphone down into Baker Street from the upstairs window, and picked up the random noise of dawn traffic in the street. I was leaning out of the window, while Gordon studied the cell. "It's growing an ear," he said solemnly (*ipsissima verba*).

A few years later Gordon was to write [Pask 1960b, 261]:

> We have made an ear and we have made a magnetic receptor. The ear can discriminate two frequencies, one of the order of fifty cycles per second and the other of the order of one hundred cycles per second. The "training" procedure takes approximately half a day and once having got the ability to recognize sound at all, the ability to recognize and discriminate two sounds comes more rapidly. . . . The ear, incidentally, looks rather like an ear. It is a gap in the thread structure in which you have fibrils which resonate at the excitation frequency.

This, then, was the truly remarkable feature of Pask's chemical computers. One way to put it is to note that the "senses" of all the cybernetic machines we have discussed so far were defined in advance. At base they were sensitive to electrical currents, and at one remove to whatever sensors and motors were hooked up to them. Pask's chemical computers, however, acquired new senses which were not designed or built into them at all—hence "growing an ear," but also acquiring a sensitivity to magnetic fields (a quite nonhuman "sense") in other experiments. If the homeostat, say, could adapt to new patterns within a fixed range of input modalities, Pask's chemical computers went decisively beyond the homeostat in searching over an open-ended range of possible modalities. If we think of the chemical computers as model brains, these brains were, at least in this one sense, superior to human brains, which have not developed any new senses in a very long time.

One possibly confusing point here is that Pask and Beer trained the computer to acquire the faculty of hearing and responding to sound—as if they somehow inserted the sense of hearing into the computer even if they did not explicitly design it in from the start. But to think that way would be to miss a key point. One should imagine the computer not in the training situation but in use—as hooked up to a factory, say—in which the coupled system was running to equilibrium. In that event, in its trial reconfigurations, a sound-sensitive thread structure might come into existence within the chemical computer and find itself reinforced in its interactions with the factory in the absence of any intervention from the experimenter whatsoever. In this scenario, the machine could thus genuinely evolve new senses in its performative interactions with its environment: hearing, a feeling for magnetic fields, or, indeed, an indefinite number of senses for which we have no name. And, as both Beer and Cariani have emphasized, no machine that could do this had been built before—or since, unless very recently: "It could well have been the first device ever to do this [develop a new sense], and no-one has ever mentioned another in my hearing" (S. Beer 2001, 555).

The Epistemology of Cybernetic Research

A CYBERNETIC HYPOTHESIS IS SOMETHING WHICH THE . . . CYBERNETICIANS DIG FROM THE OUTLANDISH SOIL OF THEIR ORGANIC ASSEMBLAGES.

GORDON PASK, "ORGANIC CONTROL AND THE CYBERNETIC METHOD" (1958, 171)

BUT, MORE IMPORTANT THAN THIS IS THE QUESTION OF WHETHER, IN SOME SENSE, THE NETWORK IS LIKE MY IMAGE OF MYSELF BEING A MANAGER (THIS PART OF THE INTERVIEW IS DIFFICULT, FOR THERE IS NO VERBAL COMMUNICATION . . .). ON THIS TEST, I SHALL ACCEPT THE NETWORK IF AND ONLY IF IT SOMETIMES LAUGHS OUTRIGHT.

GORDON PASK, *AN APPROACH TO CYBERNETICS* (1961, 113)

Throughout this book I have discussed the performative epistemology that I associate with cybernetics, the idea that representational knowledge is geared into performance, a detour away from and back to performance. It might be useful to come at this topic from a different angle here, via Pask's own epistemological reflections on his work with chemical computers, where he articulated a distinctive understanding of the "cybernetic method." Pask concluded his essay "Organic Control and the Cybernetic Method" (1958) by discussing cybernetics as a distinctive form of *practice*. He first defines a generic figure of an "observer" as "any person or appropriate mechanism which achieves a well defined relationship with reference to an observed assemblage" (Pask 1958, 172).[25] He then makes a distinction between what I call two "stances" that the observer can take up with respect to the object of inquiry (172–73):

> Any observer is limited by a finite rate at which he or it may make decisions. Since the limit exists we shall distinguish a *scientific observer* who minimises interaction with an observed assemblage and a *participant observer* who, in general, tries to maximise his interaction with an assemblage. If observers were omniscient there would be no distinction. A scientific observer decides whether or not the evidence of an observation leads him to accept each of a finite set of hypotheses, and may, as a result, determine his next observation. Since he is minimally associated with the assemblage he may determine his next observation

> precisely. . . . A scientific observer seeks to confirm as many hypotheses as possible.

Leaving aside questions of confirmation versus falsification, here we recognize a standard stereotype of the hypothesis-testing scientist. Note that on this stereotype, the scientist's access to matter passes through representations: a hypothesis is a verbal formulation—"If I do X then Y will happen"—itself embedded in all sorts of statements, theoretical and empirical. On the other hand (173; my italics),

> a cybernetician is a participant observer who decides upon a move which will *modify* the assemblage and, in general, favour his interaction with it. But, in order to achieve interaction he must be able to infer similarity with the assemblage. In the same way cybernetic control mechanisms must be similar to the controlled assemblage. The development of this similarity is the *development of a common language*. . . . [The cybernetician needs] to adopt *new languages*, in order to interact with an assemblage. [There is] an ignorance on the observer's part, about the kind of enquiry he should make. A common language is a *dynamic* idea, and once built up must be used. Thus if a conversation is disturbed it must be restarted, and one of the structured regions we have discussed must continually rebuild itself. . . . A cybernetician tries, by interaction, to bring about a state of a macrosystem which exhibits a consistent pattern of behaviour that may be represented by a logically tractable analogy.

There is, I think, an original philosophy of science adumbrated in these few sentences, which deserves a brief exegesis. Most important, the emphasis is on performative interaction with the object to be known, modification which might promote further interaction. One can recall here the earlier discussion of chemical computers and of the manager coming to terms with the controller in the same way as the controller comes to terms with the factory—by each interfering performatively with the other until some mutually acceptable, not pregiven, equilibrium is found. What Pask adds is that cyberneticians learn *in general* about their objects in just the same way: they interfere with them as much as possible in an exploratory fashion to see what they will do, with each observation provoking new, situated interferences. But what, then, of Pask's references to language and analogy? Does this return us to the hypothesis-testing model he just associated with "science"? No, because, first, Pask does not regard language as a given and stable medium in which hypotheses can be formulated and judged. The cybernetician does not know the appropriate terms—the language, the relevance conditions—for describing his or her object in advance; they have to be discovered in interaction with that object. Further, we know we have found suitable terms (not a true description) when we use them to construct a model of the object which enables us to understand its behavior when subject to additional interferences. Cybernetic interference produces *new languages* in which to address and interrogate its object.

But second, as in his usage of "conversation," it seems clear that Pask's sense of "language" is not necessarily verbal or representational in the usual sense. The model that grows in the cybernetician's interaction with some object might be nonverbal—as in the "model" of the factory that builds up in the chemical computer as it comes into equilibrium with the factory—and it may be a material object which bears no resemblance at all to the thing modelled—a thread structure does not look like Bill Smith happily making ornamental plaques; the homeostat does not look like a brain. Or it might be a conceptual construct—one of Stafford Beer's models of the Chilean economy, for example. All that matters is that the model facilitates continued commerce with the object itself.

Where does this leave us? Clearly, Pask's account of the cybernetic method indeed points to a performative epistemology. Second, we can think of his chemical computers as a vivid act of epistemological theater. The thread structures stage for us the idea that knowledge (of the factory, of the manager) need not take a representational form. Third, Pask's contrast between the scientific and the cybernetic methods warrants some brief elaboration. One could sum up the findings of the last twenty years and more of science studies as the discovery that real scientists are more like Paskian cyberneticians than his stereotype of them. They, too, struggle open-endedly with their objects and invent new languages and models to get to grips with them (Pickering 1995). But we need to think here of Pask's two kinds of observer and their different stances with respect to the world. Such open-ended struggles indeed happen in scientific practice, but this is thrust into the background of the modern sciences and forgotten, effaced, in the "hypothesis testing" model of science. And the argument I having been trying to make throughout this book—albeit with an emphasis on ontology rather than epistemology—is that these stances are consequential. Although we are all in the same boat, they make a difference: cybernetics, in its practices and in its products—chemical computers that develop new senses being a striking example of the latter—is different in its specificity from the modern sciences. This is, of course, precisely Pask's argument rehearsed earlier in the context of aesthetics and Musicolour—cybernetics suggests an unfamiliar and productive stance in science, as well the arts, entertainment, and teaching.[26]

CAs, Social Science, and F-22s

Pask discontinued his work on chemical computers in the early 1960s, and we should think about this along the lines already indicated in the previous chapter. Like Beer's biological computers, Pask's chemical ones were a valiant attempt at radical innovation undertaken with no support, more or less as a hobby, typified by "two very tired young men" trailing a microphone out of a window as the sun came up over Baker Street. We could also note that even within the cybernetics community, no one, as far as I know, sought to emulate and elaborate Pask's efforts—this in contrast, for example, to the many emulators of Walter's tortoises. Meanwhile, from the later 1950s onward typing trainers and teaching machines held out more immediate prospects of paying the rent. But one spin-off from Pask's research is interesting to follow briefly.

In an attempt to understand the dynamics of his threads, Pask undertook a series of numerical simulations of their behavior, which involved a form of idealization which is now very familiar in the sciences of complexity: he represented them schematically as two-dimensional cellular automata (chap. 4). In these simulations the dish of ferrous sulphate was replaced by a two-dimensional space, with "automata" residing at the intersections of a Cartesian grid. These automata evolved in discrete time steps according to simple rules for persistence, movement, reproduction, and death according to their success in exploiting a finite supply of "food." The early chemical-computer publications reported "hand simulations" of populations of automata, and in 1969 Pask reported on a set of computer simulations which prefigured more visible developments in work on cellular automata and artificial life a decade or two later (Pask 1969a).[27] Interestingly, however, Pask framed his account of these computer simulations not as an exploration of chemical computing but as a study of the emergence of norms and roles in social systems. Over the past decade there has been something of an explosion of social-science research on computer simulations of populations of automata.[28] It is not clear to me whether Pask's work was a formative historical contribution to this new field or whether we have here another instance of independent reinvention. What is clear is that this contemporary work on social simulation, like Pask's, can be added to our list of examples of ontology in action.[29]

Pask's interest in automata and simulation will reappear below in his work in architecture. But let me close this section with two remarks. First, following Cariani (1993, 30), we can note that the move from chemical computers to numerical simulation was not without its cost. The chemical computers found their resources for developing new senses in their brute materiality; they could find ways to reconfigure themselves that had not been designed into them. Pask's simulated automata did not have this degree of freedom; their relevance conditions were given in advance by the programs that ran them. No doubt this, too, had a bearing on the erasure of the chemical computers even from the consciousness of cybernetics.

Second, while one should beware of exaggeration, we can observe that cybernetic controllers are back in the news again. "Brain in a Dish Flies Plane" (Viegas 2004) is one of many media reports on a project strongly reminiscent of Pask, Beer, and even Ashby (who, we recall, discussed the virtues of homeostatic autopilots). In work at the University of Florida, rat neurons (in the style of Beer's biological computers) were grown in a dish and connected into the world via a sixty-channel multielectrode array (à la Pask). When this device was hooked up to an F-22 fighter jet flight simulator, "over time, these stimulations modify the network's response such that the neurons slowly (over the course of 15 minutes) learn to control the aircraft. The end result is a neural network that can fly the plane to produce relatively stable straight and level flight." Another version of the philosopher's apocryphal brain in a vat, though not so apocryphal any more, and robustly connected into the world of performance rather than seeking to represent a world of which it is not a part.[30]

The Arts and the Sixties

We have traveled a long way from Musicolour to chemical computers via typing trainers and teaching machines. For the remainder of this chapter I want to return to Pask's work in the theater, the arts, and architecture, picking up the story in the early 1960s (that is, in work that ran in parallel to his work on trainers and teaching machines). I am interested in three projects in particular: Pask's plans for a cybernetic theater; his robotic artwork, the Colloquy of Mobiles; and his contributions to architecture, beginning with the London Fun Palace. These projects are interesting in themselves as fresh instances of ontology in action, and they are also worth contemplating as yet more instances of crossovers from cybernetics to the distinctive culture of the 1960s. At an early stage in their careers, the Rolling Stones were apparently "roped in" to try out the adaptive machines at System Research (Moore 2001, 770).[31]

Pask's student Ranulph Glanville had a fleeting association with Pink Floyd, who lived nearby, and built a piece of electronic equipment for them—a ring modulator; he also did a sound mix for Captain Beefheart (Glanville, email, 16 August 2005). More consequential than such contacts with the iconic bands of the sixties, however, were Pask's contacts dating back to his undergraduate days with Cedric Price (Price 2001, 819), one of Britain's leading postwar architects, and with the radical theater director Joan Littlewood. If we pursued Pask's projects chronologically, the order would be Fun Palace, cybernetic theater, Colloquy of Mobiles, but for the purposes of exposition it is better to begin with the theater and to end with architecture.[32]

Cybernetic Theater

Joan Littlewood (1914–2002), the founder of the postwar Theatre Workshop in Britain and of the Theatre Royal in Stratford, London, writer and producer of *Oh, What a Lovely War!* and many other plays that marked an era, occupies an almost legendary place in the history of British theater (Ezard 2002).[33] She recalled that she had heard stories about Pask in the 1950s and that he had "flitted across my life from time to time like a provocative imp. . . . He had some idea of what we were up to. I wrote to him a couple of times. He seemed to be as *de trop* in English society as we were. They simply did not know how to use him—the Yanks did." The reference to the Yanks is an exaggeration, but, as usual for our cyberneticians, de trop sounds about right. Littlewood and Pask first met in person, presumably in the late 1950s, at System Research, "a normal looking house, from the outside, but we were standing in a labyrinth of wires, revolving discs of cardboard, cut from shredded wheat packets, little pots and plugs, while through it all a small, perfectly normal baby girl [Hermione Pask] was crawling in imminent danger of being electrocuted from the looks of things, though she was cooing contentedly" (Littlewood 2001, 760).

Here is Littlewood's recollection of a subsequent conversation with Pask (Littlewood 2001, 761): "I told him about two Red Indians taking their morning coffee in the Reservation Cafe and discussing last night's film. 'I thought we were going to win till that last reel,' said one. 'It would be fun,' I said, 'if the Red Indians did win for a change.' This caused a spark. He knew that I worked with inventive clowns. 'We could have a set of different endings,' he said. 'At least eight and the audience could decide which they wanted,' 'How?' 'By pressing a button attached to their seat, quite simple.' " The upshot of this conversation was a thirty-page 1964 document entitled "Proposals for a Cybernetic Theatre," written by Pask on behalf of Theatre Workshop and System

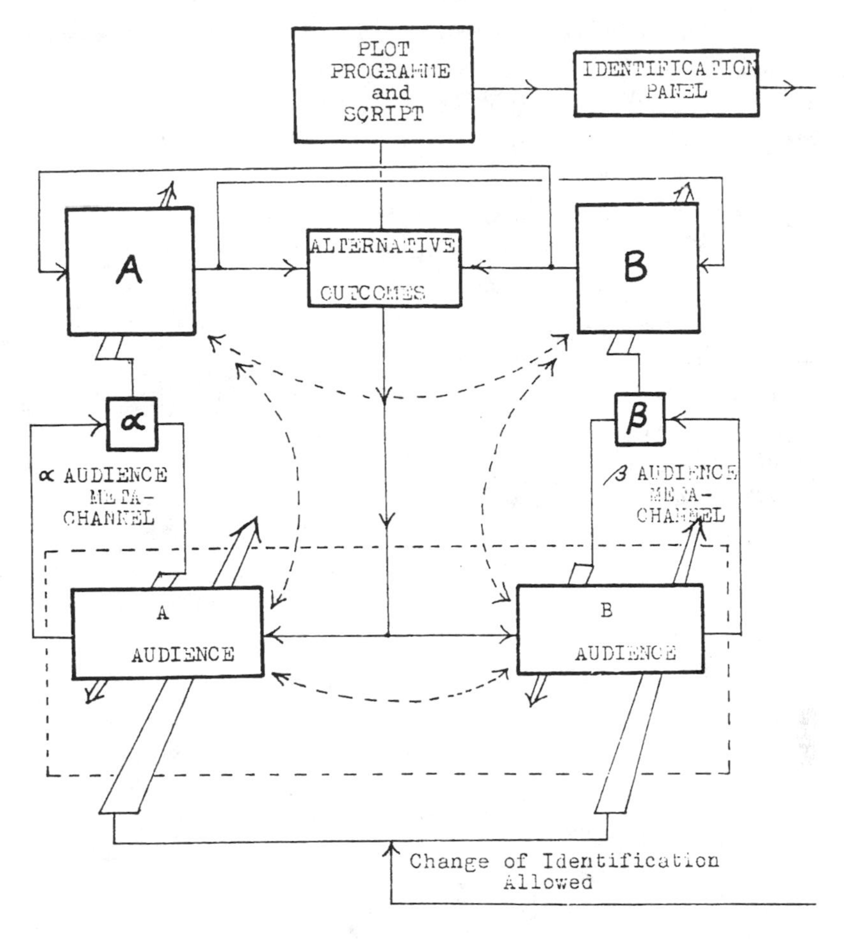

Figure 7.14. Logic diagram for a cybernetic theater. Source: Pask 1964b, 25, diagram 10. (Reproduced by permission of the North American Pask Archive.)

Research. In it Pask describes in considerable detail how audience and actors could be coupled together via feedback loops in determining the substance of any specific performance. Pask's basic idea here was that of a new kind of play, which would retain certain set-piece structural elements, specified in advance, but which would include alternative routes of plot development between the set pieces in an entailment-mesh-like structure, including the possibility that the trajectories might redefine the significance of the fixed elements.[34] During the performance, members of the audience could signal their identification with one or another of the principal actors (designated *A* and *B* in fig. 7.14). At specified branchpoints, the audience could also use levers to advocate different choices of action for their chosen character, drawing upon both their understanding of how the play had developed thus far and also upon "metainformation" on their character's thinking at this point, developed in rehearsal and provided in real time by "interpreters" (alpha and beta in fig. 7.14) via headphones or earpieces. The interpreters in turn would then use hand signals, or perhaps radio, to let the actors know their supporters' inclinations, and the play would proceed accordingly. Depending on how the play developed from these branch points, the audience was free to change identifications with actors, to make further plot decisions, and so on.

I thought this plan was totally mad when I first came across it, but when I read Littlewood's obituaries I realized that, unlike her, I still wasn't cybernetic enough (*Guardian* 2002, 12, Richard Eyre): "She didn't disrespect writers, but she had a contempt for 'text' and the notion that what was said and done on stage could become fixed and inert. She believed in 'the chemistry of the actual event,' which included encouraging the audience to interrupt the play and the actors to reply—an active form of alienation that Brecht argued for but never practised."[35] Pask's proposal indicated that in 1964 (Pask 1964b, 2)

> an initial experimental system (a physical communication system) is being constructed and will be used to determine a number of unknown values required for the efficient realisation of the mechanism. The experimental system will be used informally in Theatre Workshop and will accommodate an invited audience of between 50 and 100 people. Next it is proposed to build and install a large system accommodating an audience of between 550 and 750 people and to use it for a public presentation. . . . There are many intriguing dramatic problems that can only be solved when a suitable performance has been developed and a large system is available to embody it.

I do not know whether the experimental system was ever constructed, but it is safe to say that Pask's proposal to scale it up found no backers. A shame, but it is still instructive to reflect on these ideas.

We can see the cybernetic theater as yet another manifestation of Pask's ontology of open-ended performative engagement and the aesthetic theory that went with it. The cybernetic theater would be an "aesthetically potent environment" for both actors and audience in much the same way as Musicolour and the later training and teaching machines were. The same vision will reappear below with respect to art and architecture: it was, in fact, an enduring theme that ran through all of Pask's projects. Of course, the structural elements of the play meant that plot-development would not be fully open ended; nevertheless, Pask argued that the cybernetic theater would mark a significant departure from existing theatrical practices and experience. As in the previous discussion of Musicolour, Pask was happy to acknowledge that "a [conventional] theatrical audience is not completely passive, in which respect, amongst others, it differs from a Cinema audience or a Television audience. There is a well attested but badly defined 'Feedback' whereby the actors can sense the mood of the audience (and play their parts in order to affect it)." Thus "this control system [i.e., feedback from the audience] is embedded in the organisation of *any* dramatic presentation," but "its adequacy may be in doubt and its effectiveness is hampered by *arbitrary* restrictions. To remove these restrictions would not render a dramatic presentation something other than a dramatic presentation although it might open up the possibility for a novel art form" (Pask 1964b, 4, 5). Again, then, we have here a nice example of how ontology can make a difference, now in a new form of theater.

And, following this train of thought, it is worth remarking that Pask's cybernetic theater was literally an ontological theater, too. One might think of conventional theater as staging a representational ontology, in which the audience watches a depiction of events, known already to everyone on the other side of the curtain, suggesting a vision of life more generally as the progressive exposure of a pregiven destiny. I have repeatedly argued that a different ontological moral could be extracted from cybernetic devices, but in the case of Pask's cybernetic theater no such "extraction" is necessary—within the frame of the play's structural elements, the audience was directly confronted with and participated in an unforeseeable performative becoming of human affairs. In the cybernetic theater, then, the ontology of becoming was right on the surface.[36]

A few further thoughts are worth pursuing. One is historical. We can note a continuity running from Pask's notion of an explicit feedback channel from audience to actors to his friend Stafford Beer's experimentation with algedometers in Chile in the early 1970s. In the previous chapter we saw that Beer's devices were prone to playful misuse, and Pask was prepared for something similar in the theater, wondering if "many people will participate in a more experimental or mischievous manner"—seeking somehow to throw the actors off balance, as had Beer's subjects. Pask remarked that "unless there are statistically well defined and concerted attempts to upset the system this should not pose a real problem," but nevertheless, "various devices have been embodied in this design to avoid 'illegal' manipulation of the response boards. We assume that 'illegal' manipulation is bound to occur either mischievously or by accident" (Pask 1964b, 15, 18).

Second, we can note that, as in all of Pask's projects, the cybernetic theater undercut existing power relations. Most obviously, the audience was given a new weight in determining the substance of each performance in real time. The role of actors was likewise elevated relative to writers and directors in their responsibility for making smooth traditions from one plot trajectory to another. And, at the same, Pask's vision entailed the integration of new social roles into theatrical performances: the interpreters who provided meta-information to the audience, the technicians who would wire up the feedback channels and maintain them, even the cyberneticians as new theorists of the whole business, quite distinct from conventional theater critics.

And third, we need to think about the kind of company that Pask kept. In the 1960s, Joan Littlewood was one of the most successful directors in British theater: "She had three shows in the West End by 1963, triumph on a Lloyd Webber scale, and to incomparably higher standards" (Ezard 2002, 20). In his collaboration with Littlewood just one year later, Pask thus crossed over from the narrow world of typing trainers into one of the most lively and visible currents of British popular culture. It is therefore worth examining precisely which current he stepped into.

The key observation is that, unlike Andrew Lloyd Webber, Littlewood was an avowedly antiestablishment figure, who understood theater as one of those technologies of the self we have discussed before, aimed now at reconstituting British society. After studying at RADA (the Royal Academy of Dramatic Art) she moved first from London to Manchester, which brought her "closer to the counter-culture she sought," and where she worked for the BBC, the *Manchester Guardian*, and "small leftist agit-prop groups dedicated to taking drama to the people of the north." The Theatre Union, which she cofounded in 1936 with the folksinger Ewan McColl, "saw itself as a vanguard of theory; its productions were influenced by Vsevolod Meyerhold, the Stanislavsky disciple who was the first director of postrevolutionary Soviet drama until Stalin purged him." During World War II, her group was "often splendidly reviewed but [was] always refused grants by the Council for the Encouragement of Music and the Arts, the Arts Council predecessor. She and McColl were blacklisted by the BBC and by forces entertainment group ENSA as subversives." Her group renamed itself Theatre Workshop after the war and supported the early Edinburgh Fringe Festival—the alternative to the high-culture Edinburgh Festival—and rented the Theatre Royal on Angel Lane in London in 1953 for £20 a week—"a dilapidated palace of varieties reeking of cat urine"—before making its first breakthrough to the West End in 1956 with *The Good Soldier Schweik* (Ezard 2002, 20). "She was wholly unclubbable," wrote a fellow theater director, "a self-educated working-class woman who defied the middle-class monopoly of theatre and its domination by metropolitan hierarchy and English gentility. She believed in realising the potential of every individual, being in favour of 'that dull working-class quality, optimism,' a necessary virtue in a life dedicated to demonstrating that political theatre wasn't always an oxymoron" (*Guardian* 2002, Eyres).

Pask's involvement with the theater in the sixties did not, then, lead him into the high culture of the British establishment, but rather into the countercultural, antiestablishment milieu, here typified by Littlewood, that somehow succeeded, for a brief moment around that decade, in becoming a defining formation in British culture. We have examined before the ontological resonances between cybernetics and the counterculture—flicker and the Beats, Bateson and Laing's radical psychiatry, Beer and Eastern spirituality—and Pask's alignment with Littlewood should be understood in just the same way. We can return to this theme below, but we can note now that this alignment also doomed cybernetics to going down with the ship. Cybernetics has itself continued up to the present, but its visibility in popular culture declined with the overall decline of the counterculture. Littlewood herself seems to have become disgusted with the form of life that went with being a successful London theater director. "Success is going to kill us," she wrote in the mid-1960s. "Exhausted and miserable, she walked out at the crowning moment when she and Raffles had managed to buy the [Theatre Royal]. She disappeared alone to Nigeria to work on an abortive film project with the writer Wole Soyinka. She returned but never recaptured the momentum: if it meant diluting standards or becoming a full-time impresario, she did not want to" (Ezard 2002, 20).

Cybernetic Serendipity

In the 1960s the ICA, the Institute for Contemporary Arts, in London was Britain's center for new developments in art. If something exciting and important was happening in Britain or abroad, the ICA aimed to represent it to the British public.[37] Conversely, a show at the ICA ratified a new movement or whatever as, indeed, exciting and important. Jasia Reichardt, who had organized the first show of British Pop Art in London, *Image in Progress*, at the Grabowski Gallery in 1962, joined the ICA as assistant director in 1963, where she organized a show on concrete poetry in 1965, *Between Poetry and Painting* (Reichardt 1971, 199). In the autumn of that year she began planning "an international exhibition exploring and demonstrating some of the relationships between technology and creativity." In 1968, "there was enough financial support for it to go ahead," and her exhibition, now called *Cybernetic Serendipity*, opened at the ICA on 2 August and closed on 20 October 1968 (Reichardt 1968a, 3, 5).[38] The exhibition was divided into three parts (Reichardt 1968b, 5):

1. Computer-generated graphics, computer-animated films, computer-composed and -played music, and computer poems and texts
2. Cybernetic devices as works of art, cybernetic environments, remote-controlled robots and painting machines
3. Machines demonstrating the uses of computers and an environment dealing with the history of cybernetics

As one can gather from this list and from figure 7.15, "cybernetic" in *Cybernetic Serendipity* should be interpreted broadly, to include almost all possible intersections between computers and the arts, including, for example, computer graphics, one of Reichardt's special interests (Reichardt 1968b). But two of our

Figure 7.15. Norman Toyton, cartoon of computer confessional. Source: J. Reichardt (ed.) *Cybernetic Serendipity: The Computer and the Arts* (London: W. & J. Mackay, 1968), 8.

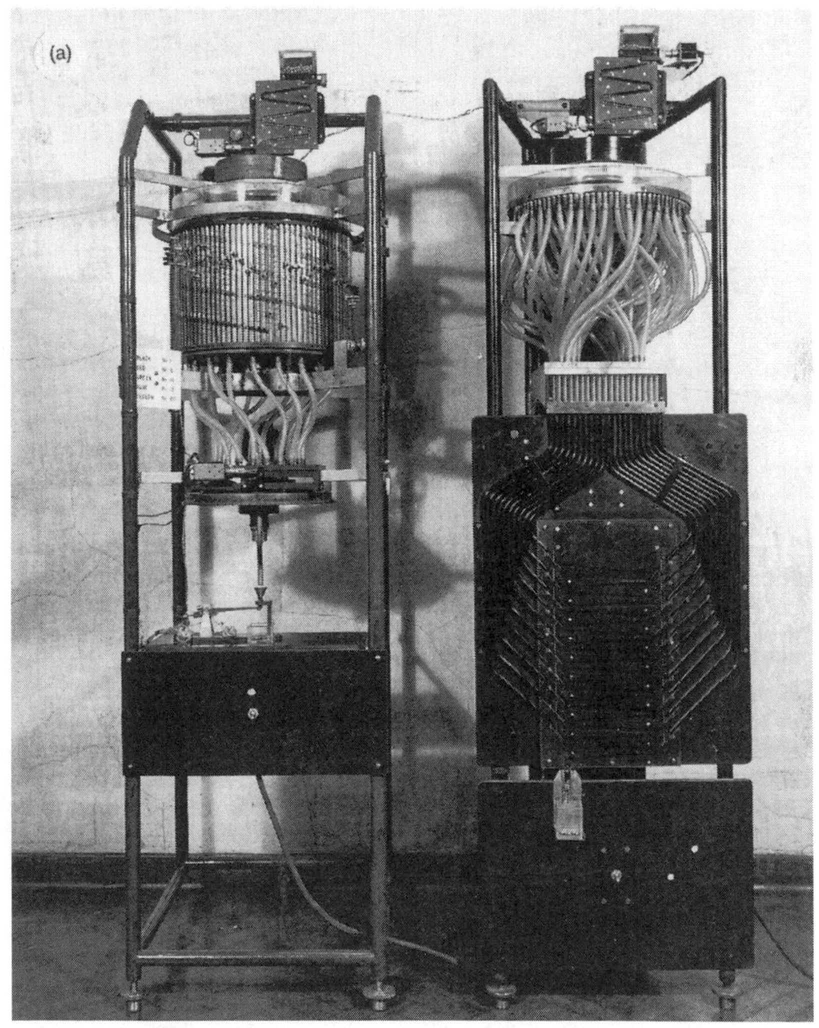

Figure 7.16. Beer's stochastic analog machine. Source: S. Beer "SAM," in J. Reichardt (ed.), *Cybernetic Serendipity: The Computer and the Arts* (London: W. & J. Mackay, 1968), 12.

cyberneticians showed their work at the exhibition, Stafford Beer and Gordon Pask.[39] Beer's contribution was a futuristic-looking electromechanical device for generating random numbers as inputs to Monte Carlo simulations of steel production, SAM, the Stochastic Analogue Machine (fig. 7.16), which was described in an accompanying poem by Beer (Beer 1968b; for more details on SAM, see Beer 1994a). I want to focus here, however, on Pask's exhibit, which he called Colloquy of Mobiles (Pask 1968, 1971; see fig. 7.17).[40]

Figure 7.17. Photo of the Colloquy of Mobiles. Source: G. Pask, "A Comment, a Case History and a Plan," in J. Reichardt (ed.), *Cybernetics, Art, and Ideas* (Greenwich, CT: New York Graphics Society, 1971), 96, fig. 40.

Like all of Pask's creations, the Colloquy was a baroque assemblage. Perhaps the best way to think of it is as a sophisticated variant of Walter's tortoises.[41] As we saw in chapter 3, the tortoises were mobile, phototropic robots which in combination engaged in complex mating dances, and just the same can be said of the components of the Colloquy. Pask's robots were, in one way, somewhat less mobile than Walter's. As shown schematically in figure 7.18, the Colloquy consisted of five robots, three designated "female" and two "male," each suspended from above. Their mobility consisted principally in their ability to rotate on their axes, driven by electric motors. The males each had two "drives," designated *O* and *P*, which built up over time (as charges on a capacitor) and were indicated by the intensity of either an orange or a puce light on the robot. These lights were reminiscent of the tortoises'

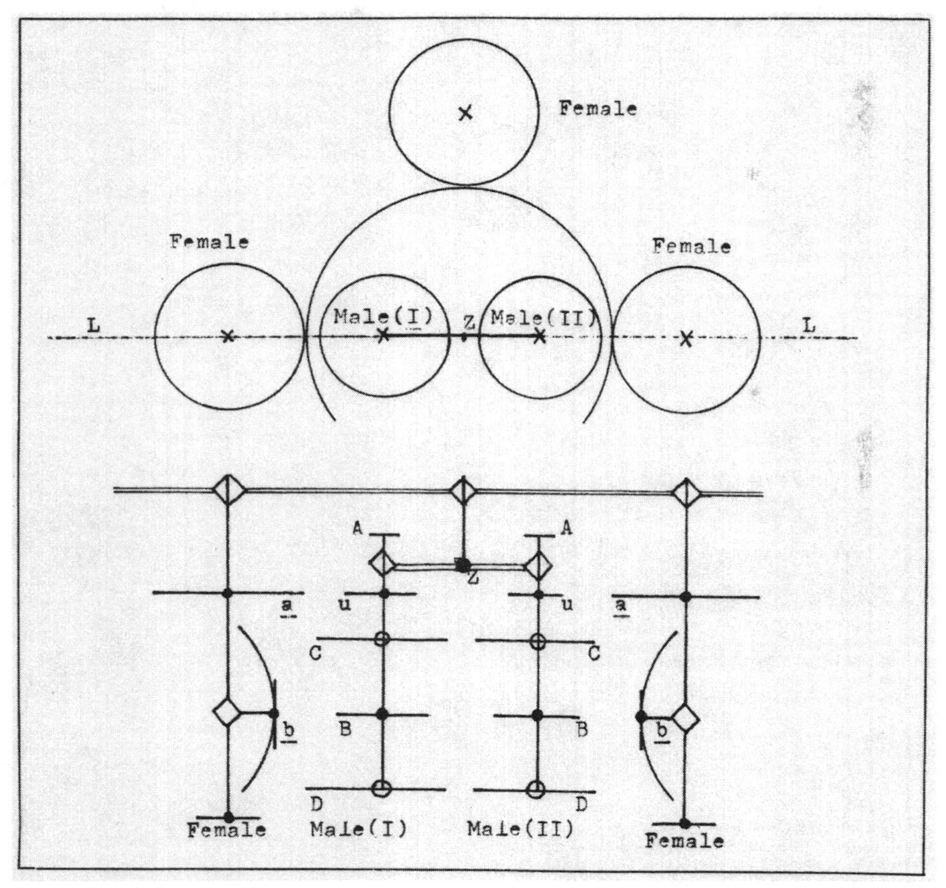

Figure 7.18. Plan of the Colloquy of Mobiles. *Top*, horizontal plan; *bottom*, vertical section taken through line *L* in horizontal plan; *A*, drive state display for male; *B*, main body of male, bearing "energetic" light projectors *O* and *P*; *C*, upper "energetic" receptors; *D*, lower "energetic" receptors; *U*, non-"energetic," intermittent signal lamp; *a*, female receptor for intermittent positional signal; *b*, vertically movable reflector of female *Z*, bar linkage bearing male I and male II; ◇, drive motor; ⊕, free coupling; ●, fixed coupling; ═, bar linkage. Source: G. Pask, "A Comment, a Case History and a Plan," in J. Reichardt (ed.), *Cybernetics, Art, and Ideas* (Greenwich, CT: New York Graphics Society, 1971), 90, fig. 34.

running lights but were not the crucial feature of the Colloquy; much more complicated signaling was involved in the latter.

Each male sought to "satisfy" its drives, first by locating a female while spinning on its axis (an equivalent of the tortoise's scanning mechanism) via an intermittent directional visual signal which indicated both its identity and its desire (*O* or *P*). If a female picked this up and was interested in *O* or *P* satisfaction herself, she would respond with an identifying sound synchronized to the male light. The male receiving this would then lock onto the female (just as the tortoise locked onto a light source) and emit an intense orange or puce light from its central part (*D* in fig. 7.18). If this fell upon the reflector of the female (*b*) she would reciprocally lock onto the male and commence a scanning motion of the reflector, up and down. The object of this was to reflect the beam back onto the appropriate part of the male, *D* or *C* in figure 7.18, depending whether the drive in question was *O* or *P*. If the female was successful in doing this, the male drive would be satisfied (temporarily, until the charge on the capacitor built up again); the male would also emit a "reinforcement" sound signal, which would discharge the female's drive. The overall behavior of the setup was controlled by purpose-built electronics, which received and instigated sensory inputs and outputs from each robot and switched the motion of the robot from one basic pattern to another in accordance with flowcharts such as that shown in figure 7.19.[42]

Thus the basic arrangement of the Colloquy of Mobiles and the principles of their mating, but we can note some further complications. First, the males hung from a common bar (fig. 7.18), which meant that they competed for females: a male in search mode could disturb the other which had locked onto a female. This made for a more lively performance and added another dimension of interest for the viewer. Second, the males could differ in which receptor (*C* or *D*) was the target for satisfaction of *O* or *P* drives, and the females could adapt to this by remembering which direction of scanning (upward or downward) was successful for which drive for each male. And third, the Colloquy was open to human intervention. As Pask wrote before the exhibition (1971, 91),

> The really interesting issue is what happens if some human beings are provided with the wherewithal to produce signs in the mobile language and are introduced into the environment. It is quite likely that they will communicate with the mobiles. . . . The mobiles produce a complex auditory and visual effect by dint of their interaction. They cannot, of course, interpret these sound and light

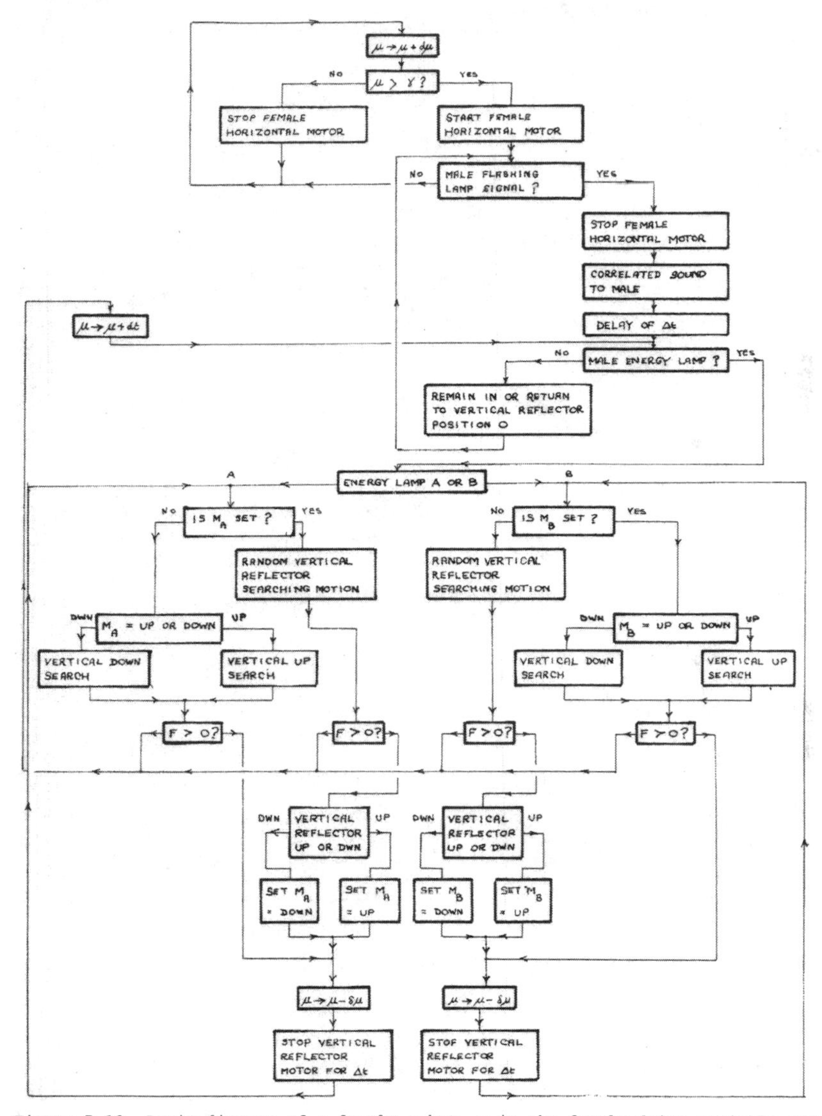

Figure 7.19. Logic diagram of a female robot. μ is the female drive variable and γ is a limit on the variable μ; each of $\Delta\mu$, $d\mu$, and $\delta\mu$ is a different increment; M_A is memory for orange (up or down vertical position); *MB* is memory for puce (up or down vertical position); *F* is a reinforcement variable, *F* = 1 or 0, evaluated by the male; *t* is a fixed delay. Source: G. Pask, "A Comment, a Case History and a Plan," in J. Reichardt (ed.), *Cybernetics, Art, and Ideas* (Greenwich, CT: New York Graphics Society, 1971), 92, fig. 35.

> patterns. But human beings can and it seems reasonable to suppose that they will also aim to achieve patterns which they deem pleasing by interacting with the system at a higher level of discourse. I do not know. But I believe it may work out that way.

In an October 1968 appendix to the same essay, Pask (1971, 98) recorded that this prediction concerning human involvement had proved to be "quite accurate, though entrainment is not nearly so effective with even moderate ambient illumination level." In other words, interaction with the Colloquy was best achieved in the dark. According to John Frazer (personal communication, 30 September 2004), people used womens' makeup mirrors to redirect the robots' light beams. One visitor to the exhibition recalled, "Some of the visitors stayed for several hours conversing with the mobiles" (Zeidner et al. 2001, 984).

- - - - -

What can we say about the Colloquy as ontological theater? Evidently it was in the same space as Musicolour, staging open-ended performative encounters between its participants, now multiple concurrent dances of agency among the five robots. In this instance, however, the participants in these dances were all machines, putting the Colloquy in the same space as the mirror and mating dances of Walter's tortoises and Ashby's multihomeostat setups. Like Walter and Ashby's machines, the Colloquy did not evolve in a fully open-ended fashion—the robots had a finite range of behaviors and fixed goals—but the introduction of human participants modified the picture, making possible a more fully open-ended range of possible performances by the human-Colloquy assemblage. As Pask noted in above quotation, the humans could engage with the robots at "a higher level of discourse," finding their own goals for the behavior of the system, just like a Musicolour performer but in cooperation with a different opto-electro-mechanical setup. It is also worth noting that the Colloquy foregrounded the role of language, communication, and signaling more sharply than the tortoise or the homeostat. One can indeed speak of signaling in connection with the tortoises, say: they responded to the presence or absence of light, and also to thresholds in light intensity. But the combination of different lights and sounds in the Colloquy (and the limited possibilities for robotic movement) brought this signaling aspect to the fore. Once more, then, we can say that the Colloquy was a piece of epistemological as well as ontological theater, and again I want to note that its epistemological aspects were geared straight into the ontological ones. The various modes of signaling in the Colloquy both were precipitated by the robots' performances and served to structure them, rather than to construct self-contained representations of the world. As usual, the Colloquy also staged a vision of a performative epistemology.[43]

The Social Basis Again

We can return to the question of the social locus of cybernetics, and the story bifurcates here. On the one hand, *Cybernetic Serendipity* was socially serendipitous for Pask. At the exhibition he met an American, Joseph Zeidner, then on the staff of the U.S. Office of Naval Research, later of the U.S. Army Research Institute. And the upshot of this meeting was that the U.S. military was among the sponsors of Pask's work on decision making and adaptive training systems over the next fifteen years. This takes us back to the lineage of training and teaching machines discussed earlier, and I will not explore the technicalities of that work further. We should bear in mind, however, that these machines were the bread and butter of Pask's life for many years. And we can also note that here we have another example of the typically nomadic pattern of propagation and evolution of cybernetics: from Musicolour and entertainment to typing trainers via the meeting with Christopher Bailey at the Inventors and Patentees Exhibition, to the Colloquy and thence to research on decision making for the U.S. military via the ICA.[44] In this section, however, I want to stay with the art world.

Nothing comes from nowhere, and we can certainly equip Pask's Colloquy with a pedigree. There is a whole history of automaton construction and machine art more generally (in which Jacques de Vaucanson's famous duck usually figures prominently) into which the Colloquy can be inserted. The Colloquy was a moment in the evolution of that tradition, distinguished (like many cybernetic artifacts) by its open-ended liveliness and interactivity. But the point I want to focus on now is that this pedigree is hardly a distinguished one and lurks, instead and as usual, in the margins of social awareness, "marginalised by both Art History and the histories of Engineering and Computer Science" (Penny 2008).[45] No doubt there are many reasons which could be adduced for this, but here we should pay attention to the oddity of machine art when seen against the backdrop of the cultural mainstream. As I stressed earlier concerning Gysin's Dream Machine and Pask's Musicolour, works like the Colloquy are odd objects that are refractory to the classifications, practices, and institutions of the modern art world. They are strange and nonmodern in just this sense. They are machines and thus, for the past couple of centuries, associated more readily with the grimy world of industry than with the lofty realms of high art; they lack the static and quasi-eternal quality of paintings and sculptures, foregrounding processes of becoming and emergence instead; as discussed before, interactive artworks tend to dissolve the primacy of the artist, thematizing real-time interplay between artwork and "user" (rather than "viewer"); they also threaten the social demarcation between artists and engineers; and, of course, they need more and different forms of curatorial attention: a sculpture just stands there, but machine art requires technological servicing to keep it going.[46]

In this sense, the marginality of machine art, including cybernetic art, is just the other side of the hegemony of modernity, and what calls for more thought is the move toward cultural centrality of works like the Colloquy. But this is no great puzzle. In earlier chapters we have seen many examples of crossovers fostered by an ontological resonance between cybernetics and the sixties counterculture, and that the Colloquy briefly positioned Pask in a sort of countercultural artistic vanguard can, I think, be similarly understood. Strange art hung together with novel forms of life in many ways.[47] The other side of this connection is that, as I said earlier, cybernetic art went down with the countercultural ship and very quickly lost its presence in the art world. "Cybernetic Serendipity might be considered the apogee of computer-aided art, considered as a mainstream art form. . . . [But] the late 1960s were both the apogee and the beginning of the end for . . . the widespread application of Cybernetics in contemporary art. . . . Cybernetic and computer art was [after the sixties], rightly or wrongly, regarded as marginal in relation to both the traditional art establishment or to avant-garde art practice" (Gere 2002, 102–3).[48]

— — — — —

After the sixties, then, "Kinetic, robotic, cybernetic and computer art practices were largely marginalized and ignored. With the odd exception . . . no major art gallery in Britain or the United States held a show of such art for the last 30 years of the twentieth century."[49] "But this does mean that . . . other kinds of art involving technology did not continue to be practised," even if they no longer commanded the heights of the art world (Gere 2002, 109,110). Much of the contemporary art discussed in chapter 6 under the heading of hylozoism—by Garnet Hertz, Eduardo Kac, Andy Gracie—is robot art, though I focused on the biological element before. Simon Penny's work is at the Paskian engineering end of the spectrum.[50] And these names are simply a random sample, examples that I have happened upon and been struck by.[51] Beyond these individual efforts, however, we can note that some sort of institutional social basis for this sort of art has also been emerging. Charlie Gere (2002, 110) mentions the Ars Electronica Centre and annual festival held in Linz, Austria, since 1979 as a key point of condensation and propagation of such work, and also that having directed the Linz festival from 1986 to 1995, Peter Weibel moved to direct the Zentrum für Kunst und Medientechnologie (ZKM) in Karlsruhe, Germany, itself a "highly funded research centre and museum dedicated to new media arts."[52] As in previous chapters, here we find traces of the emergence of a new social basis for cybernetics and its descendants, now in the sphere of art, not within mainstream institutions but in a parallel social universe (echoing the ambitions of Trocchi's sigma project).

And to round off this line of thought, it is instructive to think of the career of the British artist Roy Ascott, whom we encountered in the previous chapter as the man who first introduced the musician Brian Eno to cybernetics. Ascott was the leader in Britain in introducing cybernetics into art, having first encountered the field in 1961, reading the works of Ross Ashby, Norbert Wiener, and Frank George (Shanken 2003, 10). As head of foundation at Ealing College of Art, he introduced the Ground Course (1961–63), focused on cybernetics and behaviorism, which "fundamentally affected the work of those who taught it and of their students" (Stephens and Stout 2004, 31, 41).[53] Despite his influence on British art in the 1960s, Ascott has been "largely ignored by the British art establishment. The Tate Gallery . . . does not own any of his work. He has, however, achieved international recognition for his interactive work, and his teaching" (Gere 2002, 94). Indeed, in 2003 Ascott became the founding director of a novel pedagogical institution called the Planetary Collegium, "a world-wide transdisciplinary research community whose innovative structure involves collaborative work and supervision both in cyberspace and at regular meetings around the world." Those altered states and technologies of the nonmodern self we have been discussing also loom large in the collegium's self-description:

> The Planetary Collegium is concerned with advanced inquiry in the transdisciplinary space between the arts, technology, and the sciences, with consciousness research an integral component of its work. It sees its influence extending to new forms of creativity and learning in a variety of cultural settings. Far from eschewing the study of esoteric or spiritual disciplines, it seeks to relate ancient, exotic, even archaic knowledge and practices to radically new ideas emerging at the forward edge of scientific research and speculation, and thereby

to new forms of art and cultural expression. It seeks dynamic alternatives to the standard form of doctoral and post doctoral research while producing, if not exceeding, outcomes of comparable rigour, innovation and depth.[54]

The Fun Palace

THE HIGH POINT OF FUNCTIONALISM IS THE CONCEPT OF A HOUSE AS A "MACHINE FOR LIVING IN." BUT THE BIAS IS TOWARDS A MACHINE THAT ACTS AS A TOOL SERVING THE INHABITANT. THIS NOTION WILL, I BELIEVE, BE REFINED INTO THE CONCEPT OF AN ENVIRONMENT *WITH* WHICH THE INHABITANT COOPERATES AND *IN* WHICH HE CAN EXTERNALIZE HIS MENTAL PROCESSES.

GORDON PASK, "THE ARCHITECTURAL RELEVANCE OF CYBERNETICS" (1969A, 496)

If the sixties were the decade of interactive art, they were also the decade of interactive and adaptive architecture. In Britain, the Archigram group of architects built almost nothing, but the designs featured in *Archigram* magazine were iconic for this movement. Ron Herron's fanciful Walking City (fig. 7.20) in 1964 caught the mood, adaptive in the sense that if the city found itself somehow misfitted to its current environment, well, it could just walk off to find somewhere more congenial. Peter Cook's concept of the Plug-In City was a bit more realistic: the city as a mesh of support services for otherwise mobile units including housing—the city that could continually reconfigure itself in relation to the shifting needs and desires of its inhabitants.[55]

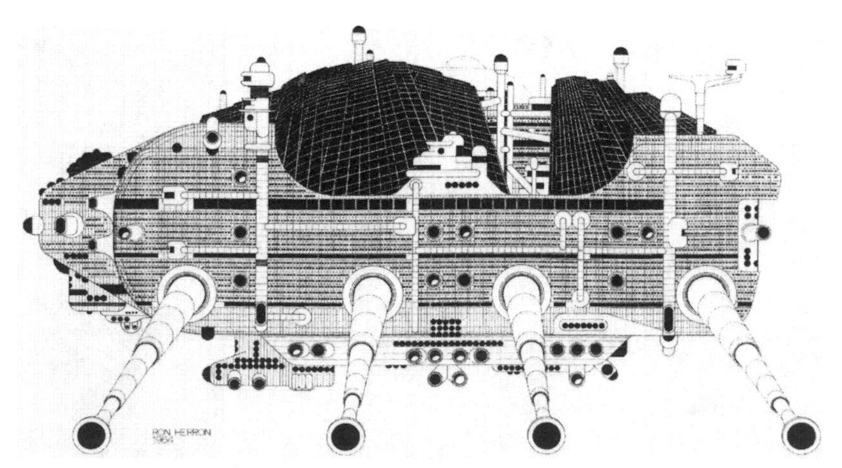

Figure 7.20. The Walking City, 1964. Source: Sadler 2005, 39, fig. 1.32.

At a relatively mundane level, the interest in adaptive architecture could be seen as a reaction to the failure of postwar urban planning for the future of London (Landau 1968). If the planners could not foresee how London would develop, then perhaps the city should become a self-organizing system able to reconfigure itself in real time in relation to its own emerging situation. This idea, of course, takes us straight back to Ross Ashby's ideas of evolutionary design and, in another register, to Beer's and Pask's biological and chemical computers that evolved and adapted instead of having to be designed in detail: the city itself as a lively and adaptive fabric for living.

At a more exalted and typically sixties level was an image of the city as a technology of the nonmodern self, a place where people could invent new ways to be, where new kinds of people could emerge. Metonymically, Archigram's Living City installation at the ICA in 1963 included a flicker machine (taking us back to Grey Walter and Bryan Gysin; fig. 7.21), Much of the inspiration for this conception of the built environment came from the tiny but enormously influential Situationist International group centered on Guy Debord in Paris, which had come into existence in the 1950s. As a founding document from 1953 put it, "The architectural complex will be modifiable. Its aspect will change totally or partially in accordance with the will of its inhabitants. . . . The appearance of the notion of relativity in the modern mind allows one to surmise the EXPERIMENTAL aspect of the next civilization. . . . On the basis of this mobile civilization, architecture will, at least initially, be a means of experimenting with a thousand ways of modifying life, with a view to mythic synthesis."[56]

Closely associated with Archigram and sharing its enthusiasm for adaptive architecture while maintaining an "avuncular" relation to it, was the architect Cedric Price, mentioned earlier as a fellow undergraduate of Pask's at Cambridge (Sadler 2005, 44), and Price was Pask's link to architecture. Around 1960, Joan Littlewood "turned . . . to a childhood dream of a people's palace, a university of the streets, re-inventing Vauxhall Gardens, the eighteenth-century Thames-side entertainment promenade, with music, lectures, plays, restaurants under an all-weather-dome" (Ezard 2002). This Fun Palace, as it was known, is one of the major unbuilt landmarks of postwar British architecture (fig. 7.22). Cedric Price was appointed as the architect for the project, and "I thought of Gordon [Pask] and Joan did too. He immediately accepted the post—unpaid as I remember—as cybernetician to the Fun Palace Trust. It was his first contact with architects and he was extremely patient. He immediately formed a cybernetic working party and attracted those he wanted to join it too. The meetings became notorious—and Trust Members attended" (Price

Figure 7.21. Flicker machine at the Living City, ICA, 1963. Source: Sadler 2005, 57, fig. 2.6.

1993, 165). "Pask agreed to join the Fun Palace team and organised the Fun Palace Cybernetics Subcommittee, and along with Littlewood and Price, he became the third major personality behind the Fun Palace" (Mathews 2007, 75).[57]

What was the Fun Palace? Like Archigram's designs, but at a much more practical level, the Fun Palace was intended as a reconfigurable adaptive space that could support an enormous variety of activities that changed over time (Landau 1968, 76):

> The activities which the Fun Palace offered would be short-term and frequently updated, and a sample suggested by Joan Littlewood included a fun arcade, containing some of the mechanical tests and games which psychologists and engineers usually play; a music area, with instruments on loan, recordings for anyone, jam sessions, popular dancing (either formal or spontaneous); a science playground, with lecture/demonstrations, teaching films, closed-circuit T.V.; an acting area for drama therapy (burlesque the boss!); a plastic area for modeling and making things (useful and useless). For those not wishing to take part, there would be quiet zones and also screens showing films or closed-circuit television of local and national happenings.

> This program called for an architecture which was informal, flexible, unenclosed, and impermanent; the architecture did not need to be simply a response to the program, but also a means of encouraging its ideas to grow and to develop further. With an open ground-level deck and with multiple ramps, moving walkways, moving walls, floors, and ceilings, hanging auditoriums, and an overall moving gantry crane, the physical volumes of the spaces could be changed as different usages were adopted. The kit of parts for these operations included charged static vapor barriers, optical barriers, warm air curtains, a fog dispersal plant, and horizontal and vertical lightweight blinds. In the Fun Palace, no part of the fabric would be designed to last for more than ten years, and parts of it for possibly only ten days.

A large number of people worked on the design of the Fun Palace, and it is impossible to spell out in detail Pask's individual contributions. At the level of content, the Cybernetics Subcommittee suggested dividing the Fun Palace into six organizational zones, and "Zone one was dedicated to the various types of teaching machines that Pask and his Systems Research had already developed." Stanley Mathews describes the Littlewood-Pask cybernetic theater as part of the overall conception of the Fun Palace (Mathews 2007, 114, 116).[58] Like the flicker machine at the Living City, the machines and the theater can be seen as metonyms for the entire building.[59] More broadly, Pask's contribution appears to have been to see the Fun Palace on the model of Musicolour—as an aesthetically potent environment that in its inner

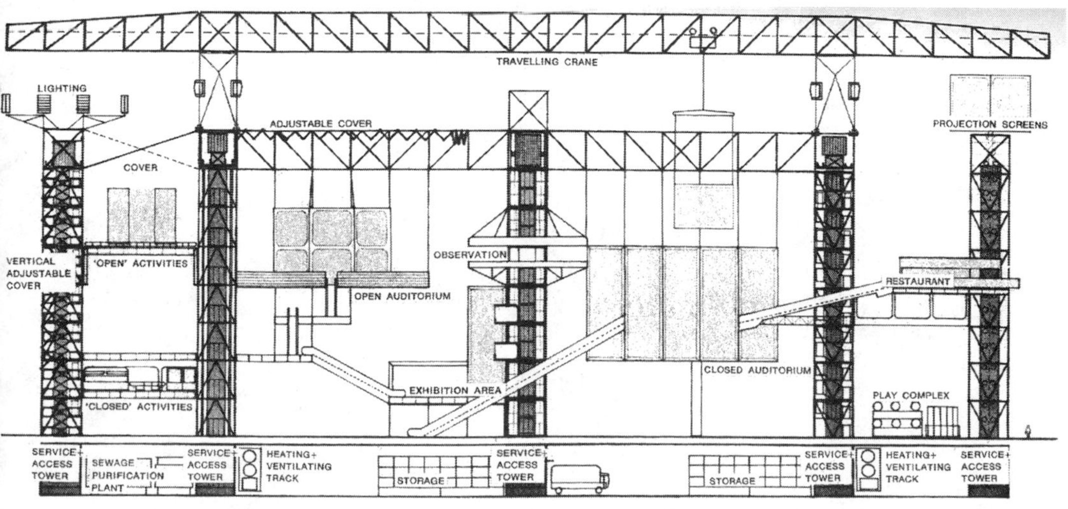

Figure 7.22. The Fun Palace. Source: Landau 1968, 79, fig. 56.

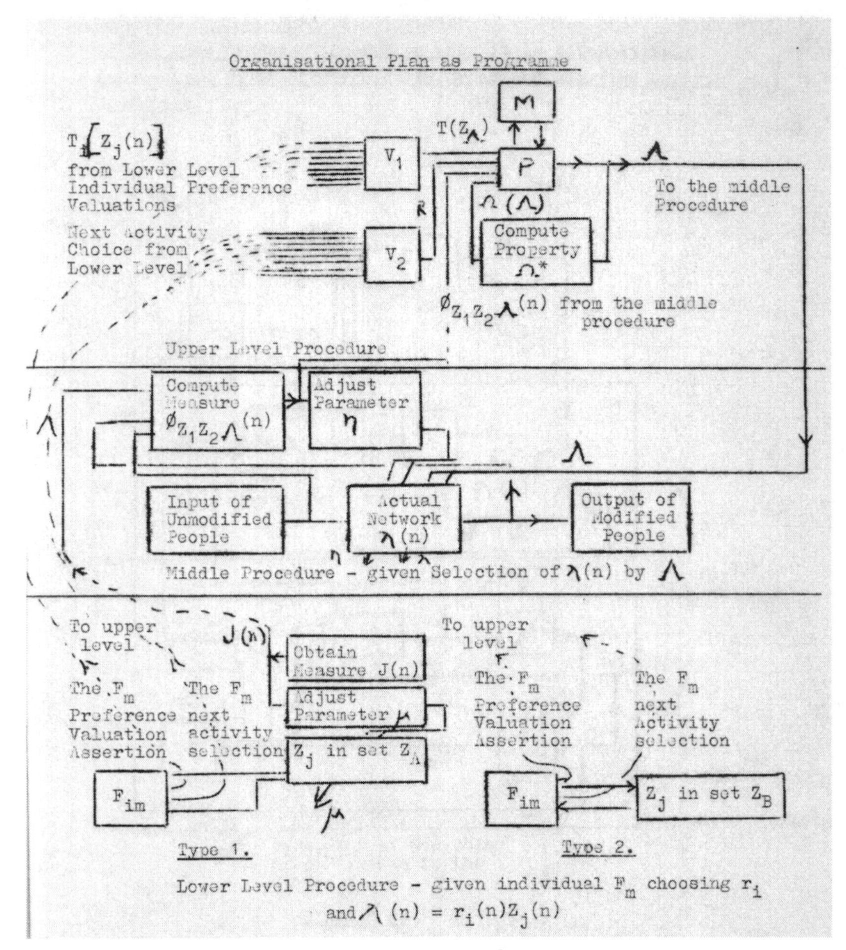

Figure 7.23. The Fun Palace's cybernetic control system. Pask 1965, 3, diagram 1. (By permission of Cedric Price Fonds, Collection Centre d'Architecture/Canadian Centre for Architecture, Montréal.)

reconfigurations both reacts to emergent patterns of use and fosters new ones.[60] Hence, I think, Roy Landau's reference to the Fun Palace as "encouraging . . . ideas to grow and to develop further." In a 1969 essay, Pask argued that cybernetic architecture would "elicit [the inhabitant's] interest as well as simply answering his queries" (Pask 1969a, 496), citing Musicolour and the Colloquy of Mobiles as examples of what he had in mind. Figure 7.23 reproduces Pask's 1965 logic diagram of the "cybernetic control system" for the Fun Palace, which features "unmodified people" as input and "modified people" as output—echoing the Situationist analysis of adaptive architecture as a transformative technology of the self.[61] Alexander Trocchi and his sigma project (chap. 5) made the connection, since he was allied with the Situationists in Paris and friends with both Price and Littlewood in London (with whom he met regularly in 1964) (Mathews 2007, 112–14).

Visitors to London will have noticed that the Fun Palace does not exist. Despite a lot of work from a lot of people, political support especially within the Labour Party, and the inclusion of such notables as Yehudi Menuhin and Lord Harewood among its trustees, the project collapsed in the second half of the 1960s, and the building was never even begun. One can cite many of the usual mundane reasons for this: the problems of finding a site, getting permissions, and, not least, raising money.[62] But another problem more germane to our theme came up again and again: the sheer difficulty of saying what the Fun Palace was. Like the Dream Machine and Musicolour before it, the Fun Palace failed to fit easily into any of the accepted architectural categories. Not only did it deliberately aim to cut across the usual demarcations—combining the arts, entertainment, education, and sport in all sorts of guises, familiar and unfamiliar, including participation in what were usually taken to be spectator activities—the broader aim was to experiment: to see what might emerge from combining these opportunities in an adaptive space. This, of course, left outsiders to the project free to project their own nightmares on it, and, as Littlewood's obituary in the *Guardian* put it, the very phrase "Fun Palace" "evoked for councillors a vision of actors copulating in the bushes," and Littlewood's "support dissipated in a fruitless search for a site" (Ezard 2002).[63]

Two thoughts before we leave the Fun Palace. The first goes back to the social basis of cybernetics. We can think once more about amateurism. I noted above that Pask's work on the Fun Palace was voluntary and unpaid, done out of interest and for fun and, no doubt, belief in the worth of the project. Here I can just add that, as Mathews (2007, 120) puts it, "like Price, Littlewood had a 'day job' and worked on the Fun Palace on the side." Again we have the sense of something welling up outside the structure of established social institutions and without support from them.

We can also think in this connection about the relation between modern architecture and buildings like the Fun Palace. The last sentence of Landau's *New Directions in British Architecture* (1968, 115) reads: "So if architecture is becoming . . . anti-building . . . perhaps it *should* be classified as not architecture . . . but this *would* signify that it had taken a New Direction." Mary

Lou Lobsinger (2000, 120) picks up the negative and describes the Fun Palace as "the quintessential anti-architectural project." We are back with the "antis"—with the Fun Palace facing contemporary architecture in much the same way as Kingsley Hall faced modern psychiatry. In neither case does the "anti" amount to pure negation. Adaptive architecture was *another* and *different* approach that crossed the terrain of established forms. If mainstream architecture aspired to permanent monuments, aesthetic and symbolic forms drenched in meaning, and fitness to some predefined function, the Fun Palace was envisaged as just a big and ephemeral rectangular box from the outside and a "kit of parts" on the inside. The heart of antiarchitecture lay in its inner dynamics and its processes of transformation in response to emergent, not given, functions—none of which existed (or, at least, were thematized) in the modern tradition. Here we have once more antiarchitecture as nomad science, sweeping in from the steppes to upset, literally, the settled lives of the city dwellers (the Walking City!), and a Situationist architecture as Heideggerian revealing—as keenly open to new ways to be—in contrast to an architecture of enframing, growing out of and reinforcing a given aesthetic and list of functions. Ontology as making a difference. No wonder that "for those who thought architecture had a visually communicative role . . . [Price's] work was anathema to everything architecture might stand for."[64]

Second, we can go back to the critique of cybernetics as a science of control. Mathews's account of Price's work in the 1960s takes a strange turn just when Pask appears at the Fun Palace. Speaking of a 1964 report from the Cybernetics Subcommittee, Mathews (2007, 119, 121) picks out what he considers a "rather frightening proposal" discussed under the heading of "Determination of what is likely to induce happiness" and continues:

> This . . . should have alerted Littlewood that the Fun Palace was in danger of becoming an experiment in cybernetic behavior-modification. However, in a 1964 letter to Pask, she actually agreed with his goals, and seemed naively oblivious to the possibility that the project might become a means of social control. . . . The idea that the Fun Palace would essentially be a vast social control system was made clear in the diagram produced by Pask's Cybernetics Subcommittee, which reduced Fun Palace activities to a systematic flowchart in which human beings were treated as data [fig. 7.23 above]. . . . Today, the concept of "unmodified or modified" people would be treated with a considerable amount of caution. Yet, in the 1960s, the prevailing and naive faith in the endless benefits of science and technology was so strong that the Orwellian implications of modification went largely unnoticed.

What can we say about this? First, this is a pristine example of the sort of critique of cybernetics that I mentioned in the opening chapters, which is why it deserves some attention. Second, the cyberneticians asked for it. They were rhetorically inept, to say the least. They went on endlessly about "control," and "modified people" in figure 7.23 sets one's teeth on edge. It invites Mathews's slide to "behavior-modification," which is a polite way to say "brainwashing." But third, of course, I think the critique is misdirected. It hinges on what I called the Big Brother sense of control—of hierarchical domination, of enframing—and nothing in Pask's work on the Fun Palace contradicts the idea that he was in the same space as Littlewood and Price (and the Situationists before them) in trying to imagine a building in which, far from being stamped by some machine, people could experiment with new and unforeseen ways to be.[65] Littlewood was not being naive in agreeing with Pask's goals. The Fun Palace, from Pask's perspective, continued the lineage of Musicolour, a machine that would get bored and encourage the performer to try something new. On a different level, as I have tried to show in this chapter and throughout the book, the cybernetic ontology was one of exceedingly complex systems which necessarily escape domination and with which we have to get along—Pask's notion of "conversation"—and the Fun Palace was just another staging of that ontology. As I also said before, the control critique might be better directed here at the established architectural tradition, which in its symbolic aspect attempts, at least, to tell us what to think and feel, and in its functional guise tries to structure what we do: the factory as a place to work (not to play games, learn or have sex), the school as a place to learn, the home as the dwelling place of the nuclear family . . . This repetitious critique of cybernetics stifles its own object.[66]

After the Sixties: Adaptive Architecture

AN EVOLUTIONARY ARCHITECTURE. . . . NOT A STATIC PICTURE OF BEING, BUT A DYNAMIC PICTURE OF BECOMING AND UNFOLDING—A DIRECT ANALOGY WITH A DESCRIPTION OF THE NATURAL WORLD.

JOHN FRAZER, *AN EVOLUTIONARY ARCHITECTURE* (1995, 103)

The social history of adaptive architecture closely mirrored that of cybernetic art, reaching a zenith in the sixties with the Fun Palace and receding into the margins thereafter, but here we can glance briefly at some postsixties developments that connect to Pask.

With the Fun Palace as the nexus, the sixties were the decade in which Pask established an enduring connection with architecture more generally. Pask (1969) described cybernetics as a theory of architecture, much as Walter and Ashby had described it as a theory of the brain. Institutionally, Pask's closest relationship was to the AA, the Architecture Association school in London. Cedric Price, who had completed his studies at the AA in 1959 (Melvin 2003), taught there part-time during the Fun Palace project and "was delighted when Gordon agreed to sit on my architectural juries" (Price 2001, 819). Thus began an association between Pask and the AA that continued for the rest of his life: "His presence and inventions within the life of the Architectural Association are both legendary and of day to day relevance" (Price 2001, 820). Pask often spoke and gave workshops at the AA, and in the late 1980s he took up a continuing position there as an assistant tutor.[67] The connection to the AA in turn proved auspicious for the propagation of Pask's cybernetics: "Of 12 successful students Pask had at Brunel University, eight were architects and six came from the Architectural Association" (Scott and Glanville 2001). Archigram's Peter Cook (2001, 571–72) speaks of "a whole generation of young architects. . . . They are, of course, the direct progeny of Gordon." To close this chapter I want to review of few examples of Paskian architecture in practice running up to the present. What these projects have in common is the Paskian idea of a dynamically evolving relation between the human and the nonhuman.

At the level of complete structures, in 1978 Cedric Price had another attempt at designing a Fun Palace–style building that was reconfigurable in use, this time for the Gilman Paper Corporation. Again, the project was never completed, but Price hired John and Julia Frazer as computer consultants, and they constructed a working electronic model of the Generator project, as it was called.[68] "It was proposed to grid the site (a clearing in a forest in Florida) with foundation pads and to provide a permanent mobile crane for moving components, allowing the users of the building to become involved in its organization. . . . We were concerned that the building would not be changed enough by its users because they would not see the potential to do so, and consequently suggested that a characteristic of intelligence, and therefore of the Generator, was that it would register its own boredom and make suggestions for its own reorganization. This is not as facetious as it may sound, as we intended the Generator to learn from the alterations made to its own organization, and coach itself to make better suggestions. Ultimately, the building itself might be better able to determine its arrangements for the users' benefit

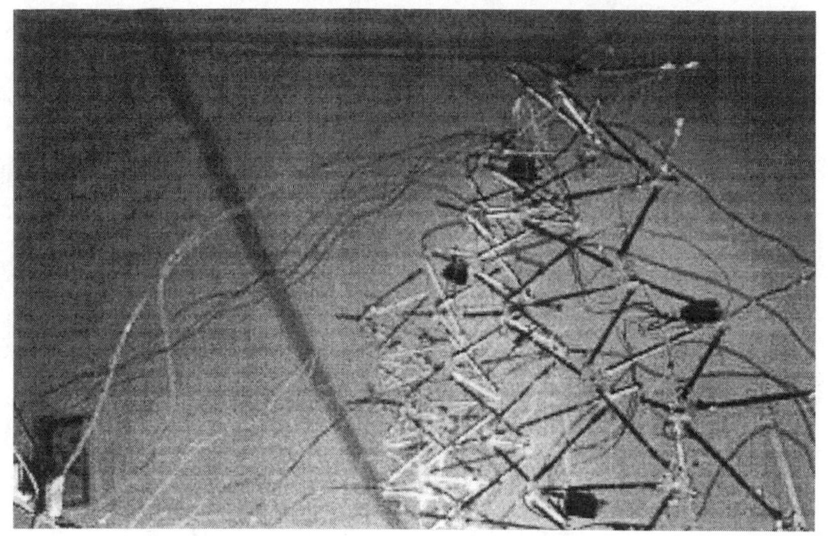

Figure 7.24. Digitally controlled architectural structure. Source: Silver et al. 2001, 907.

than the users themselves. This principle is now employed in environmental control systems with a learning capability" (Frazer 1995, 41). The reference to "boredom" here was an explicit evocation of the Musicolour machine.[69]

At a more micro level and closer to the present, a 2001 survey of projects at the Bartlett School of Architecture at University College, London, "Prototypical Applications of Cybernetic Systems in Architectural Contexts," subtitled "A Tribute to Gordon Pask," is very much in the Musicolour–Fun Palace tradition, assembling the elements for structures that can transform themselves in use (Silver et al. 2001). One project, for example, entailed the construction of a digitally controlled transformable structure—the skin of a building, say—a key element of any building that can reshape itself in use (fig. 7.24). Another project centered on communication via sounds, lights, and gestures between buildings and their users, reminiscent of the communications systems linking the robots in Pask's Colloquy of Mobiles (fig. 7.25).

In another tribute to Pask, "The Cybernetics of Architecture," John Frazer (2001) discusses the history of a big, long-term project at the AA, in which Pask participated until his death. The morphogenesis project, as it was called, ran from 1989 to 1996 and was very complex and technically sophisticated; I will simply discuss a couple of aspects of it in general terms.[70]

We have so far discussed cybernetic architecture in terms of relations between buildings and users—the former should somehow constitute an

Figure 7.25. Architectural communication device. Source: Silver et al. 2001, 911.

aesthetically potent environment for the latter. But one can also conceive of another axis of cybernetic incursion into architecture, this time concerning the relation between the architect and architectural design tools. The classic design tool in architecture is the drawing board—a passive object on which the architect inscribes his or her vision. The drawing board is thus not an aesthetically potent environment in Pask's terms. And much of Pask's involvement with architecture focused on changing that situation, via the development of tools that could adapt to and encourage the architect—again on the model of Musicolour. This was a topic on which he collaborated with Nicholas Negroponte at MIT in the development of what Negroponte called the Architecture Machine—a computerized system that could collaborate more or less symmetrically with the architect in designing buildings—turning crude sketches into plans, indicating problems with them, suggesting extensions, and so on.[71]

Frazer's morphogenesis project took this idea of creating an aesthetically potent environment for design further, along at least two axes. One was to explore new ways of communicating with computers. "Our attempts to improve the software of the user-interface were paralleled by attempts to improve the hardware. The keyboard and mouse have never seemed to me well suited to manipulating models or graphics: a digitizing tablet might be closer to a drawing board, but it is rarely used that way. In any case, we were eager to get away from conventional, drawing board dependent design approaches." Around 1980 a system of cubes was developed, each with an embedded processor. These cubes could be assembled as model structures and could be read by a computer that would build up an internal representation of structures that were somehow patterned on the arrangement of cubes (Frazer 1995, 37).

Beyond this, the morphogenesis project sought to incorporate the idea that architectural units—buildings, cities, conurbations—*grow*, quasi-biologically, and adapt to their environments in time.[72] As we have seen, in the 1950s and early 1960s, Pask had experimented with inorganic analogue models of organic growth processes—the chemical computers—but he had moved on to mathematical experimentation on cellular automata in the later sixties, and the morphogenesis project likewise took advantage of novel mathematical structures, such as genetic algorithms and cellular automata, to simulate processes of growth, evolution, and adaptation within the computer. The architect would supply the computer with a "seed" structure for a building, say, which the machine would then evolve, taking account of coevolutionary interactions with the building's environment. At the same time, the architect could interfere with this process, in the choice of seed, by selecting certain vectors of evolution for further exploration, and so on. In this way, the computer itself became an active agent in the design process, something the architect could interact with symmetrically, sailing the tides of the algorithms without controlling them (and taking us back to Brian Eno in the previous chapter)—a beautiful exemplification of the cybernetic ontology in action. Figure 7.26 is a single example of this style of coevolutionary design, a computer simulation of how the city of Groningen might develop into the future taking account of interactions between the growing city itself, its inhabitants, and its geographic environment. The quasi-organic structure is evident. As the original caption says, the generating computer model behind it was inspired by Pask's work in the 1950s, and this was, in fact, the last student project that Pask himself supervised.

One can thus trace out streams of Paskian cybernetic architecture ramify-

Figure 7.26. Groningen study. Source: J. H. Frazer, "The Cybernetics of Architecture: A Tribute to the Contribution of Gordon Pask," *Kybernetes, 30* (2001), 641–51, p. 648.

Figure 7.27. Gerbil architecture. Source: Nicholas Negroponte, *Soft Architecture Machines*, (Cambridge, MA: MIT Press, 1975), 46, fig. 1. (© 1976 Massachusetts Institute of Technology, by permission of the MIT Press.)

ing from the Fun Palace toward the present, in the development of design tools as well as building structures. It remains the case that nothing on the scale of the Fun Palace has yet been built, but perhaps the sixties might be coming back: in 2002 the Royal Institution of British Architects gave the Archigram group its gold medal (Sadler 2005, 7).

— — — — —

Let me end this chapter with one more, light-hearted, example of cybernetic architecture, an installation exhibited by Pask's collaborator, Nicholas Negroponte, at the Jewish Museum in New York from September to November 1970.[73] Close inspection of figure 7.27 reveals of a mass of small cubes inhabited by a colony of gerbils. The gerbils push the cubes around, as is their wont. At intervals, a computer scans the scene and either pushes the blocks back where they were, if they have not moved much, or aligns them to a grid in their new positions. The gerbils then go to work again, the computer does its thing once more, and thus the built environment and its inhabitants' use of it coevolve open-endedly in time in ways neither the architect, nor the computer, nor the gerbils could have foreseen—just like a Musicolour performance.

8

SKETCHES OF ANOTHER FUTURE

WHAT A LONG STRANGE TRIP IT'S BEEN.

THE GRATEFUL DEAD, "TRUCKIN'" (1970)

Writing this book has taken me to places I never expected to go. Strange worlds, new civilizations. To some I went willingly; to others, less so. I was happy to find myself in art worlds I had never imagined; I have acquired an odd interest in architecture, a field that never spoke to me before; it was fun and edifying to sit for a while at the feet of Wizard Prang. I had forgotten—and this is an important datum, I think—how literally wonderful the sixties were. On the other hand, my heart sank when I finally admitted to myself that three of my principals had written books with "brain" in the title and that I had to figure out some brain science. Likewise the realization that three of the "four founders of cybernetics" were strongly associated with psychiatry and that I needed to grapple with the history of that bleak field too. The brain and psychiatry were more daunting prospects than biofeedback and even chemical computers. Other topics just sucked me in: pursuing DAMS from a single published footnote into the labyrinths of Ashby's private journals could drive anyone mad, and the journey remains unfinished. Still, I got to visit all those places, and I am very glad I did; I thank the shades of my cyberneticians for taking me along. I am changed by the trip.

What else can one do after a journey like that but look back and look forward?

Themes from the History of Cybernetics

At different times I have imagined two different ways of organizing this book. The one I chose was a sequential exploration of the work of named individuals. Perhaps it had to be that way. The most important context for understanding the work of any individual is what they did before. If you want to get the hang of DAMS, it helps to have the homeostat fresh in your mind; if you want to understand the Fun Palace, start with Musicolour. But there is this other way. I could have arranged the material thematically and looked at how various cybernetic projects bore on each theme—and we can review some of these themes now, briefly, as another way of remembering the trip.

ONTOLOGY

Ontology is the major cross-cutting theme I announced in advance and that I have pursued pretty conscientiously as we went along, so I will not dwell on it at any length. But there are things to say. Another working title for the book was *Performance*.

The discovery of pragmatist philosophy was a major turning point in my intellectual life. Reading William James's *Pragmatism and the Meaning of Truth* (1978 [1907, 1909]) in 1985 suddenly offered me a way seeing knowledge as situated (rather than transcendentally true) while continuing to take it seriously (and not as epiphenomenal froth).[1] It sometimes seems that everywhere I have gone since then, even in this book, James was there first. And yet there is something frustrating about the pragmatist tradition. It displaces many of the standard philosophical problematics in valuable ways, but it remains representationalist, epistemological, largely centered on knowledge. The pragmatist insists that knowledge has to be understood in relation to practice, but practice always features as something ancillary, to be wheeled on as needed to combat epistemological arguments from other schools of philosophy. You can read the pragmatists forever without learning much about practice and performance apart from a few armchair examples. It is as if knowledge remains the luminous sun around which these little planets called "practice" and "performance" revolve.[2]

In retrospect, then, I can see much of my own work as an exploration of this neglected side of pragmatism, an inquiry into practice *in its own right*, without a pregiven presumption that the end of inquiry has to be an argument about knowledge. And, to put it simply, the upshot for me was a gestalt switch into what I call the performative idiom. The argument of *The Mangle of Practice* was that if there is a sun around which all else revolves, it is performance, not knowledge—knowledge is a planet or maybe a comet that sometimes participates in the dynamics of practice and sometimes does not, and the discovery, for me, was that practice has its own structure that one can explore and talk about—as a dance of agency, for example.

The modern sciences background their own practice, organizing it around a telos of knowledge production and then construing it retrospectively in terms of that knowledge (a tale of errors dispelled). We have seen that cybernetics was not like that. Cybernetics was about systems—human, nonhuman, or both—that staged their own performative dances of agency, that foregrounded performance rather than treating it as some forgettable background to knowledge. This is the primary sense in which one can read cybernetics as ontological theater—as forcibly reminding us of the domain of practice and performance and bringing that to the fore. As showing us, in a fascinating range of instances, that performance is not necessarily about knowledge, and that when knowledge comes into the picture it is as *part of* performance.

Beyond that, cybernetics helps us think further about the nature of practice and performance. The key idea in grasping many of the examples we have explored is Beer's notion of an "exceedingly complex system"—meaning a system with its own inner dynamics, with which we can interact, but which we can never exhaustively know, which can always surprise us. A world built from exceedingly complex systems would necessarily be one within which being would always center on performative dances of agency and findings-out, where neither knowledge nor anything else would constitute a still, reliable center. This, I think, is our world. It is certainly the world of science as I described it in *The Mangle*. Again, cybernetics dramatizes this vision for us, and in at least two ways. On the one hand, the cyberneticians built machines and systems that interacted with and adapted to the world as an exceedingly complex system, in a list running from the tortoise and the homeostat up to biological computers, the VSM, Musicolour, and the Fun Palace. These examples can help bring home to us what unknowability can mean (as well as "performance"). They also demonstrate that a recognition that we live in a world of exceedingly complex systems does not imply paralysis, that we can, in fact, *go on* in a constructive and creative fashion in a world of exceedingly complex systems. On the other hand, the history of cybernetics offers us many simple "toy" examples of exceedingly complex systems. The tortoise, the homeostat, DAMS, cellular automata, Musicolour can all function as ontological icons—inscrutable Black Boxes in their performance, even though one can, in fact, open these boxes and understand them at the level of parts. My argument was that if we look through the end of the telescope that picks out performance, then these can all function as instructive examples of what the world in general is like (though from the other end they look like modern science and engineering and conjure up an ontology of knowability and control).

DESIGN

A distinctive notion of design has surfaced from time to time on our journey: Ashby on DAMS and the explorer who finds Lake Chad but no longer knows where he is, Beer on biological computers and the entrainment rather than deliberate reconfiguration of materials, Frazer and Pask building performatively inscrutable cellular automata into their architectural design systems, Eno and music. I have always thought of design along the lines of rational planning—the formulation of a goal and then some sort of intellectual calculation of how to achieve it. Cybernetics, in contrast, points us to a notion of design in the thick of things, plunged into a lively world that we cannot control and that will always surprise us (back to ontology). No doubt real designers have always found themselves in medias res, continually coping with the emergent exigencies of their projects. What interests me is that cybernetics serves both to foreground these exigencies (rather than treating them as unfortunate side effects) and to make a virtue of them, to enjoy them!). Ashby came to see an evolutionary approach to design—continually taking stock and exploring possibilities—as integral to the development of truly complex systems like DAMS. Beer's idea was that there is completely another way to the construction of performative computing elements: finding some material with the appropriate liveliness rather than laboriously engineering dead matter. Eno, Frazer, and Pask *wanted* to see where their uncontrollable CAs would take them—what sort of a trip that would be. Throughout the book I have tried to show that ontology makes a difference, but most of my examples have concerned specific systems or artifacts; here we can see that it makes a difference more generally, now in an overall stance toward design.

POWER

Following this line of thought takes us to another theme that has run through the book: power. Throughout, and especially in the later chapters, I have sought to address the critique of cybernetics as a science of control. To do so, I have have found it useful to distinguish two senses of "control." The critics' sense, I think, is that of a hierarchical, linear "command and control," of a power that flows in just one direction in the form of instructions for action (from one group of people to another, or, less conventionally, from humans to matter). I have been at pains to show that the cybernetic sense of "control" was not like that. Instead, in line with its ontology of unknowability and becoming, the cybernetic sense of control was rather one of getting along with, coping with, even taking advantage of and enjoying, a world that one cannot push around in that way. Even in its most asymmetric early moments, cybernetics never imagined that the classical mode of control was in fact possible. Ashby's appalling notion of blitz therapy did not envisage any determinate result; its only aspiration was an open-ended homeostat-like reconfiguration of the mentally ill, perhaps in a beneficial direction, but usually not. Even there the fantasy of command and control was absent, and this in a principled, not incidental, way. But from chapter 5 onward we have been especially concerned with what I called the symmetric fork in the road, the branch of later cybernetics that imagined a world in which adaptation goes both ways—in psychiatry, between doctor and patient, as at Kingsley Hall, but in many other realms too.

The appeal of this symmetric branch of cybernetics is that it both adumbrates and argues for a performative form of democracy, within social organizations, between social organizations, and even between people and things. This is just the recognition that we are always in medias res put another way. But we can take the thought one stage further by referring, as I have, to the philosophy of Martin Heidegger, and his contrast between enframing and revealing. Heidegger's idea was that modernity is characterized by a stance of enframing—the stance of command and control that goes along with an ontology of knowability, and that assumes we can obtain determinate results from our initiatives in the human and material worlds. Everything I know about the history of science and technology tells me that this assumption is a mistake (though a productive one in certain circumstances); what I have learned from Heidegger and cybernetics is to see it as a sad one. It closes us off from what the world has to offer; in the mode of enframing, the unexpected appears with a negative sign in front of it, as a nuisance to be got around. The stance of revealing, in contrast, is open to the world and expects novelty, for better or for worse, and is ready to seize on the former. And what I take the history of cybernetics to show is that such words are not empty philosophical pieties. It is possible to develop rich and substantial ways of going on in the world in the mode of revealing. The tortoise, the homeostat, Kingsley Hall, the viable system model, syntegration, Musicolour, the Fun Palace—these are all *revealing machines* that in one way or another explore their worlds for what they have to give.

THE ARTS

We have explored the origins of cybernetics in science, as a science of the brain and psychiatry, but we have also seen how quickly it spilled over into all sorts of fields—robotics, complexity theory, management, politics, education, and so on. Some of these only came to the fore in specific chapters, but others appeared in several. One theme that has arisen in most concerns intersections (or fusions) of cybernetics and the arts: the Dream Machine, brainwave music, Brian Eno's music (and even Jimi Hendrix and feedback), architecture (the detailed tuning of parts in Christopher Alexander's work; the adaptive architecture of Archigram, Price, and Pask; aesthetically potent design environments), synesthesia and Musicolour, interactive theater, robotic and interactive sculpture (the Colloquy of Mobiles). The Dream Machine and brainwave music can be postponed for a moment, but the other artworks just mentioned all serve to dramatize the basic ontology of cybernetics beyond the world of science and help us grasp it. In doing so, they also echo the other themes just mentioned: an experimental approach to design as a process of revealing rather than enframing, a leveling of power relations between artists and audiences, a blurring of modern social roles. At the same time, we should recall the oddity of many of these artworks. Again and again, the question that has come up is: is it art? Does the Dream Machine count as visual art? Does brainwave music or Eno's music count as music? Did the Fun Palace count as architecture? What could Pask sell Musicolour as? The clash between these odd artifacts and modern classifications points to the more general theme of this book: that ontology makes a difference, in both practices and products.

SELVES

Another cross-cutting theme that surfaced especially in chapters 3, 5, and 6 has to do with, variously, the brain and the self, in ways that I tried to catch up in a contrast between modern and nonmodern apprehensions. The modern take on the brain can be exemplified by work in traditional AI (and its counterparts in the cognitive sciences more broadly): an image of the brain as, centrally, an organ of representation, calculation, and planning. The modern apprehension of the self I take to resonate with this: an idea of the self as a bounded locus of agency, again centering on representation, calculation, planning, and will. I have not sought to delineate the modern self at all sharply (imagine a race of accountants), but it is clear, I think, that the cybernetic take on the brain and the self was very different. Cybernetics began by imagining the brain (and later the self) as performative, and, as I said, it is possible to be curious about the performative brain in ways that hardly arise within the modern perspective. Right from the start, Grey Walter was interested in madness, epilepsy, visions, yogic feats, and nirvana—topics that escape from modern discourse, or at most appear as regrettable deviations from the norm. We could say that cybernetics had a more capacious appreciation of the possibilities of brains and selves than modern discourse sanctions—an open-ended vision of what people are like and can be, in contrast to the narrowly conceived field of the modern self.

From a cybernetic standpoint, there is always something more to be found out in exploration of the brain and the self. Hence the trajectory that led from flickering stroboscopes via dynamic visions to Gysin's Dream Machines, and another trajectory in the same line of descent leading to EEG biofeedback and brainwave music. In the same vein, Bateson and Laing (and Huxley) pointed to the possibility of the dissolution of the modern self and related that both to schizophrenia and to Buddhist enlightenment. Madness as not only a sad aberration to be stamped out by electroshock or drugs, but perhaps an opening, too—an opening seized upon by the sixties, with its "explorations of consciousness" but also long central to Eastern philosophical and spiritual traditions, as taken up in these pages and integrated with cybernetics by Stafford Beer (and others).

I have found Foucault's notion of technologies of the self useful here. Foucault used it to point to strategies deployed in the construction of varieties of the modern (in my sense) self—freestanding and self-controlled centers of will and calculation—technologies of self-enframing. In the history of cybernetics, we can apply the phrase quite literally, but now in reference to technologies ranging from flicker to meditation that somehow elicit and explore other states beyond the range of the modern, selves that are out of control, techniques that might even dissolve the modern self. Technologies of the self as technologies of self-revealing.

SPIRITUALITY

From nonstandard selves one can drift almost continuously into a discussion of nonmodern spirituality, but a few more specific points are worth making.

When I began writing this book, my interest in the spiritual realm was close to nonexistent. A compulsory school indoctrination into the worldview of the Church of England had left me with the conviction that even geography, my worst subject, had more going for it. But I was struck to find my cyberneticians dragging me back into this realm in ways that I found challenging. Walter, Ashby, Bateson, Laing, and Beer all, though in different ways, registered spiritual interests and connections (and I suspect there is more to be said about Pask than I have so far discovered). So what is it with cybernetics and spirituality, especially Eastern spirituality? My qualifications to speak in this area remain tenuous but, with that caveat in mind, here is a list of affinities (which I cannot quite make into a unity, though they all hang together with a performative understanding of the brain and the self):

(1) As just discussed, a cybernetic curiosity about the performative brain leads naturally to an interest in the sorts of strange performances and altered states characteristic of yogic traditions. (2) The performative brain is necessarily a relational brain that responds to its context, including technologies of the self that can evoke and respond to novel performances. We have seen that some of these technologies were distinctly Western (flicker, LSD), but many again made a direct connection to the East: meditation, yoga, diet, tantric exercises as practiced by Wizard Prang. (3) Cybernetic models of the brain, and understandings of the self, point immediately to a decentering of the mind and the self. From the tortoise and the homeostat onward, the cybernetic preoccupation with adaptation has continuously eroded the modern understanding of the bounded, self-contained, and self-moving individual. Instead, one has the image of the brain and the self as constitutively bound up with the world and engaged in processes of coupled becomings. And this image is not far from the ontology of Buddhism, say, with its emphasis on undoing the modern self for the sake of an awareness of being, instead, part of a larger whole ("yoga means union"). (4) A corollary of the cybernetic epistemology with its insistence that articulated knowledge is part of performance rather than its container is, I think, a hylozoist wonder at the performativity of matter and the fact that such performativity always overflows our representational abilities. I remain unsure about the spiritual lineage of hylozoism, but we have seen how Beer aligned it first with the Christian tradition and later with Eastern philosophy and spirituality.

Beyond these specifics, what interests me most in this connection is how thoroughly cybernetics elided the modern dichotomy of science and religion. Since the nineteenth century, in the West at least, a sort of precarious settlement has been reached, in which science and religion relate to two disparate realms of existence, each empowered to speak of topics in its own territory but not the other's. Again, cybernetics was not like that. As we saw in most detail in Beer's work, his science (cybernetics) and spirituality were of a piece, shading into one another without any gap or transition. I talked about the "earthy" quality of this sort of spirituality; one could just as well speak of the "elevated" quality of the science (I made up the phrase "spiritual engineering"). One could say much the same about Ashby's brief foray into the spiritual realm ("I am now . . . a Time-worshipper"). Aldous Huxley and his scientific contacts (Osmond and Smythies) stand as a beautiful example of how to think about an immanent rather than transcendent dimension of the spirit. Those of us who grew up in the Church of England (and, no doubt, other more dogmatic churches) find this a difficult position to even imagine—how can you have a religion without a transcendent God?—but, as I said, it might be worth the effort, a conceptual and spiritual breather from current agonizing about the relation between Christianity and Islam, modern science and fundamentalist Christianity.

THE SIXTIES

The sixties—in particular, the sixties of the counterculture—have skipped in and out of these chapters, too. Substantively, in the recent history of the West, the sixties were the decade when the preoccupations of cybernetics with performative experimentation came closest to popular culture—iconically in the countercultural fascination with "explorations of consciousness," but also in sixties experimentation with new identities and art forms, new forms of social and sexual arrangements, and even with new relations to matter and technology: Hendrix abusing his guitar and overloading the amps again. My suggestion is that the many crossovers from cybernetics into the world of the counterculture index a shared nonmodern ontology. Both the sixties and cybernetics can be understood as nonmodern ontological theater, nonmodern ontology in action. If cybernetics began as a science of psychiatry, it became, in its symmetric version, the science of the sixties.

ALTERED STATES

There are some places I wanted to go that cybernetics did not take me. My interest in cybernetics as ontology grew out of my earlier work in the history of physics, and my conviction is that the cybernetic ontology is illuminating across the board. But while it is true that the work of my cyberneticians span off in all sorts of directions—robots and gadgets run through all of the preceding chapters (except chap. 5); Ashby regarded his formal cybernetics as a theory of all possible machines, which could be equated with all (or almost all) of nature; cellular automata have appeared here and there; Ashby worshipped time; Beer worshipped matter and wrote hylozoist poems about the Irish Sea—still, a certain asymmetry remains. If a performative notion of the brain implies a space for curiosity that can lead into the field of altered states of consciousness, one can imagine a similar trajectory leading to a fascination with altered states of matter, especially for a hylozoist like Beer. I think here of a tradition of research into the self-organizing properties of complex inorganic and biological systems that came to life in the 1980s, the same period that saw the resurgence of Walterian robotics (and neural networks in computing)—research that focused on, for example, the emergence of structure in convection flows (Bénard cells), the uncanny quasi-organic dynamic patterns associated with the Belousov-Zhabotinsky reaction in chemistry, the tendency of slime molds to morph from unicellular entities into aggregate superorganisms and go marching off. I would have enjoyed the opportunity to tie these examples into our story, but the occasion did not arise. I once asked Stafford Beer about this line of research into complexity; he replied that the physicists and mathematicians were "re-inventing the wheel." For once, I think he was wrong.[3]

THE SOCIAL BASIS

The social basis of cybernetics is another topic I have discussed systematically in each chapter. What we found throughout are the marks of a continual social marginality of cybernetics: its hobbyist origins outside any institutional frame, its early flourishing in tenuous and ad hoc organizations like dining clubs and conference series, its continual welling-up outside established institutions and its lack of dependable support from them. We could think about this marginality more constructively as the search for an institutional home for cybernetics: individual cyberneticians often found visiting and part-time positions in universities (with Ashby at the BCL in Illinois, and the Brunel Cybernetics Department as exceptions that proved the rule, but only for a time: neither unit has lasted to the present); other cyberneticians lodged themselves in the world of business and industry (Beer and Pask as consultants; Kauffman at Bios, temporarily at least; Wolfram at Wolfram Research). We have also come across the emergence of novel institutions, from the Santa Fe Institute to the Planetary Collegium and, more radically, Kingsley Hall, the Archway communities, and the antiuniversity, with Alexander Trocchi's sigma project as a sort of overarching blueprint and vision. More generally, the counterculture, while it lasted, offered a much more supportive environment to cybernetics than did the organs of the state.

I am left with an image of the social basis of cybernetics as not just marginal but evanescent, always threatening to wink out of existence, always in need of re-creation. At times, I am inclined to see an arrow of change here, moving toward more substantial and resilient social structures as the years have gone by—an image of the emergence of a parallel social universe, as I called it, that maps only poorly and partially onto the institutional structures of modernity. At other times I wonder if the Santa Fe Institute and the Planetary Collegium will last any longer than Kingsley Hall and the BCL.

There are several ways to think about this marginality. Their descendants often blame the founders of cybernetics for caring more about themselves than the institutional future of the field, and there is something to that, though one would have to exonerate Stafford Beer (and Heinz von Foerster) on this score. But we should also think not about individuals and their personalities, but about the connections between sociology and ontology. At the simplest level, a metaphor of attraction and repulsion comes to mind. From the start, the cyberneticians were in the same ontological space as one another, and much more attracted to each other's thought and practice than that of their colleagues in their home departments and institutions, and no doubt the inverse was equally true. More materially, within the grid of institutionalized practice, doing cybernetics required different facilities and resources from conventional equivalents—building little robots as a way of doing psychiatric theory, growing biological computers as the cutting edge of management. At the limit, one finds not only incompatibility but material, practical, and theoretical collisions, battles, and warfare. Here I think especially of Villa 21 and Kingsley Hall from chapter 5: "antipsychiatry" as not just different from conventional psychiatry but institutionally incompatible with it, unable to coexist with mainstream practices within a single institution, and theoretically intensely critical of them. Here, above all, is where we find Deleuze and Guattari's notion of nomad science being staged before our eyes—and the nomads eventually driven off. We can return to the question of the social basis in the next section.

Sketches of Another Future

Spiegel: And what takes the place of philosophy now?
Heidegger: Cybernetics.

MARTIN HEIDEGGER, "ONLY A GOD CAN SAVE US" (1981)[4]

So what? What can this story do for us now? Why describe a historical study as sketches of another future?

The simple answer is that the book is an attempt to rescue cybernetics from the margins and to launder it into mainstream discourse, to make it more widely available. The other future I have in mind is "another future" for people who have not yet stumbled into this area, and for a world that seems to me presently dominated by a modern ontology and all that goes with it. By rehearsing the history of cybernetics and reading it in terms of a nonmodern ontology of unknowability and becoming, I have tried to convey my conviction that there *is* another way of understanding our being in the world, that it makes sense, and that grasping that other way can make a difference in how we go on. My idea for the future is not that we should all go out tomorrow and build robot tortoises or management consultancies based on the VSM (though it might be time to have another go at the Fun Palace). My hope is that these scenes from the history of cybernetics can function as open-ended models for future practice, and that they can help to make an endless and quite unpredictable list of future projects imaginable.

Why should we care about this? This takes us back to my thoughts at the beginning about the hegemony of modernity, as I defined it, over our works and our imaginations. It would be surprising if modernity were not hegemonic: almost all of the educational systems of the West, schools and universities, are organized around the modern ontology of knowability—the idea that the world is finitely knowable—and, indeed, around an aim of transmitting positive knowledge. One cannot get through a conventional education without getting the impression that knowledge is the thing. I know of no subjects or topics that are taught otherwise. And part of the business of conjuring up another future is, then, to suggest that this is not necessarily a desirable situation. I can think of three ways to argue this.[5]

The first is simple and relatively contentless. Variety is good: it is what helps us to adapt to a future that is certainly unknown. Better to be able to grasp our being in the world in two ways—nonmodern as well as modern—rather than one. It would be nice to have available ways of thinking and acting that stage some alternative to modernity. I have never heard brainwave music and I have never played with biological computers, but I would rather live in a world that includes rather than excludes them, and one day they might be important.

The second argument has to do with the specifics of cybernetics. Cybernetics, it seems to me, offers an alternative to modernity that has some particularly attractive features. I summarized many of them in the preceding section; here I would emphasize that in its symmetric version there is something inherently democratic about cybernetics. The cybernetic ontology, as I have said before, necessarily implies respect for the other, not because respect is nice but because the world is that way. The ontology itself evokes a democratic stance. At the same time, cybernetics offers us a peculiarly and interestingly performative take on democracy that can even extend to our relations with matter and nature (without the descent into modernist anthropomorphism). And beyond that, I think there is something very attractive about programs of action that adopt the stance of revealing that I associate with cybernetics—of openness to what the world has to offer us. Given the choice, who could possibly prefer enframing to revealing? If anyone did, I would be inclined to follow R. D. Laing and consider them mad.

These arguments are right, I believe; they are cybernetically correct; and they are sufficient to warrant the effort I have put into these pages. But there is a third argument, which refers not to the specifics of cybernetics but to the modern backdrop against which cybernetics stands out. This last argument might have more force, for some readers at least, and it requires some discussion.

— — — — —

RATIONALIZATION MEANS . . . THE EXTENSION OF THE AREAS OF SOCIETY SUBJECT TO THE CRITERIA OF RATIONAL DECISION. . . . MOREOVER, THIS RATIONALITY EXTENDS ONLY TO RELATIONS OF POSSIBLE TECHNICAL CONTROL, AND THEREFORE REQUIRES A TYPE OF ACTION THAT IMPLIES DOMINATION, WHETHER OF NATURE OR SOCIETY.

JÜRGEN HABERMAS, *TOWARD A RATIONAL SOCIETY* (1970, 81–82)

THAT THESE TRAGEDIES COULD BE SO INTIMATELY ASSOCIATED WITH OPTIMISTIC VIEWS OF PROGRESS AND RATIONAL ORDER IS IN ITSELF A REASON FOR A SEARCHING DIAGNOSIS.

JAMES SCOTT, *SEEING LIKE A STATE* (1998, 342)

NOW PEOPLE JUST GET UGLIER AND I HAVE NO SENSE OF TIME.

BOB DYLAN, "MEMPHIS BLUES AGAIN" (1966)

What's wrong with modernity? Perhaps nothing, but perhaps there is. Martin Heidegger (1976 [1954]) equated modern engineering, and modern science behind it, with enframing and a reduction of the world and ourselves to "standing reserve." Jürgen Habermas (1970) worried about a rationalization of the lifeworld, an inherently political reconstruction of selves and society via scientific planning that derives its insidious force from the fact that it can find no representation in orthodox political discourse. More materially, Bruno Latour (2004) suggests that modernity (as I have defined it, following him) is coming back to haunt us. Its dark side shows up in the "unintended consequences" of modern projects of enframing, often in the form of ecological crises. A couple of centuries of industrialization and global warming would be a stock example. More generally, James Scott's (1998) catalog of projects that he calls "high modernist"—schemes that aim at the rational reconstruction of large swathes of the material and social worlds—reminds us of their often catastrophic consequences: famine as another side of the scientific reform of agriculture, for instance.[6] I think of murder, mayhem, and torture as another side of the imposition of "democracy" and "American values" on Iraq.

If one shares this diagnosis, what might be done? The obvious tactic is resistance—the enormous and so far successful opposition to genetically modified organisms in Europe, for instance. But we can note that this tactic is necessarily a negative one; it aims to contain the excesses of modernity, but only by shifting the balance, recalibrating our ambitions without changing their character. Arguments about global warming remain within the orbit of the modern ontology; they themselves depend on scientific computer simulations which aim to know the future and hence bend it to our will. This sort of opposition is immensely important, I think, but in its negativity it also contributes to the grimness of what Ulrich Beck (1992) famously called the "risk society," a society characterized by *fear*—of what science and engineering will bring us next.

And this, of course, gets us back to the question of alternatives. Is there something else that we can do beyond gritting our teeth? Heidegger just wanted to get rid of modernity, looking back to ancient Greece as a time when enframing was not hegemonic, while admitting that we can never get back there and concluding that "only a god can save us" (1981). More constructively, there is a line of thought running from Habermas to Latour that grapples with the problem of "how to bring the sciences into democracy" (the subtitle of Latour's book)—of how, literally or metaphorically, to bring scientists, politicians, and citizens together on a single and level playing field, on which the expertise and interests of none of these groups necessarily trumps that of the others.

I admire this line of thought. If translated into novel social arrangements, it might well make the world less risky, less grim and fearful. The vision may even be coming true: scientists and politicians are meeting collectively in conferences on global warming, for example. And yet it is hard to get excited about it, so much remains unchanged. In this explicitly representationalist approach to the dark side of modernity, modern science and engineering remain hegemonic; Latour's ambition is simply to slow down the hectic pace of modernity, to give us some democratic breathing space before we rush into the next high-modernist adventure.[7] And, to get back to our topic, this, for me, is where the "political" appeal of cybernetic resides, precisely in that it promises more than a *rearrangement* of the world we already have. It offers a constructive alternative to modernity. It thematizes attractive possibilities for *acting differently*—in all of the fields we have talked about and indefinitely many more—as well as thinking and arranging political debate differently. This is my third reason—specific to our historical conjuncture—for being interested in cybernetics as a challenge to the hegemony of modernity.[8]

— — — — —

HOW RARE IT IS TO ENCOUNTER ADVICE ABOUT THE FUTURE WHICH *BEGINS* FROM A PREMISE OF INCOMPLETE KNOWLEDGE.

JAMES SCOTT, *SEEING LIKE A STATE* (1998, 343)

What might it mean to challenge the hegemony of modernity? At the ground level, I have in mind both an ability to recognize cybernetic projects and to read their moral as ontological theater and a multiplication of such projects—a "quantitative" challenge, one might say. But just where does this hegemony reside? In one sense, it resides in our imaginations; there is something natural about Scott's high-modernist schemes to dominate the world; we might worry about their specifics, but we tend not to see that there is anything generically problematic about this sort of activity. But there is another side to this, which I think of in terms of "echoing back." We are not simply taught at school to find the modern ontology natural; the "made world" of modernity continually echoes this ontology back to us and reinforces it. I grew up in a factory town that depended upon turning formless sheet metal into cars. From my earliest childhood I was plunged into a world where human agency visibly acted on apparently passive matter to accomplish its ends, and where any unintended consequences of this process were hard to see and talk about. The material form of an industrialized society in this sense echoes back the modern ontology of a passive and defenseless nature awaiting reconfiguration by humanity. From another angle, while writing this book I lived in a small midwestern town where all of the streets are laid out on a north-south, east-west grid. To get out of town, one could drive for hours down a linear and featureless freeway or take a plane and travel through an unmarked space in which a trip to San Francisco differs from one to London only in terms of the number of hours of discomfort involved. In such a geography, how else should one think of space but as Cartesian, or time as linear and uniform?

In ways like this, the made world of modernity echoes back to us the basic modern ontology (and vice versa, of course). This is a reason for thinking that perhaps the Habermas-Latour approach to reining in modernity might not be enough. It would do little, as far as I can make out, to challenge this material reinforcement of modernity's ontological stance. Conversely, I have paid great attention here to the made worlds of cybernetics—objects and artifacts that can echo back to us a nonmodern instead of a modern ontology. Part of the business of challenging modernity might entail moving these objects and artifacts from the margins toward the center of our culture, as I have tried to do here: multiplying them, appreciating them as ontological theater, taking them seriously. Heidegger's desperate dream was that artists and poets might save us from the world of enframing. My first reaction to that was incredulity, but there might be something to it. I am impressed by the examples of cybernetic art, theater, music, and architecture that we have encountered on this trip. If we could learn to see interactive robot artworks as ontological theater instead of vaguely amusing objects at the fringes of real art, the hegemony of modernity would indeed be challenged—which is not to say that art is enough or, pace Heidegger, that artists are the only people we should look to.

Another question that arises here is: how far should the challenge to modernity go? What could we imagine? A complete displacement of the modern by the nonmodern? The answer to that is no, for a couple of reasons. One goes like this: Cybernetics has often been characterized as a science of information, different in kind and having a different referent from the usual sciences of matter. I know of no way of thinking about electrical power stations other than modern physics, and no way of building and running them other than modern civil and electrical engineering. Take away the modern elements and our society would quickly collapse into a species of chaos grimmer than the grey world we already live in. Of course, this book has not leaned on this conventional information-matter contrast. I have emphasized the performative aspects of cybernetics, even when representation and information have been at stake. And the discussions of hylozoism certainly open up a space in which one can entertain the possibility of somehow entraining nature differently from Heidegger's archetypal power station straddling and enframing the Rhine. Readers of science fiction may think of lighting schemes featuring luminous lichens or bacteria. We should not, however, hold our breath; we will need modern science and engineering for some time to come.

The other reason is that I just argued that variety is good. It would be good if we could imagine the world in a nonmodern as well as a modern fashion. An extermination of modernity would be variety-reducing—that is, bad. But challenging the hegemony of modernity might also come to mean *putting modernity in its place*. At some sort of limit, this would mean coming to see modernity precisely as an *option* rather than the natural and the only possible way to go on, and a risky one at that. A nonmodern ontology would immediately and continually remind us that we never know the future and that we should always expect unexpected consequences to accompany our projects, however scientifically thought through they are. And this would imply *being careful*, in two senses: first, not rushing headlong into Scott's high-modernist adventures; and second, watching what happens if one does embark on such schemes—being very alert to how they are developing in practice, taking it for granted that reality will depart from expectations (for better or for worse), and being ready to shift our understandings and expectations in the light of emergent discrepancies between expectations and accomplishments.

This, of course, restates the cybernetic epistemology that we have encountered repeatedly in the previous chapters, especially Stafford Beer's willingness to incorporate scientific models into the VSM coupled with his suspicion of them and his insistence that they continually fail and need to be revised in practice. It also returns us to the Habermas-Latour line of thought, that we need a better and continuously watchful form of democratic participation in modern science and engineering, but now as part of a broader recognition that there are other than modern ways to go on in the world and that these are options, too.[9]

— — — — —

Just what is it that stands between us and the deluge? Is it only cybernetics? Only the work of Walter, Ashby, Bateson, Laing, Beer, and Pask? No. As I said earlier, as far as the future is concerned, their work should be seen as offering us a set of models that both conjure up a nonmodern ontology and invite endless and open-ended extension. The ontological vision, and the realization that there are real-world projects that go with it, is the important thing, not the names and historical instances. But it might be helpful to put this another way. I have been trying to argue that the history of cybernetics indeed conjures up another form of life from that of modernity, but my suggestion is not that cybernetics was some sort of brute and isolated historical singularity. I have, for example, talked about all sorts of antecedents from which cybernetics grew, starting with work in experimental psychology that included Pavlov's salivating dogs and phototropic "electric dogs." But here it might be useful to note some of the many contemporary streams of work and thought that lie in much the same space as cybernetics even though it might not make much historical sense to label them "cybernetic."

In fact, I have mentioned many of these nonmodern traditions already. In philosophy, we could think of the pragmatist tradition. William James's idea that experience is continually "boiling over" relative to our expectations is a beautiful way into an ontology of becoming. We could also think of the tradition of Continental philosophy (as it is known in the United States and Britain), including the writings of, say, Heidegger, Deleuze and Guattari, and Isabelle Stengers, with Alfred North Whitehead as an honorary Continental. This might be the place for me to remember my own field, and to point to a very active constellation of work in "posthumanist" science and technology studies as discovering the nonmodern character of the temporal evolution of modern science and engineering (Pickering 2008a).[10] Going in another direction, I have pointed to the many intersections between cybernetics and Eastern philosophy and spirituality: shared connections with the decentering of the self, the dance of Shiva as the dance of agency. We could also think of the transposition of the East to the West in the shape of New Age philosophy, with its erasure of the modern dichotomies of mind, body, and spirit. Before New Age there was the sixties. The counterculture may no longer be with us, but, like cybernetics, as an experimental form of life it offers us a whole range of models for future practices that also stage an ontology of unknowability and becoming.

The point I want to make is that cybernetics, narrowly defined as a historical entity, can be seen as part of much larger cultural assemblage. We could continue the list of its elements into the arts. We have examined many artworks that have been more or less explicitly associated with cybernetics, but there is an endless list of others that are in the same ontological space, and I will just mention a couple. Willem de Kooning's works immediately conjure up an ontology of decentered becoming. It is impossible to think of his rich, thick, and smudgy paintings as having been constructed according to some preconceived plan; one has to understand them as the joint product of a decentered and temporally emergent process involving a constitutive back and forth between the artist and the paint on the canvas. Some of Max Ernst's most haunting images began as tracings of the knots in the floorboards of his hotel room, which we can appreciate as another example of hylozoist ontological theater—another staging of the idea that it's all there already in nature, that the modern detour through detailed design can be unnecessary and can be curtailed in a process of finding out what works in the thick of things. Antonin Artaud's (1993) vision of the "Theatre of Cruelty" is just the sort of ontological theater we have been discussing here, but now instantiated literally as theater.

Moving to the sciences, we have explored some of the resonances and intersections between cybernetics and contemporary work on complexity, and I noted, at least, relations between the original cybernetic approach to understanding the brain and current work in brain science. Dating further back into history, theories of biological evolution again confront us with a spectacle of performative adaptation to an unknown future. The histories of these sciences cannot be reduced to that of cybernetics; they, too, can be thought of as part of the larger nonmodern assemblage that I am trying to delineate.

We should also think of engineering. In chapter 3 I explored connections between cybernetics and current work in situated robotics as exemplified by the work of Rodney Brooks, but I find it noteworthy that the dark side of modernity is beginning to be recognized within the field of engineering itself, and that new approaches are being developed there, so this might be the place to introduce one last example.[11]

My example concerns the civil engineering of rivers and waterways. The traditional approach is as usual a modern one, drawing upon science in seeking to make water conform to some preconceived human plan. The U.S. Army Corps of Engineers, for instance, has been fighting the Mississippi River for 150 years, seeking to contain its tendency to flood and to change direction, all in the name of maintaining the economic health of New Orleans and the Delta region (McPhee 1989). The devastating effects of Hurricane Katrina in 2005 should give us pause about this strategy (Pickering 2008b, forthcoming), but even before that some engineers had begun to think and act differently (Harden 2002): "Scientists know what is ailing the great rivers of America. They also know how to cure it. From the Columbia . . . to the Everglades . . .

they have been empowered . . . to take control of ecologically imperilled rivers that have been harnessed for decades to stop floods, irrigate farms and generate power. Instead of demolishing dams, they are using them to manipulate river flows in a way that mimics the seasonal heartbeat of a natural waterway. Scientists have discovered that a spring rise and a summer ebb can give endangered fish, birds and vegetation a chance to survive in a mechanized river." Here, then, we have a recognition within science and engineering that domination and enframing is not the one best way of proceeding, that we have other options, that it can be better to go with the flow—of water, time, and the seasons. Much of the Midwest of the United States was under water a hundred years ago. It was drained and converted to farmland by straightening the rivers and digging ditches to feed them. Now there is a "movement afoot to undo some of draining's damage," damage which includes wrecking entire ecosystems and wiping out enormous populations of fish and birds. "Even letting a short section of a ditch or channelized stream 'do what occurs naturally' and not maintain it can be very beneficial to fish and other wildlife." "This is science in its infancy," a geography professor is quoted as saying. "It's a mixture of science and trial-and-error. We're good in ways we can command and control a stream. We're not good at figuring out ways to make it a complex system in which nature can function" (Pringle 2002).

More positively, if the Army Corps of Engineers acts in a command-and-control mode, there also exists a field called adaptive environmental management which aims instead to explore and pay attention to the performative potential of rivers. Its stance toward nature, as in the above quotation, is experimental. Asplen (2008) gives the example of experimental floods staged on the Colorado River, in which scientists monitor the ecological transformations that occur when large quantities of water are released from an upstream dam—as a way of exploring the *possibilities* for environmental management, rather than trying simply to dictate to nature what it will look like.[12] Here in the heartland of modern engineering, then, we find emerging a nonmodern stance of revealing rather than enframing, which we can assimilate to the overall nonmodern assemblage that I have been sketching out.[13]

Where does this leave us? This latest list is another way of trying to foster the idea that modernity is not compulsory, that there are other ways of going on that make sense and are worth taking seriously—an attempt to put together a more encompassing gestalt than that assembled in earlier chapters, which can offer a bigger "quantitative" challenge to the hegemony of modernity (Pickering 2009). This in turn, of course, raises the question of why we should start with cybernetics in the first place? Two answers are possible. One has to do with human finitude: it would take forever to write a history of this entire assemblage; one has to start somewhere; and cybernetics seemed, and still seems to me, a perspicuous entry point. Second, I have learned something in writing this book. I did not see in advance that all these examples could be grouped into a nonmodern assemblage. This reflects my lack of imagination, but it is also a function of the relative isolation and lack of interconnection between many of the elements I have just mentioned. Buddhism is usually thought of as a philosophical and spiritual system having implications for individual practice and conduct. It is hard to imagine (though I can do it now) that it might hang together with a certain approach to engineering. De Kooning and Ernst were just painters, weren't they? What can Rodney Brooks's robots have to do with Heidegger or Deleuze?[14] And Continental philosophy is just philosophy, isn't it?—words and representations (like science studies and *The Mangle of Practice*).

From this perspective, the appeal of following cybernetics in action is that it enables us to see interconnections between all these traditions, fields, and projects; to pick out their common staging of an ontology of unknowabilty and becoming; and, indeed, to pick out as resources for the future the strands from these traditions that have this resonance. The history of cybernetics shows us how easy it is to get from little robots to Eastern spirituality, brainwave music, complexity theory, and the Fun Palace.

One last remark. I have stressed the protean quality of cybernetics, the endless multiplicity of cybernetic projects, and I want to note now that the reference to multiplicity implies a recognition that these projects are not inexorably chained together. It is entirely possible, for example, to take Beer's viable system model seriously as a point of departure for thinking further about problems of social and political organization while admitting that hylozoism and tantrism are not one's cup of tea. You might think Heidegger is a load of incomprehensible rubbish and still be interested by situated robotics (and vice versa). An interest in cellular automata does not depend on fond memories of the sixties. As a challenge to the hegemony of modernity, all that is important is the idea that a nonmodern ontology is possible and can be staged in practice, not its specific historical staging in this field or that. Readers should not be put off if they dislike de Kooning or Ernst.

I look at this the other way around. A recognition of the relation between cybernetics and current work in complexity, robotics, and the civil engineering of rivers points to a material and conceptual robustness of this entire assemblage and helps me also to take seriously the wilder projects and artifacts we have examined: flicker machines, explorations of consciousness, tantric yoga, walking cities.[15] All of these might be of a piece with the utter sobriety of Ashby's phase-space diagrams and calculations of time to equilibrium. From this perspective, too, this book has been an attempt to counteract a narrowing of our imaginations—of what there is in the world, what we are like, what we can be, and what we can do.

— — — — —

Where might an alternative to modernity flourish? Obviously, in our imaginations and in the projects that go with a nonmodern imagining. This book certainly aspires to contribute to that. But I want to return to the question of the social basis one last time. Where, institutionally, might cybernetics and its ilk grow in the future? I have two different but compatible thoughts on this. First, one cannot help but be struck by the social marginality of cybernetics throughout its history, and this has led me to an interest in new institutions, however marginal themselves, that have emerged as a social basis for cybernetics in recent decades. The cybernetic equivalents of schools and universities have turned out to be places like Wolfram Research, the Santa Fe Institute, the Planetary Collegium, and the Zentrum für Kunst und Medientechnologie, with Kingsley Hall, the Anti-University of London, and the sixties counterculture as short-lived models for something more radical, and New Age as a massive but somehow walled-off contemporary presence. However ephemeral these institutions have been or might prove to be, for most of the time I have been writing this book I have thought of them—or nonstandard institutions like them—as the future home of cybernetics. I have referred to this possibility from time to time as a parallel social universe—an institutional space where cybernetics might reproduce itself and grow, quite apart from the usual modern instutions of cultural production and transmission—much as Trocchi imagined his sigma project in the sixties.

From that perspective, one aim of this book has been to incorporate this other social world into the overall picture of the nonmodern assemblage I have been trying to put together. Just as I have been trying to show that the cybernetic ontology and cybernetic projects and objects make sense and are worth taking seriously, so my suggestion is that we should take seriously the sometimes odd institutions which have from time to time supported them. I would like to launder these institutions into mainstream discourse and consciousness as well as more specific aspects of cybernetics. The other future I am trying to imagine has this odd social aspect too; the growth of this parallel social world might indeed be an important aspect of the challenge to modernity.

But while I have been writing this chapter, another line of thought has come upon me. As summarized above, it is clear that many cybernetic endeavors are strongly incompatible with their modern equivalents. It is indeed hard to imagine Kingsley Hall not existing in tension with conventional psychiatry. But just at the moment I can see no principled reason why something like what is laid out in this book could not be taught at schools and universities and even feature prominently in their curricula. Let me end with this.[16]

It is true, I think, as I said earlier in this chapter, that Western educational systems are strongly oriented toward the transmission of positive knowledge, and that this hangs together with a modern ontology of knowability, controllability, enframability. But, oddly enough, there are few people who would defend the proposition that this is a total description of the world. Everyone knows that surprise is a distinguishing mark of science, and that unintended consequences go with engineering initiatives. At the moment, however, such knowledge is, well, marginal. It doesn't amount to much; it continually disappears out of the corner of our eyes. And my suggestion is that it does not have to be this way, even in our pedagogical institutions. We could bring unknowability into focus for our children.

How? Here I think first about what I have been calling the discovery of complexity. Cellular automata are not at all difficult to draw—by hand, never mind with a computer—and even young children can enjoy the patterns they make. I am sure that there are already schools that incorporate them in their curriculum, but it seems highly likely that they do so as a subfield of mathematics instruction—here is yet another branch of mathematics, in a sequence that would include arithmetic, geometry, groups, or whatever. My suggestion is that one could incorporate cellular automata into the curriculum differently, along the lines of how I have discussed them here, as ontological icons, little models of a world graspable in terms of a nonmodern ontology.[17] *That* would be the point of learning about them in the sort of courses I am imagining. Many of the artifacts we have examined along the way would likewise be discussable in this way at school. Gysin's Dream Machines might similarly offer a way into thinking about the explorability (rather than the givenness) of our perceptions and, by extension, our selves. School trips might include going to look at interactive art in local museums, again explicitly discussed as ontological theater, and not: here's a Rembrandt, here's a Picasso, here's a tortoise.

Beyond that, of course, the history of cybernetics (and all that I assimilated to it a few pages ago) offers a wonderful and teachable set of examples that show that we are perfectly capable of going on in a world of unknowability and becoming, that there is nothing paralyzing about it. The more musical and theatrical students might enjoy playing a Musicolour machine (simulated on a computer) more than the trombone, say. Do-it-yourself kits for making tortoiselike robots are relatively cheap, and students could explore their emergent patterns of interaction. One can easily simulate and play with a multihomeostat setup. Some sort of quasi-organic architectural design methods might be more fun than many of the uses computers are conventionally put to in school. I am in danger of rehearsing the contents of this book yet again, and, well, more advanced students could also try reading it.

The more I think about it, the more promising this idea becomes.[18] In line with my earlier remarks on variety, I am certainly not proposing the unthinkable, to rid the curriculum of its modern elements. Nothing in this book, I repeat, threatens modernity, only its taken-for-granted hegemony and universality. I am suggesting the inclusion of a series of courses in schools and universities that would figure prominently in the curriculum, explicitly conceived as relating to a nonmodern ontology. I teach in a university, not a school, and I would be prepared to argue for one such course as a requirement for all freshmen, whatever their field.[19]

I am, in the end, very attracted to this idea of systematically introducing students to a nonmodern ontology, beginning at an early age. If done right, it could easily produce a generation that would automatically say "wait a minute" when presented with the next high-modernist project of enframing, who would immediately see the point of Latour's "politics of nature," and who would, moreover, be in just the right position to come up with new projects that center on revealing rather than enframing. I would enjoy sharing a world with people like that.

NOTES

Notes to Chapter 1

1. On the Macy conferences—so called because they were sponsored by the Josiah Macy, Jr., Foundation—see especially Heims (1991) and Dupuy (2000). These meetings did much to sustain the development of American cybernetics under the permanent chairmanship of Warren McCulloch. Their proceedings were published from 1949 onward under the title *Cybernetics: Circular Causal, and Feedback Mechanisms in Biological and Social Sciences*, edited by Heinz von Foerster, joined from 1950 onward by Margaret Mead and Hans Lukas Teuber. The proceedings have recently been republished with valuable ancillary material by Pias (2003, 2004).
2. See, for example, Bowker (1993, 2005), Conway and Siegelman (2005), Cordeschi (2002), Dupuy (2000), Edwards (1996), Galison (1994), Haraway (1981–82, 1985), Hayles (1999), Heims (1980, 1991), Kay (2001), Keller (2002), Mindell (2002), Mirowski (2002), Richardson (1991), and Turner (2006). Much of this work focuses on the history of cybernetics in the United States, and there is relatively little discussion of cybernetics elsewhere.
3. On McCulloch's enduring concern with the brain see McCulloch (1988, 2004), Heims (1991, chap. 3), and Kay (2001). In the 1930s, McCulloch worked at the Rockland State Hospital, a mental institution, and from 1941 to 1952, the key years in the initial development of cybernetics, he was director of research in the Department of Psychiatry of the University of Illinois Medical School in Chicago. In 1952, he moved to the Research Laboratory of Electronics at MIT, and he remained there for the rest of his working life. Wiener had no background in brain science, but the key paper in the genesis of his cybernetics, Rosenblueth, Wiener, and Bigelow (1943), moved immediately from the

autonomous weapons systems he had worked on earlier to a general analysis of purposive behavior in animals and machines.

4. McCulloch, Verbeek, and Sherwood (1966, ix–x) align Craik with James Clerk Maxwell, who numbered, among many scientific achievements, his theoretical analysis of the steam-engine governor.
5. The Research Laboratory of Electronics was the successsor to the wartime Rad Lab at MIT; on its history see Wildes and Lindgren (1985, chap. 16, which includes much on Wiener, Shannon, and information theory but covers the work of McCulloch and his group in a single paragraph on p. 263, without using the word "cybernetics") and www.rle.mit.edu/about/about_history.html (which mentions the work of Norbert Wiener, Claude Shannon, Jerome Lettvin, and Walter Pitts but says nothing about McCulloch). The Illinois Biological Computer Laboratory and the Brunel Cybernetics Department are discussed further in chapters 4 and 6, respectively. In Britain, two other departments of cybernetics have existed, one at the University of Reading, the other at the University of Bradford (which closed down in summer 2006), but neither of these enters our story.
6. For a recent encapsulation of the critique, see Harries-Jones (1995, 104): "Social critics across the political spectrum, from the left-leaning Jürgen Habermas to the right-leaning Jacques Ellul, have identified cybernetics and the systems concepts derived from it as the foremost ideology of the military-industrial technocracy that threatens our planet." The point of Harries-Jones's summary is to exempt Bateson from the critique. If one follows Harries-Jones's citations, one finds the young Habermas (1970, 77, 117–18) bivalent: enthusiastic about cybernetics as a medium of interdisciplinary communication between otherwise isolated scientific specialties, but anxious about a "cybernetic dream of the instinct-like self-stabilization of societies." The latter follows a list of possible future technological innovations that begins with "new and possibly pervasive techniques for surveillance, monitoring and control of individuals and associations" and is thus an instance of the "control" critique discussed at some length in the next chapter. We are, of course, monitored much more intensively than we were in the late sixties, but we do not have cybernetics to thank for that. One finds the later Habermas (1987) struggling with the sociology of Talcott Parsons, whose work did incorporate a cybernetic element but is hardly exemplary of the strands of cybernetics at issue in this book. Ellul (1964) mentions cybernetics from time to time, but only within the flow of his overall critique of the "technological society" and the reduction of human life to "technique." Our examples of cybernetics fit Ellul's concept of technique exceedingly poorly.
7. Though industrial automation is often cited as a paradigmatic instance of the Marxist deskilling thesis, the situation was often seen differently, at least in postwar Britain and France. There, automation was widely imagined to hold out the welcome promise of relieving the human race of many of the demands of repetitive wage labor (while giving rise to the "leisure problem"—what would people do with all their spare time?). More on this in chapter 7. The idea seems quaint and touching now—how could anyone think that?—but it is worth remembering that automation once had this positive valence, too.

Notes to Chapter 2

1. Continuing a line of thought from chapter 1, I should note that not everything that can be counted as cybernetic speaks to the ontological concerns at issue here. The interest, for me, of the British cyberneticians is precisely that their work draws one in an ontological direction. Again, there are aspects of the work of Walter, Ashby, and others that do not obviously engage with my ontological concerns, and I will not go into them in any detail. Ontological interest is thus another principle of historical selection that informs this book.
2. "Modern" is a word with more resonances and associations than I need, but I cannot come up with a better one. This paragraph and those that follow try to define my own usage. In the history of Western philosophy, Descartes's articulation of the duality of mind and matter—the idea that they are made of different stuff—was a key moment in the history of modernity thus conceived. Likewise the Scientific Revolution: Newton's laws of motion as specifying regular properties of matter, independent of any human knower; also the Enlightenment's emphasis on reason as a special and defining property of the human. Our scientific and commonsense stances toward the world remain, I think, largely within the space thus defined, and hence they are "modern" on this definition.
3. The dominant strand in twentieth-century philosophy of science portrayed the material world as a passive substrate which was the origin of observation statements, themselves either the basis for scientific induction or a deductive constraint on theory building. On such a view, there is no space to think about the performative aspects of the world and our constitutive entanglement with them in knowledge production.
4. This characterization of modern science is crude, though not, I think, misleading. A properly nuanced and exemplified discussion would require another book. I should stress therefore that my object here is not to do full justice to modern science, but to set up an illuminating contrast that will pick out what I take to be key features of cybernetics—and these are the topic of this book. Thematizing this contrast serves to intensify my own tendency to dichotomize, while, from the cybernetic side, too, there are nuances that might be considered to blur the distinction—some of which are discussed below. Canguilhem (2005 [1983]) notes the different ontologies espoused by scientists and historians of science—one fixed, the other fluid—but does not develop this observation, treating it as simply a fact about the two camps; I thank Hans-Jörg Rheinberger for bringing this essay to my attention.
5. This echoes an argument due to van Gelder (1995).
6. From yet another angle, the suggestion in the following chapters is not that ontological claims came first and somehow gave rise to specific cybernetic projects, and neither is it the reverse. I am interested in how the ontology and these projects hung together and informed one another.
7. Thus, the argument of *The Mangle of Practice* was that the material world as explored by modern science is itself an exceedingly complex system in Beer's terms. The scientists, however, adopt a peculiar stance in that world, construing

their experience in terms of fixed entities (whose ascribed properties turn out themselves to evolve in practice). From my ontological perspective, the modern sciences read the world against the grain—and *The Mangle* undid this by reading the sciences against their grain.

8. In the second postscript to *The Mangle* I floated the idea that my analysis might be a Theory of Everything, as the physicists like to say of their theories. Later I realized that the idea of a theory of everything could be understood in two rather different ways. The physicist's sense is that of a complete mathematical theory from which all worldly phenomena can be deduced, in principle at least. This is the sense of a theory of everything as the end of science—all that remains is to fill in the details. The nonmodern ontology of *The Mangle* instead suggests an always infinite horizon of constructive engagement with the world—in modern science as a never-ending finding out; in cybernetics as an endless staging of the ontology in this situation or that, in this way or that.
9. In Walter's work especially, the interest in strange performances and altered states can be seen as part of an attempt to map input-output relations of the brain as a Black Box, and I can make another connection back to *The Mangle* here. In the first postscript to the book I commented on the apparently never-ending range of material powers and performances that have manifested themselves in the history of science, and I contrasted this with the apparently unvarying historical parameters of human agency. In an attempt to symmetrize the picture, I mentioned a couple of examples of "non-standard human agency" but was unable to offer any very convincing documentation. I did not know then that cybernetics itself had plunged into explorations in this area, and I had forgotten that the sixties were a golden age for the same sort of experimentation in everyday life.
10. Kauffman (1971) explicitly identifies this style of explanation as "cybernetic"; I return to his work below.
11. The preceding paragraphs are not meant to suggest that the cyberneticians were the first to discover the existence of unpredictably complex systems. The three-body problem was around long before cybernetics. The argument is that the discovery of complexity in cybernetics emerged in a specific way within the frame of key projects. Furthermore, we will see that the cyberneticians addressed this problematic in very different ways from the authors usually associated with "complexity," including Kauffman and Wolfram. For popular accounts of the recent history of work on chaos and complexity, see Gleick (1987) and Waldrop (1992), neither of whom mentions cybernetics.
12. For the pre-Katrina version of this story, see Pickering (2002); for a post-Katrina version, Pickering (2008b).

Notes to Chapter 3

1. I am indebted to Rhodri Hayward for much enlightenment on Walter and the relevant literature, including his own writings prior to publication. I have drawn upon two sets of archival holdings below, both in London: John Bates's papers at the Wellcome Library, and the papers of the Burden Neurological Institute at the Science Museum.

2. This biographical sketch is based on Hayward (2001a) and Holland (2003).
3. Walter spent the first six months of his Rockefeller fellowship in Germany, including a visit to Hans Berger in Jena (see below), and he was then based at the Maudsley and at the Hospital for Epilepsy and Paralysis in Maida Vale (Wellcome Library, GC/179/B.35, p W.2a; Walter 1938, 6). On Golla and the Maudsley, see Hayward (2004, forthcoming).
4. Walter (1966) acknowledges the importance of his collaboration with the electrical engineer Geoffrey Parr. At the time of Walter's earliest EEG work, Parr was "head of the special products division of Edison Swan and could let me have special valves. . . . He also had his own workshop at home and every time he came to see me he brought a little present—a specially mounted and calibrated meter, or a set of matched resistors or an engraved graticule." Parr and Walter collaborated on the development of Britain's first commercial electroconvulsive therapy machine (below). As the editor of *Electrical Engineering* it was Parr who encouraged Ashby to play up the futuristic aspects of his first publication on the homeostat (next chapter).
5. On Walter's place in the institutional history of EEG research, see Cobb (1981).
6. Besides Golla and Walter, the research staff comprised Dr. Reiss, a physiologist, and Mr. Tingey, a chemist (salaries £350 and £250, respectively), plus a laboratory mechanic, a laboratory boy, and a woman dispenser and clerk. A clinical director, Dr. E. L. Hutton, was appointed in November 1939 at £800 a year (on a par with Walter). In 1949, the Burden had eighteen beds for inpatients at eight guineas per week, and Golla himself established a clinical practice that "straddle[d] the boundary between neurology and psychiatry" (Cooper and Bird 1989, 15–16, 21, 61).
7. Sometimes referred to as *The Swerve of the Cornflake*. The snowflake of the title is actually Walter's word for what would now be known as a fractal geometry, which allows characters in the book to travel into the future.
8. Another story branches off here, that of Walter's determined attempts to correlate EEG spectra with human personality traits. See Hayward (2001a).
9. Of course, any materialist conception of the brain is liable to destabilize this split, including the prewar experimental psychology mentioned below.
10. Walter employed this tactic of electromechanical modelling frequently. Besides the tortoise, CORA, and the pattern-tracing device mentioned in note 11, *The Living Brain* includes an appendix on an artificial nerve that Walter constructed (1953, 280–86). In 1954, Walter (1971 [1954], 40–44) displayed a gadget that he call IRMA, for innate releasing mechanism analogue, that he had built following a 1953 visit to the Burden by Konrad Lorenz. IRMA, like CORA. was intended as an attachment to the tortoise, and one begins to glimpse here a more extensive cyborgian project of building a realistic synthetic animal.
11. Yet another of Walter's electrical contraptions further illuminated scanning. "The manner in which such a system might work in the brain is illustrated in a device which we had developed for quite another purpose, for reconverting line records, such as EEG's, back into electrical changes" (Walter 1953, 109–10). This device consisted of a cathode-ray oscilloscope coupled via an amplifier to a photoelectric cell which viewed the oscilloscope screen. The

spot was arranged to oscillate along the *x* axis of the screen—scanning—and the circuitry was cleverly arranged so that along the *y* axis the spot would trace out the edge of any opaque object placed in the screen. Thus, a geometric image was converted into a time-varying electrical signal.

12. Walter later interpreted the other brain rhythms as also modes of adaptive search: the delta rhythms were associated with a search for order and stability; thetas were a search for specific pleasurable entities. See Hayward (2001b).
13. In *Cybernetics* Wiener also discussed the alpha rhythms as a scanning mechanism, citing Walter, and he developed this idea further in the second edition (1948, 141; 1961, 198). See also John Stroud's contribution (1950) to the sixth Macy conference.
14. The protoypical version of this argument is to be found in Rosenblueth, Wiener, and Bigelow's foundational essay "Behavior, Purpose and Teleology" (1943). Rosenblueth et al. suggest that purposeful behavior in humans, animals, and machines can be understood in terms of a single model: the servomechanism with negative feedback. The tortoise can be seen as a specific instantiation of this idea.
15. The notion of exploration is important here and marks a difference between the cybernetic devices we will be examining and earlier ones such as the thermostat and the steam-engine governor. Those earlier devices did not interrogate their environments; they were "hard wired" to respond to certain features: if the temperature goes up, the thermostat automatically turns the heating down, and so on. The tortoise, in contrast, went looking for specific elements in the environment, like lights. This, of course, contributed to the lively and lifelike quality of its behavior, and its effectiveness in undermining the boundary between animals and machines.
16. The tortoise's predecessors and contemporaries typically did not have the opportunity to display this kind of inscrutable variability. The prewar brain models in experimental psychology discussed by Cordeschi (2002) usually aimed to emulate a narrowly circumscribed range of performances, such as the adaptation of the eye to varying light intensities or simple phototropism, and they either succeeded or failed in that predetermined task—likewise the cybernetic maze-running robots of the early 1950s mentioned above.
17. The individual mechanical and electrical components of the tortoise were likewise nonadaptive—the electronic valves, say, did not change their properties in the course of the tortoise's explorations. Adaptation consisted in variations of the interconnections of parts. It is, of course, hard to imagine building any fully adaptive system. Models for the latter might be biological evolution, including the evolution of the nonbiological environment, and the coevolution of science, technology, and society.
18. Strictly speaking, the tortoises were capable of modifying their environment, but only in a simple and mechanical fashion: small obstacles moved when the tortoises bumped into them, which helped the tortoises to navigate past them. And in multitortoise configurations, the tortoises constituted lively environments for each other, as in the mating dance. But Walter's writings thematize neither of these observations, and it is better to postpone this phase of the ontological discussion to the next chapter.

19. Craik was the author of a highly original book on philosophy and psychology and of a posthumous collection of essays (Craik 1943, 1966). Craik's essay "Theory of the Human Operator in Control Systems" (1947) derives from his wartime research and outlines a detailed vision of the mechanical re-creation of the human operator of a weapons system. On Craik and his relation to cybernetics in Britain, see Hayward (2001b, 295–99), Clark (2002), Gregory (1983), and Zangwill (1980). John Stroud's paper (1950) at the sixth Macy conference attests to the importance of Craik's work in the history of U.S. cybernetics.
20. The quote continues, "But it will be a worse 'animal' for though it will keep more closely to its beam it will have to be aimed roughly in the right direction and will not 'speculate'—that is, spy out the land—nor will it solve Buridan's dilemma [of choosing between two targets]." I think "the usual way" here harks back to the prewar tradition in experimental psychology of constructing phototropic robots. As discussed by Cordeschi (2002), such robots typically used a pair of photocells to home in on their targets.
21. For more on this, see Holland (2003, 2090–91), which includes a description of Walter's workshop.
22. In this connection it is also profitable to read Warren McCulloch's biographical account of his own route to cybernetics: McCulloch (2004).
23. On the Burden Neurological Institute's transformation during World War II, Cooper and Bird (1989, 14) record that "less than six months after the opening ceremony [at the Burden] war started and the Emergency Medical Service used the Institute as a neurological hospital for the whole of the West Country, it possessing the only neurosurgical theatre West of London. Despite the strain due to the requirements of the neurosurgical unit, the laboratories continued to function as centres for clinical research in neurology and psychiatry." Walter was presumably exempt from military service during the war by virtue of his occupation. One can get some feeling for his work during the war from an undated handwritten document by him entitled "The Genesis of Frenchay" (Science Museum, BNI papers, 6/38). Frenchay was a hospital built close to the Burden, "as a sort of reverse land-lease project to provide the U.S. Army Medical Corps with a clinical service not too far from the rapidly advancing front after D-Day. . . . My task was to help the neurologists & neurosurgeons by taking EEGs with a home-made portable in cases of head-wounds." After the Battle of the Bulge, "I can still see in vivid horror what is now Ward 2 with beds touching, and the pressure on, not only for treatment, but to get the men back to the States as soon as possible for local morale over there." Besides his EEG work, "there were also cases of 'battle fatigue' in the Hospital (what we called 'shell shock' in the first War) and I was engaged to help with these too, by giving ECT."
24. In a quite different context, the history of the electronic synthesizer, Pinch and Trocco (2002, 280) make the same point about the importance of postwar army-surplus electronics, referring specifically to Lisle Street in London, "just full from one end to the other of second-hand, ex-Army, electronic stores."
25. In another field, psychology, Richard Gregory's *Mind in Science: A History of Explanations in Psychology and Physics* (1981), refers only briefly to Walter, mainly in reproducing the circuit diagrams for the tortoises and CORA (286, 287).

(His reference to Ashby is even briefer [82], and there is no mention of Bateson, Laing, Beer, or Pask.) We should note that Walter did find some support at the Burden. As noted, it was Bunny Warren, a Burden engineer, who built the machines that Walter displayed in the 1950s. But there is no indication of any sustained support for Walter's work there. A book on the history of the Burden coauthored by Ray Cooper, who worked there from 1955 onward and was director from 1971 until his retirement in 1988, devotes a chapter to Walter's work on the electrophysiology of the brain but mentions the tortoises only in passing, with the comment: "Cybernetics suffered from the fact that it was fine in theory [*sic*] but whenever a practical application was attempted the whole thing appeared to be facile. . . . A pity, for some general theory of brain function (if one exists) would be most welcome" (Cooper and Bird 1989, 45).

26. Bates's letter and Walter's reply are in the Wellcome Archives, GC/179/B.1.
27. This information is taken from an excellent PhD dissertation, Clark (2002). See Clark (77–80) on the Ratio Club; the quotation from Walter is from a letter to Bates dated 29 September 1949 (Clark 2002, 80n221). Clark includes an appendix collecting archival documentation on the Ratio meetings (207–15). At the meeting on 22 February 1951, Walter gave the address "Adaptive Behaviour" and demonstrated the tortoises; on 31 May 1951, Ashby delivered "Statistical Machines" (the homeostat, etc.); on 6 November 1952, Walter discussed Ashby's new book, *Design for a Brain*; the meeting on 2 July 1953 again featured a visit by McCulloch; and the meeting of 6 May 1955 included a demonstration at Barnwood House (Ashby's place of work) and lunch at the Burden Institute (Walter's).
28. The Ratio Club's membership list grew somewhat relative to the people named in Bates's letter to Walter, the most eminent addition being the mathematician and computer pioneer Alan Turing, invited by Bates in a letter dated 22 September 1949 (Clark 2002, 80; on Turing, see Hodges 1983). If one focuses on the Ratio Club as a whole, the case for a military origin of British cybernetics looks much stronger, and Clark makes such a case (2002, 77): "Craik appears . . . as the lost leader of a group of physiologists and psychologists, who as a consequence of their wartime redeployment, developed an interest in electronics and control mechanisms. The interest in electrical mechanisms in the nervous system was not new, it had been established before the war. But the redirection of research imposed by necessity, moved it toward engineering technology rather than pure science, something that, but for this wartime demand, might not have had such comprehensive influence." It is worth emphasizing therefore that neither Walter nor Ashby was mobilized as part of the war effort. Walter continued his work at the Burden throughout the war (n. 23 above) and Ashby worked at Barnwood House mental hospital until June 1945, when he was called up for military service in India.
29. The lack of sociologists was in contrast to the important presence of social scientists at the Macy conferences, including Margaret Mead and Gregory Bateson. In this respect the U.S. cybernetics community was even more markedly interdisciplinary than the British one. See Heims (1991).
30. Echoing the heterogeneity of these institutions, Cooper and Bird (1989, 20) note that "throughout the history of the [Burden] Institute there have been

serious attempts to establish some formal links with the University of Bristol. For all kinds of reasons, some good some bad, often to do with finance, these never succeeded."

31. At the first Namur conference, twenty-four papers were presented in the session "Cybernetics and Life," at which Walter presided. Several of the authors are listed in the *Proceedings* simply as doctors of medicine; others as neuropsychiatrists and neurosurgeons; Reginald Goldacre gave the Royal Cancer Hospital in London as his affiliation. Academics came from departments including electrical engineering, physics, physiology, and philosophy. Two authors were based in the cybernetics group of the Max Planck Institute for Biology in Tübingen. Albert Uttley listed the British National Physical Laboratory; French authors recorded affiliations to the Fédération Nationale de l'Automation, the Conseil en Organisation électronique et nucléaire pour l'Industrie, and the CNRS.
32. Other robots modelled on the tortoise included an American robot squirrel called Squee, built by Edmund Berkeley (Berkeley 1952), dubbed *M. speculatrix berkeleyi* by Walter (1953, 132), and "la tortue du Vienne," built by Heinz Zemanek, an engineer at the Technical University in Vienna, and exhibited at the first Namur conference (Zemanek 1958). Zemanek also exhibited a copy of Ashby's homeostat. I thank Garnet Hertz for making me aware of Berkeley's manuscript.
33. In his early work in science studies, Harry Collins (1974) emphasized how difficult it is to replicate an experiment with only a published description to rely upon and how important personal contacts are. My point concerning cybernetics is this obverse of this.
34. See people.csail.mit.edu/brooks/.
35. On Allen, see Brooks (2002, 32–44). For a technical discussion of Brooks's robotics at this time, including subsumption architecture and Allen (not named), see Brooks (1999 [1986]).
36. Though not central to our story, a key event here was the extinction in the 1960s of funding for neural-network research stemming from the work of McCulloch and Pitts (1943), in favor of support for symbolic AI: see Olazaran (1996). Like Brooks's robotics, neural networks, too, emerged from the shadow of representational AI in the 1980s. The contrast between cybernetics and AI is sometimes phrased as that between analog and digital computing. Historically this stands up, but the continuity between Walter's analog machines and Brooks's digital ones suggests that the key difference does not lie there, but rather in the overall nature of the projects and their respective emphases on adaptive performance or symbolic representation.
37. The standard AI response is, of course, wait for the next generation of processors.
38. Brooks explicitly conceived the contrast between his approach and mainstream robotics as that between performative and representational approaches: see his essays "Intelligence without Reason" (1999 [1991]) and "Intelligence without Representation" (1999 [1995]).
39. At the time of writing, details of the meeting are still to be found online at www.ecs.soton.ac.uk/~rid/wgw02/first.html. The proceedings were published

as Damper (2003) but do not cover the full diversity of presentations (a keynote talk by Brooks is missing, for example). It is significant to note that the revival of this style of robotics coincided in the mid-1980s with a revival of neural-network research in AI (n. 36 above), and, as is evident in the proceedings, neural networks have become an important feature of robot development. Here two branches of the original cybernetic synthesis as laid out by Wiener have been reunited in a symbiosis of research in engineering and brain science that echoes but goes far beyond Walter's work. A glance at the proceedings of the conference makes clear that the resurrection of cybernetics in robotics was not the work of Brooks alone; I discuss him as a conspicuous and illuminating example of this process. At much the same time as Brooks changed his style in robotics, Valentino Braitenberg published a book called *Vehicles: Experiments in Synthetic Psychology* (1984), which reviewed the imagined performances of fourteen different conceptual variants of tortoise-style robots, each of which could mimic distinct psychological performances (from "getting around," "fear and aggression," and "love" up to "egotism and optimism"). When he wrote the book, Braitenberg was director of the Max Planck Institute of Biological Cybernetics in Germany.

40. More could be said on how Brooks and others managed to redirect their research fields, never an easy task in academia. The simple answer is that the performativity of Brooks's robots was attractive to outside funding agencies, which, in the United States at least, is a powerful argument in the university and makes it possible to support and train graduate students and postdoctoral fellows. Thus, the long list of acknowledgments in Brooks (2002, vii–viii) begins with "all my students over the years who have contributed to building robots and helping to solve the problems of making life-like behavior from nonlifelike components," immediately followed by "my sponsors at DARPA (the Defense Advanced Research Projects Agency) and the Office of Naval Research, and more recently from NTT (Nippon Telegraph and Telephone Corporation), who have had patience and faith over the years that something good would come of my crazy ideas." Information on Brooks's company, iRobot, can be found at www.irobot.com/home.cfm. Until recently, one product of iRobot was a robotic doll for children, but, echoing the failure to turn the tortoise into a commercial toy, this no longer appears on the website. The products now listed are Roomba, a robot vacuum cleaner, and PackBot, "a portable unmanned vehicle [which] is helping to protect soldiers." Exploration of this website reveals that much of the research and development at iRobot is funded by U.S. military agencies and directed toward military ends.

41. To take this thread of the story a bit further, we could note that much current research on the mind, brain and consciousness is recognizably in the Walter-Ashby model-building tradition. See, for example, Edelman (1992), who, despite considerable historical erudition, manages not to mention cybernetics.

42. Walter (1951, 62): "This process may of course be accelerated by formal education: instead of waiting for the creature to hit a natural obstacle the experimenter can blow the whistle and kick the model. After a dozen kicks the model will know that a whistle means trouble, and it can thus be guided away from danger by its master."

43. The "great Pavlov" quotation is from Walter (1966, 10), which speaks of "a period when I was working in Cambridge under the direction of the great I. P. Pavlov."
44. There are some further interesting ideas about CORA and memory in Walter (1951, 63): "In *M. Docilis* the memory of association is formed by electrical oscillations in a feedback circuit. The decay of these oscillations is analogous to forgetting; their evocation, to recall. If several learning pathways are introduced, the creature's oscillatory memory becomes endowed with a very valuable feature: the frequency of each oscillation, or memory, is its identity tag. A latent memory can be detected and identified by a process of frequency analysis, and a complex of memories can be represented as a synthesis of oscillations which yields a characteristic wave pattern. Furthermore a 'memory' can be evoked by an internal signal at the correct frequency, which resonates with the desired oscillation. The implications of these effects are of considerable interest to those who study the brain, for the rhythmic oscillation is the prime feature of brain activity."
45. At the first Namur conference in 1956, Heinz Zemanek's replication of the tortoise included a very much simplified version of CORA, which used a resistance with a negative temperature coefficient instead of the sophisticated differentiating and integrating circuits devised by Walter (Zemanek 1958).
46. We could return here to the cybernetic discovery of complexity. Like the tortoise, the CORA-equipped tortoise displayed emergent properties that Walter had not designed into it, some of which he was not able to explain (1953, 180–82). He constructed a complicated after-the-fact explanation of *M. docilis*'s display of different patterns of conditioning, analogous to "appetitive and defensive reflexes" in animals, when sounds were associated with lights or with obstacles, but also found that "such models show inexplicable mood changes. At the beginning of an experiment the creature is timid but accessible, one would say, to gentle reason and firm treatment; later, as the batteries run down there is a paradoxical reversal of attitude; either the reflex or the acquired response may be lost altogether, or there may be swings from intractability to credulity. Such effects are inevitable; however carefully the circuits are designed, minute differences and changes are cumulatively amplified to generate temperaments and tempers in which we can see most clearly how variations in quantity certainly do, in such a system, become variations in quality." *Docilis*, like *speculatrix*, can thus be seen in itself as an instructive ontological icon, demonstrating again that systems of known parts can display unexpected behavior, remaining, in this sense, Black Boxes knowable only in their performances.
47. "Of course," he added, "were the model an engine for guiding a projectile or regulating the processes of an oil refinery, this tendency to neurotic depression would be a serious fault, but as an imitation of life it is only too successful."
48. In 1954 Walter recalled that in Cambridge in 1934, working with Pavlov's student Rosenthal, he had mistakenly imposed inconsistent conditioning regimes on experimental dogs: "One of the five dogs retained a quite reasonable conditioned reflex system, two of them became inert and unresponsive, and two became anxious and neurotic" (1971 [1954]. 54).

49. Thus, as one would expect, cybernetics emerged from a distinctively performative approach to psychiatry (in contrast, say, to the inherently representational approach of psychoanalysis). On the history of psychiatric therapies in this period, see Valenstein (1986) and Shorter (1997). Sleep therapy in fact developed earlier than the others and was hardly as benign as Walter suggests. It consisted in inducing sleep lasting for days or a week, using first bromides and then barbiturates. Shorter mentions a study in the late 1920s that found a 5% mortality in sleep cures. William Burroughs (who reappears below) includes sleep therapy in his catalog of the cures he had attempted for heroin addiction: "The end result was a combined syndrome of unparalleled horror. No cure I ever took was as painful as this allegedly painless method. . . . After two weeks in the hospital (five days sedation, ten days 'rest') I was still so weak that I fainted when I tried to walk up a slight incline. I consider prolonged sleep the worst possible method of treating withdrawal" (Burroughs 2001 [1956]. 219). In British psychiatry, the most important of the desperate cures was ECT, first used by the Italian psychiatrist Ugo Cerletti in 1938.
50. The report acknowledges that Walter's attempts at EEG research had not borne fruit in the form of scientific publications and focuses on clinical applications of EEG techniques in the localization of brain tumors and explorations of epilepsy, from which "disappointingly meagre conclusions can be drawn" (Walter 1938, 11). One of Walter's "first jobs" at the Burden "was to set up a clinical EEG department—the first in the country—in which epilepsy could be further studied" (Cooper and Bird 1989, 67).
51. "Golla wishing to assess the method [ECT], got Grey Walter to make the necessary equipment. . . . After trying it out on a sheep which had 'neat fits' the first ECT in Great Britain was done in the Institute in 1939" (Roy Cooper, "Archival Material from the Burden Neurological Institute, Bristol," May 2000, 31: Science Museum, BNI archive §6).
52. The author unmentioned so far, Gerald William Thomas Hunter Fleming, was physician superintendent of Barnwood House, Gloucester, a mental hospital close to the Burden, where Ashby was employed from 1947 to 1959 and did much of his foundational work in cybernetics. Fleming reappears in the next chapter in connection with Ashby's brief and disastrous appointment as director of the Burden.
53. See note 23 above. Walter (1938, 12–13) also mentions an experimental lobotomy. "Aware at the same time how flimsy was the excuse for intervention, Dr. Golla, Mr. Wylie McKissock and myself decided to operate on one of Dr. Golla's patients at Maida Vale." Part of the superior frontal gyrus of an epileptic patient was removed, and the patient was then free from fits for six weeks, after which they returned. Walter comments that "these cases are reported here . . . to illustrate the disappointments which are likely to attend the extension of electro-encephalography from the purely diagnostic to the therapeutic field, and also to indicate the futility of forming pathological hypotheses while the electro-physiology of the cortex is still in such an undeveloped state."
54. More narrowly, as we have seen, we also need to think about a tradition of electromechanical modelling in experimental psychology, Pavlovian behavior-

ism, Walter's specific interest in EEG research, and so on—and, indeed, World War II appears in this story, too, though not as the key element which it was for Wiener.

55. Chlorpromazine was approved for sale under the name Thorazine by the U.S. FDA in May 1954, and Starks and Braslow (2005, 182) provide revealing statistics on the use of different therapeutic regimes before and after this date at Stockton State Hospital, California's oldest mental institution. "In fact, antipsychotic drugs proved so effective in therapeutically eliminating recalcitrant, hostile, and violent behaviors that lobotomy disappeared almost overnight." Up to 1953, 63% of psychotic patients were treated with ECT; after that, 9% received ECT only, 28% a combination of ECT and drugs, and 47% drugs only.
56. This work might be described as subcybernetic, since it thematized performance but not necessarily adaptation. Often, though, performance and adaptation turn out to be entangled in the developments at issue, so I will not dwell on this distinction in what follows. As mentioned below, for example, Walter's understanding of yogic feats depended on a notion of willful cerebral adaptation to otherwise autonomous bodily functions; likewise, his interpretation of the effects of flicker went back to the idea of scanning as a modality of adaptation to the world. I should mention that a curiosity about more mundane performances also characterized the early years of cybernetics. Norbert Wiener and Kenneth Craik were both intensely interested in how human beings operated complex tracking and gun-aiming systems (Galison 1994; Craik 1947), while Borck (2000) describes the origins of cybernetics in German aviation medicine in World War II. Again, adaptation was a key feature of this work—to the swerving of a target aircraft, or to the novel conditions of flight in high-performance aircraft.
57. Berger's pioneering EEG work was itself inspired by an interest in finding a physiological basis for psychic performances such as telepathy (Robbins 2000, 18; Borck 2001). Walter (1953, 269) also floated a more generic conception of an altered state: "Amplified by understanding of the basic functions involved, the physiological training of unusual brains may have results that are quite unforeseeable. We are so accustomed to mediocrity . . . that we can scarcely conceive of the intellectual power of a brain at full throttle." In fiction this was taken up by Colin Wilson in relation to his idea that humanity has thus far failed to exploit anything but a small fraction of its mental abilities, and elaborated in relation to all sorts of technologies of the self (below), including psychoactive drugs, sensory deprivation chambers, and EEG machines: see, for example, Wilson (1964, 1971). "My dynamo was suddenly working at full power" (1964, 211).
58. His discussion of all the items on the list is entirely nonsceptical, with the possible exception of extrasensory perception.
59. This notion of the brain is central to the distinctive analysis of cognition in autopoietic systems: see Maturana and Varela (1980, 1992).
60. This discussion of the modern self (like the earlier discussion of modern science) is meant to be suggestive rather than exhaustive, as a counterpoint to the nonmodern pole that is the focus of interest. On the historical construction of the modern self see, besides Foucault, Rose (1989).

61. Phenomenologically, the nonmodern forms of selfhood adumbrated by Walter display either a surrender of rational control (madness, nirvana) or a gain in control of performances not recognized in the modern body (yogic feats). For an insightful discussion of technologies of the self that entail a surrender of control (drugs and music), see Gomart and Hennion (1999) and Gomart (2002), who draw upon Jullien's (1999) analysis of Chinese understandings of decentered agency.
62. It might be that an interest in strange performances was not as suspect in science in the mid-twentieth century as it subsequently became. In 1942, for example, Harvard physiologist Walter B. Cannon (the man who gave homeostasis its name) published an article entitled "Voodoo Death" (Cannon 1942), a topic he had been working on since the 1930s (Dror 2004). On the historical origins of Walter's list, I am struck by its coincidence with the topics and approaches that featured at this time in the *Journal of the Society for Psychical Research*, a society founded in the late nineteenth century for the scientific exploration of psychic phenomena. In the pages of this journal one indeed finds discussions of ESP alongside yogic feats, clairvoyant visions, and so on, and the idiom of these discussions—seeking scientific explanations of such phenomena in terms of the properties of the brain—is the same as Walter's. Walter in fact joined the Society for Psychical Research in 1962 (having delivered the society's Myers Memorial Lecture in 1960), though I have no concrete evidence to connect him with it before the publication of *The Living Brain*. Geiger (2003, 40) notes that John Smythies (see below) visited Walter at the Burden, where they "discussed theories about the cause of the flicker experience" (the date of this meeting is not clear), and Smythies published a major article on psi phenomena in the SPR journal in 1951, discussing flicker as a key piece of evidence and citing Walter (J. Smythies 1951). (In the same volume, Smythies's father published an account of witnessing an instance of levitation in India [E. Smythies 1951].) Reports of EEG research into ESP and "mediumistic trance" appeared in the SPR's journal in the early 1950s—Wallwork (1952) and Evans and Osborn (1952), for example—and Wallwork (699) acknowledges assistance in his research from Smythies. The medium who was the subject of these EEG experiments was Eileen Garrett. Garrett knew Smythies, and Geiger (2003, 84–88) argues that Walter's interest in the paranormal was inspired by her (though this seems to refer to the 1960s). Walter attended one of Garrett's parapsychology conferences in 1961, and Garrett took part in one of Walter's experiments which involved the conjunction of flicker and LSD. Walter related Garrett's EEG record to "expectancy waves" (the phenomenon of contingent negative variation, mentioned above), an idea taken up and amplified in the 1970s by ESP researchers Russell Targ and Harold Putthof at the Stanford Research Institute. Stafford Beer is also recorded as a member of the Society for Psychical Research in 1962 and 1971, and George Spencer Brown (chap. 5, n. 25) joined the society as early as 1949 (and held the Perrott Studentship in Psychical Research at Trinity College, Cambrudge, in 1951–52). Despite his 1930 decision to "accept spiritualism" (chap. 4, n. 19), Ashby does not appear to have been a member of the SPR, and neither does Pask. For this information

on the society's membership, I thank its secretary, Peter Johnson (email, 19 June 2007).

63. One difference between Huxley and Walter has to do with first-person experience. Walter's stance was "scientific" inasmuch as he simply treated it as a matter of observable fact that people had visions, experienced hallucinations, and so on. In *The Doors of Perception*, Huxley offers a lyrical phenomenological account of what it is like to experience the world after ingesting mescaline. Another contrast is that Walter's scientific explanations of altered states usually hinged on the electrical properties of the brain, whereas Huxley's science was largely chemical.
64. On the history of the electronic stroboscope see Wildes and Lindgren (1985, chap. 8). "Looking" at a strobe with closed eyes reduces dazzle and homogenizes the ilumination of the retina; also, as mentioned earlier, the alpha rhythms of the brain are typically present when the eyes are closed but disappear when they are open. Adrian and Matthews (1934, 378) discussed the effects of flicker in their pioneering paper on the alpha rhythms of the brain, and observed that if the light is too bright, "the [visual] field may become filled with coloured patterns, the sensation is extremely unpleasant and no regular waves are obtained." Walter's 1938 report to the Rockefeller Foundation mentions attempts "to 'drive' the alpha rhythms with a flickering light" (4), but these were abandoned because of technical difficulties with the apparatus.
65. Walter's early technical publications on flicker were Walter, Dovey, and Shipton (1946), Walter and Walter (1949), and Walter (1956b). Norbert Wiener offered a mathematical analysis of flicker in the second edition of *Cybernetics* in terms of "frequency pulling," offering the synchronization of firefly flashing and the coupling of alternating electrical generators as examples of related phenomena. He also mentioned experiments in Germany, in which subjects were exposed to electrical fields alternating around the alpha frequency, where the effect was reported to be "very disturbing" (Wiener 1961, 197–202).
66. "The moment such diagreeable sensations were reported the flicker was of course turned off; recruitment of normal volunteers is not encouraged by stories of convulsions which also might unjustly impair the good repute of electroencephalography as a harmless experience" (Walter 1953, 97).
67. Evans was a Welsh novelist who suffered from epilepsy and attended the Burden Neurological Institute for treatment, where a romance developed between her and Frederick Golla (Hayward 2002). Besides moving patterns, Walter (1953, 250) mentions the induction of "vivid visual hallucination: 'a procession of little men with their hats pulled down over their eyes, marching diagonally across the field.' "
68. I am grateful to Rhodri Hayward for telling me of this work, and even more so to Geiger for sending me a copy when it was still unavailable outside Canada.
69. The key figures in British research were Humphrey Osmond, a psychiatrist, and John Smythies (n. 62 above), a qualified doctor with a PhD in neuroanatomy (Geiger 2003, 27–45). Osmond coined the words "hallucinogen" and "psychedelic" and provided Huxley with the famous 0.4 g of mescaline, the effects of which Huxley described in *The Doors of Perception*. Osmond's work

on LSD led to his involvement with the CIA and MI6, "which were interested in LSD as a possible 'truth drug' to make enemy agents reveal secrets" (Tanne 2004). Heims (1991, 167–68, 224–30) discusses parallel research in the United States, where the psychologist Heinrich Klüver was a key figure and a regular participant in the Macy cybernetics conferences. Heims again discusses the CIA's interest in "mind control" drugs, especially LSD, and possibilities of "behaviour modification," and notes that in the 1950s "the Macy Foundation was for a time used as a conduit for CIA money designated for LSD research." The key figure here was Frank Fremont-Smith, who organized both the cybernetics conferences and three other conference series which "brought leading contractors for CIA-sponsored drug work together with government people concerned with its application." Fremont-Smith thus had an early opportunity to try out LSD for himself.

70. Burroughs connected the effects of flicker to Walter's explorations of synesthetic overflow between different areas of the brain (Geiger 2003, 47; Walter 1953, 72).
71. Geiger (2003, 49): "Sommerville had also read *The Living Brain*, and he and Burroughs sought out Walter, attending a lecture and speaking with him afterwards."
72. There are many echoes of Walter's work in Burroughs's early masterpiece *Naked Lunch* (2001 [1959]). In the section entitled "benway," for example, Dr. Benway is described as "an expert on all phases of interrogation, brainwashing and control," who uses various regimes of Walter-Pavlov-style cross-conditioning to reduce the population of Annexia not to madness, but to gibbering docility. Burroughs invents a technology of the self called the "Switchboard" for this purpose: "Electric drills that can be turned on at any time are clamped against the subject's teeth; and he is instructed to operate an arbitrary switchboard, to put certain connections in certain sockets in response to bells and lights. Every time he makes a mistake the drills are turned on for twenty seconds. The signals are gradually speeded up beyond his reaction time. Half an hour on the Switchboard and the subject breaks down like an overloaded thinking machine" (21–22). Immediately after the description of the Switchboard, Benway offers the opinion quoted in part in chapter 1, that "the study of thinking machines teaches us more about the brain than we can learn by introspective methods," and he continues with a classically Walterian, if fictional, materialist account of the phenomenology of cocaine that I quoted in chapter 1: "Ever pop coke in the mainline? It hits you right in the brain, activating connections of pure pleasure. . . . C is electricity through the brain, and the C yen is of the brain alone, a need without body and without feeling. The C-charged brain is a berserk pinball machine, flashing blue and pink lights in electric orgasm. C pleasure could be felt by a thinking machine, the first hideous stirrings of insect life. . . . Of course the effect of C could be produced by an electric current activating the C channels." One thinks of the mirror and mating dances of Walter's tortoises souped up with random jolts of current surging through their rudimentary neurons, now as a materialist model for getting high.
73. Geiger (2003, 55–61). In *The Politics of Ecstasy* (1970) Leary also discusses Walter's work with implanted electrodes and the manipulation of consciousness.

74. Geiger (2003, 64, 67, 91, 95 97) mentions exhibitions in Paris (the first), Rome (1962, where Gysin constructed his Chapel of Extreme Experience), Tangiers (1964), and Basel (1979, the opening attended by Burroughs and Albert Hofmann, the chemist who synthesized LSD in 1938). Beyond the immediate orbit of the Beats, the 1963 *Living City* exhibition at the ICA in London was lit by a flicker machine (Sadler 2005, 55–57). This exhibition was put on by the Archigram group of architects, who will reappear in chapter 7. In 1979, the Pompidou Centre acquired a Dream Machine for its permanent collection. Gysin died in Paris in July 1986, but the Dream Machine continued to appear in exhibitions: in Los Angeles in 1996, and at the Hayward Gallery in London in 2000.
75. The story of flicker machines as consumer goods does not end in the sixties. In January 2005, under the headline "Décor by Timothy Leary, Dreams by You," the *New York Times* reported that David Woodard, a California composer and conductor, had made and sold more than a thousand Dreamachines since the early nineties, based on Gysin's original templates (Allen 2005). On first exposure to a Dreamachine, the reporter saw "colorful undulating patterns"; on his second experience he had visions of a campfire in a forest and a large auditorium, and had sensations of following someone down a basement hallway and of sharing his visions with someone else in the room (who was not there). The association with the counterculture continues: the Dream Machine in question belonged to Kate Chapman, "a former neuroscience researcher for the Multidisciplinary Association for Psychedelic Studies." One can also buy commercially produced flicker glasses today, consisting of LEDs mounted inside the frame of a pair of sunglasses. The flash rate of the LEDs is electronically controlled and can be programmed to flash at different rates in different sequences. I thank David Lambert for lending me his glasses. I enjoyed using them very much. I saw only moving geometric patterns, but I was struck by the objective quality of the images and their aesthetic appeal, and also by visceral sensations akin to falling when the flicker frequency changed. My son Alex tried them out, with no background explanation from me, and immediately said, "I can see people playing table tennis." (The glasses I tried are made by Synetic Systems International: www.syneticsystems.com.)
76. Geiger (2003, 72–77) also discusses the deliberate production of flicker effects in film, especially Tony Conrad's 1966 film *The Flicker*, and mentions a flicker sequence in Stanley Kubrick's *2001: A Space Odyssey*. Theodore Roszak wrote a fabulous novel, also called *Flicker* (1991), in which a Manichaean sect develops very sophisticated flicker techniques in cinematography to act in sinister ways on audiences. I thank Fernando Elichirigoity for bringing this book to my attention.
77. Here and elsewhere "the sixties" refers to a specific cultural formation rather than a well-defined chronological period. Writing this book has brought home to me how much of what one associates with "the sixties" in fact originated in the 1950s (Huxley's books, *Naked Lunch*) or even earlier (Walter's work on flicker, for instance). Further exemplifications of this observation in the chapters to follow. The chronological sixties were, of course, the period when the developments at issue became very widely known and practiced.

78. "Trepanning" here refers specifically to the practice of removing part of one's skull with the object of achieving some form of enlightenment. It appears that very few people actually went through with this procedure, but it was certainly part of the cultural imaginary of the sixties. See J. Green (1988, 67, 97); and, for more detail, www.noah.org/trepan/people_with_holes_in_their_heads.html (I thank Peter Asaro for directing me to this website). Walter mentions that "some ancient skulls do in fact show trephine holes where there is no evidence of organic disease" (1953, 228) but assimilates this to the prehistory of leucotomy.
79. On later experiments at the Burden Neurological Institute using theta feedback in a "vigilance task"—spotting vehicles that would occasionally pass across a TV screen—see Cooper and Bird (1989, 39–40). Borck (2000) discusses biofeedback experimentation in German aviation medicine in World War II, with the object, for example, of flashing a warning signal to pilots that they were about to faint due to lack of oxygen before they were consciously aware of the fact.
80. I am especially interested here in brainwave biofeedback, but such techniques have also been applied to other bodily parameters—the heartbeat, for example (and hence a possible connection to yogic feats). Butler (1978) lists 2,178 scholarly publications covering all areas of biofeedback research.
81. Returning to a familar theme in the history of cybernetics, Robbins also discusses the continuing marginalization of biofeedback in psychiatric research and practice—"part of the problem is the fact that biofeedback doesn't fit neatly into any category" (2000, 144).
82. I am very grateful to Henning Schmidgen and Julia Kursell for alerting me to this cybernetic connection and pointers to the literature. Rosenboom (1974, 91–101) is a bibliography of books and articles on biofeedback and the arts, listing around two hundred items.
83. Lucier (1995, 46) ascribes the original inspiration for this work to conversations with Edmond Dewan when they were both teaching at Brandeis University. Dewan was then working for the U.S. Air Force, who were interested in the possibility that flickering reflections from propellors were inducing blackouts in pilots prone to epilepsy.
84. Teitelbaum (1974, 62–66) also discusses works in which he "played" the synthesizer used to convert the EEG readout into an audible signal, referring to his role as that of "a guide" who "gathers together the subject's body rhythms and helps make a sound image of the electronically extended organism. The sound image thus created becomes an object of meditation which leads the subject to experience and explore new planes of Reality. . . . Even the audience seemed to enter into a trance like state, thereby entering into the feedback loop and lending positive reinforcement to the whole process. Describing her own subjective response later, Barbara Mayfield said it was 'like having an astral body through the wires.'" He also mentions that in 1969 the group used another technology of the self, yogic breathing, to control heart rates and alpha production. Evidently my imputation of a simple hylozoism to such work requires some qualification: turning brain states into music was highly technologically mediated (as in the setup depicted in fig. 3.15). The contrast I want to

emphasize is that between composing a musical piece note by note in advance and seeing what emerges in real time from a biological system such as the human body. In this sense, one might see Lucier and Teitelbaum's compositions as a hybrid form of ontological theater. As usual, I am trying get clear on the less familiar aspect of the assemblage. I can note here that Teitelbaum (1974, 69) mentions a dark side of biofeedback: "With some of the most technically 'advanced' psychology work currently being carried out in our prisons [under] the guise of aversion therapy and the like, there is clearly great cause for concern." One thinks of the graphic portrayal of aversion therapy, complete with EEG readout, in Stanley Kubrick's film of Anthony Burgess's novel *A Clockwork Orange* (1963). We can explore this "double-valuedness" of technologies of the self further in chapter 5, on Bateson and Laing.

85. One tends to imagine that biofeedback music died some time in the sixties like other cybernetic projects, but, as usual, a quick search of the Web proves the opposite to be true. "Brainwave music" turns up many sites, often with a distinctly New Age flavor. Feedback music is still performed; one can buy EEG headsets and PC software that converts brain rhythms into MIDI output that can be fed directly into a synthesizer. For a recent report on the state of the art, which includes an excellent overview of postsixties scientific EEG research, see Rosenboom (1997).

Notes to Chapter 4

1. This biographical sketch is based upon a typescript biography dating from the late 1960s in the Biological Computer Laboratory Archives, University of Illinois at Urbana-Champaign, S.N. 11/6/26; B.N. (cited below as "BCL archive"), much amplified by information from Jill Ashby and John Ashby, for which I am very grateful. I also thank Peter Asaro for many conversations about Ashby over the years, Amit Prasad for research assistance, and Malcolm Nicolson and Steve Sturdy for their help in understanding medical careers.
2. In a biographical notebook, Ashby (1951–57) notes that his father wanted him to grow up to be "a famous barrister or a famous surgeon" and that he was "savage when annoyed" and "so determined and forceful" (10, 27). He recalls Hans Christian Anderson's story of the little mermaid who suffered horribly in leaving the sea for love of her prince, and he identifies the prince with his father (and presumably himself with the mermaid), continuing, "I learned to hate him, & this held from about 16 to 36 years of age. Now, of course, as a person, he's just an elderly relative" (28–31). "To sum up . . . one could only describe [my life], so far, as thirty years of acute unhappiness, ten years of mild unhappiness, & (so far) a few years that might be described as breaking even" (22).
3. Some measure of Ashby's financial circumstances can be gauged by his only known self-indulgence: sports cars. When he went to Palo Alto, he took one of the first Triumph TR2s with him. When he moved to Illinois it was a Jaguar XK120 (John Ashby, personal communication).
4. Since I began my research a wealth of archival material has been deposited at the British Library in London by his daughters, Jill Ashby, Sally Bannister, and Ruth Pettit, and I thank them for early access to this material, which is

now also available on the Web at http://rossashby.info/. Jill Ashby's sons, Ross's grandsons, John and Mick Ashby, have been very active in making Ashby's papers and works publicly available, and I thank them both for much assistance. The most valuable archival resource is a complete set of Ashby's journal running from 1928 to 1972, some 7,189 pages in all, cited below as "journal" by page number and date (sometimes only approximate). I have also drawn upon a biographical notebook that Ashby wrote between 1951 and 1957 under the title of "Passing through Nature" (Ashby 1951–57); another notebook titled "The Origin of Adaptation," dated 19 November 1941 (Ashby 1941); copies of Ashby's correspondence, cited by correspondent and date; Ashby's list of publications, as updated by Peter Asaro; and a family biography of Ashby including many photographs written by Jill Ashby. For a collection of papers presented at a conference marking the centenary of Ashby's birth, see Asaro and Klir (2009).

5. See Hayward (2004) on Golla's central role in British psychiatry.
6. There is an interesting parallel here with the work of Walter Freeman in the United States. After extensive unsuccessful attempts to identify organic correlates of insanity, Freeman went on to become one of the key figures in the development of lobotomy: Pressman (1998, 73–77).
7. I am grateful to Jill Ashby for providing me with copies of annual reports from St. Andrew's, and for the information that Ashby joined the hospital as a pathologist, bacteriologist, and biochemist on 27 March 1936 at a salary of £625 per annum.
8. Mrs. Ashby to Mai von Foerster, 5 August 1973; BCL archive: S.N. 11/6/26; B.N. 1.
9. The quotation is an edited extract from pp. 6–13.
10. Letter from BBC to Ashby, 21 February 1949.
11. We can make this machine seem a little less strange. A 1932 paper by L. J. Henderson (1970 [1932], 163, fig. 1) includes a diagram showing four rigid bodies attached to a frame and to one another by elastic bands, as an exemplification of a system in the kind of dynamic equilibrium discussed in the following paragraph. Ashby drew much the same diagram to make much the same point nine years later, in his notebook "The Origin of Adaptation" (1941, 35). What Ashby then added to Henderson's picture is that the elastic bands can break, as discussed in the next paragraph but one. I have not been able to trace any reference to Henderson in Ashby's journals or published writings. Henderson was a Harvard colleague of Walter B. Cannon, on whom see the following note.
12. Ashby's journal, pp. 2072–81. A note on 28 December 1946 (p. 2094) records, "I have devised a Unit which conforms to the requirements of pp. 2079 and 2083," and includes a sketch of the wiring diagram (reproduced here as fig. 4.3). On 3 May 1947, Ashby wrote: "Triumph! A unit has been made to work," and a note added records that the machine was demonstrated at a meeting of the Royal Medico-Psychological Association held at Barnwood House in May 1947 (p. 2181). On 3 March 1948 Ashby wrote, "Have completed my new four-unit machine," discussed below, and drew a more detailed circuit diagram (pp. 2431–32), and on 16 March 1948, he could again exclaim: "Triumph! The machine of p. 2432 the 'automatic homeostat' was completed today. . . . After all

the trouble, it works" (p. 2435). This is the first occasion on which I have found Ashby calling the machine a "homeostat." It is clear from his journal that he got the word from Walter B. Cannon's 1932 book *The Wisdom of the Body*, but it is also clear that Ashby's cybernetics owed little to Cannon (and that Ashby had trouble spelling "homeostasis" for many years). The first mention of Cannon that I have found in Ashby's journal is in an entry dated 8 October 1937 (or just later) (p. 365). (This is at the end of Ashby's volume 2 of the journal. The earlier entries in this volume are from 1931, but it seems clear that Ashby went back later to fill this volume up. Other notes from 1937 appear in volume 3.) Here, in a long sequence of notes on the literature that Ashby had been reading we find under "Cannon, Walter B. (ref. 399)" this entry: "He spells it homeostasis. Richert (Ref. 414) said 'The living being is stable. It must be in order not to be destroyed.' Nothing much." Reference 399 is a 1929 paper by Cannon which Ashby lists with the title "Organisation for Physiological Homoe[o]stasis." The third "o" is an insertion by Ashby into a word already mispelled. All the evidence, including that "nothing much," suggests that Ashby did not draw any inspiration directly from Cannon. Earlier in the same series of notes on the literature, we find "Edgell continues: '[To Prof. Jennings] an organism does not reply to its environment by a simple reflex which is at once relevant to the situation. On the contrary, stimulation is followed by many & varied movements from which the successful movement is selected by a process of trial and error. It will be that movement which relieves the organisation with respect to the stimulation in question'" (p. 341, dated 8 October 1937). This passage anticipates the key features of the homeostat; the book under discussion is Edgell (1924). Cannon's name next appears in Ashby's journal within a set of notes discussing a 1933 book by George Humphrey, where the sole comment is "Cannon (Ref. 399) calls all stabilising processes 'hom[o]eostatic' (note the title of Cannon's paper) (to be read)" (p. 793, dated 26 January 1940). Two pages later, we find a striking series of protocybernetic remarks. Still discussing Humphrey (1933): "He considers the animal as a system in mechanical eq[uilibriu]m. . . . Ostwald (Ref. 402) 'A petrol motor which regulates its petrol supply by means of the ball governor in such a way that its velocity remains constant, has exactly the same property as a living organism.'" Ashby then goes on to discuss an early twentieth-century phototropic robot which homed in on a light, much like Walter's tortoises: the citation is Loeb (1918) (and see Cordeschi 2002). We can note that Ashby (1940), his first cybernetic paper, was received at the *Journal of Mental Science* on 25 February 1940, just one month after this passage in his journal, though, typically, Ashby makes no mention of this publication there. Cannon (1929) reappears briefly in an entry made on 30 July 1944, where Ashby copied out a quotation (as well as one from Claude Bernard), and ended: "Summary: Some other people's quotations on equilibrium" (p. 1721). Finally, on 9 June 1947 Ashby noted that "I have just read Cannon, W. B. The Wisdom of the Body. London.1932. He points out that the recognition of stability is as old as Hippocrates' recognition of the 'vis medicatrix naturae' four hundred years B.C. In order to avoid misunderstandings he coins a new word 'homeostasis' (note spelling). His definition is vague and unhelpful." Following this Ashby transcribed four pages of quotations by hand from Cannon plus two pages of typescript inserted in his journal—transcriptions interspersed with comments such as "Why in Heaven's name doesn't he conclude. . . . How can anyone be so blind?" (p. 2196) and "He gives Four Principles, obvious enough to me but quotable" (p. 2197) (pp. 2195–98).

13. On the Barnwood House demonstration, see the previous note; on the Burden Neurological Institute demonstration, see Ashby (1948, 383n4).
14. Ashby (1948) cites Rosenblueth, Wiener, and Bigelow's key 1943 essay and mentions radar-controlled antiaircraft guns, but he does not cite Wiener's book or use the word "cybernetics." He does cite his own 1940 essay, in effect asserting his priority in the discussion of negative-feedback mechanisms. In *Design for a Brain* (1952, 154), Ashby does use the word "cybernetics," crediting it to Wiener, again in a discussion of antiaircraft gunnery, but without citing Wiener's book. This passage disappeared in the second edition (Ashby 1960) as did "cybernetics" from the index, though a citation to Wiener's *Cybernetics* was added.
15. This concern with what the world must be like if we can learn to adapt to it runs throughout Ashby's work. One can find it clearly expressed in his journal at least as early as 15 January 1940 (p. 482).
16. Ashby's four-homeostat setups could be more or less symmetric. The fully symmetric version was the one in which in all four homeostats were uniselector controlled and thus able to adapt to one another, and this is the version referred to here. The less symmetric version was one in which the parameters of the three world homeostats were fixed and the brain homeostat did all the adapting. Even in this second version the world remains intrinsically dynamic and performative, responding in specific ways to specific outputs from homeostat 1. Nevertheless, the question of whether a given situation should be modelled by the symmetric or asymmetric configuration could be highly consequential. In psychiatry, for example, this marked the point of divergence between Ashby and Bateson, as discussed below.
17. "Twenty-five positions on each of four uniselectors means that 390,625 combinations of feedback pattern are available" (Ashby 1948, 381). Something does not quite make sense here. Figure 4.4c indicates that each homeostat unit contained three uniselectors, *U*, one for each of the input lines from the other units (but not on the feedback input from itself). If each of these had twenty-five settings, then each unit could be in 25^3 possible states. And then if one wired four units together, the number of possible states would be this number itself raised to the fourth power, that is, 25^{12}. I have no convincing explanation for the discrepancy between my calculation and Ashby's, except that perhaps he was thinking of each unit as having just a single uniselector switched into its circuits, but nothing in what follows hinges on a precise number.
18. In the same notebook Ashby (1951–57) also mentions the "emphasis laid in my early theorising on 'nirvanophilia,'" and in a journal entry on 8 October 1937 he remarks that "Jennings may be added to the list of 'nirvanophiliacs'" (p. 341). In the previous chapter I mentioned the connection Walter made between homeostasis and nirvana, and Ashby clearly had a long-standing interest in this topic, identifying the brain's achievement of dynamic equilibrium with its environment and the experience of Buddhist detachment from the world:

"Nirvanophilia is identical with stable equilibrium" (journal entry dated 4 May 1939, p. 586); "the adaptation is perfect, intelligence infallible, all in Nirvana" (13 June 1959, p. 6118). One might wonder whether Ashby's military posting to India bore upon his spiritual views, but the interest in nirvana predates that, and "Passing through Nature" makes no explicit connections to spirituality in connection with his time in India. He remarks instead that "the army and India released something in me. (. . . In India I first played like a child, & enjoyed it all with immense delight, packing into a year all the emotions of a decade.) Since then I have become steadily more active, tending to emerge from the shadow in which I used to live" (p. 56, written between September and December 1954). In the passage immediately preceding the "Time-worshipper" quotation, Ashby recalls something like a spiritual awakening early in World War II. He was overcome by a mysterious illness which resulted in admission to Northampton General Hospital for several days at the time of the German air raids on Coventry in November 1940, which he heard from his hospital bed (pp. 31–36). "What this 'illness' was I've never discovered. . . . I felt, in Stephen Leacock's words, as if I had 'swallowed a sunset.' . . . What did the old writers mean when they said that God threw . . . into a deep sleep, and spoke to him, saying '. . .'—I have a very open mind on the question of what that illness really was" (1951–57, 35–36; the last two groups of three dots are Ashby's; one should read them as "X").

19. There is yet another angle on Ashby's spirituality, which I discovered as I was completing this book and have been unable to take further. In his journal on 5 August 1930 Ashby wrote: "Today, after reading de Brath (ref 88) I have made my decision, and will accept spiritualism. I think that to refuse to take the plunge when one is convinced is mere cowardice" (164). The book he cites is de Brath (1925), *Psychical Research, Science, and Religion*. The entry continues, "I have just read a statement by Oliver Lodge," and goes on to discuss the interaction of spirit with matter, and the brain as an "elaborate and exceedingly delicate trigger mechanism . . . the "obvious place for spirits . . . to step in and seize control" (165). On 18 October 1931, Ashby made several pages of journal notes from his readings on clairvoyance and hypnotism, reproducing quotations such as "Experiment seems to show that thoughts and words leave traces in the super-individual mind which can be recognised by the seer" (280). Thirty pages later he made more notes on his readings on clairvoyance, mediums, possession, psychiatric patients, and "supernormal knowledge" (318–23). This interest evidently persisted. On 18 February 1950 Ashby sent John Bates a list of thirty-four possible topics for discussion at future meetings of the Ratio Club, which included "18. Telepathy, extra-sensory perception, and telekinesis." (The list ended with "If all else fails: 34. The effect of alcohol on control and communication with practical work.") On 26 April Ashby sent Bates a revised list of twenty-eight suggestions, which included "12. Is 'mind' a physical 'unobservable'? If so, what corollaries may be drawn? . . . 27. Can the members agree on a conclusion about extra-sensory perception? 28. What would be the properties of a machine whose 'time' was not a real but a complex variable? Has such a system any application to certain obscure, ie spiritualistic, properties of the brain?" (Wellcome Library, Bates papers, GC/179/B.5). I thank

Mick Ashby for pointing out that in a 1972 biographical sketch Ashby listed a consultancy with the Institute for Psychophysical Research in Oxford (www.rossashby.info/index.html), and that a 1968 book on out-of-the-body experiences published by the Institute thanks Ashby, Walter, and many others for advice and assistance (C. Green 1968). With spiritualism, mediums, telekinesis, and so on, we are back in the space of strange performances and altered states, as discussed in the previous chapter on Walter. The idea of the brain as a material detector for a universal field of mind and spirit runs through the work of John Smythies (chap. 3, nn. 62, 69). In the present context, we can also note that such a view can hang together with a notion of biological/spiritual evolution in which the brain latches onto and adapts to the universal field of mind, as in the work of Michael Murphy (1992) (mentioned in chap. 6, n. 69, below).

20. Conant's obituary rightly conveys the amateur flavor of Ashby's early cybernetics, though the details should be taken with a pinch of salt. I am grateful to Jill Ashby for consulting Denis Bannister on her father's work and letting me know what she found (email, 15 April 2005). Bannister worked for Ashby at Barnwood House for just over a year and married his daughter Sally (subsequently divorced). Bannister's opinion is that the kitchen table story was one of Ashby's jokes, though he recalls that some parts of the homeostat were purchased from the post office. Bannister and Ashby's principal project was "researching enzymes concerned with ECT therapy. Only when they had spare time did they build the Homeostat which was done definitely as a sideline and definitely at Barnwood House. It took about three months after which they took it round to meetings. It was very bulky and heavy. They mixed frequently with those at the Burden and Grahame White from the Burden (also mentioned in the notebooks with Denis) also helped with the building." The journal entry in question is dated 3 March 1948 and reads: "My two lads [White and Bannister] have been most useful & did a good deal of the wiring up" (p. 2433). The only other reference I have found to Ashby's source of parts for the homeostat is to the detritus of World War II: "It has four ex RAF bomb control switch gear kits as its base, with four cubical aluminium boxes" (3 March 1948, p. 2341).

21. Jill Ashby (email, 8 March 2005) pointed out to me a journal entry for 1 February 1943, pp. 1181–82, which refers to Ashby's key 1943 essay that was eventually published in 1945 (Ashby 1945a): "I have rewritten the previous paper, restricting it to an explanation of 'adaptation by trial and error,' keeping the maths in Part II, & have sent it to the Brit. J. Gen. Psychol. And I just don't give a damn what happens. I'm just losing interest. (Am doing astronomy at the moment). Now that my previous childish phantasies have been shown to be untrue I have lost my drive. And without something to give a drive one just can't move." The continuation of this passage nicely documents the professional precariousness of Ashby's cybernetics: "Tennent has given me to understand quite clearly that he wants nothing to do with it. It is now impossible for me to move further unless I can feel that there will be some benefit somewhere. As it is, it merely leads to trouble. *Summary*: Theory has been submitted for publication for the third time." Jill Ashby adds that "Dr Rambout sadly suddenly died and his position was filled by Dr Tennent who I can remember my mother saying did not like my father. Dr Tennent did not mention my father's work in the next four

annual reports." Rambout was the medical superintendent when Ashby was appointed to St. Andrew's Hospital in Northampton, where he worked from 1936 to 1947. Number 12 in the "Letters and Documents" section of the Ashby archive is a letter from N. Rashevsky, 8 January 1947, rejecting a paper Ashby had submitted to the *Bulletin of Mathematical Biophysics*, entitled "Dynamics of the Cerebral Cortex, VII."

22. It is interesting to contrast cybernetics' institutional travails in the West with its enormous success in the Soviet Union (Gerovitch 2002). Although much remains to be explored about the substance of Soviet cybernetics, it is tempting to reflect that, ontologically, cybernetics was much more congenial to a materialist Marxist ideology than to the taken-for-granted dualism of the Western academy.
23. On the history of the BCL and its closure, see Müller and Müller (2007); Umpleby (2003); and Franchi, Güzeldere, and Minch (1995), where Franchi recalls that when he visited the University of Illinois in the mid-1980s, "I couldn't find a single person who knew of the BCL" (301). See also the BCL website, which includes valuable information and links to other sites, including several festschrifts for von Foerster: bcl.ece.uiuc.edu. On the distinctive research program of the BCL, see Asaro (2007); on von Foerster, see Brier and Glanville (2003).
24. As far as I can make out, Ashby makes the mistake of dividing by 2 instead of 1/2 when he evaluates T_2 and T_3, but nothing hinges on this.
25. One gets a feeling for this kind of world in some of Stanisław Lem's fiction.
26. This last example strikes me as assuming an unwarranted Platonism about the structure of mathematics, but it still counts as a nice example of Ashby's train of thought.
27. Think, for example, of Newtonian physics or quantum field theory. All of the possible entities and their interrelations are already prescribed by the theory; there is no room even to raise the sort of question that Ashby addresses. Of course, questions of weak and strong couplings do arise in such sciences, but as matters of epistemological convenience and possibility. Coupling constants are what they are. Weakly coupled systems can be analyzed by the methods of perturbation theory; strongly coupled ones are resistant to such methods (and this gets us back, of course, to questions of complexity and predictive transparency). From another angle, Ashby's analysis reminds me of arguments in science and technology studies. There the humanist critique is that posthumanist analyses are irrevocably useless because, lacking the principle of humanist centering, they amount to a brute assertion that everything is coupled to everything else. My own reply to this critique is that units of empirical analysis are not given but have to be found in any specific instance. It turns out that one finds that some cultural elements are strongly coupled to each other—constituting the object of analysis—while indefinitely many are weakly coupled at most. For more on this, see Pickering (1997, 2005b, 2005c).
28. Ashby seems to have finally concluded that the task was hopeless by 1957. In a journal entry dated 23 February 1957 he recorded, "Now we come to the (most unpleasant) deduction of p. 5539 [seven pages earlier]—that there is no secret trick to be discovered, no gimmick, no Ashby's principle, no ultimate ruling

law:—there is only information, knowledge of what is best in each particular case. So the idea that has been nagging me ever since the 'Red mass' of p. 2957—find once for all how the discriminative feedback is to find the right step-mechanism—is now shown to be mistaken. The feedback can be directed with discrimination only after the necessary information has been collected. If then I (or other designing factor) am to get the feedback right without personal attention, I must achieve a supplementation. This is not systematic: it just means looking for any trick, using the peculiarites of the material at hand, that will meet the case. Thus, if I build a new machine & want discriminative feedback, I may use any trick that happens to be convenient (for there is no way better), but I must admit that it is ad hoc, and not claim any wider validity for it. Ingenuity—it is now up to you! Summary: 'Thinking things over' in a multi-stable system. Discriminative feedback requires mere opportunism [marginal notes forward to pp. 5549, 5584, 5601]" (pp. 5546–47, and see also following pages). "Summary: No general principle can be sufficient guide when selection must be done; some actual channel is necessary" (27 February 1957, p. 5549).

29. Ashby's journal makes clear the extent to which his thinking was importantly informed by trying to build DAMS and finding that his expectations were frustrated. An entry from 15 May 1951: "Saw Wiener today at the B.N.I., & asked him whether he could give any help towards getting a mathematical grasp of DAMS, towards solving the problem of p. 3151 (lower half). He was firm in his opinion. There is no branch of mathematics that can, at present, be of much use. The work I am doing is quite new, and the mathematician has no experience to guide his intuitions. Therefore, says Wiener, I must develop my machine empirically until I have established a clear & solid structure of factual material. It will then become clearer what are the important factors, & what are the significant questions. The mathematical attack must come after the empirical exploration" (p. 3304).
30. In the discussion of the homeostat I emphasized that Ashby's approach required to him to model the world as well as the brain, and the same went for DAMS. The earliest explicit discussion I have found of DAMS's environment comes from a journal entry on 23 September 1951. "A much more practical idea for the 'env[ironmen]t' of DAMS. My previous idea was to have parts that can be joined together in a variety of ways. I now see a simpler way of getting variety:—Build a 'machine' of, say, coils and magnets, making the relations permanent. But provide it with, say, eight inputs & eight outputs that can be joined to the machine [i.e., DAMS] at random." But by the end of this note, Ashby had gone around in a circle and concluded, "This leads naturally to the conception of just thinking of the last row of valves [in DAMS] as the 'envt' [of the rest of DAMS]" (pp. 2482–83). Just as the world of the homeostat was more homeostats, so the world of DAMS was a subset of DAMS. The brain-world symmetry is even more evident in this latter setup.
31. More than a year later, on 20 October 1951 (p. 3152) Ashby recorded: "DAMS is working quite well now with 40 valves for simple jobs. It could probably be made to work like a simple ultrastable system by just feeding back some disturbance to shatter the neons. . . . On the other hand it is showing little sign

of the characteristic multistable system's power of adapting by parts and of accumulating patterns. I presume that if the machine demonstrates clearly that something can not be done with 100 valves, then that will be worth knowing!"

32. A version of the idea that DAMS should evolve by trial and error reappears in Ashby's *Thalès* essay under the heading "'Darwinian' Processes in Machinery" (Ashby 1951, 5), though there the emphasis is on spontaneous reconfigurations within the machine (and Ashby acknowledges a private communication from Wiener for his use of "Darwinian"). The perspective of the human designer returns in Ashby's notes for 20 October 1951 (pp. 3512–17). "Here is a simple and, I think, potential method for improving DAMS—a sort of sequential analysis" (p. 3512). Ashby then discusses regulating a watch by fiddling with its "index" (while not understanding how that enters into the mechanism). "The method proposed is not new: it has been used in endless trades throughout the history of civilisation for the improvement of products and processes when the conditions were too complex to allow of scientific analysis" (21 October 1951, p. 3516). He gave as examples Morocco leather, white lead, linseed oil, the motor car, and different ignition systems. "Summary: Improvement by the purely empirical is as old as industry. Corollary: If I can make substantial improvements in DAMS by purely empirical process, I shall provide facts that will give a solid basis on which later workers may found a science—just as the rubber technologists enabled the scientists eventually to build a science of high polymers" (p. 3517). Ashby's remark about the necessity of new techniques in new developments in science, which seem "plain crazy" in the old framework, are echoes of Kuhn's (1962) thoughts on the incommensurability of paradigms more than a decade *avant la lettre* and expressed in a more thoroughgoingly performative register. Ashby's recognition of the problems raised by machines like DAMS for notions of understanding and explanation likewise presage much later philosophical discussions of complex systems more generally (e.g., Kellert 1993, Pickering 1995).
33. In the quoted passages from both *Thalès* and *Design for a Brain*, Ashby cites Humphrey (1933), the work discussed in note 12 above in connection with Ashby's earlier steps toward the homeostat.
34. Habituation actually surfaced in DAMS as a block toward the overall goal of accumulating adaptations. Ashby wrote in a journal note dated 1 May 1951, "During the last two or three months I have been testing the first 10-valve block of DAMS Mark 13. Whereas I expected it to be remarkably unstable I have found it to be remarkably, almost undesirably, tenacious; so much so that I had doubts whether I could ever make it learn."
35. Ashby seems to have saved this argument up for six years, before presenting it in public at the 1958 "Mechanisation of Thought Processes" conference discussed below.
36. In the next chapter we will find Stafford Beer eloquently lamenting the same situation a few years later.
37. The only such machine that I have seen written up is the one Ashby demonstrated at the "Mechanisation of Thought Processes" conference in 1958 which simulated the production and cure of mental pathologies, discussed in the next section.

38. This review appeared as an undated special issue of the *Journal of Mental Science*, but internal evidence suggests that it appeared in 1950. It is a measure of the visibility of cybernetics in British psychiatry by this date that both Ashby and Walter were invited to contribute to it (Walter with the entry "Electro-Encephalography"). Frederick Golla's entry "Physiological Psychology" begins with the remark "The past five years have seen an attempt to reformulate the basic conceptions . . . in terms of neural mechanics . . . Cybernetics," and then discusses Ashby's work at length, followed by that of McCulloch, Pitts, and Walter on "scanning" and Masserman's cats, before moving on to more prosaic topics (Golla 1950, 132–34).
39. One might read Ashby as asserting that the undoing of habituation by any large and random disturbance is discussed in *Design for a Brain*, but it is not. As far as I know, Ashby's only attempt to argue this result in public was at the "Mechanisation of Thought Processes" conference in 1958, in a paper titled "The Mechanism of Habituation" (Ashby 1959a). This includes an extended discussion of "de-habituation" (109ff.) without ever mentioning electroshock. At the same meeting, Ashby also exhibited "A Simple Computer for Demonstrating Behaviour" (Ashby 1959b). According to its settings, Ashby claimed that it could display "over forty well known pieces of biological behaviour," including various simple reflexes, "accumulation of drive," "displacement activity," and "conflict leading to oscillation" or "compromise." As exhibited, it was set to show "conflict leading to catatonia, with protection and cure" (947–48). It might be significant that in the passage quoted Ashby uses the memory loss clinically associated with ECT to make a bridge to his cybernetic analysis of dehabituation. Though I will not explore them here, Ashby developed some very interesting cybernetic ideas on memory (Bowker 2005). It was in this period that the first discussions of memory appear in Ashby's journal. The suggestion is, then, that the concern with the mode of action of ECT might have been a surface of emergence also for Ashby's cybernetics of memory (and see, for example, the long discussion of memory that appears in Ashby 1954).
40. Walter's article on CORA appeared in the August 1951 issue of *Scientific American*, just one month before Ashby made this note.
41. Besides his journal and the published literature, Ashby also raised the topic of possible links between cybernetics and psychiatry at the Ratio Club. The Bates papers at the Wellcome Archive in London include a three-page typescript by Ashby entitled "Cybernetics and Insanity" (Contemporary Medical Archives Centre, GC/179/B.2a), cataloged as "possibly prepared for Ratio Club meeting 1952." It develops the idea mentioned in the following note that the brain consists of a hierarchy of homeostatic regulators and that mental illness might be identified with malfunctions at the highest-but-one level. There are also two lists of possible topics for discussion at the Ratio Club sent earlier by Ashby to Bates (GC/179/B5), which include "3. To what extent can the abnormalities of brains and machines be reduced to common terms?" (26 April 1950) and "26. The diagnosis and treatment of insanity in machines" (18 February 1950).
42. The paper was read to a Royal Medico-Psychological Association meeting in Gloucester (presumably at Barnwood House) in July 1953. For the sake of com-

pleteness I can mention two more connections that Ashby made from cybernetics to psychiatry. First, the essay suggests a quite novel general understanding of mental illness. Ashby supposes that there exist in the brain homeostatic regulators ancillary to the cortex itself, and that mental pathologies might be due to faults in the ancillary systems. He argues that if that were the case, "we would see that the details of the patient's behaviour were essentially normal, for the details were determined by an essentially normal cortex; but we would find that the general tenor was essentially abnormal, a caricature of some recognisable temperament. . . . Thus we might see the healthy person's ability to think along new and original lines exaggerated to the incomprehensible bizarreness of the schizophrenic" and likewise for the "maniac" and the "melancholic" (123). Second, Ashby returns to the question of the organic basis for mental illness, but from a distinctively cybernetic angle. This hinged, in the first instance, on the question of the brain's essential variables. Ashby had already suggested that biological limits on essential variables were set by heredity, and here he supposed that sometimes heredity would go wrong. He mentioned a child born unable to feel pain, who thus "injures himself seriously and incessantly," and imagined that "the mental defective who is self-mutilating . . . may be abnormal in the same way" (121).

43. One thinks here of the "sensitive dependence on initial conditions" later thematized in complex systems theory.
44. Ashby discontinued his specifically psychiatric research after leaving England in 1960, but he continued to contribute to the psychiatric literature after moving to Urbana. See, for example, Ashby (1968c). These later works uniformly seek to educate psychiatrists about cybernetics and, especially, information theory, and the interest in ECT disappears completely. Ashby was happy to talk about psychiatric therapy, as in "The theory of machines . . . may well provide the possibility of a fully scientific basis for the very high-level interactions between patient and psychotherapist" (1968c, 1492). On the other hand, one can still detect understated connections between the degree of connectivity within the brain and lobotomy. Having discussed his student Gardner's discovery of a threshold in "connectance," he remarks that "it is obvious that a system as complex and dynamic as the brain may provide aspects at which this 'mesa' phenomenon may appear, both in aetiology and therapy. There is scope for further investigation into this matter, both in its theory and its application. . . . All the studies of the last twenty years . . . show that systems should be only moderately connected internally, for in all cases too rich internal connection leads to excessive complexity and instability. The psychiatrist knows well enough that no one can produce associations so quickly or so wide-ranging as the acute maniac; yet his behaviour is inferior, for knowing what associations to avoid, how to stick to the point, is an essential feature for effective behaviour" (1968c, 1494, 1496). As it happens, the very last entry Ashby made in his journal was about schizophrenia: he noted that in an article in the *British Journal of Psychiatry*, vol. 120 (1972), "Schizophrenics were classified by handedness . . . dominant eye . . . and six [sex]" (8 March 1972, pp. 7188–89).
45. There are many more discussions of texts in Ashby's journal than records of conversations, though see below on Mrs. Bassett.

46. This observation reflects back on Ashby's understanding of the prototypical four-homeostat setup. While that setup can help us imagine a fully symmetric performative relation between entities and their world, the present discussion suggests that Ashby's basic model was one in which the brain homeostat was uniselector controlled and adaptive but the other three homeostats had their properties fixed: a brain that adapts to the world but not vice versa. He mentions somewhere that it is a good thing that cars and roads do not change their properties when we are learning to drive, otherwise the process would be even more fraught, and it is hard to disagree with that. On the other hand, it is integral to the history of science that the material world can perform in new and surprising ways when subjected to novel trials in the laboratory, and this seems to me to be modelled better by a symmetric field of adaptive homeostats, each capable of taking on new states in response to novel stimuli. More generally, biological species, including the human race, do not typically take the material world as given; they transform it instead, and, again, such transformations may elicit new worldly performances—think of global warming.
47. We might think again about the standard origin story, of cybernetics from warfare. Against that, we could observe that Ashby's first cybernetic publication appeared in 1940, eleven years before he got Clausewitz out of the public library. From another angle, Ashby's discussion of wrestling and the throat-gripping hands strike me as wonderful quick analysis of why our brave leaders prefer to bomb their latest enemies from high altitude (at night, with vision-enhancing technologies, after destroying their radar) rather than fighting them in the street.
48. While most of Ashby's discussions of state-determined machines can be seen as integral to the development of his cybernetics and psychiatry, this letter to *Nature* was part of a deliberate strategy to draw attention to his work, recorded in his journal in early June 1944 (p. 1666): "My plan is to write articles on political & economic organisation to try and make a stir there, knowing that then I can say that it is all based on my neuro-physiology. Another line is to watch for some dogmatic statement which I can flatly contradict in public, the bigger the authority who makes it the better." Of course, that he even thought about applying his mathematical analysis to economics is enough to illustrate the instability of its referent.
49. *An Introduction to Cybernetics* displays no interest at all in psychiatry, apart from a remarkable paragraph at the end of the preface. After thanking the governors of Barnwood House and Fleming, its director, for their support, Ashby continues: "Though the book covers many topics, these are but means; the end has been throughout to make clear what principles must be followed when one attempts to restore normal function to a sick organism that is, as a human patient, of fearful complexity. It is my faith that the new understanding may lead to new and effective treatments, for the need is great" (1956, vii).
50. Thanks to the efforts of Ashby's grandson, Michael Ashby, *An Introduction to Cybernetics* is now online as a PDF file at the *Principia Cybernetica* website at http://pcp.vub.ac.be/IntroCyb.pdf. The scanning process has left its marks in some easily spotted typographical errors.
51. The book is indeed written as a textbook, including many practice questions, with answers at the back. The question for which I have a particular fondness

is example 2 in section 2.9 (1956, 15): "In cricket, the runs made during an over transform the side's score from one value to another. . . . What is the cricketer's name for the identical transformation?" The Englishness of it all. I wonder how many readers of the Bulgarian translation, say, or the Japanese, would have known that the answer is "a maiden over" (1956, 274), and how many promising careers in cybernetics ended abruptly with that question.

52. On 15 October 1951, Ashby made some notes on extracts from a symposium on information theory held in London in September 1950. "A most important (but obscure) contribution by Shannon. . . . I cannot, at present, follow it. I must find out more about this sort of thing" (p. 3510).

53. Recall how readily hyperastronomical numbers were liable to appear in the kinds of combinatorics that cybernetics led into, as in Ashby's estimates of how long a multielement homeostat setup would take to reach equilibrium. On Bremermann's limit see, for example, Ashby (1969). Ashby mentions there his work on deriving a measure of "cylindrance," a number that characterizes the extent to which systems of interconnected variables can be decomposed into independent sets, and thus how likely our thinking on such systems is to confront Bremermann's limit. The problems of DAMS were clearly still on Ashby's mind in the mid-1960s; Ashby says of cylindrance that "it treats not only the fairly obvious case in which the relation consists of *k* wholly independent subrelations but also the much more interesting case in which the whole relation has something of the simplicity of a *k*-fold division while being in fact still connected. (An elementary example is given by a country's telephone communications, in that although all subscribers are joined potentially to all, the actual communications are almost all by pairs)" (Ashby 1968a, 74).

54. I cannot resist one last observation on *An Introduction to Cybernetics*, which is that in it Ashby talked himself yet again into the problem of accumulating adaptations, which we discussed earlier. At the end of the penultimate chapter (Ashby 1956, 260–62), he turned to the question of the "duration of selection"—how long would take it take to pick the right element out of an enormous array of possibilities? This was an immediate translation of the question in *Design for a Brain* of how long it would take a multihomeostat setup to find its equilibrium state. Again in *An Introduction to Cybernetics* he concluded that a direct search would take a hyperastronomical time and remarked that, in contrast, the search would be quick if the system were "reducible," the equivalent of letting each homeostat come to equilibrium independently of the others. Of course, the problem of getting around this assumption of absolute independence was the rock on which DAMS was already, by 1956, foundering and which eventually the gating mechanism of the second edition of *Design for a Brain* magically conjured away. But *An Introduction to Cybernetics* was an introductory text, so after a simple discussion of reducibility Ashby felt at liberty to remark: "The subject must be left now, but what was said in *Design for a Brain* on 'iterated systems,' and in the chapters that followed, expands the thesis" (1956, 262). The circle is squared; the unfortunate reader is sent off to wander endlessly in a hall of mirrors.

55. This was earlier argued by Kenneth Craik (1943). Having read his book, Craik was, as mentioned earlier, one of the first people Ashby wrote to about his nascent cybernetics.

56. A model is a representation of its object, so one wonders what sort of a mechanism Ashby conceived for representation. If one thinks of the homeostat and DAMS, this is a puzzle, but Ashby's thinking here was distinctly undramatic—he was happy to conceive representation in terms of programs running on digital computers. "The Big Computer—how much difference will be made by its advent may well be a matter of opinion. I rather lean to that of Martin Shubik who suggested that its impact may eventually be as great as that of the telescope, opening up entirely new worlds of fact and idea (after we have learned to use it appropriately)" (Ashby 1966, 90; see also Ashby 1970). Even so, Ashby had a distinctive understanding of how computers might be used. Ashby (1968c) mentions Illiac IV as a parallel processing machine which will need a new style of programming, and continues: "Little, however, is being done in the direction of exploring the 'computer' that is brainlike in the sense of using nearly all its parts nearly all the time" (1493). The discussion that follows includes habituation, which points to DAMS as a referent for the latter style of machine.
57. Ashby's *An Introduction to Cybernetics* culminates with the epistemology of selection: " 'Problem solving' is largely, perhaps entirely, a matter of appropriate selection. . . . It is, in fact, difficult to think of a problem, either playful or serious, that does not ultimately require an appropriate selection as necessary and sufficient for its solution" (1956, 272) The book ends within two paragraphs, in a discussion of intelligence amplification, which Ashby discusses elsewhere in terms of the use of a random source to generate a list of possibilities which can then be checked automatically against some well-defined goal.
58. To put some flesh on this critique, I can note that in his 1970 essay on models, discussed above, Ashby pondered the question of finding the right model for some system in much the same terms as he had with Walker, invoking many of the same historical examples: Newton, Gauss, Mozart. His pièce de résistance, however, was the nineteenth-century mathematician Sir William Rowan Hamilton (Ashby 1970, 109), about whose construction of the "quaternion" system I have, as it happens, written at length (Pickering 1995, chap. 4). My argument there was that Hamilton visibly shifted the "goal" of his research in midstream, starting with the intention of constructing a three-place analogue of complex numbers but ending up with a four-place one instead, and *made up* all sorts of mathematical possibilities (including ones like noncommuting variables, explicitly disallowed in the mathematics of the day) and tried them out to see what worked. Ashby and Walker would have to regard the former as blatant cheating and lack any language to conceptualize the latter.
59. This biographical information is taken from Alexander's website: www.patternlanguage.com. Besides architecture, Alexander's work has also been very influential in software design, but I will not follow that here—see the "wiki" link at the website just cited. I thank Brian Marick for an introduction to this aspect of Alexander's work.
60. Alexander's notions on design match Ashby's thoughts on evolutionary design and the blueprint method, but the latter were unpublished, and Alexander arrives at them from a different route.

61. One can find instances of Ashby talking about arrays of lightbulbs as a way of conjuring up the horrendous numbers one can generate by combinatorics, but all the examples that I have yet found date from after Alexander's book was published. See, for example, Ashby (1970, 99).
62. As usual, bringing the ontology down to earth involved adding something by way of specification: in this instance, the assumption that the matrix of interconnecting misfits indeed falls into weakly coupled sectors. Alexander thus "solved" the problem of complexity by fiat. DAMS, in contrast, was intended to solve this problem for itself. As analogues to the brain, Alexander's designs were not as adaptable as Ashby had hoped DAMS would be.
63. I am grateful to Stuart Kauffman for conversations about his life and work when he visited the University of Illinois in March 2004.
64. Ashby, von Foerster, and Walker (1962) had earlier discussed the instability of a network of idealized neurons. They showed analytically that the network would settle down either to zero or full activity depending upon the intensity of the input stimulation. In a further continuation of this work, Gardner and Ashby (1970) used computer calculations to explore the stability of linear networks as a function of the density of their interconnection and discovered a discontinuity with respect to this variable: below some value (dependent on the total number of elements in the system) the system would be stable; above that it would be unstable. Gardner's research was for a master's thesis at the University of Illinois, cited in Ashby (1968b).
65. Waldrop (1992, chap. 3) covers the trajectory of Kauffman's research career. He became interested in the problematic of genetic circuits and embryogenesis as a graduate student in 1963 on reading the work of Jacob and Monod on gene-switching, and tried to explore the behavior of small, idealized, genetic circuits in pencil-and-paper computations, before paying for the computer simulations in which the results discussed below first emerged. In 1966, he contacted the doyen of American cybernetics, Warren McCulloch, who was well known for his work on neural networks (dating from McCulloch and Pitts's foundational paper in 1943, and identical to genetic nets at a binary level of abstraction). At McCulloch's invitation, Kauffman spent the period from September to December of 1967 with McCulloch's group at the MIT Research Laboratory of Electronics (living at McCulloch's house). Kauffman's first publication on genetic networks was written jointly with McCulloch as an internal report of that laboratory: Kauffman and McCulloch (1967). Kauffman's first refereed publications appeared in 1969: Kauffman (1969a, 1969b). In 1969 at the University of Chicago, "Kauffman heard about Ashby's *Design for a Brain*, and "I got in touch with him as soon as I found out about it" (Waldrop 1992, 121). The only citation of prior work on networks in Kauffman (1969b) is of Walker and Ashby (1966), and the paper thanks Crayton Walker (and others) for encouragement and criticism. In 1971 the American Cybernetics Society awarded Kauffman its Wiener Gold Medal for the work discussed here. As a sidelight on the continuing marginalization of cybernetics, Waldrop manages to tell this whole story without ever using the word "cybernetics" (and there is no entry for it in the index to his book). McCulloch is described as "one of the grand old men of neurophysiology—not to mention computer science, artificial

intelligence, and the philosophy of mind" (Waldrop 1992, 113), and Ashby is an "English neurophysiologist" (120).

66. For earlier and explicitly cybernetic work in this area, including an electrical model of cell differentiation, see Goldacre's presentation at the second Namur conference in 1958 (Goldacre 1960a, also Goldacre and Bean 1960). At the same meeting Goldacre presented the paper "Can a Machine Create a Work of Art?" based on the same electrical model (Goldacre 1960b). Goldacre's affiliation was to the National Cancer Hospital in London. For a continuing strand of self-consciously cybernetic biology, see the journal *Biological Cybernetics* (originally entitled *Kybernetik*). On the history of developmental biology in the twentieth century, see Keller (2002).

67. On self-organization as a key concern at the BCL, see Asaro (2007). The idea that organisms have structures that can be understood without reference to the specific vicissitudes of evolution echoes most notably the work of D'Arcy Thompson (1961 [1917]). Though Kauffman's style of theoretical biology remained marginal to the field while the emphasis was on the reductive unraveling of genomes, in these days of postgenomics it appears now to be coming to the fore: see Fujimura (2005) (making interesting connections between theoretical biology and Walter/Brooks-style robotics) and O'Malley and Dupré (2005).

68. As usual I, am extracting the ontological features I want to draw attention to. The behavior of any finite state-determined network must eventually be cyclical, but this is not a feature I would recommend for ontological generalization. Neither should we generalize the idea that objects can exist in a denumerably finite number of states (or that such states possess a knowable matrix of transition probabilities). Kauffman (n.d.) himself reads the moral of his later work in terms of an ontology of unknowability: "This truth is a radical departure from our image of science from physics. It literally means that we cannot know beforehand how the biosphere will evolve in its ceaseless creativity. . . . The growth of the economic web may NOT be algorithmic. It may be fundamentally unknowable, but nevertheless livable. Life, after all, is not deduced, it is lived."

69. www.santafe.edu/sfi/People/kauffman/.

70. We can note also some of the tensions within science that have led to the establishment of institutions like the SFI. Wise and Brock (1998, 386) quote remarks on complexity from orthodox physicists at a meeting at Princeton University in 1996: "One really can't help feeling childish fascination looking at this picture of different beautiful systems. But switching to my adult mode, I start thinking about what I can really do as a theorist apart from going to my kitchen and trying to repeat these experiments"; "It seems to me that you are viewing the patterns in non-equilibrium systems like a zoo, where we view one animal at a time, admire it and describe it, and then go on to the next animal." The opposition between "adult" and the string "child-kitchen-zoo-animal" is interesting in many ways, but, at the least, it registers the difficulty that modern scientists have in taking seriously what lies outside the modern circle, and it functions as a warning not to venture beyond that circle.

71. For a popular account of the SFI and complexity theory, including Kauffman's work, see Waldrop (1992), and on work there on artificial life see Helmreich

(1998). For an overview of the SFI, see www.santafe.edu/aboutsfi/mission.php. "About a quarter of our activities are funded through the corporate affiliates program, another quarter through private donations, and the remaining half via government and foundational grants. We do have a small amount of endowed funds, and would warmly welcome anyone wishing to make that a large amount. We also welcome smaller private donations. Ultimately, we want to define and understand the frontiers of science, and the very nature of such a quest often requires us to rely on non-traditional funding sources." www.santafe.edu/aboutsfi/faq.php. The SFI does offer postdoctoral fellowships, which enable young researchers to work with established faculty members. The SFI website also currently lists two graduate students working with SFI faculty on their dissertation research.

72. www.biosgroup.com/company_history.asp (25 April 2007).
73. This information is taken from an unpublished document, "Stephen Wolfram: A Time Line," which was on Wolfram's website in 1999 but is no longer to be found there. I am not going to reconstruct Wolfram's historical route to CAs and his "new kind of science" here; unlike Alexander's and Kauffman's, Wolfram's early intellectual development did not pass through cybernetics (interview, 19 May 1999). Many different trajectories have fed into current work on complexity theory. I am grateful to Stephen Wolfram for several opportunities to discuss his work over recent years; I regret that I cannot go further into its substance here.
74. "Sometimes I feel a bit like a naturalist, wandering around the computational world and finding all these strange and wonderful creatures. It's quite amazing what's out there" (Wolfram 2005, 8).
75. The source on all of these applications and more is Wolfram (2002).
76. www.wolframscience.com/summerschool/2005/participants/.

Notes to Chapter 5

1. On Bateson, see Harries-Jones (1995), Lipset (1980), M. C. Bateson (1984), and Heims (1991, chap. 4); for a variety of contemporary perspectives on and extensions of Bateson work, see Steier and Jorgenson (2005). This outline of Bateson's life is drawn from Harries-Jones (1995, xi–xiii), and I thank him for very useful comments on an earlier draft of this chapter. Bateson overlapped with Ashby as an undergraduate at Cambridge; there is no evidence that they knew each other then, but Bateson was one of the first people Ashby wrote to when attempting to make contacts about his cybernetics in the mid-1940s. Bateson described his early entanglement with cybernetics thus (G. Bateson 2000, xix–xx): "In 1942 at a Macy Foundation conference, I met Warren McCulloch and Julian Bigelow, who were then talking excitedly about 'feedback.' The writing of *Naven* had brought me to the very edge of what later became cybernetics, but I lacked the concept of negative feedback. When I returned from overseas, I went to Frank Fremont-Smith of the Macy Foundation to ask for a conference on this then-mysterious matter. Frank said that he had arranged such a conference with McCulloch as chairman. It thus happened that I was privileged to be a member of the famous Macy conferences on cybernetics. My debt to Warren McCulloch, Norbert Wiener, John von Neumann, Evelyn

Hutchinson, and other members of these conferences is evident in everything I have written since World War II." I thank David Hopping for encouraging me to take an interest in Bateson, and him and Judith Pintar for stimulating discussions of psychotherapy and madness. Before Hopping, Emily Ignacio and Richard Cavendish tried unsuccessfully to interest me in Bateson.

2. For extended accounts of this project and the contribution of different members, see Haley (1976) and Lipset (1980, chap. 12). On its empirical aspect: "We have studied the written and verbal reports of psychotherapists who have treated such [schizophrenic] patients intensively; we have studied tape recordings of psychotherapeutic interviews; we have interviewed and taped parents of schizophrenics; we have had two mothers and one father participate in intensive psychotherapy; and we have interviewed and taped parents and patients seen conjointly" (Bateson et al. 1956, 212).
3. Thus Bateson (1959) introduces a discussion of schizophrenia and the double bind with a discussion of Pavlovian learning.
4. Part of Bateson's understanding of the double bind was also that discussion of it was somehow impossible for those involved in it.
5. One might think here of J. L. Austin on "speech acts" as "performative utterances," but Bateson's distinctly cybernetic take on this was to recognize the interactive and dynamic aspects of performative language.
6. Harries-Jones (1995, 111, 114) notes that Bateson was "so excited" by the homeostat that "he made it the focus of a correction of his former ideas . . . a sort of auto-critique of his prior belief in mechanistic versions of social change," and that he "tried to develop his own homeostat." Shorter (1997, 177) suggests that mothers tend to get demonized in family approaches to madness, and this example from the first schizophrenia paper certainly points in that direction. But the symmetric image of mutually adapting homeostats implies that double binds should not be seen as originating in specific individuals within any system. Bateson was later criticized by feminists and others for refusing to ascribe causality to specific family members: see Dell (1989).
7. Strictly speaking, a koan is a paradoxical verbal formulation upon which one meditates—"the sound of one hand clapping," or whatever—a representation of what seems to be an impossible referent. It is significant that Bateson favors a performative version here.
8. In a 1964 essay Bateson (2000, 304) quotes a Zen master as stating that "to become accustomed to anything is a terrible thing" and continues: "To the degree that a man . . . learns to perceive and act in terms of the contexts of contexts, his 'self' will take on a sort of irrelevance. The concept of 'self' will no longer function as a nodal argument in the punctutation of experience." For an extended and systematic exposition of the Buddhist notion of losing the self, see Varela, Thompson, and Rosch (1991). They discuss a technology of the self which they call mindfulness/awareness meditation, which can afford direct access to the nonexistence of any enduring self (and likewise the nonexistence of any enduring outer world). Francisco Varela was one of the founders (with Humberto Maturana) of the branch of cybernetics concerned with the autopoiesis of living systems and was also science adviser to the Dalai Lama. The stated aim of Varela et al.'s book is to open up a "dialogue between

science and experience" (xix). "Experience" here refers to the mind as known in the Buddhist tradition, and "science" to cognitive science. Within the latter, the book is a critique of mainstream AI research and makes alliances with the branches concerned with performance and embodiment, mentioning the work of Walter and Ashby and concentrating on Rodney Brooks (see chap. 3 above; Varela et al. 1991, 208–12). A discussion of the Buddhist "aggregates" is followed by one of EEG readings in a class of flicker experiments pertaining to the alpha rhythms of the brain, which point to a kind of temporal "chunking" of experience. We saw in chapter 3 that the cyberneticians saw this as pointing to a scanning mechanism in the brain; Varela et al. (1991, chap. 4) take it as evidence for the Buddhist notion that our experience of both the inner and outer worlds is ephemeral, discontinuously arising and passing away.

9. Huxley was closer to Walter than Bateson in looking for material explanations of the go of both transcendence and madness. *The Doors of Perception* mentions two possible mechanisms. One is that our evolutionary history has fitted us to engage with the world from a goal-oriented standpoint; our very senses function as a "reducing valve" appropriate to an alertness to dangers and opportunities but not to other properties and relations. The modern self is thus the product of adaptation over evolutionary time, and Huxley's argument is that psychedelic drugs somehow undercut our innate tendencies to enframe the world, in Heidegger's sense. In terms of inner mechanisms, Huxley refers to ideas on brain chemistry taken from Humphrey Osmond and John Smythies (chap. 3) rather than to electrical properties of the brain. Osmond supplied the mescaline on which *Doors* was based.
10. In the late 1960s, Bateson found a model for this higher level of adaptation in experiments on dolphins. In 1969, he referred to a series of experiments in which a dolphin was trained to perform specific tricks for rewards. The experimenter then decided that what was required next was simply to produce some trick the dolphin had never done before. In succeeding trials the dolphin went through its existing repertoire and became exceedingly agitated on getting no reward. Finally, its behavior changed between runs, and at the next opportunity it performed a spectacular series of new displays—as if it had figured out that the required response was something new, and had become a happier dolphin in the process (Bateson 1969,241–42). For more on this work see Lipset (1980, 249–51).
11. Including the analyst within the scope of cybernetic systems is one definition of second-order cybernetics, though, as discussed in chapter 2, second-order cybernetics usually stresses epistemological questions while I want to highlight the performative aspect of Bateson and Laing's work here.
12. The first schizophrenia paper also mentions the therapist Frieda Fromm-Reichmann, who obtained positive results by imposing a "therapeutic double bind" upon her patients—one that would challenge the conditions of the original double bind (in this case, manifest in the patient's conviction that she was being ordered to perform certain acts by an array of powerful deities), and encourage finding a different avenue of escape (Bateson et al. 1956, 226–27).
13. They also remind me of the octopus which appears at key junctures in Thomas Pynchon's masterpiece, *Gravity's Rainbow* (1975). We can now appreciate an

allusion that has escaped literary scholars: the name of this numinous creature is Grigori. For a list of pages on which it appears, see www.hyperarts.com/pynchon/gravity/alpha/o.html#octopus.

14. On the Wenner-Grenn conferences, see Lipset (1980, 26–68), and for a brief survey of cybernetic approaches to ecology, including Bateson's work, and a nuanced discussion of the critique of cybernetics as a machinelike, command-and-control approach to nature, see Asplen (2005).
15. "My own slight experience of LSD led me to believe that Prospero was wrong when he said, 'We are such stuff as dreams are made on.' It seemed to me that pure dream was, like pure purpose, rather trivial. It was not the stuff of which we are made, but only bits and pieces of that stuff. Our conscious purposes, similarly, are only bits and pieces. The systemic view is something else again" (Bateson 1968, 49). The Mental Research Institute in Palo Alto, at which Allen Ginsberg first took LSD (chap. 3), is probably the same as that established by Don Jackson, a member of Bateson's schizophrenia group. Bateson was invited to join the latter MRI, but declined (Lipset 1980, 227).
16. Sensory deprivation is another technology of the nonmodern self. A sensory deprivation tank is a large container full of water to which salts have been added so that the human body achieves neutral buoyancy. The tank also excludes noise and light, so that floating in it one is largely deprived of any sensory input. Scientific research on sensory deprivation began with the work between 1951 and 1954 of D. O. Hebb's group at McGill University in Montreal, which "began, actually, with the problem of brainwashing. We were not permitted to say so in the first publishing. What we did say, however, was true—that we were interested in the problem of the effects of monotony. . . . The chief impetus, of course, was the dismay at the kind of 'confessions' being produced at the Russian Communist trials. 'Brainwashing' was a term that came a little later, applied to Chinese procedures. We did not know what the Russian procedures were, but it seemed that they were producing some peculiar changes of attitude. How?" (Hebb, quoted in Heron 1961, 6). For reports on current research from a 1958 symposium, see Solomon et al. (1961). Like LSD, sensory deprivation proved to be a bivalent technology, leading to mental breakdowns in "scientific" settings but, as Lilly and others discovered, giving rise to transcendental states in more congenial ones. For a characteristically sixties version of the effects of sensory deprivation, one has only to look at the account given by Bateson's wife, Lois, of her hour in Lilly's tank: "Roamed and sauntered through a kind of cosmic park, full of density but infinite boundaries . . . a kind of total consciousness. . . . Sudden enlightenment—there is no such thing as separate consciousness. My roamings were a kind of total consciousness of all that was. The dense bits here and there—I was it—it was me—the people—same—there was no boundary between me and them—pronouns are only illusions!" (Lilly 1977, 190). Enlightenment as the loss of the self. For more on Lois Bateson and the counterculture (not named as such), see Lipset (1980). Bateson married Lois Cammack in 1961, and it would be interesting to know more about her biography and relation to the counterculture. Bateson's life certainly seems to have taken a countercultural turn after this date. Lilly developed a theory of mind that was explicitly "cybernetic"—see, for example,

chapter 7 of *The Deep Self* (Lilly 1977), "The Mind Contained in the Brain: A Cybernetic Belief System." I thank Mike Lynch for suggesting that I take an interest in Lilly. Two movies fictionalizing dramatic effects of sensory deprivation are Basil Dearden's *The Mind Benders* (1963) and Ken Russell's *Altered States* (1980). I thank Peter Asaro for tracking down both of these. In literature, see Colin Wilson, *The Black Room* (1971). There the Black Room is a device to break down spies, and the central character discovers that one can defeat this via another technology of the self, developing the will, which leads to a new state of "heightened consciousness," which the human race has always been capable of but never before systematically achieved. EEG readings and flicker also put in appearances in the plot.

17. If the focus of this book were on the United States rather than Britain, Brand would be a key figure. He often lurks in the shadows here: he wrote a wonderfully cybernetic book on architecture called *How Buildings Learn* (1994) and another (Brand 1987) on the MIT Media Lab, founded by Nicholas Negroponte, who appears in chapter 7 as a collaborator with Gordon Pask. In the next chapter I discuss Brian Eno's relation to Stafford Beer, and Brand appears in Eno's diary as his principal interlocutor (Eno 1996a). For much more on Brand, see Turner (2006).
18. In a phrase, the New Age movement is based on a nonmodern ontology in which mind, body and spirit are coupled to one another in a decentered fashion, much like cybernetics, and hence the crossover. The history of Esalen deserves more attention than I can give it here; see Anderson (2004 [1983]), Kripal and Shuck (2005), and Kripal (2007). Richard Price, one of the two founders of Esalen (the other was Michael Murphy), took anthropology courses with Bateson at Stanford University (Kripal 2007, 79; Bateson held the position of visiting professor at Stanford while he was working on schizophrenia in Palo Alto, and his grants were administered by the university: Lipset 1980, 196, 237). Price and Murphy consulted Bateson in their planning for the institute (Anderson 2004 [1983], 49). The first seminar at Esalen was offered by Murphy, the second by Bateson (and Joe Adams) under the title "Individual and Cultural Definitions of Reality." "What they were actually up to was an explicit comparison between the present state of the mental health profession and theInquisition of the late medieval period" (Kripal 2007, 101, 170). Erickson describes Esalen's founding in 1962 as very much in the spirit of "antipsychiatry": "Esalen is Price's revenge on mental hospitals" (Erickson 2005, 155, quoting Murphy). A series of informal talks that Bateson gave at Esalen between 1975 and 1980 is available on audiotape at www.bigsurtapes.com/merchant.mv36.htm. On Laing's connection to Esalen, see note 43 below. Stafford Beer was also invited to visit Esalen (Allenna Leonard, email, 8 May 2006).
19. Harries-Jones (1995, 11) remarks that Bateson's readers "can see that Bateson points towards an entirely different set of premises about a science of ecology, but find it difficult to distinguish his radical thinking about holistic science from the communal or mother-earth spiritualism of the counter-culture—the 'New Age' approach made familiar through the mass media. Bateson rejected the latter as anti-intellectual." More issues open up here than I can explore, but briefly, if we see the Esalen Institute as an early home of New Age, it is hard to

see Bateson's choice to spend his last two years there as a rejection; it is also hard to see Esalen's enfolding of Bateson as a display of anti-intellectualism. But Harries-Jones is right if we take his reference to the mass media as pointing to a subsequent commodification of New Age. New Age has become an industry and a market as well as a form of life in which the intellectual concerns of people like Bateson, Laing, Huxley, Walter, and Beer are conspicuously absent. We could see this, in turn, as a symptom of the walling-off from mainstream culture of the sixties curiosity about strange performances and altered states—a reciprocal purification that mirrors the expunging of the nonmodern from the world of modern science, capital, and militarism. In this history of cybernetics I am trying to remember that our possibilities are not confined to these two alternatives, an anti-intellectual New Age and the world of modern science. To take this line of thought a little further, I could note that New Age has not become entirely anti-intellectual—see, for example, the discussions at Esalen of science and spirituality reproduced in Abraham, McKenna, and Sheldrake (1992). These discussions, however, remain largely at the level of ideas; the performative dimension is largely absent. I thank Jan Nederveen Pieterse for alerting me to the work of these authors.

20. On Laing, see, for example, Howarth-Williams (1977), Laing (1985), Kotowicz (1997), Miller (2004), Burston (1996), and a biography written by his son, Adrian: A. Laing (1994). I am indebted to Malcolm Nicolson and, especially, Ian Carthy for guidance on Laing and the relevant literature.
21. Laing trained as a psychoanalyst in London. "This shows in his ample use of psychoanalytical language although it seems that it did not inform the way he worked very much" (Kotowicz 1997, 74).
22. The term "antipsychiatry" seems to have been put into circulation by Laing's colleague David Cooper in his book *Psychiatry and Anti-psychiatry* (1967), but Laing never described himself as an "antipsychiatrist."
23. Laing traveled to the United States in 1962 for discussions with Bateson (as well as Erving Goffman and others; Howarth-Williams 1977, 4–5). Lipset (1980) includes quotations from interviews with Laing that make it clear that he and Bateson became friends. Laing quotes Warren McCulloch disapprovingly on the cybernetics of the brain; a later passage makes it clear that it is the vivisectionist aspects of cybernetics that he cannot abide (which we could associate here with the asymmetric psychiatry of Walter and Ashby; Laing 1976, 107–8, 111ff.).
24. Likewise, "the 'double-bind' hypothesis . . . represented a theoretical advance of the first order" (Laing 1967, 113). We can note that, like Bateson, Laing understood the double bind in a decentered fashion. Thus, "One must remember that the child may put his parents into untenable positions. The baby cannot be satisfied. It cries 'for' the breast. It cries when the breast is presented. It cries when the breast is withdrawn. Unable to 'click with' or 'get through' mother becomes intensely anxious and feels hopeless. She withdraws from the baby in one sense, and becomes over-solicitous in another sense. Double binds can be two way" (Laing 1961, 129).
25. Laing and Pask eventually met, years after Pask had begun citing Laing, at a 1978 conference in Sheffield on catastrophe theory. "That night Ronnie, Gordon

Pask and I [Adrian Laing] got totally inebriated while intermittently ranting and raving. . . . Ronnie had met a soul mate in Gordon Pask—his bifurcation talk stayed in Ronnie's mind for many years" (A. Laing 1994, 203–4). One can identify at least one further personal connection between Laing and the British cybernetics community. Adrian Laing (1994, 33) refers to his father's friendship later in his life with "the mathematician and author David George Spenser-Brown," and he must be referring to George Spencer Brown (hyphenated by some, à la Grey-Walter), the author of a book on a nonstandard approach to logic and mathematics, *Laws of Form* (1969). "During the 1960s, he [Brown] became a disciple of the maverick British psychiatrist R. D. Laing, frequently cited in *Laws of Form*" (en.wikipedia.org/wiki/G._Spencer-Brown [accessed 28 November 2006]), The first edition of *Laws* in fact cites Laing's *The Politics of Experience* just once, but a later book by Brown, writing as James Keys (1972), links Eastern and Western philosophy and includes a preface by Laing (twenty-nine words, stretched out over six lines covering a whole page). *Laws of Form* attained almost cult status in second-order cybernetics as offering a formalism for thinking about the constructedness of classifications: "a universe comes into being when a space is severed or taken apart" (Brown 1969, v). On the Web one can find the proceedings of the American University of Masters Conference, held at the Esalen Institute in California in March 1973, in which Brown held forth over two days to an audience including Gregory Bateson, Alan Watts, John Lilly, and Heinz von Foerster. In the first session he derided Russell and Whitehead's theory of logical types, central to Bateson's understanding of the double bind, as contentless (www.lawsofform.org/aum/session1.html [accessed 28 November 2006]). Stafford Beer (phone interview, 23 June 1999) told me that he twice found paid employment for Brown in the 1960s while he was writing the book, at his SIGMA consultancy and at the International Publishing Company. Brown sued him for wrongful dismissal at the latter but was "seen off by [IPC] lawyers." Beer also told me that Brown was unable to find a publisher for *Laws of Form* until Bertrand Russell put his weight behind it. Beer and Heinz von Foerster reviewed the book for the *Whole Earth Catalog*, founded by Stewart Brand (n. 17 above), thus bringing it to the attention of the U.S. counterculture.

26. Laing (1985, 143), referring to his first university appointment in Glasgow. The quotation continues: "In this unit all psychiatric social-work students have a standing order not to permit any schizophrenic patient in the wards to talk to them."
27. In 1953, Osmond and Smythies (n. 9) had proposed that psychiatrists take mescaline "to establish a rapport with schizophrenic patients. . . . 'No one is really competent to treat schizophrenia unless he has experienced the schizophrenic world himself. This it is possible to do quite easily by taking mescaline'" (Geiger 2003, 29).
28. On the history of LSD in British psychiatry, see Sandison (1997). Barnes and Berke (1971) list Sigal (spelled "Segal") among the past and present members of the Philadelphia Association, and Mary Barnes recalled that he was the first person she met at Kingsley Hall (Barnes and Berke 1971, 5, 95). Kotowicz (1997, 87) says that Sigal's novel was not published in Britain for fear of libel action.

29. One can complicate this story further. Laing incorporated LSD into his private psychiatric practice in the sixties and "preferred to take a small amount of the LSD with the patient, and for the session to last not less than six hours. . . . Throughout the early sixties, Ronnie's practice in Wimpole Street gained a reputation verging on the mythological, principally due to his use of LSD in therapy" (A. Laing 1994, 71–72). In these sessions LSD functioned as a technology of the nonmodern self for both psychiatrist and patient.
30. Shorter (1997, 229–38) gives a brief history, emphasizing the importance of the Tavistock Insitute, where Laing and his colleagues worked.
31. Ken Kesey's (1962) *One Flew over the Cuckoo's Nest* is a biting portrayal of a pseudoimplementation of social psychiatry in the United States and as first a novel and then a movie became a key document in the critique of orthodox psychiatry in the 1960s. See also Joseph Berke's account of "social psychiatry" in a U.S. mental hospital in Barnes and Berke (1971, 89–92).
32. Cooper (1967) is his own account of the Villa 21 experiment. The problems arising from attempts to embed a bottom-up unit within a top-down structure were not peculiar to mental hospitals. I think of the Pilot Program installed at a General Electric plant in the United States in the 1960s, as documented by Noble (1986) and discussed in Pickering (1995, chap. 5).
33. Adrian Laing (1994, 101) describes these as the core group, and adds Joan Cunnold and Raymond Blake.
34. Kotowicz (1997, 79–87) discusses attempts to set up institutions analogous to Villa 21 and Kingsley Hall at the Heidelberg University psychiatric clinic in Germany and at the Gorizia mental hospital in Italy, where similar tensions also surfaced. "The hospital embodies the basic contradiction that is at the root of psychiatry—the contradiction between *cura* (therapy, treatment) and *custodia* (custody, guardianship). The only solution was to dismantle the institution altogether" (82). The Italian experiment was relatively successful. The German story became entangled with that of the Baader-Meinhof Red Army Faction; some of the leaders of the German project received prison sentences, others went on the run and fled the country.
35. *Asylum* was released in 1972, with introductory exposition by Laing, and is available as a DVD distributed by Kino on Video (www.kino.com). I thank David Hopping for showing me Burns (2002) and Judith Pintar for bringing *Asylum* to my attention and lending me her DVD. There is an extensive Laing archive at the University of Glasgow, but historians of psychiatry are only beginning to explore this source and seem bent on situating Laing in the history of ideas rather than practices.
36. Laing's only substantive writing about Kingsley Hall that I have been able to find is a short talk from 1968, Laing (1972), which mentions Mary Barnes's "voyage" at Kingsley Hall and an incident in which one resident (with a bird tied to his head) shot another resident (David, naked and obsessed with fears of castration) in the genitals with a Luger. It turned out the gun was loaded with blanks, and Laing emphasized the performative aspect of this interaction: "David looked down and saw his genitals were still there. . . . He lost as much of his castration anxiety in that incident as he had done in the four years that I had been seeing him in analysis. No interpretations could be as primitive as that dramatic action, completely unpredictable and unrepeatable. At Kingsley

Hall we have hoped to have a place where such encounters could occur" (Laing 1972, 21).

37. There is a good description of the building and its history in Barnes and Berke (1971, 215–18). Kingsley Hall's most illustrious previous inhabitant was Gandhi, who lived there for six months in 1931 and met representatives of the British government there, while sharing his room with a goat and living on its milk. See also Kotowicz (1997, 76).
38. Burns (2002) goes into the mundane details. One important asymmetry concerned money: the mad, who typically could not hold down jobs, had less of it than the others. See also Barnes and Berke (1971) on a split between Laing and Esterson concerning the need for organizational structure at Kingsley Hall. Here and below I follow the standard usage of Laing and his colleagues and refer to "the mad" instead of "the mentally ill" or similar formulations. This seems appropriate, as part of what was contested at Kingsley Hall was whether "the mad" were indeed "ill."
39. A central character in the *Asylum* documentary talks almost continuously in a way that relates only obscurely, if at all, to remarks addressed to him, frequently driving the other residents to distraction in their attempts to communicate both with him and each other.
40. Howarth-Williams (1977, 5) notes that Laing moved into Kingsley Hall on his own, without his wife and five children, and that his return home was a period of family and personal crisis for him, coinciding with his separation from his family. Kotowicz notes, "Laing left after a year, Esterson did not last any longer, and the maximum any therapist stayed was two years. Does it say something that the only person who stayed from the beginning to the end was Mary Barnes?" I am grateful to Henning Schmidgen for an illuminating personal account of his stay at a comparable French institution, the Clinique de La Borde at Courvenchy.
41. Burns (2002) begins: "For more than fifteen years an experiment has been carried out in London. . . . I lived in these communities for five years, from early 1970 until late 1975 and was in association with them until late 1977." We can thus date his manuscript to around 1980. Of the period of transition, Burns (1–2) recalls, "the group living at Kingsley Hall toward the end had lost cohesiveness and the therapists known as the Philadelphia Association decided, after some hesitation, to continue the experiment with a new group. Only three of us, therefore, moved into the new community. This consisted of two houses in a deteriorating neighbourhood. Since these houses were scheduled for demolition we had to move to new homes of the same nature, again and again. This was difficult and painful but the advantages were that any damage done to the structures was relatively unimportant and that there was somewhat less than the usual necessity for residents to keep up normal standards of behavior on the streets."
42. Despite its asymmetry in singling out specific individuals at any given time, on Burns's account it could be applied to or invoked by all of the members of the community, so the underlying symmetry remains.
43. A biography of Laing records that "Kingsley Hall had a dreadful record when it came to documenting therapeutic sucesses. The whole ethos of the experiment was against documentation." On the other hand, Loren Mosher implemented

"a rigorous research program" in two therapeutic households in the United States between 1971 and 1983 and "demonstrated that therapeutic households that make minimal use of medication and extensive use of suitably trained paraprofessionals are just as effective as standard psychiatric facilities—sometimes more effective—and can circumvent the toxic side effects of neuroleptic drugs or electroshock." Returning to the topic of marginality, the same work continues: "Mosher's experiment was inspired by his brief stay at Kingsley Hall and is well documented elsewhere. But . . . it has been studiously ignored by the psychiatric community" (Burston 1996, 244–45). We can also return to the Esalen Institute at this point (nn. 18 and 19 above). At Esalen in the midsixties, Richard Price "was beginning to think about starting, perhaps as an annex of Esalen, a treatment center modelled on Kingsley Hall," and was joined in this by Julian Silverman, a clinical psychologist from the National Institute of Mental Health in Bethesda, Maryland. Murphy, Price, and Silverman were interested in Laing's work. Laing visited Esalen in 1967 and was expected to figure centrally in a major Esalen seminar series in 1968, although in the event he did not appear. Silverman subsequently left Bethesda for a position at the Agnews Hospital in San Jose, California, where he attempted to explore the efficacy of the Kingsley Hall approach to psychiatry in a project sponsored by the state, the National Institute of Mental Health, and Esalen (Laing was invited to join this project and again refused to travel to the United States). Like Mosher's, the findings of Silverman's three-year project were positive: "Follow-up research . . . reported that those who had been allowed to recover without medication showed more long-term clinical improvement, lower rates of rehospitalization, and 'better overall functioning in the community between one and three years after discharge'" (Anderson 2004 [1983], 214–18). Silverman became the director of Esalen from 1971 to 1978 (Kripal 2007, 179).

44. Burns (2002) mentions that residents would sometimes get into trouble with the police—for wandering naked down the street, for example—who would turn them over to the local mental hospital. But after a brief stay there, they would often elect to return to their communities. Sigal (1976) helps make such a choice very plausible in relation to Villa 21 in the early 1960s.
45. www.philadelphia-association.co.uk/houses.htm.
46. "A new Philadelphia Association [had] virtually emerged (most of the original members having left and gone their separate ways), one which [was] somewhat less focussed on families and schizophrenia and much more organized, with a wide-ranging study programme" (Ticktin n.d., 5).
47. In a 1974 interview on "radical therapy," Laing described his work in almost entirely performative terms. In rebirthing sessions, "People would start to go into, God knows what, all sorts of mini-freak-outs and birth-like experiences, yelling, groaning, screaming, writhing, contorting, biting, contending." He continued, "I should mention massage, bodily sculpture, improvised games, etc, are all part of our ordinary, ongoing culture: wearing masks, dancing. . . . We are not identified with any special technique but we are into it, as the saying goes, for me particularly, music rhythm and dancing. When I go into one of our households for an evening usually music, drumming, singing, dancing starts up" (quoted in A. Laing 1994, 180).

48. For two U.S. projects modelled on Kingsley Hall, see note 43 above.
49. By 1971, *Politics* had sold four hundred thousand copies in paperback (A. Laing 1994, 161).
50. Laing published his first account of Watkins's voyage in 1965. Watkins appears as a visitor to Kingsley Hall in Barnes and Berke (1971).
51. J. Green (1988,178–80, 167) quoting Alan Marcuson and Peter Jenner; Leary (1970), quoted by Howarth-Williams (1977, 1).
52. Laing first met Ginsberg in New York in October 1964, where he also met Joe Berke (for the second time), Leon Redler, and Timothy Leary (A. Laing 1994, 98).
53. Before the Anti-University, there was the Notting Hill Free School, founded in 1965 and inspired by similar institutions in the United States. Joe Berke, newly arrived in London, was among the organizers. It was "a scam, it never really worked," but it somehow mutated first into the Notting Hill Fayre and then into the Notting Hill Carnival, which is still very much with us (J. Green 1988, 95–103; the quotes are from John Hopkins on p. 96).
54. The Huxley in question here is Francis, son of Sir Julian and nephew of Aldous, who met Laing around this time and whom Laing invited to become a member of the Philadelphia Association (F. Huxley 1989). An anthropologist, Francis studied Amazonian Indians in the early 1950s, "developed an interest in LSD and mescaline" through his uncle, and worked with Humphrey Osmond in Saskatchewan in the late 1950s and early 1960s (Melechi 1997, 48).
55. Byatt's (2002) fictional evocation of an antiuniversity in the north of England likewise includes a commune at its core. As the Anti-University became a commune—"dosshouse" was Jeff Nuttall's word—Berke, worried about who would pay the bills, "blew his top one night. He picked up one of these guys bodily and threw him out. Then he did this strange boasting thing: 'Who's the biggest guy on the block, who can throw anybody here? I can beat you! I'm the biggest guy on this block!' It was really dippy. He just regressed. Very sad" (Nuttall, quoted in J. Green 1988, 238–39).
56. D. Cooper (1968) reproduces some of the papers given at the conference.
57. The "numero uno" quote is from Alan Marcuson (J. Green 1988, 209). The list of attendees is taken from Miles and Alan Marcuson (J. Green 1988, 208, 209). The Berke quotation is from Berke (1970). The argument with Carmichael seems to have been a key moment in the political history of the sixties, though people recall it differently. Sue Miles: "It was quite frightening. Stokely Carmichael started this tirade against whitey. Though one could see perfectly well why he had this point of view, it was completely unworkable. Then there was this meeting afterwards back at this house where he was staying and there was extreme bad feeling and a huge argument and split between them all. Allen [Ginsberg] was going, 'This is dreadful. We have not argued this long for everyone to start getting at each other's throats and getting divided. This is not going to get us anywhere.'" Alternatively, Alan Marcuson again: "There was this wonderful dinner at Laing's. Laing had all the superstars there for dinner and he was very into being honest and he said to Stokely Carmichael, 'The thing is, Stokely, I like black people but I could never stand their smell,' and Stokely didn't like that and left" (J. Green 1988, 209, 210).

58. The quotations are from Trocchi's "Invisible Insurrection of a Million Minds" (1991a [1962], 178, 186) and "Sigma: A Tactical Blueprint (1991b [1962], 199). Trocchi was a key node linking several countercultural networks running from the British underground to the Beats and the French Situationist International (with situationism as central to the works just cited). For more on situationism, see Marcus (1989) and situationist international online: www.cddc.vt.edu/sionline/index.html. On Trocchi's part in the 1965 Albert Hall poetry reading, see J. Green (1988, 67–71).
59. A great deal of the discourse of social science, at least in the United States, circles around issues of equality and inequality, power and hierarchy, but I find it hard to think of any such work that thematizes issues of enframing and revealing. The organizing problematic is usually that inequality is simply bad and that we owe it to this group or that not to oppress them. The idea is almost never that we (the oppressors) might be able to learn something positive from the others. Think, for example, of the social science discourse on race.

Notes to Chapter 6

1. My primary sources on Beer's biography are a CV that he provided me, dated 1 January 1998, and a biographical letter Ian Beer sent to all of Stafford's children on 25 August 2002, immediately after Stafford's death. Otherwise unidentified quotations below are from this letter, and I thank Ian Beer for permission to reproduce them. Beer was the only one of the principles of this book still alive when I began my research. I talked to him at length on the phone twice: on 23 June 1999 and 22 December 1999. Various attempts to arrange to meet stumbled geographically: Beer was in Britain when I was in North America and vice versa. I went to Toronto hoping to see him in June 2002 but Beer was terminally ill and unable to speak while I was there. I met Beer's partner, Allenna Leonard, and his daughter Vanilla during that visit. I am very grateful to them for talking to me at length at that difficult time, and for many subsequent communications which I draw on throughout this chapter. I also thank David Whittaker for an informative conversation on Beer's poetry and spirituality and his relation to Brian Eno (Woodstock, Oxon, 24 August 2004). I should mention that Liverpool John Moores University has extensive holdings of Beer's papers: "Ninety Two Boxes of Beer" is the subtitle of the catalog. I have not consulted this archive: the present chapter is very long already.
2. The only use of Stafford's first name that I have found in print is in his dedication to his mother of his contribution to Blohm, Beer, and Suzuki (1986), which he signed "Tony."
3. Beer (1994b, viii): "Neurophysiologists did not talk much about logic at that time . . . and logicians were completely indifferent to the ways brains work."
4. Telephone conversation, 23 June 1999.
5. For more on the history of OR, see Fortun and Schweber (1993), Pickering (1995a), and Mirowski (2002).
6. Telephone conversation, 23 June 1999. There is actually a psychiatric connection here at the beginning of Beer's career. According to Beer (1989a, 11–12), "At the end of my military service, I spent a year from the autumn of 1947

to that of 1948 as an army psychologist running an experimental unit of 180 young soldiers. . . . All these men were illiterate, and all had been graded by a psychiatrist as pathological personalities. . . . I had a background in philosophy first and psychology second; the latter school had emphasized the role of the brain in mentation and of quantitative approaches in methodology. The analytical models that I now developed, the hypotheses set up and tested, were thus essentially neurophysiological in structure and statistical in operation. I had a background in the Gurkha Rifles too. What made these people, unusual as they were, tick—and be motivated and be adaptive and be happy too . . ? And how did the description of individuals carry over into the description of the whole unit, for it seemed to do so: every one of many visitors to the strange place found it quite extraordinary as an organic whole. . . . This was the start of the subsequent hypothesis that there might be *invariances* in the behaviour of individuals . . . and that these invariances might inform also the peer group of individuals, and even the total societary unit to which they belong."

7. On the unpredictability of the firm's environment: "The first kind of regulation is performed in the face of perturbations introduced by the environmental economy, both of the nation and of competition in the money market. The second is performed in the face of market perturbations, which may be due to the aggressive marketing politicies of competitors, but which are fundamentally caused by the rate of technological innovation" (Beer 1981, 186–87).
8. Noble (1986) discusses U.S. visions of the automatic factory and lays out the Marxist critique, but it is worth noting that effects of automation were viewed with less suspicion in Britain and Western Europe. In the early 1960s it was taken for granted that automation would make it possible for people to work less, but the principal concern, especially on the left, was with the so-called leisure problem: what would people do with their spare time? Would the young dissolve into deliquency while their parents sat at home watching the television all day (as in Anthony Burgess's *A Clockwork Orange*)? The optimistic British and European response was that this was an opportunity for a new kind of people to emerge, no longer exhausted by labor, and both Trocchi's sigma project (in the preceding chapter) and the Fun Palace project (next chapter) should be seen in relation to this—as ways to foster the emergence of new postindustrial selves (likewise the activities of the Situationist International in France).
9. The painting also includes an R-machine operating at a lower cerebral level than the others. For simplicity I have skipped over this.
10. "The research is under the direction of the author, but the detailed results given . . . were obtained by a project team consisting of three operational research workers: Mr. T. P. Conway, Miss H. J. Hirst and Miss M. D. Scott. This team is led by Mr. D. A. Hopkins, who is also the author's chief assistant in this field" (Beer 1962a, 212).
11. This idea clearly goes back to Norbert Wiener's work on predictors and learning (Galison 1994), and one can easily see how it could be automated, though it had not, in fact, yet been automated by Beer's team.
12. Beer (1959, 138–41) discusses a simple, realistic example of how this might go.

13. For a clear statement, see Beer (1975a [1973]).
14. The most detailed information I have been able to obtain is from D. J. Stewart (email, 8 June 2006), and I am grateful for his assistance. Stewart, another British cybernetician and friend of Walter, Beer, and Pask, had been employed at Brunel since 1965, before it was granted university status, and he was also centrally involved in the establishment of the Cybernetics Department. The fund-raising dinner was held at Claridge's on 7 June 1968, and the department began operation in that year. "Frank George was appointed Director, and the initial staff consisted of Gordon Pask part time and me [Stewart] full time. Ross Ashby was in Illinois at the time but was expected to return and join us in 1970. In the event he never did." In 1981 "serious financial troubles hit all British universities [and] [f]rom then on further financial restrictions, together with this rather unsatisfactory structure, caused both the research and the teaching [in cybernetics] gradually to wither away."
15. As mentioned earlier, from the 1970s onward Beer held several visiting academic appointments which made it possible for him to propagate his vision of cybernetic managagement within the system of higher education, but part-time positions are themselves marginal to the academic establishment and cannot be seen as an institutional solution to the problem of the social basis of cybernetics.
16. In the *Hitchhiker's Guide*, the earth is a giant analog computer built by mouse-like beings to answer the Ultimate Question. On the earth as an analog computer, see Blohm, Beer, and Suzuki (1986).
17. For more on this robot, see www.conceptlab.com/control/. I am grateful to Ellen Fireman for first telling me about this project, and to Garnet Herz for telling me more about it when I visited Irvine in October 2005. When I first visited this website (21 July 2005), the project was entitled "Control and Communication in the Animal and the Machine," a direct quotation of the subtitle of Wiener's *Cybernetics*. The title has since changed to "Cockroach Controlled Mobile Robot," though the previous title is still listed too.
18. The Phumox project of Andy Gracie and Brian Lee Yung Rowe does thematize adaptation, aiming at coupling an AI system into the evolution of biological systems, but it is not clear to me how far this project has got (www.aminima.net/phumoxeng.htm). In a related project, "Small Work for Robots and Insects," a neural network analyzes the song of crickets and aims at a reciprocally adaptive sonic coupling. Gracie says that "Phumox frames our interest in exploring connections between machine and nature that are outside the typical areas covered by cybernetics," but his work strikes me as directly in the line of cybernetic inheritance from Beer and Pask. I thank Guillermo Santamaria for telling me about Gracie's work.
19. System 3 "is ideally placed to use every kind of optimizing tool in its direction of current operations, from inventory theory to mathematical programming. A dynamic, current model of the firm's internal workings must in fact emerge at this level, and offers the ideal management tool for the control of internal stability" (Beer 1981, 178).
20. Beer had sketchily included these parasympathetic signals in his 1960 cybernetic factory essay in association with the equally sketchy R-machine. In effect,

levels 1–3 of the VSM were elaborations of Beer's vision of the R-machine. As an example of what might be at stake here, think of a situation in which management calls for a sudden increase in production. Left to themselves, systems 1–3 might simply try to implement this demand even if production quality went down, the machines started breaking, and the workers went on strike.

21. Beer contrasted his vision of higher management as a densely connected network of neurons with the traditional conception of a hierarchical pyramid, as respectively adaptive and nonadaptive (Beer 1981, 201, fig. 39; 204, fig. 40). Beer connected this to McCulloch's notion of the "redundancy of potential command" (232ff.)—the idea that control switches between structures in the brain as a function of the different situations encountered. Beer's idea was that higher management should function likewise.
22. Beyond multiplicity, the recursive aspect of the VSM also implies a notion of scale invariance: whatever the scale of analysis, one finds the same structures: viable systems. Both of these ideas strike me as ontologically valuable (Pickering 1995). On the other hand, Beer's tidy form of recursion, with layers of viable systems neatly stacked within each other, is less cogent. My own studies have never turned up anything like this regular structure, and for this reason I am inclined to discount this aspect of the VSM as ontological theater. Beer sometimes claimed to have demonstrated logically that all viable systems have to have such a structure, but I have not been able to find a proof that I can understand. At other times, he noted that it was *useful* to think of viable systems as recursive—"In order to discuss the organization of vast institutions as well as small ones, the principle of recursiveness was invoked. We should depict the organization as a set of viable systems within a set of viable systems, and so on. That decision was perhaps not a necessity; but it did offer a convenient and powerful convention for our work" (Beer 1979, 199). Recursivity can clearly be a "convenient and powerful convention" in getting to grips with complex biological and quasi-biological organizations even if it does not reflect a necessary feature of the world.
23. This idea also runs through the development of game-playing computer programs.
24. For an extended discussion of the system 4 model and its adaptability, see Beer (1981, 183–92). My listing of variables is an abbreviated selection from the variables that appear in Beer's fig. 36 (p. 188). As far as I know, the best-developed system 4 model was that constructed for the Chile project (below). This had the form described above, drawing upon Jay Forrester's "Systems Dynamics" approach to modelling complex systems (on the history, substance and applications of which, see Elichirigoity 1999).
25. "These charts are, or more usually are not but could be, supported by detailed job descriptions intended to show how the whole thing works. So the charts themselves specify an anatomy of management, while the job descriptions specify its physiology" (Beer 1981, 77). Beer (79) describes these organizational charts as "arbitrary" and "frozen out of history." The rhetorical other to the VSM was always this vision of linear hierarchy of command. In recent years, of course, organizational theory and practice have moved away from that model, often, in fact, in rather cybernetic directions. See, for instance, the

work of David Stark (Neff and Stark 2004; Stark 2006) on flat organizations, self-conscious concerns for adaptation, and the redundancy of potential command (mentioned in n. 21 above). Stark lists the Santa Fe Institute (chap. 4) among his affiliations.

26. In the language of contemporary science and technology studies, Beer here addresses topics concerned with "distributed cognition"—the idea that relevant knowledge of complex organizations is spread throughout the organization rather than fully centralized. See Hutchins and Klausen (1996) and Star (1991) on automation as severing important social connections, and Hutchins (1995) for an extended treatment. Beer (1981, 109–10) also mentions the converse problem: "In fact, one of the key problems for scientists installing such systems [computers] in industry is that the connections they *wish* to cut are not always successfully cut." He gives an example relating to foremen and chargehands who continue to operate the old system using "little books of private information." "Surgeons have encountered a precisely similar phenomenon when performing trunk sympathectomies. . . . The surgeon does not expect the feedback circuits involved to operate any longer—but sometimes they do."
27. This produces a Beer/cybernetics-centric account of developments within an exceptionally turbulent and eventful period of Chilean history. Medina (2006) does an excellent job of situating Beer's project within the wider frame of social, economic, and political developments in Chile, but in the present section I am principally concerned with the project as an exemplification of the VSM in action. I am grateful to Eden Medina for discussion of her work prior to publication, and for her detailed comments on the present account of Cybersyn.
28. This was Beer's understanding of what engagement with individual enterprises would look like, but, as Medina (personal communication, 21 September 2007) notes, the situation on the ground was rather different: "It would be more accurate to describe management as politically appointed interventors. The OR team from the Cybersyn project rarely had any interactions with the workers at any level. In the case of the Yarkur textile mill . . . the OR scientists worked exclusively with the interventor in charge of finances. Stafford may not have realized this. Moreover, longstanding class prejudices also kept Cybersyn scientists and engineers from interacting with the rank and file."
29. Medina (personal communication, 21 September 2007) points out that a government report in 1973 gave a figure of 27% for the number of enterprises connected to Cybernet. Nothing hinges here on this figure, though it suggests that Beer's account of the rate of progress of Cybernsyn might be overly optimistic.
30. Meadows et al. (1972). For the history of Forrester's work and its part in an emerging "discourse of globality" at the heart of "planet management" see Elichirigoity (1999).
31. Von Foerster papers, University of Illinois archives, box 1, Beer folder.
32. Flores completed a PhD at the University of California, Berkeley, under the direction of Hubert Dreyfus, whom I thank for an illuminating conversation. Flores's management consultancy draws directly on insights from the philosophy of Martin Heidegger rather than the VSM (Rubin 1999). Traces of the ontology of unknowability and the stance of revealing are evident: "It is the third

realm of Flores's taxonomy to which people should aspire: What You Don't Know You Don't Know. To live in this realm is to notice opportunities that have the power to reinvent your company, opportunities that we're normally too blind to see. In this third realm, you see without bias: You're not weighted down with information" (Rubin 1999).

33. The quotation is from an email message to the author, 3 April 2003. I had not consciously encountered the word "metanoia" before. Besides this usage, it also turns out to have been one of R. D. Laing's favorite words for tranformative voyages through inner space (see Laing 1972, for example). The word also turns up in Pearce (2002 [1971]), a minor classic of the sixties countercultural literature. (I thank Fernando Elichirigoity for bringing this book to my attention.) The *Oxford English Dictionary* defines "metanoia" as "the setting up [of] an immense new inward movement for obtaining the rule of life; a change of the inner man."
34. I. Beer (2002); email messages from Vanilla Beer to the author, 3 April 2003, and Allenna Leonard, 5 April 2003. The only published discussion of Marx that I have found in Beer's writing is the following passage, which concerns Beer's work in Chile in the early 1970s (below): "President Allende was a Marxist-Leninist who did not accept the model now in use in the USSR. . . . Allende was well aware that the Hegelian concept of the dialectic, used by Marx, was paralleled in the ubiquitous biological mechanism of homeostasis [citing Ashby]. . . . My idea was to replace the Marxist 'classes' (where the ruling class exploits the proletariat) with a richer and less tendentious categorization based on shared information. 'Exploitation' then becomes the deprivation of information. . . . What are (laughably) called the 'mass media' very often carry not zero, but *negative* information" (Beer 1994b, 11).
35. The two most extended pieces of critical writing are a long essay by Ulrich (1981) and a review of both technical and political critiques by Jackson (1989). Beer (1983) gives a short reply to Ulrich but focuses on a difference of paradigms between his own cybernetics and Ulrich's Kantianism. I find it more useful to think here about the details of the VSM, which is why I formulate a response to the criticisms myself. We could note in passing that a few decades down the line we are all enmeshed to an unprecedented degree in a vast multiplicity of noncybernetic, hierarchical systems of surveillance and control, and that most of them are simply taken for granted.
36. Beer called this the *gremio* strike, but it was commonly referred to as the October Strike, the *Paro de Octubre* (Medina, personal communication, 21 September 2007).
37. Beer understood this change in the mode of operation of Cybersyn as an instance of the "redundancy of potential command" (n. 21 above).
38. A standing concern of cybernetics from its earlier days was that feedback systems can show pathological behavior when responding to out-of-date data—the thermostat that turns the heating up after the temperature has already risen for other reasons.
39. The cybernetic innovation here hinged on the usual move from representation to performance. On general election nights, Bob McKenzie's swingometer would display voting trends relative the previous election, but after the votes

had been cast and not in an attempt to influence the process of voting. In contrast, Beer's algedometers were designed to make possible an emergent real-time interplay between the parties they coupled.

40. Presumably this is why Beer chose to reply to Ulrich in rather rarefied philosophical terms rather than responding to the details of Ulrich's charges (n. 35 above).
41. See also Bula (2004) and Donoso (2004) on applications of Beer's cybernetics in Colombia.
42. Crises in the environment and the third world are mentioned in many of Beer's post-1970 writings; see, for example, Beer (1975 [1970]). Much of Beer's reflection on Chile and Cybersyn took the form of an analysis of "the cybernetics of crisis" (Beer 1981, 351–78).
43. The Syntegrity Group based in Toronto and Zurich offers to help "organizations gain clarity and conviction as they tackle complex, multi-faceted challenges and opportunities" and lists a long string of clients running from Canadian Blood Services via multinationals such as IBM and Monsanto to the World Wildlife Fund: www.syntegritygroup.com (accessed 12 April 2005). Syncho is a British-based consultancy specializing in both the VSM and team syntegrity. It was founded in 1985 by Raul Espejo, who was earlier the senior project manager on Project Cybersyn and is currently both director of Syncho and a visiting professor at University College, Worcester, England: www.syncho.com (12 April 2005). The inside cover of Beer (1994b) notes that he was then chairman of Syncho and of Team Syntegrity (Canada). Another management consultancy firm, Phrontis—"a registered partner of Microsoft"—lists team syntegrity in its repertoire: www.phrontis.com (12 April 2005). The director of Phrontis, Anthony Gill, was a director of Syncho from 1990 until 1996 (see the Syncho website, above).
44. *Beyond Dispute* also includes a "Collaborators' Surplus" comprising seven essays on syntegrity by other authors variously involved in the project.
45. The origins of team syntegrity lay in Beer's reflections on what he later came to conceive as the adaptive connections between levels in the VSM. Allenna Leonard told me (22 June 2002) that he would try to gather representatives of the unions and management after work at United Steel for glasses of whisky, in the hope that this would precipitate open-ended discussions about the state of the company and its future contours, and this was the approach that was formalized as syntegration. Beer (1994b, 9) describes syntegration as "a means of capturing the informal talk [at a meeting] 'Later in the Bar.'"
46. Beer understood the self-organizing quality of infosets by analogy to McCulloch's notion of the redundancy of potential command in the brain (mentioned above) and as an argument against fixed hierarchical structures in organizations. Thus, *Beyond Dispute* (Beer 1994b, 148–61) includes a very interesting discussion of the 3-4 homeostat in the VSM, suggesting that its constituents in practice are not necessarily the people one would imagine, and that syntegration might be an effective way to bring them together: "Take, for example, the leading directors of a company board; add the most respected staff aides; include (possibly) representatives of workers, clients, and the community: here are 30 people strongly committed by a motive, a collegiate pur-

pose. Can they afford to meet for an intensive 5-day exploration of the future of their enterprise, using the Team Syntegrity model and protocol? If not, they are probably condemning themselves to years of orthodox, strung-out committee work that ties up thinking time, exhausts patience, frustrates innovation—and may be too late" (159–60). At a more macro scale, Beer likewise argued against a fixed hierarchy of government in human affairs running from local through national to transnational (e.g., the UN) and possibly global levels. Instead, self-organized infosets could themselves be seen as "centres of potential command" at any level of governance in respect of the specific issues in which they especially engaged and on which they were especially knowledgeable.

47. Fuller analyzed the peculiar properties of his domes in terms of a notion of tensile integrity, or "tensegrity" for short. It turned out that "tensegrity" "had been appropriated for commercial use by architects," so Beer adopted the suggestion that his approach should be called "syntegrity," once more invoking the notion of synergy that had gone into naming Project Cybersyn (Beer 1994b, 13–14). Fuller constitutes another link in our story to the counterculture, especially in the United States: he was a role model for key figures such as Stewart Brand, and his domes were the preferred architectural forms for communes (Turner 2006). Like the cyberneticians, Fuller espoused a strange and nonmodern ontology, this time a non-Cartesian spatial geometry, but I cannot pursue that here.

48. In *Beyond Dispute*, Beer mentions the 1970 syntegration devoted to the formulation of a new constitution for the OR society (Beer 1994b, 9), and another in 1987 devoted to the political future of Ontario (10). Among the 1990 experiments was a syntegration on "world governance" held at the Manchester Business School (chap. 3) and another on "the future" in Toronto (chap. 6; discussed in the text above). The book includes the chapter "Governance or Government?" (chap. 10), in which Beer contrasts "government" as an enduring entity endowed with specific powers with "governance" as the open-ended deliberations of concerned citizens—organized here as syntegrations on "world governance," ad hoc assemblages transverse to global entities such as the United Nations. The point of such syntegrations, as I understand it, would be to articulate a range of issues and concerns that might not otherwise figure in the political discourse of nations and their aggregates—another attempt to open up a space for disussion outside the frame of established politics (cf. Beer's contrast between algedometers and opinion polls).

49. This quotation and the following are from I. Beer (2002).

50. Phone interview with Stafford Beer, 23 June 1999; conversation with Allenna Leonard, 22 June 2002.

51. Beer (1994b, 227): "In my teens I had set down this statement: 'There is only one mystery: why or how is there anything.'" In an interesting echo of Grey Walter, Beer also remarks that "there appears to be a complicated set of rules for computing with neurons which prevents many of them from working at once. The neurons are electrically triggered, and if the rules are broken we get an electrical overload. This is the cybernetic explanation (in brief) of what we usually call epilepsy, or (perhaps) what our forefathers called 'possession'" (Beer 1965, 294).

52. I am not sure how meaningful these words might be for readers, but to inject a personal note, they ring true for me. They remind me of my days as a postdoctoral researcher in theoretical particle physics, spending weeks and months trying and failing to understand mathematically how quarks interact, while being painfully aware that the quarks themselves were doing their own thing all the time, in real time, "getting the answer continuously right," throughout the cosmos (or so I believed at the time). That kind of experience leaves one with a feeling for scientific knowledge as a pale simulacrum of the world (or not even that in my case), a simulacrum one nevertheless finds it hard not to mistake for the thing in itself. That, I think, is the origin of Beer's awe and wonder at the indefinite excess of the world itself in relation to our representational capacity.
53. Beer never used the word "hylozoism," as far as I know, though the protocyberneticist Kenneth Craik (chap. 3) did. Speaking of his philosophy of mind, Craik (1943, 58) remarks: "It would be a hylozoist rather than a materialistic scheme. It would attribute consciousness and conscious organisation to matter whenever it is physically organised in certain ways." The cybernetic artist David Medalla also described himself as a hylozoist (chap. 7, n. 51).
54. This is my performative gloss. In second-order cybernetics, Spencer Brown's *Laws of Form* (chap. 5, n. 25) is often invoked in this context, the idea that drawing a distinction creates two terms and a relation between them from what was originally a unity.
55. I do not mean to suggest here that the Christian tradition is entirely lacking in resources for articulating a spiritual stance like Beer's. Beer (like Laing) was happy to draw on a whole variety of traditions including Christianity in mapping out his own spiritual path, but the emphasis was always on mystical experiences and practices, and Eastern traditions thematize these in a way that modern Christianity does not.
56. I am grateful to Allenna Leonard for giving me a copy of the manuscript. Page citations below are from this version. An unpaginated version is now available online at www.chroniclesofwizardprang.com.
57. Thus, in chapter 17 Prang contests a returning disciple's Buddhist understanding of time and reincarnation, suggesting that she is trapped within a "paradigm," meaning a "not negotiable model" (Beer 1989b, 149).
58. Kripal's (2007, 18–21) history of the Esalen Institute (see above, chap. 5) includes extended discussions of tantra, which he associates with Hinduism, Buddhism, and Taoism. "Whereas ascetic Asian traditions . . . tend to privilege strongly the transcendent order . . . and consequently denigrate or renounce the everyday world (*samsara*) as illusory (*maya*) or as impermanent (*anitya*), the Tantric traditions tend to insist rather on the essential unity of the transcendent and immanent orders and in fact often privilege the immanent over the transcendent in their rituals and meditations." Hence Kripal's conception of "the enlightenment of the body" and the fact that a Web search for "tantra" leads largely to sources on tantric sex. Eliade (1969) discusses tantric yoga at great length, emphasizing its embodied and performative aspects, as well as its connections to magic and alchemy. All of this helps to illuminate what I describe below as Beer's "earthy" form of spirituality.

59. Prang puts this phrase in quotes, and continues. "You couldn't define that, and besides the phrase was associated with strong drugs. Candles and incense have effects on the nervous system too, but you never have a 'bad trip' " (Beer 1989b, 41). One can think of Beer's demonstrations of immunity to pain, mentioned earlier by his brother, as also hinging on a technology of the self.
60. When Beer described himself as teaching meditative yoga on the inside covers of his later books, it was the transmission of this form of knowledge and practice that he was referring to. Most of the conversations and interactions in *Prang* take place between Prang/Beer and his "shishyas," female apprentices, especially Perny, his current apprentice in the stories.
61. Beer 1989b, chap. 2, p. 15; chap. 5, p. 36; chap. 15, p. 116; chap. 6, p. 51; chap. 15, p. 119; chap. 15, p. 124; chap. 16, p. 142; chap. 5, p. 42; chap. 18, p. 166.
62. On *siddhis*, see, for example, Eliade (1969), Kripal's (2007) history of the Esalen Institute, and an enormous compendium of strange performances published by one of Esalen's founders: Murphy (1992).
63. He adds that he used the enneagram in devising the layout of his set of paintings relating to the Requiem Mass that were exhibited at the Metropolitan Cathedral in Liverpool in 1992 and again in 1993. For more on this exhibition, see Beer (1993b).
64. I am grateful to Joe Truss for showing me the enneagrammatic trajectories in a model icosahedron (there turn out to be many of them), though I doubt whether I could find them again unaided. Beer (1994b) offers a textual and diagrammatic description of how to find them, but I cannot claim to follow it. Properties of three-dimensional geometries are extremely hard to grasp without a three-dimensional object to refer to; one could describe Beer and Truss's explorations of the icosahedron as genuine research in this sense.
65. Beer also discusses other mandalas that can be associated with the icosahedron. One can, for example, construct a dodecahedron from twelve pentagons within the icosahedron, and Beer comments that the "central dodecahedron thus became inviolate in my mind, and because for years I had been using these forms as mandalas in meditation, it acquired the private name of the 'sacred space.' . . . I shall add that the sacred space 'breathes' with the cosmos through invisibly fine tubes connecting the centre of each dodecahedral face to its orthogonal vertex. The bridge from this mystical to the normal description lies in Aristotle's pneuma, the chi of Chinese medicine, and ends up (quite safely) with the segments of polar axes which subtend the dodecahedron" (1994b, 192–93). Discussing certain planar projections of the icosahedron, he remarks that "many Indian mandalas reflect this configuration" (195).
66. Beer develops a formalism to explain this theory of consciousness in *Beyond Dispute*, chapter 13, "Self-Reference in Icosahedral Space," with acknowledgement of inspiration from Heinz von Foerster. Beer, Pask, and von Foerster were among the leading contributors to the development of a cybernetic theory of consciousness, also elaborated in Humberto Maturana and Francisco Varela's theory of "autopoiesis." Beer wrote a preface for their *Autopoiesis and Cognition* (1980; as mentioned before, Varela was a Buddhist). I have two reasons for not going further into the cybernetic analysis of consciousness. One is that

it seems to me largely theoretical, rather than connecting to novel domains of worldly practice. The other is that I do not understand it (or have perhaps failed to get some key point). I am reminded of the idea common to early cybernetics that closed loops of neurons might be entailed in memory, and I can see how that works, but I cannot see how reentrant loops connect to any idea of consciousness that I can grasp.

67. To make the contrast it might help to go back again to AI as a model of the modern brain and the rational self. AI has few if any resources to make bridges to the spiritual.
68. Leadbeater (1990, 40): "The radiating spokes of the chakras supply force to these sympathetic plexuses to assist them in their relay work; in the present state of our knowledge it seems to me rash to identify the chakras with the plexuses, as some writers appear to have done." Evidently Beer was rash enough. For some background on the British occult tradition in which Leadbeater's writings can be situated, see Owen (2004).
69. In this conection, I think of *The Web of Life* (1996) by one of the leading thinkers of the New Age movement, Fritjof Capra. This book is one of the best popular introductions to contemporary work on complexity and self-organization (see chap. 4 above), tying them into a Buddhist perspective on being while acknowledging Capra's many personal contacts with key figures in cybernetics and related fields, including Ilya Prigogine, Francisco Varela, Humberto Maturana and Heinz von Foerster. Beer had no active connection to the New Age movement as far as I know (though perhaps he might have if "Wizard Prang" had been published). His tantric teaching in Wales appears to have been a distinctly old-fashioned operation, not, for example, advertised in the New Age literature. On the other hand, a few leaps of association are enough to bridge the gap. Beer (1994b, 203) mentions that the monk who gave him his enneagrammatic mandala (above) was based in Santiago at a "mystical mission known as Arica," and that after leaving Chile he learned that Arica's founder, Oscar Ichazo, used enneagrams in his teaching on "how to break the tyranny of ego." One can learn more about Ichazo in John Lilly's book *The Center of the Cyclone* (1972), which recounts Ichazo's attempts to develop courses in his own esoteric system and describes Lilly and Ichazo's early involvement with the Esalen Institute (see also Kripal 2007, 177–78). I discussed Esalen as an epicenter of the New Age movement briefly in chapter 5 (nn. 18, 19, 25, 43), including tentative connections to Beer and Laing, and Laing met Ichazo during his trip to the United States in 1972, the trip on which he also met Elizabeth Fehr (see chap. 5 and Burston 1996, 121). Esalen was also one of the origins in the United States of the "human potential" movement growing out of the writings of Huxley and others, and, as mentioned above, Michael Murphy shared the cybernetic and spiritual fascination with strange performances and wrote a striking book, *The Future of the Body* (Murphy 1992), very much in the tradition of William James (1902) and Aldous Huxley but thematizing more strongly the supernormal powers that accompany spiritual practice.
70. Whittaker (2003, 47) includes a select discography of Eno's music. One can access short extracts from Eno's recordings via a link at www.inmotionmagazine

.com/enol.html. According to Eno, David Bowie listed *Brain of the Firm* as a "desert island book," presumably for BBC radio's never-ending *Desert Island Discs* series (Whittaker 2003, 51). Eno (1996a) reproduces Eno's diary for 1995, together with a selection of his short essays which I draw on below. Stewart Brand, whom we encountered in the previous chapter in connection with Bateson, features frequently in this diary as an email correspondent. Eno explains his current connection to Brand via GBN, the Global Business Network, "a futures scenario development group" (on which see Turner 2006), closing a loop between Eno and Beer's world of management consultancy.

71. The essay was published with the title "Generating and Organizing Variety in the Arts" in 1976 in *Studio International* and is reprinted as "The Great Learning" in Eno (1996a, 333–44). I thank Henning Schmidgen for providing me with a copy of this. Since we were just discussing Beer on Eastern spirituality and philosophy, it is relevant to note that the essay focuses on a 1967 piece by Cornelius Cardew called *The Great Learning*, which is based on Confucian texts. The essay ends on an explicitly ontological note very reminiscent of Beer: "As the variety of the environment magnifies in both time and space and as the structures that were thought to describe the operation of the world become progressively more unworkable, other concepts of organization must become current. These concepts will base themselves on the assumption of change rather than stasis and on the assumption of probability rather than certainty. I believe that contemporary art is giving us the feel for this outlook" (Eno 1996a, 344).
72. This is almost the same phrasing as in the *Studio International* essay, which cites p. 69 of the first edition of *Brain of the Firm* (Eno 1996a, 339). Eno's essay begins with a discussion of the cybernetic notion of variety, citing Ashby's *An Introduction to Cybernetics* (Eno 1996a, 334–35).
73. Both of these pieces are discussed in historical context in Pinch and Trocco (2002, 37).
74. Another short essay makes it clear that Eno was using software called Koan from a company called Sseyo (Eno 1996a, 330–32).
75. Needless to say, feedback figured in much the same way in the history of the electric guitar. The solid body of the electric guitar was originally conceived as a way to minimize feedback effects, but as the sixties drew on, rock guitarists, most notably Jimi Hendrix, learned how to make music from loops running through loudspeakers and guitar pickups (McSwain 2002). One could thus see performances such as Hendrix's rendering of "The Star-Spangled Banner" at the Woodstock festival—almost entirely constituted from feedback effects—as themselves cybernetic ontological theater. Unlike Eno, Hendrix did not, I think, read Beer. We should see his explorations of what a guitar can do as part of the performative experimentalism of the sixties that echoed the cybernetic ontology.
76. And having found a desirable configuration, they sometimes had to struggle to hold onto it. "Brian Eno had to leave a little note on his VCS_3 synthesiser telling his technician, 'Don't service this part. Don't change this'—he preferred the sound the ring modulator produced when it was 'broken'" (Pinch and Trocco 2002, 223).

77. Eno: "So I wanted the Staffordian approach to do two things: to pitch me into aesthetic areas beyond where my taste would normally take me. That's one of the things you find working with systems, that they throw up configurations that you couldn't have thought of. I wanted the system to confront me with novelty; but I did also want to say 'I prefer this part of it to that part, this part doesn't make sense, that part does.' . . . The systemic approach . . . is certainly very good at imagination expanding" (Whittaker 2003, 59). Wolfram now markets CA-generated sounds as ringing tones for cellphones: tones.wolfram.com/generate/.
78. Whittaker is paraphrasing Eno from the sleeve notes of *Music for Airports* (1978), a "manifesto" for ambient music: "Ambient Music must be able to accommodate many levels of listening attention without enforcing one in particular; it must be as ignorable as it is interesting" (Eno 1996a, 295, 296). Eno connects the substance and mode of production of ambient music with an emergent style of consumption. "In 1978 I released the first record which described itself as Ambient Music, a name I invented to describe an emerging musical style. It happened like this. In the early seventies, more and more people were changing the way they were listening to music. Records and audio had been around long enough for some of the novelty to wear off, and people were wanting to make quite particular and sophisticated choices about what they played in their homes and workplaces, what kind of sonic mood they surrounded themselves with. The manifestation of this shift was a movement away from the assumptions that still dominated record-making at the time—that people had short attention spans and wanted a lot of action and variety, clear rhythms and song structures and, most of all, voices. To the contrary, I was noticing that my friends and I were making and exchanging long cassettes of music chosen for its stillness, homogeneity, lack of surprises and, most of all, lack of variety. We wanted to use music in a different way—as part of the ambience of our lives—and we wanted it to be continuous, a surrounding" (Eno 1996a, 293).
79. This variation is a degree of magnitude greater than the inevitable variation between performances of a traditional piece of music. One might think of improvisational jazz as being in the same space as ambient and generative music, and there is something right about this. Generative music, however, stages an overt *decentering* of composition between the musician and the dynamic system with which he or she interacts in way that jazz does not.
80. Other artists in Eno's pantheon include Cornelius Cardew, John Cage, and Christian Wolff, all of whom were "inventing systems that produced music" (Whittaker 2003, 57). Another musical admirer of Beer is Robert Fripp, guitarist and founder of the band King Crimson (see his letters reproduced in Whittaker 2003, 52). There are significant echoes of Beer's spirituality in this list: Cage was deeply engaged with Zen Buddhism, and Fripp "has also been involved with running J. G. Bennett's International Society for Continuous Education in Sherborne, which is based on the teachings of G. I. Gurdjieff" (Whittaker 2003, 47). Cage (1991, 2): "I could not accept the academic idea that the purpose of music was communication. . . . I determined to give up composition unless I could find a better reason for doing it than communica-

tion. I found this answer from Gira Sarabhai, an Indian singer and tabla player: The purpose of music is to sober and quiet the mind, thus making it susceptible to divine influences. I also found in the writings of Ananda K. Coomaraswamy that the responsibility of the artist is to imitate nature in her manner of operation. I became less disturbed and went back to work."

81. Eno: "Now that's a total break from the Western classical idea. . . . So you forego the thing that composers usually do which is design music in detail, so that you're no longer exactly the architect of a piece of work but more the designer of a musical ecosystem. You put a few things in place and see how they react or what they do to each other" (Whittaker 2003, 57). This reference to ecosystems reminds us of another shift in power relations. As the above examples make clear, Eno is interested in music generated by a multiplicity of agents, humans and machines. In the *Studio International* essay he begins by discussing the traditional hierarchical structure of an orchestra, running from the conductor and the leader down to rank-and-file musicians, and emphasizes that this produces a structured listening experience—some elements of the music are deliberately foregrounded for the audience; others constitute the background against which the foreground stands out. This arrangement depends on skilled musicians, who can be relied upon to produce a certain kind of sound, and, importantly, "it operates accurately and predictably for one class of task but it is not *adaptive*. It is not self-stabilizing and does not easily assimilate change or novel environmental conditions" (Eno 1996a, 342; my emphasis). The music dominates its environment, one can say, or else the performance is a failure. We can then note that the compositional systems described by Eno flatten out the performative space, with all of the contributing elements of the music-generating system interacting symmetrically with one another. In this essay, Eno is also at pains to evoke the ways in which the specific character of the environment in which it is performed serves to influence any particular rendering of Cardew's *The Great Learning*. The resonant frequencies of the room, for example, pull the performers toward certain frequencies rather than others (338). This connection between organizational form and adaptability is, as we have seen, also classically Beerian, though Eno's geometries are that of neither the VSM nor syntegration.

Notes to Chapter 7

1. Biographical information on Pask can be found in two festschrifts: Glanville (1993) and Glanville and Scott (2001a). In this section I also draw on unpublished biographical writings on Pask by his wife, Elizabeth (E. Pask n.d.); I thank Amanda Heitler for showing me these notes and her own, and for permission to quote from them (email from A. Heitler, 2 May 2003). More broadly, I am very grateful to the late Elizabeth Pask, Amanda Heitler, Peter Cariani, Paul Pangaro, Ranulph Glanville, Bernard Scott, Jasia Reichardt, Yolanda Sonnabend, and John Frazer for conversations and email messages about Pask. Glanville commented in detail on drafts of this chapter, but the errors remain mine. Pangaro has an extensive collection of Pask's writings, cataloged at pangaro.com/Pask-Archive/; Amanda Heitler has an extensive but disorganized

collection of Pask's papers. For videoclips of Pask in the sixties and seventies, see cyberneticians.com/index.html#pan.

2. I thank Malcolm Nicolson for bringing Pain's article to my attention. Gar's self-experimentation takes us back to the realm of strange performances and altered states, but now in a military context. For an account of parallel proto-cybernetic wartime research in Germany, see Borck (2000). One of the characters in Grey Walter's novel (1956a) sounds a lot like Gar.
3. See the previous chapter for the establishment of the Brunel department.
4. On the interactions of these three, see E. Pask (1993), McKinnon-Wood (1993), and Glanville (1996).
5. System Research endured until the early 1980s, the Thatcher years (B. Scott 1982, 486): "The non-profit organisation, System Research Ltd, no longer exists. Pask continues to write, teach and consult, based partly in London (the Architectural Association) and Holland (the University of Amsterdam). The research team is dispersed: Kallikourdis has returned to Athens, Bailey is a successful entrepreneur in microelectronics, Lewis has long been with the Open University, Mallen with the Royal College of Art, Richards is a commercial systems analyst, Scott is currently a teacher of mathematics. The whereabouts of others is unknown. One thing is certain, all who passed through System Research Ltd were deeply affected by their time there. Its spirit lives on in other conversations." After the demise of System Research Pask was left to improvise his career even more than hitherto. Pask (1982), for example, lists the Architecture Association as his primary affiliation but also mentions the Department of Cybernetics at Brunel University, Concordia University, Montreal, the Institute for Applied System Research, in the Netherlands, and System Research Developments, in Britain.
6. S. Beer (2001, 551): "People started telling me colourful stories about Gordon when he was still at Cambridge and rather precocious to be a legend. Maybe he was still there when we first met. At any rate, our truly collaborative friendship lasted through the 1950s. We remained close for the rest of his life." S. Beer (2001, 552), speaking of the early 1950s: "Gordon was driving. And he was conversing in his usual lively fashion. This meant he was looking intently at me, and waving his expressive hands under my nose. It follows that he was steering the car with his elbows, intuition and a large slice of luck. It was, I confess, the last time that I drove with him. Hair-raising stories about his driving followed his reputation around for years, until he finally gave up."
7. In the early 1990s, Pask also worked on a novel, *Adventures with Professor Flaxman-Low* (Choudhury 1993). The book was never published, but some background information and audiofiles of extracts can be found at www.justcontract.org/flax.htm. The extracts are read by Pask's assistant and later collaborator, Nick Green, and I thank him for telephone conversations and email messages about Pask and the spiritualist dimension of cybernetics more generally. Pask's novel was modelled on a series of Flaxman Low stories by E. and H. Heron that appeared in *Pearson's Monthly Magazine* in 1898 and 1899 (vols. 5 and 7), some of which are available at gaslight.mtroyal.ca/prchdmen.htm. Pask's hero, like his predecessor, was a spiritual detective, exploring spirit phenomena like hauntings in a materialist vein reminiscent of the Society for Psychical Research

(see chap. 3, n. 62). "'I hold,' Mr. Flaxman Low, the eminent psychologist, was saying, 'that there are no other laws in what we term the realm of the supernatural but those which are the projections or extensions of natural laws'" (*The Story of Konnor Old House*, 1899, gaslight.mtroyal.ca/flaxmnXJ.htm).

8. Grey Walter was interested in the neurophysiology of synesthesia, which he understood in terms of electrical spillovers from one region of the brain to another (a less drastic form of epilepsy) and explored using sophisticated EEG apparatus (discussed at length in Walter 1953). With synesthesia we are back in the realm of altered states and strange performances, but Pask did not thematize this fact, so I will not pursue it further here, except to note that this is another aspect of Walter's work that interested William Burroughs (chap. 3), and synesthesia has often been associated with a mystical spirituality in art (Tuchman 1986). I thank Jan Pieterse for alerting me to the latter connection.
9. Two points need clarification here. First, this notion of achieving a satisfying dynamic equilibrium with some changing other runs through Pask's work, though he never gave any easily graspable description of its character and perhaps such description is not to be had. Pask (1971, 78) invokes a game metaphor: "Given a suitable design and a happy choice of visual vocabulary, the performer (being influenced by the visual display) could become involved in a close participant interaction with the system. . . . Consequently . . . the machine itself became reformulated as a game player capable of habituating at several levels, to the performer's gambits." I think of this on the model of finding an enjoyable opponent for repeated games of chess, say, or squash. One somehow knows that each game stands a good chance of being fun, though the specific details and outcome of any given game remain always to be found out. Second, my text is a drastically oversimplified account of the circuitry shown in fig. 7.2. From our perspective, the details of the circuitry and Musicolour's functioning do not matter; the important thing is simply that the machine parameters changed in a way the performer could not control. But in practice much of the work entailed in building such a machine no doubt went into finding out just what kinds of circuitry would be propitious in use. A performer could not engage with changes that were too fast or too slow on a human time scale, for example. The most accessible technical description of how Musicolour worked that I have found is in Pask and McKinnon-Wood (1965), which mentions a "slightly more elaborate version" in which the performer had more control over the machine, being able to "'reinforce' or 'reward' part of the machine's 'learning' process, by indicating his approval or disapproval of the prevailing characteristics on a foot switch. . . . Initially, each trigger circuit to spotlamp connection occurs with equal probablility. But any connection that meets with the performer's approval (as indicated by pressing a foot switch) becomes more likely" (955).
10. Elizabeth Pask (n.d.) added that Musicolour "was very badly affected by the wine waiters pouring the odd glass of wine down it."
11. The wiring diagram invites us to think about simulating Musicolour on a computer, and this has been done. But such simulations leave one having to find out what the behavior of the simulated machine will be in relation to a given set of inputs.

12. See chap. 5, n. 25, on Pask's personal relationship with Laing.
13. Pask was not entirely alone in the 1950s in his interest in cybernetics and the arts. He later commented that Musicolour "has much in common with Nicolas Schöffer's artifacts which I learned about many years later" (1971, 78n2). Schöffer (1912–92) has sometimes been called the "father of cybernetic art"; a discussion of his work would lead us into the history of cybernetics in France. On Schöffer, see Burnham (1968).
14. In a story by Lucian, ca. AD 150, Eucrates tells the story of the sorcerer's apprentice. This was popularized by Disney's *Fantasia* (1940), which was in turn Pask's point of comparison for Musicolour. Wiener often invoked the sorcerer's apprentice in conjuring up the uncanny quality of cybernetic machines (chap. 1).
15. For the details of SAKI's construction and discussion of its functioning, see Pask (1960a, 1961, 67–70). We should note that, like Musicolour, the training machines also functioned as test beds for the experimental investigation of the interaction between humans and adaptive machines. Such investigations were the topic of Pask's 1964 PhD in psychology, which acknowledges support for experimental work from the Aerospace Medical Research Laboratories of the U.S. Air Force and for "abstract model building" from the air force's Office of Scientific Research, both via the air force's European Office of Aerospace Research; "the special purpose computor [*sic*] and other adaptive control machines were made available for these experiments by System Research" (Pask 1964a, i).
16. This passage continues: "When Gordon took the machines to market, in the form of SAKI, . . . the engineers somehow took the cybernetic invention away. I suspect that they saw themselves as designing a machine to achieve the content-objective (learn to type), instead of building a Paskian machine to achieve the cybernetic objective itself—to integrate the observer and the machine into a homeostatic whole. Machines such as these are not available to this day, because they are contra-paradigmatic to engineers and psychologists alike."
17. On Thoughtsticker from the early 1980s to the present, see Pangaro (2001). Pangaro is an American who moved to Britain to work with Pask. He records that, having been funded by the British Social Science Research Council, the project was later continued with support from the Admiralty. He also comments on the improvised nature of the computer hardware that typified System Research: "This THOUGHTSTICKER used display hardware that was obsolete at Negroponte's lab [see below] and had been shipped to System Research for dynamic displays of a Paskian nature. An incessant software bug, which Pask contended was a 'feature,' led to extraneous lines in these displays, but did little to discourage the imagination that, with decent funding, something really amazing could be done here" (Pangaro 2001, 794). Pangaro himself, with Pask as consultant and adviser, established a consultancy to build a more up-to-date version of the machine with a hypertext interface. This fell victim to a software platform battle in the mid-1980s, though Pangaro continues to imagine some version of Thoughtsticker as a more active and intelligent Web browser than any currently available. The affinity between Pask's interface techniques and hypertext was noted at the time by Ted Nelson (1987), quoted as "Ted Nelson on Gordon Pask," www2.venus.co.uk/gordonpask/clib.html.

18. Philosophers of science will recognize an affinity between Pask's entailment meshes and Mary Hesse's (1966, 1974) network theory of knowledge. Pask argued that entailment meshes close in on themselves (for more on this, including the idea that the topology of closure is a torus, see B. Scott 2001b), which is obviously a useful approximation in the construction of teaching machines, but surely wrong in general.
19. Other recollections and historical accounts of Pask's educational work at Brunel and the OU include Thomas and Harri-Augstein (1993) and Laurillard (2001). The most successful pedagogical machine in the Paskian tradition up to the present appears to be a system called Byzantium, developed in the late 1990s by a team from a consortium of British universities including Bernard Scott (Patel, Scott, and Kinshuk 2001).
20. Leaving the realm of machines, one could explore the cybernetics of education further by following the lead of another of Pask's Brunel students, Ranulph Glanville. Glanville (2002a, 2002b) takes seriously the idea of children as exceedingly complex systems with their own dynamics and connects it to the idea that education should be understood as a constructive fostering of that dynamics rather than any simple transmission of information. This idea returns to the work of Friedrich Froebel (1782–1852) who, according to Glanville, invented the metaphor of the child as a growing plant—a very cybernetic image. Concretely, Glanville points to Froebel kindergartens, Montessori schools, and a form of studio education common in teaching design as examples of cybernetic pedagogic institutions. (Grey Walter was also keen on Montessori schools: Walter 1953, 269.) I would be tempted to add the Anti-University of London (discussed in chap. 5) and Joan Littlewood's Fun Palace from the same period (discussed below) to the list, with Kingsley Hall as a sister institution in the realm of psychiatry. As usual, we can note that the cybernetic approach entails a symmetrizing shift away from the familiar teacher-student hierarchy, and that Glanville's examples of his preferred forms of pedagogy remain marginal to the educational establishment.
21. If we wanted to follow this line of thought further, we could note that Pask is seen is one of the founders of second-order cybernetics and a significant contributor to Maturana and Varela's "autopoiesis" tradition and recall that Francisco Varela was scientific adviser to the Dalai Lama.
22. Pask's publication list begins with eighteen papers written between 1957 and the publication of *An Approach to Cybernetics* in 1961. Ten are about teaching machines; the other eight are about cybernetic controllers such as chemical computers.
23. On genuinely biological computers, Pask (1960b, 258) says that he and Beer "have examined models, where currency is food supply and unicellulars like paramecium are active elements, sufficiently to show that such colonial organization may be coupled to a real process."
24. As usual, Pask was himself not very forthcoming on the practical details. Pask (1958, 166–67) sketches out a means for monitoring the development of threads in some region using four ancillary electrodes. Two electrodes emit current periodically (presumably to inhibit growth of threads from them) and the others register some response. The trick would be to look for changes in response correlated with sounds in the required range (as detected by a filter

attached to the microphone). Positive correlations could then be encouraged by injecting more current into the assemblage as they occur.

25. See also the distinction between a cybernetic "natural history" approach and the traditional scientific method in Pask (1960b).
26. For further development of Pask's ideas on epistemology in relation to adaptive and self-organizing systems see the work of Peter Cariani, available at his website: www.cariani.com. I am very grateful to Cariani for discussions about Pask's chemical computers, and responsibility for inadequacies in my account rest with me. Isabelle Stengers's (1997) critical philosophy of science is an attempt to push science in the direction of a cybernetic and "risky" method. Looking in a different direction, the "grounded theory" approach to social-scientific research foregrounds the kind of dense engagement with the object that Pask took to characterize cybernetics. Genuinely "risky" research conducted under the aegis of grounded theory thus constitutes yet another of example of ontology in action.
27. Pask (1961, plate IV, facing p. 65) reproduces images of the evolution of an "activity surge in 2-dimensional cell array of neurone-like units, simulated on a computer by R. L. Beurle." On the later simulations discussed below: "The programs . . . were written by D. J. Feldman in the machine language of the ICT 1202 computer and the programs were run on this machine. (A good deal of reprocessing of cards was needed to augment the limited storage capacity of this machine.)" Pask (1969a, 106).
28. See, for example, the online journal, first published in 1998, the *Journal of Artificial Societies and Social Simulation*, jasss.soc.surrey.ac.uk/JASSS.html.
29. Pask was not alone in the 1960s in his interest in cellular automata as models for the biological and social sciences. He cited the work of five other individuals and groups (Pask 1969a, 102). In chapter 4 I discussed the way in which aspects of Ashby's cybernetics were elaborated in the work of Christopher Alexander, Stuart Kauffman, and Stephen Wolfram in later developments centering on cellular automata. In a 1969 paper on architecture, Pask included Alexander in a list of people with whom he had personal contact (1969b, 496). The hand calculations were presented by Pask in talks and publications in the early 1960s. Pask (1962) discusses models of slime molds, which makes a bridge to the work on self-organization by Ilya Prigogine (and thus back to Stengers): see Prigogine and Stengers (1984) and Pickering (2005a).
30. I thank Peter Asaro for bringing this work to my attention.
31. I continue to suspect that Pask was the model for one of the BBC's Dr. Who's (Patrick Troughton—the perky little one who played the penny whistle), but the only documented connection to Dr. Who that I can find is this (Moore 2001, 770): "One evening [in the 1970s]. prior to our meeting [of the Cybernetics Society, of which Pask was Chairman], I recognised at the bar Tom Baker, famous for his performances on television as Dr Who, all round scientist and 'Time Lord.' I invited him to come upstairs and join our monthly meeting. Alas he declined saying that he was only an actor and did not understand such high-level science."
32. I thank Sharon Irish for many conversations on what follows and its wider context, and for access to much relevant literature, especially on architecture.

33. For a fuller account of Littlewood's role in what follows, see Littlewood (1994).
34. Pask (1964b, 10): "The structural organisation of a dramatic presentation suitable for this system closely resembles the branching programmes used extensively in teaching machines." Pask's diagram 2 (1964b, 9) shows a monitoring device which displays possible lines of plot development as a network of interconnected branches, and ends with a skeletal example of the possible forms of such a play (28–30). One of Pask's suggestion was that in rehearsal actors could work back from various nodes in the action, thus developing the overall network of possible trajectories for the play and the metainformation which would be provided to the audience at the branch points (29–30). I thank Paul Pangaro for a copy of this proposal.
35. The *Guardian* obituary also recalled that "the company's exceptional flair for improvisation and rewriting—Behan's script was chaotic—drew full houses" (Ezard 2002).
36. Pask's cybernetic theater was by no means unique in the sixties in its interactive and emergent aspects: one thinks of "happenings" and performance art. My suggestion is that we can see the latter, too, as exemplifications of the cybernetic ontology in action, somewhat differently staged from Pask's project. I thank Ranulph Glanville for pointing out to me that in 2000 Jeffrey Archer wrote a West End play about his own trial for perjury in which the audience voted on Archer's innocence or guilt at the end (see Wikipedia on Archer). This is a heavily watered down version of what Pask had in mind.
37. For a historical overview of the British art scene in the 1960s, see Stephens and Stout (2004).
38. The exhibition had a relatively small budget of only £20,000; Reichardt was employed part time at the ICA at a salary of £30 per week; artists were not paid to exhibit their work; and the ICA did not produce a catalog for the exhibition (Reichardt, personal communication, 21 February 2002). Instead, a special issue of *Studio International* was published to coincide with the exhibition: Reichardt (1968a).
39. Robin-McKinnon Wood and Margaret Masterman (1968) contributed a short poem that had been generated by a computer at the Cambridge Language Research Institute: a bug in a language analysis program had randomized the output to produce something resembling a Burroughs style cut-up. The only roboticist to appear in histories of the British underground, Bruce Lacey, also contributed to the exhibition. As discussed in Lacey (1968), his exhibits at the ICA were interestingly cybernetic, and able to interact with their environment (*Owl*, 1967), with humans (*Rosa Bosom (R.O.S.A.—Radio Operated Simulated Actress)*, 1965), and with other robots (*Mate*, 1967). Nuttall (1968, 125) refers to Lacey's "magnificent hominoids, sick, urinating, stuttering machines constructed of the debris of the century, always with pointed socialist/pacifist overtones but with a profound sense of anger, disgust and gaiety that goes far beyond any political standpoint."
40. Pask (1968, 35) acknowledges "Maurice Hyams in cooperation with System Research Ltd" as "patron of the project."
41. Speaking of their undergraduate days in the early 1950s, Harry Moore (2001, 769) recalled that "Grey Walter's experiments on 'artificial neurones' and

tortoise models in his book *The Living Brain* . . . also provided Gordon with additional stimulation."

42. As Gere (2002, 95) points out, this "computer" is visible in the background of fig. 7.17. Pask (1971, 98) credits Mark Dowson for constructing the electronics, Tony Watts for the electromechanical construction, Yolanda Sonnabend (an eminent theater designer) for the design of the female robots, and himself for the male design and the overall setup.
43. One can also think about the Colloquy from a different angle. The idea of robots mating in public seems to have escaped explicit discussion from 1968 to the present, but it is hard not to see the Colloquy as some sort of reflection on the "permissive society."
44. Zeidner et al. (2001) recalls Zeidner's first meeting with Pask and is a good place to begin an inquiry into this strand of Pask's work. "He infused our research with new concepts and paradigms for understanding decision making in complex, poorly-structured systems. He also introduced us to the use of formal theory and logic to better understand and predict interpersonal or person-machine communications" (984–85). Pask organized four conferences for Zeidner's Army Research Institute in Richmond (close to his home and System Research) on decision making in complex systems, in 1975, 1976, 1978, and 1983. The aim of the first was "to elicit a fair picture of the state of the art in decision making in Europe; the status of decision oriented disciplines; and on-going or contemplated lines of research" (Pask 1976c, i). One can get a feeling for the overall problematic from the first paper at the meeting, "SIMTOS: A Review of Recent Developments," by J. Baker, the supervisory project director of the Organisations and Systems Research Laboratory of the Army Research Institute in Washington (Baker 1976). SIMTOS was a computerized battle simulation in which military personnel conducted operations against a simulated enemy. "Decision making" for the Army Research Institute thus referred to decisions made in the flow of battle, though neither the academics at the meeting nor Pask referred to this context. The second conference (Pask 1978) was devoted to the problematic of training "decision makers." I am grateful to Joseph Zeidner for providing me with copies of the proceedings of these conferences. There is more on this work in Pask and Curran (1982, 130–31): "Conference discussions stimulated research into a different type of decision aid, a sort of on-going model which System Research has since updated. This model was an evolving computer system called TDS, or Team Decision System." This system featured a space battle scenario, in which "TDS can aid decision making by allowing the commander to interrogate the system; by giving information; by presenting the tactics currently in use; by doing calculations. But it also has an extra ingredient: there are real emergencies which parallel the unexpected emergencies of real life. . . . However much the commanders interact with TDS as a decision making aid, they have no chance to deal with emergencies, simply because these happen too fast. . . . TDS takes over and makes the choice on the basis of what it has learned about the strategies commanders have chosen in the past." Pask also carried out research for the U.S. Air Force in the periods 1961–65 and 1975–77. The first report on that work

was "Research on the Design of Adaptive Training Systems with a Capability for Selecting and Altering Criteria for Adaptation" (Pask et al. 1965).

45. Simon Penny is an interactive artist and theorist based at the University of California, Irvine. I thank him for enlightening discussions on the past and present of his field. Penny (1999) discusses the writings of sculptor Jack Burnham (1968) on the history and future of sculpture. In *Beyond Modern Sculpture* Burnham devotes the first half of the book to "sculpture as object," meaning the modern tradition, the past. The second half is on "sculpture as system" running from "sculpture and automata" up to "robot and cyborg art," machine art, and the future. Cybernetics is central to Burnham's discussion of the last of these, beginning with the tortoise, the homeostat, and the maze-running mice built by Claude Shannon and continuing with Pask's Eucrates (204, 337) (the book was written before the Colloquy was exhibited).
46. Several of these points are made by Penny (2008), who also alerted me to the difficulty of keeping these machines going in an art museum (the Science Museum in London has a motionless tortoise displayed in a glass case). The threat to the distinctive identity of the artist was explicit at the *Cybernetic Serendipity* exhibition: "Two aspects of this whole project are particularly significant. The first is that at no point was it clear to any of the visitors . . . which of the various drawings, objects and machines were made by artists and which were made by engineers; or, whether the photographic blow-ups of texts mounted on the walls were the work of poets or scientists. There was nothing intrinsic in the works themselves to provide information as to who made them. Among the contributors . . . were forty-three composers, artists and poets, and eighty-seven engineers, doctors, computer systems designers and philosophers. The second significant fact is that whereas new media inevitably contribute to the changing forms of the arts, it is unprecedented that a new tool should bring in its wake new people to become involved in creative activity. . . . Graphic plotters, cathode-ray tube displays and teleprinters have enabled engineers, and others, who would never even have thought of putting pen to paper, to make images for the sheer pleasure of seeing them materialize" (Reichardt 1971, 11). Artists might well have been concerned for their already tenuous social status.
47. A growing if ambiguous fascination with technology is visible in the history of British art in the sixties (Stephens and Stout 2004). One indicator of the cultural centrality achieved by machine art is that *Cybernetic Serendipity* was opened by Anthony Wedgewood-Benn, then minister of technology, with Lord Snowdon (Anthony Armstrong-Jones, photographer) and Princess Margaret also in evidence (Reichardt, interview, 21 February 2003).
48. We can see this as a return to "normal" in the art world, reinforced in the late sixties and early seventies by a critique of science and technology *within* the counterculture. The critique grew out of disgust at the Vietnam War and the complicity of scientists and engineers with the military-industrial complex. Sadler (2005) discusses this in relation to adaptive architecture (below).
49. In 1972, Jasia Reichardt tried to put together another project fusing art and science, but was unable to find support for it: "Every proposal was dismissed." Later she sought to develop a project called "Fantasia Mathematica" on the

arts and mathematics for the BBC but was again unable to attract a sufficient budget. She told me that no one in Britain was interested in archival papers on *Cybernetic Serendipity*, and that the papers are now in Japan. Pask led a discussion at "Event 1" of the Computer Art Society, 29–30 March 1969; "Event 2" never happened (Reichardt, interview, 21 February 2002).

50. For examples of Penny's work, see ace.uci.edu/penny/. Penny knew Pask personally; Hertz was working in Penny's Masters Program in Arts, Computation, and Engineering at the University of California, Irvine, when I met him (24 October 2005).

51. Gere (2002, 110) gives a different list of names, including Roy Ascott (below), David Medalla ("who refers to himself as a 'hylozoist,' a philosopher of the pre-Socratic Ionian school devoted to the belief that all matter and life are inseparable"; Burnham 1968, 345) and the ubiquitous Stelarc. For a recent ill-tempered critique of interactive art from the mainstream as represented by the *New York Times*, see Boxer (2005).

52. Another circle closes here. Latour and Weibel (2002, 2005) have recently organized two exhibitions at the ZKM which attempted, like this book, to bring science studies, engineering, art, and politics together in new ways: "Iconoclash: Beyond the Image-Wars in Science, Religion and Art" in 2002, and "Making Things Public: Atmospheres of Democracy" in 2005. The latter included a recreation of Stafford Beer's control room for the Cybersyn project in Chile (overseen by Eden Medina), and the opening sentences of the online description of the exhibition are eerily Paskian: "As soon as you enter the show, you feel that something odd is happening: lights, sound and labels seem to react to your presence as a visitor in some invisible and yet palpable manner. You have just encountered the atmospheric conditions of democracy. Soon you will discover that the whole space of the show is embedded in the PHANTOM PUBLIC, a work of art that aims to lend a different, emotional colour to political involvement and political envelopment": www.ensmp.fr/~latour/expositions/002_parliament.html (accessed 3 March 2005).

53. In London, Ascott also taught Pete Townshend of The Who; his contact with Eno dated to the midsixties, when he had moved to Ipswich: Gere (2002, 94). Irish (2004) discusses the work and career of Stephen Willats, another cybernetic artist and student of the lived environment (see, for example, Willats 1976). In the early 1960s Willats was a student at Ealing, where he was taught by Ascott (Stephens and Stout 2004, 111), and he was associated later in the sixties with System Research. Since 1965 he has been editor and publisher of the unfortunately titled *Control Magazine*. A glance at its contents reveals that "control" here should be read in the Paskian sense of "conversation" and not in the authoritarian sense beloved of the critics of cybernetics (more on this below re architecture).

54. www.planetary-collegium.net/about/ (accessed 3 March 2005). Besides Ascott himself, Brian Eno and Ranulph Glanville appear on the collegium's list of supervisors and advisers. Centered on the University of Plymouth in England, and with "nodes" in Zurich, Milan, and Beijing, the collegium echoes the trans-institutional constitution of early cybernetics (chap. 3), but now as a possibly enduring basis for graduate training and advanced research. The Arts, Com-

putation, and Engineering graduate program at U.C. Irvine (n. 50 above) is an attempt to reconfigure the inner structure of the university as a social basis for cybernetic art (broadly construed).

55. Sadler (2005) is a detailed and beautifully illustrated history of Archigram.
56. Gilles Evain (Ivan Chtcheglov), quoted in Heynen (1999, 152). Heynen discusses Constant's New Babylon project as one of the most worked out visions of Situationist adaptive architecture. Behind much of this imagining of architecture as a technology of the self lay "the leisure problem," as it was known in Britain in the 1960s (chap. 6, n. 8). Many people took the leisure problem seriously, and the Situationist dream was that it presented a revolutionary opportunity for a new kind of people to emerge. An excellent source of information on Situationism is the Situationist International online website: www.cddc.vt.edu/sionline/index.html. Marcus (1989) is a beautiful account of the Situationists (and much else). On the connection between Situationism and antipsychiatry, again mediated by Alexander Trocchi (below), see chap. 5, n. 58.
57. Mathews (2007) is the key source on the history of the Fun Palace. Mathews (2007, 274, app. C) lists twenty-six members of the Cybernettics Subcommittee: among those one might recognize, besides Littlewood and Price, are Roy Ascott, Stafford Beer, Tom Driberg (MP), Dennis Gabor, Frank George, Reginald Goldacre, Richard Gregory, A. R. Jonckheere (Pask's supervisor in his London PhD research), Brian Lewis, Robin McKinnon-Wood and his wife, and Ian Mikardo (MP). Several members of the subcommittee were important in the subsequent creation of the Open University, and one wonders if these connections were important to Pask's affiliation to the OU (discussed above).
58. Pask's (1964a) proposal for a cybernetic theater does not mention the Fun Palace.
59. As a member of the Cybernetics Subcommittee, Roy Ascott proposed a "Pillar of Information" for the Fun Palace, an electronic kiosk that would respond to queries in an adaptive fashion: "Based on patterns of user interaction, the Pillar of Information would gradually develop an extensive network of cognitive associations and slippages as a kind of non-hierarchical information map, both allowing and provoking further inquiry beyond the user's initial query" (Mathews 2007, 119).
60. At the reactive level, "the Cybernetics Subcommittee . . . outlined plans to use the latest computerized punch card system to track and allot resources for various activities." But beyond that, the committee also noted that the list of activities and zones could never be complete because "the variety of activities could never be entirely forecast" (Mathews 2007, 116, 118).
61. I thank Howard Shubert and Anne-Marie Sigouin at the Collection Centre d'Architecture/Canadian Centre for Architecture, Montréal, for providing me with a copy of the Cybernetics Subcommittee minutes from which fig. 7.23 is taken and permission to reproduce the figure. This figure has been reproduced at least twice before, in Lobsinger (2000, 131, fig. 5.8) and Mathews (2007, 120). Lobsinger labels it "diagram for a cybernetics theater," but this is a mistake. I return to Lobsinger and Mathews's commentary on the figure below. The Cybernetics Subcommittee "also suggested methods of identity-shifting

and role-playing. . . . Roy Ascott proposed an 'identity bar' which would dispense paper clothing, enabling people to try on different and unfamiliar social personae or even gender roles" (Mathews 2007, 118).

62. Mathews (2007, chap. 4) explores many of these difficulties, including, for example, a shift in the pattern of local government from the London County Council to the Greater London Council engineered by the Conservative government at a key moment in 1964.
63. Mathews (2007) offers much documentation of these kinds of fears circling around the Fun Palace. Littlewood (2001, 761) later wrote: "Alas, for those of us who had given our lives to an idea, the powers that be wouldn't let us have the land for the new Vauxhall Gardens—any land! A bowdlerized version of the structure was erected in Paris but, without activists skilled in managing such activities, as we had foreseen, it became merely a rather pleasant empty space" (Littlewood 2001, 761). This watered-down and flexible but noninteractive version of the Fun Palace lives on as Renzo Piano and Richard Rogers's Pompidou Centre, which "would have been inconceivable without the Fun Palace" (Melvin 2003). (Piano was awarded the highest architectural honor, the Pritzker Prize, in 1998; Rogers the same in 2007: Pogrebin 2007.) Mathews (2007, 232–35) also notes a connection between another of Cedric Price's unbuilt sixties projects, the Potteries Thinkbelt (an adaptive university on wheels, running on disused railway lines in the north of England) and Rem Koolhaas and Bernhard Tschumi's Parc de la Villette.
64. The architect George Baird "stated that Price's refusal to provide 'visually recognizable symbols of identity, place, and activity' and his reduction of architecture to a machine for 'life-conditioning' displayed a gross misconception of architecture's place in human experience. For Baird, Price's architecture-as-servicing mechanism was equivalent to architecture as 'a coffee-vending machine'" (Lobsinger 2000, 126, 134).
65. Thus, the Cybernetics Subcommittee minutes to which Mathews refers, drafted by Pask himself, speak of maintaining "the environment of the individual varied or novel enough to sustain his interest and attention but not so varied that it is unintelligible" (Pask 1965, 7), of a "consensus of opinion . . . in favour of a Fun Palace which stimulated people to think for themselves and to engage in creative activities [and] strong resistance to the view that the Fun Palace should have a specifically educational function" (10), of a pilot project as "possibly modified by the community" (11), and of there being "many legitimate objectives for we do not know, at the outset, the character of Fun" (14). "In a conventional or arbitrary concatenation these facilities [fixed ones, like cinemas and restaurants] appear as objects that satisfy a need. In a Fun Palace they function as operations that catalyse further activity, in particular and according to the defined objectives [of?] participant, co-operative and creative activity" (17). "Mr Pinker also warned against the dangers of moralising, It was not the function of the Fun Palace to turn out a 'Participant-Citizen,' or to give them spiritual uplift. Its job, vis-à-vis the external environment was simply to open up new vistas" (17–18). Pask presented fig. 7.23 as his summary of the committee's deliberations along these lines and added, "If you find this picture confusing, please neglect it" (1).

66. The critique does appear to propagate by repetition (rather than, for instance, reflective thought) and to have an oddly ritualized quality. Mathews's 2007 version is isomorphous to Lobsinger's from 2000 (which Mathews cites), both taking off from fig. 7.23 and saying much the same things. Lobsinger remarks, "At the mention of control systems and the lax behaviorist psychologizing to produce happiness, one is inclined to recoil in amused disdain," and goes on to cite Alvin Toffler's much-read *Future Shock* from 1970: "Toffler himself cites the Fun Palace as an instance of technocratic thought and the impoverishment of the most significant part of human experience, the built environment." This is itself a revealingly misleading gloss on what Toffler says about the Fun Palace (Toffler 1970, 54–57)—and actually the central problematic of *Future Shock* is a supremely cybernetic one: adaptation to an unknown future; we might also note that Toffler later wrote the foreword for the English translation of Prigogine and Stenger's *Order out of Chaos* (Toffler 1984)—but, anyway, half a page later Lobsinger's essay ends on an approving if not very precise note with "In the 1960s, as today, the Fun Palace offers architects a challenging conception of architecture that privileges organization and idea over architecture as built form" (134). It might help to note that Lobsinger also invokes Deleuze on forms of social control that go beyond familiar disciplinary mechanisms. This line of argument might be productive in thinking about some of Pask's teaching machines. SAKI aimed at entraining people in the world of typing in a way that went beyond simple punishments and rewards, as did the later pedagogical machines, and we could certainly ask ourselves whether we are in favor of developing new ways of teaching people to type. But the difference between those machines and the Musicolour–Fun Palace line is that the former had extrinsically defined goals while the latter did not.
67. John Frazer (interview, London, 3 September 2004). Pask's public talks at the AA were popular cultural events, and both Jasia Reichardt and Yolanda Sonnabend told me that they would go along to the AA whenever Pask was speaking there (interviews, London, 21 and 22 February 2002). Pask also taught workshops at the AA in a series organized by his student Ranulph Glanville. One of the first people to invite Pask to speak at the AA was Peter Cook, around 1965, for an "event day" at which interesting people from outside the world of architecture presented their work (John Frazer, interview, 3 September 2004).
68. For a collection of drawings and illustrations of the Generator project, including photographs of the prototype control system, see Riley et al. (2002). I am very grateful to Molly Wright Steenson for sharing her extensive knowledge of the Generator project with me (telephone conversation, 2 February 2007).
69. John Frazer, interview, London, 3 September 2004. John Frazer studied at the AA from 1963 to 1969, and "right from the outset Gordon Pask was a source of inspiration and soon became directly involved" (Frazer 2001, 641). Frazer also taught at the AA from 1973 to 1977 and 1987 to 1996. A couple of the threads of our story converge on Frazer: as mentioned in chapter 3, at an impressionable age in the early 1950s he encountered Grey Walter's tortoises at the Radio Show in Earls Court, London, and subsequently tried to build one; and he worked on cellular automata while at Cambridge (1969–73) at much the same time as John Conway was developing the Game of Life. Both men used the

Atlas Titan computer at Cambridge, desirable for its graphics display, but they worked on different shifts at night (Frazer, interview, 3 September 2004). I am grateful to John Frazer for our conversation and the provision of biographical information.

70. For an extended account, see Frazer (1995), which includes a foreword written by Pask and a 1990 photograph of Pask with the "Universal Constructor" (p. 7).
71. For an early discussion of the MIT Architecture Machine, see Negroponte (1969), who cites Warren McCulloch and Pask. See also Negroponte (1970, 1975); the latter discusses work at MIT between 1968 and 1972 and includes a twenty-six-page introduction written by Pask. For later developments at MIT, see Brand (1987). Ranulph Glanville points out that a drawing board is "a conversational mirror (for the designer) and a source of idea theft in the studio. It's not nearly so static/passive as you think!" (email, 18 Aug 2005). This echoes Pask's thought that any competent work of art is an aesthetically potent environment. But still, a drawing board does not thematize and foreground possibilities for dynamic interaction in the same way as the systems discussed in the text. Frazer's critique (1995, 60) of CAD exactly parallels Beer's of conventional uses of information systems—here, as introducing new technology but leaving existing design practices unchanged.
72. Brand (1994) is a very nice study of how conventional buildings actually evolve over time despite being set in bricks and mortar and concludes with ideas on adaptive architecture. Again, the distinctly cybernetic take on this was to thematize evolution in both the design process and its products.
73. The installation was called *Seek*; the show was Software, Information Technology: Its New Meaning for Art, curated by Jack Burnham (Gere 2002, 107).

Notes to Chapter 8

1. I thank Leo Marx and Philip Fisher for talking me into reading James. More recently, I thank Michael Friedman for a pointed question about the relation between my work and pragmatism.
2. The words "practice" and "performance" point in the same direction, though in different ways. "Practice" refers to human activity in the world, while "performance" is the "doing" of any entity or system, human or nonhuman. "Practice" is thus a subset of "performance." In *The Mangle of Practice* I focused on the relation between the practice of scientists and the performance of machines and instruments. In this book I have paid little attention to the practice of cyberneticians (in the sense of analyzing the day-by-day construction of tortoises or DAMS, say), though I have no doubt that it looks just like the practice of the physicists discussed in *The Mangle*. I have focused instead on the performance of the machines and systems they built.
3. The classic popular introduction to this line of research is Prigogine and Stengers (1984); see also Pickering (2005a). In philosophy, a hylozoist fascination with striking self-organizing properties of matter runs through Deleuze and Guattari (1987) and continues in the work of Manuel DeLanda—for example, the series of columns "Matter Matters," beginning with DeLanda (2005) (also DeLanda 2002). I thank DeLanda for very illuminating conversations and

correspondence (and arguments). One might also think here of the emergent beauty of fractal mathematics (Mandelbrot 1983). Beer incorporated fractal imagery in some of his paintings, and Walter's novel (1956a) includes a fractal (*avant la lettre*) time machine, but there is little beyond that that could figure in our history.

4. I cannot resist reproducing this quotation, though it does not mean what I would like. Heidegger thought of cybernetics as a universal science that could embrace and unify all the other sciences, thus displacing philosophy from one of its traditional roles, and hence his attempted reformulation of the "end of philosophy" in Heidegger (1976). For a historical account of cybernetics as a universal science, see Bowker (1993). Evidently that is not the version of cybernetics that emerges from the history of British cybernetics. It would be nice to know what Heidegger would have made of the story told here; Carol Steiner (2008) stages a fictional conversation between the great philosopher and myself!
5. The question (put to me most forcefully by Ezekiel Flannery) of who the "we" is in this paragraph and below arises here. I am content to leave this open: readers can decide for themselves whether they are in or out. "We" certainly includes an earlier me (before writing *The Mangle of Practice* and beginning the research for this book), and, faute de mieux, I take myself to be representative in the relevant respects of, at least, the contemporary West. From another angle, there are many religions that take for granted nonmodern ontologies, but I don't imagine that many Buddhists, say, often find themselves talking a lot about robotics or cellular automata. They, too, might find the history of cybernetics striking; they, too, could be part of this "we."
6. Scott's examples of high-modernist projects include scientific forestry, the Soviet collectivization of agriculture, "villagisation" in Tanzania, Lenin's conception of the Bolshevik Revolution, city planning, and Le Corbusier's design for Brazilia. I thank David Perkins for referring me to Scott's book.
7. Latour's political message is that we need to *think* differently about science, technology, and society: "We have simply to *ratify* what we have always done, provided we *reconsider* our past, provided that we *understand* retrospectively to what extent we have never been modern, and provided that we rejoin the two halves of the *symbol* broken by Hobbes and Boyle as a sign of recognition. Half of our politics is constructed in science and technology. The other half of Nature is constructed in societies. Let us patch the two back together, and the political task can begin again. Is it asking too little simply to ratify in public what is already happening?" (Latour 1993, 144; emphases added). I discuss Latour's conservatism further in Pickering (2009).
8. Neither Heidegger nor Latour addresses the possibility of different sorts of practice in science and engineering (and beyond). Habermas (1970, 87–88) considers the possibility (which he associates with Marcuse), but only to reject it: "Technological development lends itself to being interpreted as though the human species had taken the elementary components of the behavioral system of purposive-rational action, which is primarily rooted in the human organism, and projected them one after another onto the plane of technical instruments. . . . At first the functions of the motor apparatus (hands and legs) were augmented and replaced, followed by energy production (of the human

body), the functions of the sensory apparatus (eyes, ears, and skin), and finally the governing center (the brain). Technological development thus follows a logic that corresponds to . . . the structure of *work*. Realizing this, it is impossible to envisage how, as long as the organization of human nature does not change . . , we could renounce technology, more particularly *our* technology, in favour of a qualitatively different one. . . . The idea of a New Science will not stand up to logical scrutiny any more than that of a New Technology, if indeed science is to retain the meaning of modern science inherently oriented to possible technical control. For this function, as for scientific-technological progress in general, there is no more 'humane' substitute." From the present perspective this is a circular argument; my argument is that things look different if one takes the adaptive rather than the laboring body as a point of departure in thinking about what science and technology might be. J. Scott (1998) offers *metis*—local, situated knowledge grounded in experience—as an antidote to scientific high modernism. This is an interesting proposal but leaves little if any space for the experimental stance that I associate with cybernetics.

9. Latour's (2004) institutional blueprint for a politics of nature includes mechanisms for continual monitoring and reassessment of how plans are working out. J. Scott's (1998, 345) recommendations for development planning and practice are "take small steps," "favor reversibility," "plan on surprises," and "plan on human inventiveness."
10. This field now extends in many and sometimes surprising directions. This might be the place to mention Donna Haraway's (2003) brilliant, performative, and very cybernetic analysis of love. She focuses on relations between humans and dogs, but the analysis is readily extended to purely human relations, relations with nature, and so on.
11. A range of examples different from that discussed below can be found in software engineering and developments in information technology: see Marick (2008) and Neff and Stark (2004). We could also think of the literature on engineering design: see McGrail (2008) for an analysis that connects design and ontology along the lines discussed here.
12. There is a subtlety that needs to be thought through here concerning the meaning of "experiment." In the modcrn sciences, this refers to a detour away from and back to the world as found, isolating specific segments of the world and producing knowledge of them in small-scale laboratory experiments which can then be transferred back to the world in the fashion of enframing. Latour's (1983) essay on Pasteur and anthrax remains the canonical study of this maneuver. Latour does not, however, discuss the fact that sometimes the translation back into the world works and sometimes it fails. The nonmodern sense of "experiment" we need here is that of performative experimentation on the thing itself (e.g., the Colorado River) without any detour through the laboratory. The contrast is between experiment as part of a strategy of enframing and experiment as revealing.
13. J. Scott (1998, 327) quotes from a 1940 account of water management in Japan: "Erosion control is like a game of chess [or a dance of agency]. The forest engineer, after studying his eroding valley, makes his first move, locating and building one or more check dams. He waits to see what Nature's response is. This determines the forest engineer's next move, which may be another dam or

two, an increase in the former dam, or the construction of side retaining walls. Another pause for observation, the next move is made, and so on, until erosion is checkmated. The operations of natural forces, such as sedimentation and re-vegetation, are guided and used to the best advantage to keep down costs and to obtain practical results. *No more is attempted than Nature has already done in the region.*"

14. In a discussion of situated robotics, Brooks (1999 [1991], 97) includes a heading "It Isn't German Philosophy": "In some circles much credence is given to Heidegger as one who understood the dynamics of existence. Our work has certain similarities to work inspired by this German philosopher . . . but our work was not so inspired. It is based purely on engineering considerations. That does not preclude it from being used in philosophical debate as an example on any side of any fence, however."
15. Jeff Nuttall's (1968, 253–55) classic report from the British underground ends with the topic of robustness, lamenting a lack of lasting material achievements by the counterculture and looking forward to the construction of more enduring cultural elements, including some that are familiar from chapter 7: "It's time to come away from the mobile arts, poetry, jazz, theatre, dance, clothes. Too great a preoccupation with mobility constitutes a refusal of existence. Movement, like drugs is good tactics but a poor alternative to the established culture. . . . Can we build and think and organize with the passions of perpetual inner illumination? Of course we can. . . . Let us turn away from the contemplators and listen to the architects, the activists, the engineers, the Archigram Group with their Plug-In City scheme, Cedric Price the Fun Palace designer, Geoffrey Shaw and his constructions in plastic, Keith Albarn and his furniture sculpture. . . . Let's . . . build our own damn future." It is symptomatic of the structural weakness of the sixties that Nuttall never once mentions cybernetics in his book—history might have been different.
16. I thank Frederick Erickson and Kris Gutierrez, professors of education at UCLA, for a brief conversation about the ideas discussed below. Each thought such an addition to the school curriculum would be possible and desirable. Kris mentioned that the No Child Left Behind Act might be an obstacle in the United States, because she thought of this sort of course as unexaminable. Actually, if it were taught in terms of concrete examples I think there would be no problem in constructing an assessment system and even in setting formal examinations.
17. "Ontology" is an intimidating word, but "how the world is" is plain English.
18. The truly difficult intellectual maneuver is coming to see modern science and engineering as a particular stance in the world. It has taken me a long time to be able to do so, and I am not sure how one would begin to teach young children about this.
19. The question of who would teach such a course (besides me) then arises, and it takes us back to the question of the social basis. I think one can find scholars with the right sort of interests and expertise scattered around most universities, and the course itself might serve both to bring them together and to reduce their marginality. This kind of initiative in undergraduate education might even act as a center of condensation for what I once referred to as a department of emergency studies (Pickering 1995, chap. 7).

REFERENCES

Abraham, R., T. McKenna, and R. Sheldrake (1992) *Trialogues at the Edge of the West: Chaos, Creativity and the Resacralization of the World* (Santa Fe, NM: Bear & Co.).

Adams, D. (1979) *The Hitchhiker's Guide to the Galaxy* (London: Pan).

Adrian, E. D., and B. H. C. Matthews (1934) "The Berger Rhythm: Potential Changes from the Occipital Lobes in Man," *Brain*, 57, 355–85.

Alexander, C. (1964) *Notes on the Synthesis of Form* (Cambridge, MA: Harvard University Press).

Alexander, C. (1983) "Linz Café: The Architect's Evaluation," and "Contrasting Concepts of Harmony in Architecture: Debate between Christopher Alexander and Peter Eisenman," *Lotus International*, 40, 45–73.

Alexander, C., et al. (1977) *A Pattern Language: Towns, Buildings, Construction* (New York: Oxford University Press).

Allen, M. (2005) "Décor by Timothy Leary, Dreams by You," *New York Times*, 20 January, D1, 7.

Anderson, W. T. (2004 [1983]) *The Upstart Spring: Esalen and the Human Potential Movement: The First Twenty Years* (Lincoln, NE: Authors Guild Backinprint.com).

Artaud, A. (1993) *The Theatre and Its Double* (London: Calder).

Asaro, P. (2007) "Heinz von Foerster and the Bio-computing Movements of the 1960s," in A. Müller and K. H. Müller (eds.), *An Unfinished Revolution? Heinz von Foerster and the Biological Computer Laboratory, 1958–1976* (Vienna: Edition Echoraum).

Asaro, P. (2008) "W. Ross Ashby: From Mechanisms of Adaptation to Intelligence Amplifiers," in M. Wheeler, P. Husbands, and O. Holland (eds.), *The Mechanical Mind in History* (Cambridge, MA: MIT Press), pp. 149–84.

Asaro, P., and G. Klir (eds.) (2009) *The Intellectual Legacy of W. Ross Ashby*, special issue of *International Journal of General Systems*, 38, no. 2.

Ashby, W. R. (1933) "The Physiological Basis of the Neuroses," *Proceedings of the Royal Society of Medicine*, 26, 1454–60.

Ashby, W. R. (1935) "The Thickness of the Cerebral Cortex and Its Layers in the Mental Defective" (MA thesis, University of Cambridge).

Ashby, W. R. (1937a) "An Inquiry into the Incidence of 'Neuropath' Conditions in the Relatives of Normal Persons," *Journal of Mental Science, 83*, 247.

Ashby, W. R. (1937b) "Tissue Culture Methods in the Study of the Nervous System," *Journal of Neurology and Psychopathology, 17*, 322.

Ashby, W. R. (1940) "Adaptiveness and Equilibrium," *Journal of Mental Science, 86*, 478–84.

Ashby, W. R. (1941) "The Origin of Adaptation," unpublished notebook, British Library, London.

Ashby, W. R. (1945a) "The Physical Origin of Adaptation by Trial and Error," *Journal of General Psychology, 32*, 13–25.

Ashby, W. R. (1945b) "Effect of Controls on Stability," *Nature, 155*, 242–43.

Ashby, W. R. (1947) "The Existence of Critical Levels for the Actions of Hormones and Enzymes, with Some Therapeutic Applications," *Journal of Mental Science, 93*, 733–39.

Ashby, W. R. (1948) "Design for a Brain," *Electronic Engineering, 20* (December), 379–83.

Ashby, W. R. (1949a) "Effects of Convulsive Therapy on the Excretion of Cortins and Ketasteroids," *Journal of Mental Science, 95*, 275–324.

Ashby, W. R. (1949b) "Adrenal Cortical Steroids and the Metabolism of Glutamic Acid in Gluconeogenesis," *Journal of Mental Science, 95*, 153–61.

Ashby, W. R. (1950a) "The Stability of a Randomly Assembled Nerve-Network," *Electroencephalography and Clinical Neurophysiology, 2*, 471–82.

Ashby, W. R. (1950b) "Cybernetics," in G. W. T. H. Fleming (ed.), *Recent Progress in Psychiatry*, special issue of *Journal of Mental Science*, 2nd ed. (London: J. & A. Churchill), pp. 93–110.

Ashby, W. R. (1951) "Statistical Machinery," *Thalès, 7*, 1–8.

Ashby, W. R. (1951–57) "Passing through Nature" (unpublished notebook).

Ashby, W. R. (1952) *Design for a Brain* (London: Chapman & Hall).

Ashby, W. R. (1953a) "The Mode of Action of Electro-convulsive Therapy," *Journal of Mental Science, 99*, 202–15.

Ashby, W. R. (1953b) "Homeostasis," in H. von Foerster, M. Mead, and H. L. Teuber (eds.), *Cybernetics: Circular Causal and Feedback Mechanisms in Biological and Social Systems: Transactions of the Ninth Conference (March 20–21, 1952)* (New York: Josiah Macy, Jr., Foundation), pp. 73–106. Reprinted in C. Pias (ed.), *Cybernetics-Kybernetik: The Macy Conferences, 1946–1953*, vol. 1 (Zurich: diaphanes), pp. 593–619.

Ashby, W. R. (1954) "The Application of Cybnernetics to Psychiatry," *Journal of Mental Science, 100*, 114–24.

Ashby, W. R. (1956) *An Introduction to Cybernetics* (New York: Wiley).

Ashby, W. R. (1959a) "The Mechanism of Habituation," in *Mechanisation of Thought Processes: Proceedings of a Symposium Held at the National Physical Laboratory on 24th, 25th, 26th and 27th November 1958*, 2 vols. (London: Her Majesty's Stationery Office), pp. 93–118.

Ashby, W. R. (1959b) "A Simple Computer for Demonstrating Behaviour," in *Mechanisation of Thought Processes: Proceedings of a Symposium Held at the National Physical*

Laboratory on 24th, 25th, 26th and 27th November 1958, 2 vols. (London: Her Majesty's Stationery Office), pp. 947–49.

Ashby, W. R. (1960) *Design for a Brain: The Origin of Adaptive Behaviour*, 2nd ed. (London: Chapman & Hall).

Ashby, W. R. (1962) "Simulation of a Brain," in H. Borko (ed.), *Computer Applications in the Behavioral Sciences*, pp. 452–67.

Ashby, W. R. (1963) "Induction, Prediction and Decision-Making in Cybernetic Systems," in H. E. Kyburg and E. Nagel (eds.), *Induction: Some Current Issues* (Middletown, CT: Wesleyan University Press), pp. 55–65. Reprinted in Ashby, *Mechanisms of Intelligence: Ross Ashby's Writings on Cybernetics*, R. Conant (ed.) (Seaside, CA: Intersystems Publications, 1981), pp. 275–86.

Ashby, W. R. (1966) "Mathematical Models and Computer Analysis of the Function of the Central Nervous System," *Annual Review of Physiology, 28*, 89–106. Reprinted in Ashby, *Mechanisms of Intelligence: Ross Ashby's Writings on Cybernetics*, R. Conant (ed.) (Seaside, CA: Intersystems Publications, 1981), pp. 325–42.

Ashby, W. R. (1968a) "Some Consequences of Bremermann's Limit for Information Processing Systems," in H. Oestreicher and D. Moore (eds.), *Cybernetic Problems in Bionics* (New York: Gordon & Breach), pp. 69–76.

Ashby, W. R. (1968b) "The Contribution of Information Theory to Pathological Mechanisms in Psychiatry," *British Journal of Psychiatry, 114*, 1485–98. Reprinted in Ashby, *Mechanisms of Intelligence: Ross Ashby's Writings on Cybernetics*, R. Conant (ed.) (Seaside, CA: Intersystems Publications, 1981), pp. 343–56.

Ashby, W. R. (1968c) Review of *The Politics of Experience*, by R. D. Laing, in *Philosophy Forum, 7* (September 1968), 84–86.

Ashby, W. R. (1970) "Analysis of the System to Be Modeled," in R. M. Stogdill (ed.), *The Process of Model-Building in the Behavioral Sciences* (Columbus: Ohio State University Press), pp. 94–114. Reprinted in Ashby, *Mechanisms of Intelligence: Ross Ashby's Writings on Cybernetics*, R. Conant (ed.) (Seaside, CA: Intersystems Publications, 1981), pp. 303–23.

Ashby, W. R., G. H. Collins, and N. Bassett (1960) "The Effects of Nicotineic Acid, Nicotinamide, and Placebo on the Chronic Schizophrenic," *Journal of Mental Science, 106*, 1555–59.

Ashby, W. R., and R. M. Stewart (1934–35) "The Brain of the Mental Defective: The Corpus Callosum in Its Relation to Intelligence," "The Brain of the Mental Defective: A Study in Morphology in Its Relation to Intelligence," and "The Brain of the Mental Defective," pt. 3, "The Width of the Convolutions in the Normal and Defective Person," *Journal of Neurology and Psychopathology, 13*, 217–26; *14*, 217–26; *16*, 26–35.

Ashby, W. R., R. M. Stewart, and J. H. Watkin (1937) "Chondro-osteo-dystrophy of the Hurler Type: A Pathology Study," *Brain, 60*, 149–79.

Ashby, W. R., H. von Foerster, and C. C. Walker (1962) "Instability of Pulse Activity in a Net with Threshold," *Nature, 196*, 561–62. Reprinted in Ashby, *Mechanisms of Intelligence: Ross Ashby's Writings on Cybernetics*, R. Conant (ed.) (Seaside, CA: Intersystems Publications, 1981), pp. 81–84.

Ashby, W. R., and C. C. Walker (1968) "Genius," in P. London and D. Rosenhan (eds.), *Foundations of Abnormal Psychology* (New York: Holt, Rinehart and Winston), pp. 201–25.

Asplen, L. (2005) "Cybernetics, Ecology, and Ontology in the Writings of G. E. Hutchison, E. P. Odum, and H. T. Odum" (unpublished manuscript, University of Illinois).

Asplen, L. (2008) "Going with the Flow: Living the Mangle through Acting in an Open-Ended World: Nature, Culture, and Becoming in Environmental Management Practice," in A. Pickering and K. Guzik (eds.), *The Mangle in Practice: Science, Society and Becoming* (Durham, NC: Duke University Press), pp. 163–84.

Baker, J. (1976) "SIMTOS: A Review of Recent Developments," in G. Pask (ed.), *Current Scientific Approaches to Decision Making in Complex Systems: Report of a Conference Held at Richmond, Surrey, Eng;and, 14–15 July 1975*, ARI Technical Report TR-76-B1 (Arlington, VA: U.S. Army Research Institute for the Behavioral and Social Sciences), pp. 6–11.

Barnes, B., D. Bloor, and J. Henry (1996) *Scientific Knowledge: A Sociological Analysis* (Chicago: University of Chicago Press).

Barnes, M., and J. Berke (1971) *Two Accounts of a Journey through Madness* (New York: Harcourt Brace Jovanovich).

Bateson, G. (1942) "Social Planning and the Concept of Deutero-Learning," in *Science, Philosophy and Religion: Second Symposium* (New York: Conference on Science, Philosophy, and Religion), chap. 4. Reprinted in Bateson, *Steps to an Ecology of Mind*, 2nd ed. (Chicago: University of Chicago Press, 2000), pp. 159–76.

Bateson, G. (2000 [1959]) "Minimal Requirements for a Theory of Schizophrenia," the second annual Albert D. Lasker Lecture, *A.M.A. Archives of General Psychiatry, 2* (1960), 477–91. Reprinted in Bateson, *Steps to an Ecology of Mind*, 2nd ed. (Chicago: University of Chicago Press, 2000), pp. 244–70.

Bateson, G. (ed.) (1961) *Perceval's Narrative: A Patient's Account of His Psychosis, 1830–1832* (Stanford, CA: Stanford University Press).

Bateson, G. (1968) "Conscious Purpose versus Nature," in D. Cooper (ed.), *To Free a Generation: The Dialectics of Liberation* (New York: Collier), pp. 34–49.

Bateson, G. (1969) "Double Bind, 1969," paper presented at a Symposium on the Double Bind, American Psychological Association, Washington DC, August 1969. Reprinted in C. Sluzki and D. Ransom (eds.), *Double Bind: The Foundation of the Communicational Approach to the Family* (New York: Grune & Stratton, 1976), pp. 237–42; and in Bateson, *Steps to an Ecology of Mind*, 2nd ed. (Chicago: University of Chicago Press, 2000), pp. 271–78.

Bateson, G. (1972) *Steps to an Ecology of Mind* (New York: Ballantine).

Bateson, G. (2000) *Steps to an Ecology of Mind*, 2nd ed. (Chicago: University of Chicago Press).

Bateson, G., D. Jackson, J. Haley, and J. Weakland (1956) "Towards a Theory of Schizophrenia," *Behavioral Science, 1*, 251–64. Reprinted in Bateson, *Steps to an Ecology of Mind* (New York: Ballantine, 1972), pp. 201–27.

Bateson, M. C. (1984) *With a Daughter's Eye: A Memoir of Margaret Mead and Gregory Bateson* (New York: HarperCollins).

Bateson, M. C. (2000) "Foreword by Mary Catherine Bateson, 1999," in G. Bateson, *Steps to an Ecology of Mind*, 2nd ed. (Chicago: University of Chicago Press), pp. vii–xv.

Bateson, M. C. (2005) "The Double Bind: Pathology and Creativity," *Cybernetics and Human Knowing, 12*, 11–21.

Bear, G. (1997) *Slant* (New York: Tor).

Beck, U. (1992) *Risk Society: Towards a New Modernity* (London: SAGE).
Beer, I. D. S. (2001) *But, Headmaster! Episodes from the Life of an Independent School Headmaster* (Wells: Greenbank Press).
Beer, I. D. S. (2002) Letter to Stafford Beer's children, 25 August.
Beer, S. (1953) "A Technique for Standardized Massed Batteries of Control Charts," *Applied Statistics, 2*, 160–65. Reprinted in Beer, *How Many Grapes Went into the Wine? Stafford Beer on the Art and Science of Holistic Management* (New York: Wiley, 1994), pp. 35–42.
Beer, S. (1954) "The Productivity Index in Active Service," *Applied Statistics, 3*, 1–14. Reprinted in Beer, *How Many Grapes Went into the Wine? Stafford Beer on the Art and Science of Holistic Management* (New York: Wiley, 1994), pp. 43–58.
Beer, S. (1956) "The Impact of Cybernetics on the Concept of Industrial Organization," in *Proceedings of the First International Congress on Cybernetics, Namur, 26–29 June 1956* (Paris: Gauthier-Villars). Reprinted in Beer, *How Many Grapes Went into the Wine? Stafford Beer on the Art and Science of Holistic Management* (New York: Wiley, 1994), pp. 75–95.
Beer, S. (1959) *Cybernetics and Management* (London: English Universities Press).
Beer, S. (1994 [1960]) "Retrospect—American Diary, 1960," in Beer, *How Many Grapes Went into the Wine? Stafford Beer on the Art and Science of Holistic Management* (New York: Wiley), pp. 229–309.
Beer, S. (1962a) "Towards the Automatic Factory," in H. von Foerster and G. Zopf (eds.), *Principles of Self-Organization: Transactions of the University of Illinois Symposium on Self-Organization, Robert Allerton Park, 8 and 9 June, 1961* [sic: actually 1960] (New York: Pergamon), pp. 25–89. Reprinted in Beer, *How Many Grapes Went into the Wine? Stafford Beer on the Art and Science of Holistic Management* (New York: Wiley, 1994), pp. 163–225.
Beer, S. (1962b) "A Progress Note on Research into a Cybernetic Analogue of Fabric," *Artorga*, Communication 40, April 1962. Reprinted in Beer, *How Many Grapes Went into the Wine? Stafford Beer on the Art and Science of Holistic Management* (New York: Wiley, 1994), pp. 24–32.
Beer, S. (1965) "Cybernetics and the Knowledge of God," *Month, 34*, 291–303.
Beer, S. (1967) *Cybernetics and Management*, 2nd ed. (London: English Universities Press).
Beer, S. (1968a) *Management Science: The Business Use of Operations Research* (Garden City, NY: Doubleday).
Beer, S. (1968b) "SAM," in J. Reichardt (ed.), *Cybernetic Serendipity: The Computer and the Arts* (London: W. & J. Mackay), pp. 11–12.
Beer, S. (1969) "The Aborting Corporate Plan," in *Perspectives on Planning* (Paris: OECD).
Beer, S. (1975 [1970]) "The Liberty Machine," keynote address to the American Society of Cybernetics conference "The Environment," Washington DC, 8–9 October 1970, in Beer, *Platform for Change: A Message from Stafford Beer* (New York: John Wiley), pp. 307–21.
Beer, S. (1972) *Brain of the Firm* (London: Penguin).
Beer, S. (1975a [1973]) "The World We Manage," *Behavioral Science, 18*, 198–209. Reprinted as "The Surrogate World We Manage," in Beer, *Platform for Change: A Message from Stafford Beer* (New York: John Wiley, 1975), pp. 397–416.

Beer, S. (1975b [1973]) "Fanfare for Effective Freedom: Cybernetic Praxis in Government," in Beer, *Platform for Change: A Message from Stafford Beer* (New York: John Wiley, 1975), pp. 421–52.

Beer, S. (1973a) "Beer Replies," *New Scientist*, 57, 449.

Beer, S. (1973b) "Beer to Grosch," *New Scientist*, 57, 685–86.

Beer, S. (1974a) "Cybernetics of National Development," Zaheer Foundation Lecture, New Delhi, India. Reprinted in Beer, *How Many Grapes Went into the Wine? Stafford Beer on the Art and Science of Holistic Management* (New York: Wiley, 1994), pp. 316–43.

Beer, S. (1974b) *Designing Freedom* (Toronto: CBC Publications).

Beer, S. (1975) *Platform for Change: A Message from Stafford Beer* (New York: John Wiley).

Beer, S. (1977) *Transit* (Dyfed, Wales: CWRW Press). 2nd ed., 1983 (Charlottetown, PEI, Canada: Mitchell Communications).

Beer, S. (1979) *The Heart of the Enterprise* (New York: Wiley).

Beer, S. (1994 [1980]) "I Said, You Are Gods," second annual Teilhard Lecture. Reprinted in Beer, *How Many Grapes Went into the Wine? Stafford Beer on the Art and Science of Holistic Management* (New York: Wiley, 1994), pp. 375–96.

Beer, S. (1981) *Brain of the Firm*, 2nd ed. (New York: Wiley).

Beer, S. (1983) "A Reply to Ulrich's 'Critique of Pure Cybernetic Reason: The Chilean Experiment with Cybernetics,'" *Journal of Applied Systems Analysis*, 10, 115–19.

Beer, S. (1985) *Diagnosing the System for Organizations* (New York: Wiley).

Beer, S. (1989a) "The Viable System Model: Its Provenance, Development, Methodology and Pathology," in R. Espejo and R. Harnden (eds.), *The Viable System Model: Interpretations and Applications of Stafford Beer's VSM* (New York: Wiley), pp. 11–37.

Beer, S. (1989b) "The Chronicles of Wizard Prang" (unpublished manuscript, Toronto), available at www.chroniclesofwizardprang.com.

Beer, S. (1990a) "Recursion Zero: Metamanagement," *Systems Practice*, 3, 315–26.

Beer, S. (1990b) "On Suicidal Rabbits: A Relativity of Systems," *Systems Practice*, 3, 115–124.

Beer, S. (1993a) "World in Torment: A Time Whose Idea Must Come," *Kybernetes*, 22, 15–43.

Beer, S. (1993b) "Requiem," *Kybernetes*, 22, 105–8.

Beer, S. (1994a) *How Many Grapes Went into the Wine? Stafford Beer on the Art and Science of Holistic Management*, R. Harnden and A. Leonard (eds.) (New York: Wiley).

Beer, S. (1994b) *Beyond Dispute: The Invention of Team Syntegrity* (New York: Wiley).

Beer, S. (1994c) *The Falcondale Collection: Stafford Beer Initiates an Audience into the World of Systems and Managerial Cybernetics*, videotape and transcript (Liverpool: JMU Services).

Beer, S. (2000) "Ten Pints of Beer: The Rationale of Stafford Beer's Cybernetic Books (1959–94)," *Kybernetes*, 29, 558–72.

Beer, S. (2001) "A Filigree Friendship," *Kybernetes*, 30, 551–59.

Beer, S. (2004 [2001]) "What Is Cybernetics?" acceptance speech for an honorary degree at the University of Valladolid, Spain, 21 October 2001 in *Kybernetes*, 33, 853–63.

Berke, J. (ed.) (1970) *Counter Culture: The Creation of an Alternative Society* (London: Peter Owen).

Berkeley, E. C. (1952) "The Construction of Living Robots" (unpublished manuscript, New York), available at www.purplestatic.com/robot/index.htm.
Blohm, H., S. Beer, and D. Suzuki (1986) *Pebbles to Computers: The Thread* (Toronto: Oxford University Press).
Borck, C. (2000) "Aiming High: Cybernetics in Wartime Germany," paper delivered at colloquium, Max Planck Institute for History of Science, 9 May 2000.
Borck, C. (2001) "Electricity as a Medium of Psychic Life: Electrotechnological Adventures into Psychodiagnosis in Weimar Germany," *Science in Context*, *14*, 565–90.
Bowker, G. (1993) "How to Be Universal: Some Cybernetic Strategies, 1943–70," *Social Studies of Science*, *23*, 107–27.
Bowker, G. (2005) *Memory Practices in the Sciences* (Cambridge, MA: MIT Press).
Boxer, S. (2005) "Art That Puts You in the Picture, Like It or Not," *New York Times*, 27 April 2005.
Braitenberg, V. (1984) *Vehicles: Experiments in Synthetic Psychology* (Cambridge, MA: MIT Press).
Brand, S. (1987) *The Media Lab: Inventing the Future at MIT* (New York: Viking).
Brand, S. (1994) *How Buildings Learn: What Happens After They're Built* (New York: Viking).
Brier, S., and R. Glanville (eds.) (2003) *Heinz von Foerster, 1911–2002*, special issue of *Cybernetics and Human Knowing*, *10*, no. 3–4.
Brooks, R. (1999 [1986]) "A Robust Layered Control System for a Mobile Robot," *IEEE Journal of Robotics and Automation*, *RA-2*, 14–23. Reprinted in Brooks, *Cambrian Intelligence: The Early History of the New AI* (Cambridge, MA: MIT Press), pp. 3–26.
Brooks, R. (1999 [1991]) "Intelligence without Reason," in *Proceedings of the 1991 Joint Conference on Artificial Intelligence*, 569–95. Reprinted in Brooks, *Cambrian Intelligence: The Early History of the New AI* (Cambridge, MA: MIT Press, 1999), pp. 133–86.
Brooks, R. (1999 [1995]) "Intelligence without Representation," *Artificial Intelligence Journal*, *47*, 139–60. Reprinted in Brooks, *Cambrian Intelligence: The Early History of the New AI* (Cambridge, MA: MIT Press, 1999), pp. 79–101.
Brooks, R. (1999) *Cambrian Intelligence: The Early History of the New AI* (Cambridge, MA: MIT Press).
Brooks, R. (2002) *Robot: The Future of Flesh and Machines* (London: Allen Lane).
Brown, G. S. (1969) *Laws of Form* (London: George Allen & Unwin).
Buchanan, M. (2007) "The Messy Truth about Drag," *Nature Physics*, *3* (April), 213.
Bula, G. (2004) "Observations on the Development of Cybernetic Ideas in Colombia: A Tribute to Stafford Beer," *Kybernetes*, *33*, 647–58.
Burgess, A. (1963) *A Clockwork Orange* (New York: Norton).
Burnham, J. (1968) *Beyond Modern Sculpture: The Effects of Science and Technology on the Sculpture of This Century* (New York: George Braziller).
Burns, D. (2002) *The David Burns Manuscript*, Brent Potter (ed.) (unpublished manuscript), available at laingsociety.org/colloquia/thercommuns/dburns1.htm#cit.
Burroughs, W. S. (2001 [1956]) "Letter from a Master Addict to Dangerous Drugs," *British Journal of Addiction*, *53*. Reprinted in *Naked Lunch: The Restored Text*, J. Grauerholz and B. Miles (eds.) (New York: Grove Press). pp. 213–29.
Burroughs, W. S. (2001 [1959]) *Naked Lunch: The Restored Text*, J. Grauerholz and B. Miles (eds.) (New York: Grove Press).

Burston, D. (1996) *The Wing of Madness: The Life and Work of R. D. Laing* (Cambridge, MA: Harvard University Press).
Butler, F. (1978) *Feedback: A Survey of the Literature* (New York: Plenum).
Byatt, A. S. (2002) *A Whistling Woman* (London: Chatto & Windus).
Cage, J. (1991) "An Autobiographical Statement," www.newalbion.com/artists/cagej/autobiog.html.
Cameron, J. L., R. D. Laing, and A. McGhie (1955) "Patient and Nurse: Effects of Environmental Change in the Care of Chronic Schizophrenics," *Lancet, 266* (31 December), 1384–86.
Canguilhem, G. (2005 [1983]) "The Object of the History of Sciences," in G. Gutting (ed.), *Continental Philosophy of Science* (Malden, MA: Blackwell), pp. 198–207.
Cannon, W. B. (1929) "Organisation for Physiological Homeostasis," *Physiological Reviews, 9*.
Cannon, W. B. (1932) *The Wisdom of the Body* (London).
Cannon, W. B. (1942) "Voodoo Death," *American Anthropologist, 33*, 169–81.
Capra, F. (1996) *The Web of Life: A New Scientific Understanding of Living Systems* (New York: Anchor Books).
Cariani, P. (1993) "To Evolve an Ear: Epistemological Implications of Gordon Pask's Electrochemical Devices," *Systems Research, 10*, 19–33.
Castaneda, C. (1968) *The Teachings of Don Juan: A Yaqui Way of Knowledge* (Harmondsworth: Penguin).
Choudhury, R. R. (1993) "Black Holes, Time Bandits and the Gentleman's Code," *Systems Research, 10*, 209–10.
Clark, D. J. (2002) "Enclosing the Field: From 'Mechanisation of Thought Processes' to 'Autonomics'" (PhD thesis, University of Warwick).
Cobb, W. A. (1981) "E.E.G.: Historical, Personal and Anecdotal," second Grey Walter Memorial Lecture, Runwell Hospital, Essex, 27 October 1979, *Journal of Electrophysiological Technology, 7*, 58–66.
Collins, H. M. (1974) "The TEA Set: Tacit Knowledge and Scientific Networks," *Science Studies, 4*, 165–86.
Conant, R. (1974) "W. Ross Ashby (1903–1972)," *International Journal of General Systems, 1*, 4–5.
Conant, R. C., and W. R. Ashby (1970) "Every Good Regulator of a System Must Be a Model of That System," *International Journal of Systems Science, 1*, 89–97. Reprinted in Ashby, *Mechanisms of Intelligence: Ross Ashby's Writings on Cybernetics*, R. Conant (ed.) (Seaside, CA: Intersystems Publications, 1981), pp. 187–95.
Conway, F., and J. Siegelman (2005) *Dark Hero of the Information Age: In Search of Norbert Wiener, the Father of Cybernetics* (New York: Basic Books).
Cook, P. (2001) "The Extraordinary Gordon Pask," *Kybernetes, 30*, 571–72.
Cooper, D. (1967) *Psychiatry and Anti-psychiatry* (London: Tavistock Institute). Reprinted 1970 (London: Paladin).
Cooper, D. (ed.) (1968) *To Free a Generation: The Dialectics of Liberation* (New York: Collier).
Cooper, R. (1993) "Walter, (William) Grey (1910–1977)," in C. S. Nicholls (ed.), *Dictionary of National Biography: Missing Persons* (Oxford: Oxford University Press). pp. 699–700.
Cooper, R., and J. Bird (1989) *The Burden: Fifty Years of Clinical and Experimental Neuroscience at the Burden Neurological Institute* (Bristol: White Tree Books).

Cordeschi, R. (2002) *The Discovery of the Artificial: Behaviour, Mind and Machines before and beyond Cybernetics* (Dordrecht: Kluwer).

Craik, K. J. W. (1943) *The Nature of Explanation* (Cambridge: Cambridge University Press).

Craik, K. J. W. (1947) "Theory of the Human Operator in Control Systems," pt. 1, "The Operator as an Engineeering System"; pt. 2, "Man as an Example in a Control System," *British Journal of Psychology, 38*, 56–61; *39*, 142–48. Reprinted in C. R. Evans and A. D. J. Robertson (eds.), *Key Papers: Cybernetics* (Manchester: University Park Press, 1968), pp. 120–39.

Craik, K. J. W. (1966) *The Nature of Psychology: A Selection of Papers, Essays and Other Writings by the Late Kenneth J. W. Craik*, S. L. Sherwood (ed.) (Cambridge: Cambridge University Press).

Damper, R. I. (ed.) (1993) *Biologically-Inspired Robotics: The Legacy of W. Grey Walter*, Proceedings of an EPSRC/BBSRC international workshop, Bristol, 14–16 August 2002, *Philosophical Transactions of the Royal Society, 361*, 2081–2421.

Davenport, M.L. (2005) "A Home Is Broken: The Closure of a Philadelphia Association Household," www.philadelphia-association.co.uk/archive/shirland_closure.pdf.

de Brath, S. (1925) *Psychical Research, Science and Religion* (London: Methuen).

DeLanda, M. (2002) *Intensive Science and Virtual Philosophy* (Continuum Books: London).

DeLanda, M. (2005) "Matter Matters," *Domus*, no. 884 (September), 94.

de Latil, P. (1956) *Thinking by Machine: A Study of Cybernetics* (London: Sidgwick & Jackson).

Deleuze, G., and F. Guattari (1987) *A Thousand Plateaus: Capitalism and Schizophrenia* (Minneapolis: University of Minnesota Press).

Dell, P. F. (1989) "Violence and the Systemic View: The Problem of Power," *Family Process, 28*, 1–14.

de Zeeuw G. (1993) "Improvement and Research: A Paskian Evolution," *Systems Research , 10*, 193–203.

Donoso, R. (2004) "Heinz and Stafford: A Tribute to Heinz von Foerster and Stafford Beer," *Kybernetes, 33*, 659–67.

Dror, O. (2004) " 'Voodoo Death': Fantasy, Excitement, and the Untenable Boundaries of Biomedical Science," in R. D. Johnston (ed.), *The Politics of Healing: Essays on the Twentieth-Century History of North American Alternative Medicine* (New York: Routledge), pp. 71–81.

Dupuy, J.-P. (2000) *The Mechanization of the Mind: On the Origins of Cognitive Science* (Princeton, NJ: Princeton University Press).

Edelman, G. (1992) *Bright Air, Brilliant Fire: On the Matter of the Mind* (New York: Basic Books).

Edgell, B. (1924) *Theories of Memory* (Oxford: Oxford University Press).

Edwards, P. N. (1996) *The Closed World: Computers and the Politics of Discourse in Cold War America* (Cambridge, MA: MIT Press).

Eliade, M. (1969) *Yoga: Immortality and Freedom* (Princeton, NJ: Princeton University Press).

Elichirigoity, I. (1999) *Planet Management: Limits to Growth, Computer Simulation, and the Emergence of Global Spaces* (Evanston, IL: Northwestern University Press).

Ellul, J. (1964) *The Technological Society* (New York: Knopf).

Eno, B. (1996a) *A Year with Swollen Appendices* (London: Faber & Faber).

Eno, B. (1996b) "Generative Music," talk delivered at the Imagination Conference, San Francisco, 8 June 1996. Reproduced in *In Motion Magazine*, www.inmotionmagazine.com/eno1.html.

Erickson, B. (2005) "The Only Way Out Is In: The Life of Richard Price," in J. Kripal and G. Shuck (eds.), *On the Edge of the Future: Esalen and the Evolution of American Culture* (Bloomington, IN: Indiana University Press), pp. 132–64.

Espejo, R. (ed.) (2004) *Tribute to Stafford Beer*, special issue of *Kybernetes*, *33*, nos. 3–4, 481–863.

Espejo, R., and R. Harnden (eds.) (1989) *The Viable System Model: Interpretations, and Applications of Stafford Beer's VSM* (New York: Wiley).

Evans, C. C., and E. Osborn (1952), "An Experiment in the Electro-Encephalography of Mediumistic Trance," *Journal of the Society for Psychical Research*, *36*, 588–96.

Ezard, J. (2002) "Joan Littlewood," *Guardian*, September, 20, 23.

Feyerabend, P. K. (1993) *Against Method*, 3rd ed. (New York: Verso).

Fleming, G. W. T. H. (ed.) (1950) *Recent Progress in Psychiatry*, special issue of *Journal of Mental Science* (London: Churchill).

Fleming, G. W. T. H., F. L. Golla, and W. Grey Walter (1939) "Electric-Convulsive Therapy of Schizophrenia," *Lancet*, *64*, no. 6070 (December), 1353–55.

Fortun, M., and S. Schweber (1993) "Scientists and the Legacy of World War II: The Case of Operations Research," *Social Studies of Science*, *23*, 595–642.

Foucault, M. (1988) *Technologies of the Self: A Seminar with Michel Foucault*, L. H. Martin, H. Gutman, and P. H. Hutton (eds.) (Amherst, MA: University of Massachusetts Press).

Franchi, S., G. Güzeldere, and E. Minch (1995) "Interview with Heinz von Foerster," in Franchi and Güzeldere (eds.), *Constructions of the Mind: Artificial Intelligence and the Humanities*, *Stanford Humanities Review*, 4, no. 2, 288–307.

Frazer, J. H. (1995) *An Evolutionary Architecture* (London: Architectural Association).

Frazer, J. H. (2001) "The Cybernetics of Architecture: A Tribute to the Contribution of Gordon Pask," *Kybernetes*, *30*, 641–51.

Fujimura, J. H. (2005) "Postgenomic Futures: Translations across the Machine-Nature Border in Systems Biology," *New Genetics and Society*, *24*, 195–225.

Galison, P. (1994) "The Ontology of the Enemy: Norbert Wiener and the Cybernetic Vision," *Critical Inquiry*, *21*, 228–66.

Gardner, H. (1987) *The Mind's New Science: A History of the Cognitive Revolution* (New York: Basic Books).

Gardner, M. R., and W. R. Ashby (1970) "Connectance of Large Dynamic (Cybernetic) Systems: Critical Values for Stability," *Nature*, *228*, 784. Reprinted in Ashby, *Mechanisms of Intelligence: Ross Ashby's Writings on Cybernetics*, R. Conant (ed.) (Seaside, CA: Intersystems Publications, 1981), pp. 85–86.

Geiger, J. (2003) *Chapel of Extreme Experience: A Short History of Stroboscopic Light and the Dream Machine* (New York: Soft Skull Press).

Gere, C. (2002) *Digital Culture* (London: Reaktion Books).

Gerovitch, S. (2002) *From Newspeak to Cyberspeak: A History of Soviet Cybernetics* (Cambridge, MA: MIT Press).

Glanville, R. (ed.) (1993) *Festschrift for Gordon Pask*, special issue of *Systems Research*, *10*, no. 3.

Glanville, R. (1996) "Robin McKinnon-Wood and Gordon Pask: A Lifelong Conversation," *Cybernetics and Human Knowing*, 3, no. 4.
Glanville, R. (2002a) "A (Cybernetic) Musing: Cybernetics and Human Knowing," *Cybernetics and Human Knowing*, 9, 75–82.
Glanville, R. (2002b) "A (Cybernetic) Musing: Some Examples of Cybernetically Informed Educational Practice," *Cybernetics and Human Knowing*, 9, 117–26.
Glanville, R., and B. Scott (eds.) (2001a) *Festschrift for Gordon Pask*, special issue of *Kybernetes*, 30, no. 5/6.
Glanville, R., and B. Scott (2001b) "About Gordon Pask," *Kybernetes*, 30, no. 5/6.
Gleick, J. (1987) *Chaos: Making a New Science* (New York: Penguin).
Goldacre, R. (1960a) "Morphogenesis and Communication between Cells," in *Proceedings of the 2nd International Congress on Cybernetics, Namur, 3–10 September 1958* (Namur: Association Internationale de Cybernétique), pp. 910–23.
Goldacre, R. (1960b) "Can a Machine Create a Work of Art?" in *Proceedings of the 2nd International Congress on Cybernetics, Namur, 3–10 September 1958* (Namur: Association Internationale de Cybernétique), pp. 683–97.
Goldacre, R., and A. Bean (1960) "A Model for Morphogenesis," *Nature, 186* (23 April), 294–95.
Golla, F. L. (1950) "Physiological Psychology," in G. W. T. H. Fleming (ed.), *Recent Progress in Psychiatry*, 2nd ed. (London: J. & A. Churchill), pp. 132–47.
Gomart, E. (2002) "Methadone: Six Effects in Search of a Substance," *Social Studies of Science*, 32, 93–135.
Gomart, E., and A. Hennion (1999) "A Sociology of Attachment: Music Amateurs, Drug Users," in J. Law and J. Hassard (eds.), *Actor Network Theory and After* (Oxford: Blackwell), pp. 220–47.
Gray, J. A. (1979) *Pavlov* (London: Fontana).
Green, C. (1968) *Out-of-the-Body Experiences* (Oxford: Institute of Psychophysical Research).
Green, J. (1988) *Days in the Life: Voices from the English Underground, 1961–1971* (London: Heinemann).
Gregory, R. L. (1981) *Mind in Science: A History of Explanations in Psychology and Physics* (Cambridge: Cambridge University Press).
Gregory, R. L. (1983) "Editorial," *Perception, 12*, 233–38.
Gregory, R. L. (2001) "Memories of Gordon," *Kybernetes*, 30, 685–87.
Grosch, H. (1973) "Chilean Economic Controls," *New Scientist*, 57, 626–67.
Guardian (2002) "Oh, What a Lovely Woman," 25 September, G2, pp. 12–13.
Guattari, F. (1984 [1973]) "Mary Barnes, or Oedipus in Anti-psychiatry," *Le Nouvel Observateur*, 28 May 1973. Translated in Guattari, *Molecular Revolution: Psychiatry and Politics* (Harmondsworth: Penguin, 1984), pp. 51–59.
Habermas, J. (1970) *Toward a Rational Society: Student Protest, Science, and Politics* (Boston: Beacon Press).
Habermas, J. (1987) *The Theory of Communicative Action*, vol. 2, *Lifeworld and System: A Critique of Functionalist Reason* (Boston: Beacon Press).
Haley, J. (1976) "Development of a Theory: A History of a Research Project," in C. Sluzki and D. Ransom (eds.), *Double Bind: The Foundation of the Communicational Approach to the Family* (New York: Grune & Stratton), pp. 59–104.
Hanlon, J. (1973a) "The Technological Power Broker," *New Scientist*, 57, 347.

Hanlon, J. (1973b) "Chile Leaps into a Cybernetic Future," *New Scientist*, 57, 363–64.

Haraway, D. (1981–82) "The High Cost of Information in Post–World War II Evolutionary Biology: Ergonomics, Semiotics, and the Sociobiology of Communication Systems," *Philosophical Forum*, 13, 244–78.

Haraway, D. (1985) "A Manifesto for Cyborgs: Science, Technology, and Socialist Feminism in the 1980s," *Socialist Review*, 80, 65–107. Reprinted in Haraway, *The Haraway Reader* (New York: Routledge. 2004), pp. 7–45.

Haraway, D. (2003) *The Companion Species Manifesto: Dogs, People, and Significant Otherness* (Chicago: Prickly Paradigm Press).

Harden, B. (2002) "Dams and Politics Channel Mighty River," *New York Times*, 5 May, 1.

Harnden, R., and A. Leonard (1994) "Introduction," in Beer, *How Many Grapes Went into the Wine? Stafford Beer on the Art and Science of Holistic Management* (New York: Wiley, 1994), pp. 1–11.

Harries-Jones, P. (1995) *A Recursive Vision: Ecological Understanding and Gregory Bateson* (Toronto: University of Toronto Press).

Harries-Jones, P. (2005) "Understanding Ecological Aesthetics: The Challenge of Bateson," *Cybernetics and Human Knowing*, 12, 61–74.

Hawkridge, D. (2001) "Gordon Pask at the Open University," *Kybernetes*, 30, 688–90.

Hayles, N. K. (1999) *How We Became Posthuman: Virtual Bodies in Cybernetics, Literature, and Informatics* (Chicago: University of Chicago Press).

Hayward, R. (2001a) "The Tortoise and the Love-Machine: Grey Walter and the Politics of Electroencephalography," in C. Borck and M. Hagner (eds.), *Mindful Practices: On the Neurosciences in the Twentieth Century, Science in Context, 14*, no. 4, 615–41.

Hayward, R. (2001b) "'Our Friends Electric': Mechanical Models of the Mind in Postwar Britain," in G. Bunn, A. Lovie, and G. Richards (eds.), *A Century of Psychology* (Leicester: British Psychological Society & The Science Museum), pp. 290–308.

Hayward, R. (2002) "Frederick Golla and Margiad Evans: A Neuropsychiatric Romance," paper presented at the 3rd INTAS Conference on the History of the Human Sciences, Groningen, May.

Hayward, R. (2004) "Golla, Frederick Lucien," in B. Harrison (ed.), *Oxford Dictionary of National Biography* (Oxford: Oxford University Press).

Hayward, R. (forthcoming) "Making Psychiatry English: The Maudsley and the Munich Model," in V. Roelke, P. Weindling, and L. Westwood (eds.), *Inspiration, Co-operation, Migration: British-American-German Relations in Psychiatry, 1870–1945* (Rochester: University of Rochester Press).

Heidegger, M. (1976 [1954]) "The Question Concerning Technology," in D. Krell (ed.), *Martin Heidegger: Basic Writings* (New York: Harper & Row), pp. 287–317.

Heidegger, M. (1976) "The End of Philosophy and the Task of Thinking," in D. Krell (ed.), *Martin Heidegger: Basic Writings* (New York: Harper & Row), pp. 431–49.

Heidegger, M. (1981) "'Only a God Can Save Us': The *Spiegel* Interview (1966)," in T. Sheehan (ed.), *Heidegger: The Man and the Thinker* (Chicago: Precedent Publishing), pp. 45–67.

Heims, S. J. (1980) *John von Neumann and Norbert Wiener* (Cambridge, MA: MIT Press).

Heims, S. J. (1991) *The Cybernetics Group* (Cambridge, MA: MIT Press).

Helmreich, S. (1998) *Silicon Second Nature: Culturing Artificial Life in a Digital World* (Berkeley: University of California Press).

Henderson, L. J. (1970 [1932]) "An Approximate Definition of Fact," *University of California Publications in Philosophy, 14* (1932), 179–99; reprinted in Henderson, *On the Social System: Selected Writings*, B. Barber, (ed.) (Chicago: University of Chicago Press), pp. 159–80.

Heron, W. (1961) "Cognitive and Physiological Effects of Perceptual Isolation," in P. Solomon et al. (eds.), *Sensory Deprivation: A Symposium Held at Harvard Medical School* (Cambridge, MA: Harvard Univesity Press), pp. 6–33.

Hesse, M. B. (1966) *Models and Analogies in Science* (Notre Dame, IN: University of Notre Dame Press).

Hesse, M. B. (1974) *The Structure of Scientific Inference* (London: Macmillan).

Heynen, H. (1999) *Architecture and Modernity* (Cambridge, MA: MIT Press).

Hodges, A. (1983) *Alan Turing: The Enigma* (New York: Simon & Schuster).

Holland, O. (1996) "Grey Walter: The Pioneer of Real Artificial Life," in C. G. Langton and K. Shimonara (eds.), *Artificial Life V* (Cambridge, MA: MIT Press), pp. 33–41.

Holland, O. (2003) "Exploration and High Adventure: The Legacy of Grey Walter," *Philosophical Transactions of the Royal Society of London A, 361*, 2085–2121.

Howarth-Williams, M. (1977) *R. D. Laing: His Work and Its Relevance for Sociology* (London: Routledge & Kegan Paul).

Humphrey, G. (1933) *The Nature of Learning in Relation to the Living System* (London).

Hutchins, E. (1995) *Cognition in the Wild* (Cambridge, MA: MIT Press).

Hutchins, E., and T. Klausen (1996) "Distributed Cognition in an Airline Cockpit," in Y. Engeström and D. Middleton (eds.) *Cognition and Communication at Work* (Cambridge: Cambridge University Press), pp. 15–34.

Huxley, A. (1954) *The Doors of Perception* (New York: Harper).

Huxley, A. (1956) *Heaven and Hell* (New York: Harper).

Huxley, A. (1963) *The Doors of Perception, and Heaven and Hell* (New York: Harper & Row).

Huxley, F. (1989) "The Liberating Shaman of Kingsley Hall," *Guardian*, 25 August.

Irish, S. (2004) "Tenant to Tenant: The Art of Talking with Strangers," *Places, 16*, no. 3, 61–67.

Jackson, M. C. (1989) Evaluating the Managerial Significance of the VSM," in R. Espejo and R. Harnden (eds.), *The Viable System Model: Interpretations and Applications of Stafford Beer's VSM* (New York: Wiley), pp. 424–39.

James, W. (1902) *The Varieties of Religious Experience: A Study in Human Nature* (New York: Longmans, Green & Co.).

James. W. (1978 [1907, 1909]) *Pragmatism and the Meaning of Truth* (Cambridge, MA: Harvard University Press).

James, W. (1943 [1909, 1912]) *Essays in Radical Empiricism. A Pluralistic Universe* (New York: Longmans, Green & Co.).

Jullien, F. (1999) *The Propensity of Things: Toward a History of Efficacy in China* (New York: Zone Books).

Kauffman, S. (1969a) "Homeostasis and Differentiation in Random Genetic Control Networks," *Nature, 224*, 177–78.

Kauffman, S. (1969b) "Metabolic Stability and Epigenesis in Randomly Constructed Genetic Nets," *Journal of Theoretical Biology, 22*, 437–67.

Kauffman, S. (1971) "Articulation of Parts Explanation in Biology and the Rational Search for Them," *Boston Studies in the Philosophy of Science, 8*, 257–72.

Kauffman, S. (1995) *At Home in the Universe: The Search for Laws of Self-Organization and Complexity* (New York: Oxford University Press).

Kauffman, S. (2002) *Investigations* (New York: Oxford University Press).

Kauffman, S. (n.d.) "Comments on presentation at World Science Forum," www.ucalgary.ca/files/ibi/wsfblog.pdf (read April 2007).

Kauffman, S., and W. McCulloch (1967) "Random Nets of Formal Genes," *Quarterly Research Report, Research Laboratory of Electronics, 88*, 340–48.

Kay, L. E. (2001) "From Logical Neurons to Poetic Embodiments of Mind: Warren S. McCulloch's Project in Neuroscience," *Science in Context, 14*, 591–614.

Keller, E. F. (2002) *Making Sense of Life: Explaining Biological Development with Models, Metaphors and Machines* (Cambridge, MA: Harvard University Press).

Kellert, S. H. (1993) *In the Wake of Chaos: Unpredictable Order in Dynamical Systems* (Chicago: University of Chicago Press).

Kesey, K. (1962) *One Flew over the Cuckoo's Nest* (New York: New American Library).

Keys, J. [George Spencer Brown] (1972) *Only Two Can Play This Game* (New York: Julian Press).

Kotowicz, Z. (1997) *R. D. Laing and the Paths of Anti-psychiatry* (London: Routledge).

Kripal, J. (2007) *Esalen: America and the Religion of No Religion* (Chicago: University of Chicago Press).

Kripal, J., and G. Shuck (eds.) (2005) *On the Edge of the Future: Esalen and the Evolution of American Culture* (Bloomington, IN: Indiana University Press).

Kuhn, T. S. (1962) *The Structure of Scientific Revolutions* (Chicago: University of Chicago Press). 2nd ed. 1970.

Lacey, B. (1968) "On the Human Predicament," in J. Reichardt (ed.), *Cybernetic Serendipity: The Computer and the Arts* (London: W. & J. Mackay), pp. 38–39.

Laing, A. (1994) *R. D. Laing: A Biography* (New York: Thunder's Mouth Press).

Laing, R. D. (1961) *The Self and Others* (London: Tavistock Publications, 1961). 2nd ed. 1969.

Laing, R. D. (1967) *The Politics of Experience* (New York: Pantheon).

Laing, R. D. (1968a) "Author's Response," *Philosophy Forum, 7*, 92–94.

Laing, R. D. (1968b) "The Obvious," in D. Cooper (ed.), *To Free a Generation: The Dialectics of Liberation* (New York: Collier), pp. 13–33.

Laing, R. D. (1972) "Metanoia: Some Experiences at Kingsley Hall, London," in H. Ruitenbeek (ed.), *Going Crazy: The Radical Therapy of R. D. Laing and Others* (New York: Bantam), pp. 11–21. Reprinted from *Recherches*, December 1968.

Laing, R. D. (1976) *The Facts of Life: An Essay in Feelings, Facts, and Fantasy* (New York: Pantheon).

Laing, R. D. (1985) *Wisdom, Madness and Folly: The Making of a Psychiatrist* (New York: McGraw-Hill).

Laing, R. D., M. Phillipson, and R. Lee (1966) *Interpersonal Perception* (London: Tavistock).

Landau, R. (1968) *New Directions in British Architecture* (London: Studio Vista).

Latour, B. (1983) "Give Me a Laboratory and I Will Raise the World," in K. D. Knorr-Cetina and M. Mulkay (eds.) *Science Observed: Perspectives on the Social Study of Science* (Beverly Hills, CA: Sage), pp. 141–70.

Latour, B. (1993) *We Have Never Been Modern* (Cambridge, MA: Harvard University Press).

Latour, B. (2004) *Politics of Nature: How to Bring the Sciences into Democracy* (Cambridge, MA: Harvard University Press).
Latour, B., and P. Weibel (eds.) (2002) *Iconoclash: Beyond the Image Wars in Science, Religion, and Art* (Cambridge, MA: MIT Press).
Latour, B., and P. Weibel (eds.) (2005) *Making Things Public: Atmospheres of Democracy* (Cambridge, MA: MIT Press).
Laughlin, R. B., and D. Pines, "The Theory of Everything," *Proceedings of the National Academy of Science*, 97, 28–31.
Laurillard, D. (2001) "Gordon Pask Memorial Speech: Contributions to Educational Technology," *Kybernetes*, 30, 756–59.
Leary, T. (1970) *The Politics of Ecstasy* (London: Paladin).
Leadbeater, C. W. (1990) *The Chakras* (Wheaton, IL: Theosophical Publishing House, Quest Books).
Lécuyer, C., and D. Brock (2006) "The Materiality of Microelectronics," *History and Technology*, 22, 301–25.
Lettvin, J. (1988) "Foreword," in W. S. McCulloch, *Embodiments of Mind*, 2nd ed. (Cambridge, MA: MIT Press), pp. v–xi.
Lilly, J. (1972) *The Center of the Cyclone: An Autobiography of Inner Space* (New York: Julian Press).
Lilly, J. (1977) *The Deep Self: Profound Relaxation and the Tank Isolation Technique* (New York: Warner Books).
Lipset, D. (1980) *Gregory Bateson: The Legacy of a Scientist* (Boston: Beacon Press).
Littlewood, J. (1994) *Joan's Book: Joan Littlewood's Peculiar History as She Tells It* (London: Methuen).
Littlewood, J. (2001) "Gordon Pask," *Kybernetes*, 30, 760–61.
Lobsinger, M. (2000) "Cybernetic Theory and the Architecture of Performance: Cedric Price's Fun Palace," in S. Goldhagen and R. Legault (eds.), *Anxious Modernisms: Experimentalism in Postwar Architectural Culture* (Cambridge, MA: MIT Press), pp. 119–39.
Loeb, J. (1918) *Forced Movements, Tropisms and Animal Conduct* (Philadelphia).
Lucier, A. (1995) *Reflections: Interviews, Scores, Writings*, G. Gronemeyer and R. Oehlschlägel (eds.) (Köln: MusikTexte).
Mandelbrot, B. (1983) *The Fractal Geometry of Nature* (New York: W. H. Freeman).
Marcus. G (1989) *Lipstick Traces: A Secret History of the Twentieth Century* (Cambridge, MA: Harvard University Press).
Marick, B. (2008) "A Manglish Way of Working: Agile Software Development," in A. Pickering and K. Guzik (eds.), *The Mangle in Practice: Science, Society and Becoming* (Durham, NC: Duke University Press), pp. 185–201.
Masserman, J. H. (1950) "Experimental Neuroses," *Scientific American, 182* (March), 38–43.
Mathews, S. (2007) *From Agit-Prop to Free Space: The Architecture of Cedric Price* (London: Black Dog).
Maturana, H. R., and F. J. Varela (1980) *Autopoiesis and Cognition: The Realization of the Living* (Dordrecht: Reidel).
Maturana, H. R., and F. J. Varela (1992) *The Tree of Knowledge: The Biological Roots of Human Understanding*, rev. ed. (Boston: Shambala).
McCulloch, W. S. (1988) *Embodiments of Mind*, 2nd ed. (Cambridge, MA: MIT Press).

McCulloch, W. S. (2004) "The Beginnings of Cybernetics," undated manuscript, reproduced in Claus Pias (ed.), *Cybernetics-Kybernetik: The Macy-Conferences, 1946–1953*, vol 2, *Essays and Documents* (Zürich: Diaphanes), pp. 345–60.
McCulloch, W., and W. Pitts (1943) "A Logical Calculus of the Ideas Immanent in Nervous Activity," *Bulletin of Mathematical Physics*, 5, 115–33. Reprinted in McCulloch, *Embodiments of Mind* (Cambridge, MA: MIT Press, 1988), pp. 19–39.
McCulloch, W. S., L. Verbeek, and S. L. Sherwood (1966) "Introduction," in Sherwood (ed.), *The Nature of Psychology: A Selection of Papers, Essays and Other Writings by the Late Kenneth J. W. Craik* (Cambridge: Cambridge University Press), pp. ix–xi.
McGrail, R. (2008) "Working with Substance: Actor-Network Theory and the Modal Weight of the Material," *Techne*, *12*, no. 1, 65–84.
McKinnon-Wood, R. (1993) "Early Machinations," *Systems Research*, *10*, 129–32.
McKinnon-Wood, R., and M. Masterman (1968) "Computer Poetry from CLRU," in J. Reichardt (ed.), *Cybernetic Serendipity: The Computer and the Arts* (London: W. & J. Mackay), p. 56.
McPhee, J. (1989) "Atchafalaya," in *The Control of Nature* (New York: Farrar, Straus, Giroux), pp. 3–92.
McSwain, R. (2002) "The Social Reconstruction of a Reverse Salient in Electrical Guitar Technologies: Noise, the Solid Body, and Jimi Hendrix," in H.-J. Braun (ed.), *Music and Technology in the Twentieth Century* (Baltimore: Johns Hoplins University Press), pp. 186–98.
Meadows, D. H., D. L. Meadows, J. Randers, and W. W. Behrens III (1972) *The Limits to Growth* (New York: Universe Books).
Medina, E. (2006) "Designing Freedom, Regulating a Nation: Socialist Cybernetics in Allende's Chile," *Journal of Latin American Studies*, *38*, 571–606.
Melechi, A. (1997) "Drugs of Liberation: Psychiatry to Psychedelia," in Melechi (ed.), *Psychedelia Britannica: Hallucinogenic Drugs in Britain* (London: Turnaround), pp. 21–52.
Melvin, J. (2003) "Cedric Price," *Guardian*, 15 August 2003.
Miller, G. (2004) *R. D. Laing* (Edinburgh: Edinburgh University Press).
Mindell, D. (2002) *Between Human and Machine: Feedback, Control, and Computing before Cybernetics* (Baltimore: Johns Hopkins University Press).
Mirowski, P. (2002) *Machine Dreams: Economics Becomes a Cyborg Science* (Cambridge: Cambridge University Press).
Moore, H. (2001) "Some Memories of Gordon," *Kybernetes*, *30*, 768–70.
Müller, A., and K. H. Müller (eds.) (2007) *An Unfinished Revolution? Heinz von Foerster and the Biological Computer Laboratory, 1958–1976* (Vienna: Edition Echoraum).
Murphy, M. (1992) *The Future of the Body: Explorations into the Further Evolution of Human Nature* (Los Angeles: Tarcher).
Neff, G., and D. Stark (2004) "Permanently Beta: Responsive Organization in the Internet Era," in P. N. Howard and S. Jones (eds.), *Society Online: The Internet in Context* (Thousand Oaks, CA: SAGE), pp. 173–88.
Negroponte, N. (1969) "Architecture Machine" and "Towards a Humanism through Machines," *Architectural Design*, *39*, 510–13.
Negroponte, N. (1970) *The Architecture Machine: Toward a More Human Environment* (Cambridge, MA: MIT Press).
Negroponte, N. (1975) *Soft Architecture Machines* (Cambridge, MA: MIT Press).

Nelson, T. (1987) *Computer Lib: Dream Machines*, rev. ed. (Redmond, WA: Tempus Books).

Noble, D. F. (1986) *Forces of Production: A Social History of Industrial Automation* (Oxford: Oxford University Press).

Nuttall, J. (1968) *Bomb Culture* (New York: Delacorte Press).

Olazaran, M. (1996) "A Sociological Study of the Official History of the Perceptrons Controversy," *Social Studies of Science, 26*, 611–59.

O'Malley, M., and J. Dupré (2005) "Fundamental Issues in Systems Biology," *BioEssays, 27*, 1270–76.

Owen, A. (2004) *The Place of Enchantment: British Occultism and the Culture of the Modern* (Chicago: Universioty of Chicago Press).

Pain, S. (2002) "Inactive Service," *New Scientist, 176*, no. 2373 (14 December), 52–53.

Pangaro, P. (2001) "Thoughtsticker 1986: A Personal History of Conversation Theory in Software, and Its Progenitor, Gordon Pask," *Kybernetes, 30*, 790–806.

Pask, E. (1993) "Today Has Been Going On for a Very Long Time," *Systems Research, 10*, 143–47.

Pask, E. (n.d.) Notes on a biography of Gordon Pask (unpublished manuscript).

Pask, G. (1958) "Organic Control and the Cybernetic Method," *Cybernetica, 1*, 155–73.

Pask, G. (1959) "Physical Analogues to the Concept of Growth," in *Mechanisation of Thought Processes: Proceedings of a Symposium Held at the National Physical Laboratory on 24th, 25th, 26th and 27th November 1958*, 2 vols. (London: Her Majesty's Stationery Office), pp. 877–928.

Pask, G. (1960a) "Teaching Machines," in *Proceedings of the 2nd International Congress on Cybernetics, Namur, 3–10 September 1958* (Namur: Association Internationale de Cybernétique), pp. 962–78.

Pask, G. (1960b) "The Natural History of Networks," in M. Yovits and S. Cameron (eds.), *Self-Organizing Systems: Proceedings of an Interdisciplinary Conference, 5 and 6 May* (New York: Pergamon), pp. 232–63.

Pask, G. (1961) *An Approach to Cybernetics* (London: Hutchison).

Pask, G. (1962) "A Proposed Evolutionary Model," in H. von Foerster and G. W. Zopf (eds.), *Principles of Self-Organization: Transactions of a Symposium on Self-Organization, Robert Allerton Park, 8 and 9 June 1961* (New York: Pergamon), pp. 229–53.

Pask, G. (1964a) *An Investigation of Learning under Normal and Adaptively Controlled Conditions* (doctoral dissertation, University of London).

Pask, G. (1964b) "Proposals for a Cybernetic Theatre" (unpublished manuscript), available at www.pangaro.com.

Pask, G. (1965) "Fun Palace Cybernetics Committee: Minutes of the Meeting Held at the Building Centre, Store Street, London WC1, 27 January, 1965," Cedric Price Fonds, Collection Centre d'Architecture/Canadian Centre for Architecture, Montréal, DR1995:0188:525:003.

Pask, G. (1968) "The Colloquy of Mobiles," in J. Reichardt (ed.), *Cybernetic Serendipity: The Computer and the Arts* (London: W. & J. Mackay), pp. 34–35.

Pask, G. (1969a) "The Architectural Relevance of Cybernetics," *Architectural Design, 39* (September), 494–96.

Pask, G. (1969b) "The Computer Simulated Development of Populations of Automata," *Mathematical Biosciences, 4*, 101–27.

Pask, G. (1971) "A Comment, a Case History and a Plan," in J. Reichardt (ed.), *Cybernetics, Art, and Ideas* (Greenwich, CT: New York Graphics Society), pp. 76–99.

Pask, G. (1975a) *The Cybernetics of Human Learning and Performance* (London: Hutchinson).

Pask, G. (1975b) *Conversation, Cognition and Learning: A Cybernetic Theory and Methodology* (Amsterdam: Elsevier).

Pask, G. (1976a) *Conversation Theory: Applications in Education and Epistemology* (Amsterdam: Elsevier).

Pask, G. (1976b) "Conversational Techniques in the Study and Practice of Education," *British Journal of Educational Psychology*, 46, 12–25.

Pask, G. (ed.) (1976c) *Current Scientific Approaches to Decision Making in Complex Systems: Report of a Conference Held at Richmond, Surrey, England, 14–15 July 1975*, ARI Technical Report TR-76-B1 (Arlington, VA: U.S. Army Research Institute for the Behavioral and Social Sciences).

Pask, G. (1977) "Minds and Media in Education and Entertainment: Some Theoretical Comments Illustrated by the Design and Operation of a System for Exteriorizing and Manipulating Individual Theses," in R. Trappl (ed.), *Proceedings of the Third European Meeting on Cybernetics and Systems Research* (Amsterdam: North-Holland), pp. 38–50.

Pask, G. (ed.) (1978) *Current Scientific Approaches to Decision Making in Complex Systems: II: Report of a Second Conference Held at Richmond, Surrey, England, 10–12 May 1976*, ARI Technical Report TR-78-B4.

Pask, G. (1981) "Organizational Closure of Potentially Conscious Systems," in M. Zeleny (ed.), *Autopoiesis: A Theory of the Living* (New York: North Holland), pp. 263–308.

Pask, G. (1982) "SAKI: Twenty-Five Years of Adaptive Training into the Microprocessor Era," *International Journal of Man-Machine Studies*, 17, 69–74.

Pask, G. (1992) "Interaction of Actors, Theory and Some Applications," vol 1, "Outline and Overview" (unpublished manuscript), available online at www.cybsoc.org/PasksIAT.PDF.

Pask, G., with S. Curran (1982) *Micro Man: Computers and the Evolution of Consciousness* (New York: Macmillan).

Pask, G., R. Glanville, and M. Robinson (1980) *Calculator Saturnalia; or, Travels with a Calculator: A Compendium of Diversions and Improving Exercises for Ladies and Gentlemen* (New York: Vintage Books).

Pask, G., B. Lewis, et al. (1965) "Research on the Design of Adaptive Training Systems with a Capability for Selecting and Altering Criteria for Adaptation," Miscellaneous Reports under USAF Contract No. AF61(052), 402, 1961–65.

Pask, G., and T. R. McKinnon-Wood (1965) "Interaction between Man and an Adaptive Machine," in *Proceedings of the 3rd International Congress on Cybernetics, Namur, 11–15 September 1961* (Namur: Association Internationale de Cybernétique), pp. 951–64.

Pask, G., and B. Scott (1973) "CASTE: A System for Exhibiting Learning Strategies and Regulating Uncertainty," *International Journal of Man-Machine Studies*, 5, 17–52.

Patel, A., B. Scott, and Kinshuk (2001) "Intelligent Tutoring: From SAKI to Byzantium," *Kybernetes*, 30, 807–18.

Pavlov, I. P. (1927) "Lecture 23: The Experimental Results Obtained with Animals in Their Application to Man," G. V. Anrep (trans.), www.ivanpavlov.com/lectures/ivan_pavlov-lecture_023.htm.

Pearce, J. C. (2002 [1971]) *The Crack in the Cosmic Egg: New Constructs of Mind and Reality*, 2nd ed. (Rochester, VT: Park Street Press).

Penny, S. (1999) "Systems Aesthetics and Cyborg Art: The Legacy of Jack Burnham," *Sculpture, 18*, no. 1.

Penny, S. (2008) "Bridging Two Cultures: Towards an Interdisciplinary History of the Artist-Inventor and the Machine-Artwork," in D. Daniels and B. Schmidt (eds.), *Artists as Inventors/Inventors as Artists* (Ostfildern: Hatje Cantz), pp. 142–58.

Pias, C. (ed.) (2003) *Cybernetics-Kybernetik: The Macy-Conferences, 1946–1953*, vol 1, *Transactions* (Zürich: Diaphanes).

Pias, C. (ed.) (2004) *Cybernetics-Kybernetik: The Macy-Conferences, 1946–1953*, vol 1, *Essays and Documents* (Zürich: Diaphanes).

Pickering, A. (1995) *The Mangle of Practice: Time, Agency, and Science* (Chicago: University of Chicago Press).

Pickering, A. (1995a) "Cyborg History and the World War II Regime," *Perspectives on Science*, 3, no. 1, 1–48.

Pickering, A. (1997) "History of Economics and the History of Agency," in J. Henderson (ed.), *The State of the History of Economics: Proceedings of the History of Economics Society* (London: Routledge), pp. 6–18.

Pickering, A. (2002) "New Ontologies," paper presented at "Real/Simulacra/Artificial: Ontologies of Post-modernity," the Eighth International Conference on Agendas for the Millenium, Candido Mendes University, Rio de Janeiro, Brazil, 20–22 May 2002.

Pickering, A. (2005a) "A Gallery of Monsters: Cybernetics and Self-Organisation, 1940–1970," in Stefano Franchi and Güven Güzeldere (eds.), *Mechanical Bodies, Computational Minds: Artificial Intelligence from Automata to Cyborgs* (Cambridge, MA: MIT Press), pp. 229–45.

Pickering, A. (2005b) "Decentring Sociology: Synthetic Dyes and Social Theory," in Ursula Klein (ed.), *Technoscientific Productivity*, special issue of *Perspectives on Science, 13*, 352–405.

Pickering, A. (2005c) "From Dyes to Iraq: A Reply to Jonathan Harwood," in Ursula Klein (ed.), *Technoscientific Productivity*, special issue of *Perspectives on Science, 13*, 416–25.

Pickering, A. (2006) "Science and Technology Studies," lectures at the University of Aarhus, Denmark, 15–18 April.

Pickering, A. (2008a) "Culture, Science Studies and Technoscience," in T. Bennett and J. Frow (eds.), *Handbook of Cultural Analysis* (London: SAGE), pp. 291–310.

Pickering, A. (2008b) "New Ontologies," in A. Pickering and K. Guzik (eds.), *The Mangle in Practice: Science, Society and Becoming* (Durham, NC: Duke University Press), pp. 1–14.

Pickering, A. (2009) "The Politics of Theory: Producing Another World, with Some Thoughts on Latour," *Journal of Cultural Economy*, 2, 197–212.

Pickering, A. (forthcoming) "After Dualism," in the proceedings of the Gulbenkian Foundation Conference "Challenges to Dominant Modes of Knowledge: Dualism,"

SUNY Binghamton, 3–4 November 2006, edited by Aviv Bergman, Jean-Pierre Dupuy, and Immanuel Wallerstein.

Pinch, T., and F. Trocco (2002) *Analog Days: The Invention and Impact of the Moog Synthesiser* (Cambridge, MA: Harvard University Press).

Pitts, W., and W. S. McCulloch (1947) "How We Know Universals: The Perception of Auditory and Visual Forms," *Bulletin of Mathematical Biophysics*, 9, 127–47. Reprinted in McCulloch, *Embodiments of Mind* (Cambridge, MA: MIT Press, 1988), pp. 46–66.

Pogrebin, R. (2007) "Top Prize for Rogers, Iconoclastic Architect," *New York Times*, 29 March.

Poundstone, W. (1985) *The Recursive Universe: Cosmic Complexity and the Limits of Scientific Knowledge* (New York: Morrow).

Pressman, J. (1998) *Last Resort: Psychosurgery and the Limits of Medicine* (Cambridge: Cambridge University Press).

Price, C. (1993) "Gordon Pask, *Systems Research*, 10, 165–66.

Price, C. (2001) "Gordon Pask," *Kybernetes*, 30, 819–20.

Prigogine, I., and I. Stengers (1984) *Order out of Chaos: Man's New Dialogue with Nature* (New York: Bantam Books).

Pringle, K. (2002) "Movement Afoot to Undo Some of Draining's Damage," *Champaign-Urbana News Gazette*, 3 March, E1, E7.

Pynchon, T. (1975) *Gravity's Rainbow* (London: Picador).

Reichardt, J. (ed.) (1968a) *Cybernetic Serendipity: The Computer and the Arts* (London: W. & J. Mackay).

Reichardt, J. (1968b) "Computer Graphics," in J. Reichardt (ed.), *Cybernetic Serendipity: The Computer and the Arts* (London: W. & J. Mackay), pp. 70–71.

Reichardt, J. (ed.) (1971) *Cybernetics, Art, and Ideas* (Greenwich, CT: New York Graphics Society).

Richardson, G. P. (1991) *Feedback Thought in Social Science and Systems Theory* (Philadelphia: University of Pennsylvania Press).

Riley, T., et al. (2002) *The Changing of the Avant-Garde: Visionary Architecture from the Howard Gilman Collection* (New York: Museum of Modern Art).

Robbins, J. (2000) *A Symphony in the Brain: The Evolution of the New Brain Wave Biofeedback* (New York: Grove Press).

Rorvik, D. (1970) *As Man Becomes Machine: The Evolution of the Cyborg* (New York: Pocket Books).

Rose, N. (1989) *Governing the Soul: The Shaping of the Private Self* (New York: Routledge).

Rose, N. (2003) "Becoming Neurochemical Selves," in N. Stehr (ed.), *Biotechnology between Commerce and Civil Society* (Piscataway, NJ: Transaction Publishers), pp. 1–48.

Rosenblueth, A., N. Wiener, and J. Bigelow (1943) "Behavior, Purpose and Teleology," *Philosophy of Science*, 10, 18–24.

Rosenboom, D. (ed.) (1974) *Biofeedback and the Arts: Results of Early Experiments* (Vancouver: Aesthetic Research Centre of Canada).

Rosenboom, D. (1997) *Extended Musical Interface with the Human Nervous System: Assessment and Prospects* (San Francisco: Imternational Society for the Arts, Sciences and Technology). Available at mitpress2.mit.edu/e-journals/LEA/MONOGRAPHS/ROSENBOOM/rosenboom.html.

Roszak, T. (1991) *Flicker* (New York: Summit Books).

Rubin, H. (1999) "The Power of Words," *FastCompany*, issue 21, p. 142. Available at www.fastcompany.com/online/21/flores.htm (accessed 4 April 2005).

Ruesch, J., and G. Bateson (1951) *Communication: The Social Matrix of Psychiatry* (New York: Norton).

Sadler, S. (2005) *Archigram: Architecture without Architecture* (Cambridge, MA: MIT Press).

Sandison, R. (1997) "LSD Therapy: A Retrospective," in A. Melechi (ed.), *Psychedelia Britannica: Hallucinogenic Drugs in Britain* (London: Turnaround), pp. 53–86.

Scott, B. (1982) "The Cybernetics of Gordon Pask," pt. 2, *International Cybernetics Newsletter*, 24, 479–91.

Scott, B. (2001a) "Conversation Theory: A Constructivist, Dialogical Approach to Educational Technology," *Cybernetics and Human Knowing*, 8, no. 4, 2.

Scott, B. (2001b) "Gordon Pask's Conversation Theory: A Domain Independent Constructivist Model of Human Knowing," *Foundations of Science*, 6, 343–60.

Scott, B., and R. Glanville (2001) "Introduction," *Kybernetes*, 30, no 5/6, no page number.

Scott, J. (1998) *Seeing like a State: How Certain Schemes to Improve the Human Condition Have Failed* (New Haven, CT: Yale University Press).

Shanken, E. A. (ed.) (2003) *Telematic Embrace: A Love Story? Roy Ascott's Theories of Telematic Art* (Berkeley: University of California Press).

Shannon, C. E., and W. Weaver (1963 [1949]) *The Mathematical Theory of Communication* (Urbana, IL: University of Illinois Press).

Shipton, H. W. (1977) "Obituary: W. Grey Walter M.A., Sc.D. Cantab., Hon. M.D. Marseilles," *Electroencephalography and Clinical Neurophysiology*, 43, iii–iv.

Shorter, E. (1997) *A History of Psychiatry: From the Era of the Asylum to the Age of Prozac* (New York: Wiley).

Sigal, C. (1976) *Zone of the Interior* (New York: Thomas Y. Cromwell).

Silver, P., F. Dodd, T. Holdon, C. Lang, and J. Pletts (2001) "Prototypical Applications of Cybernetic Systems in Architectural Contexts: A Tribute to Gordon Pask," *Kybernetes*, 30, 902–20.

Smythies, E. A. (1951) "A Case of Levitation in Nepal," *Journal of the Society for Psychical Research*, 36, 415–26.

Smythies, J. R. (1951) "The Extension of Mind: A New Theoretical Basis for Psi Phenomena," *Journal of the Society for Psychical Research*, 36, 477–502.

Solomon, D. (ed.) (1964) *LSD: The Consciousnesss-Expanding Drug* (New York: G. P. Putnam's Sons).

Solomon, P., et al. (eds.) (1961) *Sensory Deprivation: A Symposium Held at Harvard Medical School* (Cambridge, MA: Harvard Univesity Press).

Star, S. L. (1991) "The Sociology of the Invisible: The Primacy of Work in the Writings of Anselm Strauss," in D. R. Maines (ed.), *Social Organization and Social Process: Essays in Honor of Anselm Strauss* (Hawthorne, NY: Aldine de Gruyter), pp. 265–83.

Stark, D. (2006) "Appello per una sociologia della grandezza," *Sociologia del Lavoro*, no. 104, 200–223. Available in English at www.sociology.columbia.edu/pdf/stark_fsw2.pdf.

Starks, S. L., and J. T. Braslow (2005) "The Making of Contemporary Psychiatry," pt. 1, "Patients, Treatments, and Therapeutic Rationales before and after World War II," *History of Psychology*, 8, 176–93.

Steier, F., and J. Jorgenson (eds.) (2005) *Gregory Bateson: Essays for an Ecology of Ideas*, special issue of *Cybernetics and Human Knowing, 12*, nos. 1–2, 5–158.

Steiner, C. (1999) "Constructivist Science and Technology Studies: On the Path to Being?" *Social Studies of Science, 29*, 583–616.

Steiner, C. (2008) "Ontological Dance: A Dialogue between Heidegger and Pickering," in A. Pickering and K. Guzik (eds.), *The Mangle in Practice: Science, Society and Becoming* (Durham, NC: Duke University Press), pp. 243–65.

Stengers, I. (1997) *Power and Invention: Situating Science* (Minneapolis: University of Minnesota Press)

Stephens, C., and K. Stout (eds.) (2004) *Art and the 60s: This Was Tomorrow* (London: Tate).

Stroud, J. (1950) "The Psychological Moment in Perception," in H. von Foerster (ed.), *Cybernetics: Circular Causal, and Feedback Mechanisms in Biological and Social Systems: Transactions of the Sixth Conference, March 24–25, 1949, New York, N.Y.* (New York: Josiah Macy, Jr., Foundation), pp. 27–63.

Suchman, L. (2005) "Demonically Wild Contingencies: Disciplining Interactivity in Ethnomethodology and Cybernetics," paper presented at the annual meeting of the Society for Social Studies of Science, Pasadena, CA, 20–22 October 2005.

Tanne, J. (2004) "Obituary: Humphry Osmond," *British Medical Journal, 328*, 713.

Teitelbaum, R. (1974) "In Tune: Some Early Experiments in Biofeedback Music," in D. Rosenboom (ed.), *Biofeedback and the Arts: Results of Early Experiments* (Vancouver: Aesthetic Research Centre of Canada), pp. 55–70.

Tenney, J. (1995) "The Eloquent Voice of Nature," in A. Lucier, *Reflections: Interviews, Scores, Writings*, G. Gronemeyer and R. Oehlschlägel (eds.) (Köln: MusikTexte), pp. 12–18.

Thomas, L., and S. Harri-Augstein (1993) "Gordon Pask at Brunel: A Continuing Conversation about Conversations," *Systems Research, 10*, 183–92.

Thompson, D'A. W. (1961 [1917]) *On Growth and Form*, abridged ed. (Cambridge: Cambridge University Press).

Ticktin, S. (n.d.) "Biography," laingsociety.org/biograph.htm.

Toffler, A. (1970) *Future Shock* (New York: Random House).

Toffler, A. (1984) "Science and Change," foreword to I. Prigogine and I. Stengers, *Order out of Chaos* (New York: Bantam Books), pp. xi–xxvi.

Trocchi, A. (1991a [1962]) "The Invisible Insurrection of a Million Minds," in A. Scott (ed.), *Invisible Insurrection of a Million Minds: A Trocchi Reader* (Edinburgh: Polygon), pp.177–91. Reprinted from *New Saltire Review* (1962).

Trocchi, A. (1991b [1962]) "Sigma: A Tactical Blueprint," in A. Scott (ed.), *Invisible Insurrection of a Million Minds: A Trocchi Reader* (Edinburgh: Polygon), pp.192–203. Reprinted from *New Saltire Review* (1962).

Tuchman, M. (1986) *The Spiritual in Art: Abstract Painting, 1890–1985* (New York: Abbeville).

Turner, F. (2006) *From Counterculture to Cyberculture: Stewart Brand, the Whole Earth Network, and the Rise of Digital Utopianism* (Chicago: University of Chicago Press).

Ulrich, W. (1981) "A Critique of Pure Cybernetic Reason: The Chilean Experience with Cybernetics," *Journal of Applied Systems Analysis, 8*, 33–59.

Umpleby, S. (2003) "Heinz von Foerster and the Mansfield Amendment," *Cybernetics and Human Knowing, 10*, 161–63.

Valenstein, E. S. (1986) *Great and Desperate Cures: The Rise and Decline of Psychosurgery and Other Radical Treatments for Mental Illness* (New York: Basic Books).

van Gelder, T. (1995) "What Might Cognition Be, If Not Computation?" *Journal of Philosophy*, *92*, 345–81.

Varela, F., E. Thompson, and E. Rosch (1991) *The Embodied Mind: Cognitive Science and Human Experience* (Cambridge, MA: MIT Press).

Viegas, J. (2004) "Brain in a Dish Flies Plane," discoverychannel.com news, 22 October 2004, news.ufl.edu/2004/10/21/braindish.

Waldrop. M. (1992) *Complexity: The Emerging Science at the Edge of Order and Chaos* (New York: Simon & Schuster).

Walker, C. C., and W. R. Ashby (1966) "On Temporal Characteristics of Behavior in Certain Complex Systems," *Kybernetik*, 3,100–108. Reprinted in R. Conant (ed.), *Mechanisms of Intelligence: Ross Ashby's Writings on Cybernetics*, R. Conant (ed.) (Seaside, CA: Intersystems Publications, 1981), pp. 89–97.

Wallwork, S. C. (1952) "ESP Experiments with Simultaneous Electro-Encephalographic Recordings," *Journal of the Society for Psychical Research*, *36*, 697–701.

Walter, V. J., and W. G. Walter (1949) "The Central Effects of Rhythmic Sensory Stimulation," *Electroencephalography and Clinical Neurophysiology*, *1*, 57–86.

Walter, W. G. (1938) "Report to the Rockefeller Foundation," unpublished manuscript, BNI archive, Science Museum, London.

Walter, W. G. (1950a) "An Imitation of Life," *Scientific American*, *182* (May), 42–45.

Walter, W. G. (1950b) "The Twenty-fourth Maudsley Lecture: The Functions of Electrical Rhythms in the Brain," *Journal of Mental Science*, *96*, 1–31.

Walter, W. G. (1950c) "Electro-Encephalography," in G. W. T. H. Fleming (ed.), *Recent Progress in Psychiatry*, 2nd ed. (London: J. & A. Churchill), pp. 76–93.

Walter, W. G. (1951) "A Machine That Learns," *Scientific American*, *185* (August), 60–63.

Walter, W. G. (1953) *The Living Brain* (London: Duckworth).

Walter, W. G. (1971 [1953]) "Electroencephalographic Development of Children," in J. M. Tanner and B. Infelder (eds.), *Discussions on Child Development* (New York: International Universities Press), pp. 132–60.

Walter, W. G. (1971 [1954]) "Presentation: Dr. Grey Walter," in J. M. Tanner and B. Infelder (eds.), *Discussions on Child Development* (New York: International Universities Press), pp. 21–74.

Walter, W. G. (1956a) *The Curve of the Snowflake* (New York: W. W. Norton).

Walter, W. G. (1956b) "Colour Illusions and Aberrations during Stimulation by Flickering Light," *Nature*, *177* (14 April), 710.

Walter, W. G. (1961) *The Living Brain*, 2nd ed. (Harmondsworth: Penguin).

Walter, W. G. (1966) "Traps, Tricks and Triumphs in E.E.G.: 1936–1966," the second Geoffrey Parr Memorial Lecture, BNI archive, 6/31 Science Museum, London.

Walter, W. G., V. J. Dovey, and H. Shipton (1946) "Analysis of the Response of the Human Cortex to Photic Stimulation," *Nature*, *158* (19 October).

Watts, A. (1957) *The Way of Zen* (New York: Pantheon).

Watts, A. (1976 [1961]) "A Fragment from *Psychotherapy East and West*," in C. Sluzki and D. Ransom (eds.), *Double Bind: The Foundation of the Communicational Approach to the Family* (New York: Grune & Stratton, 1976), pp. 167–60. Reprinted from Watts, *Psychotherapy East and West* (New York: Mentor Books, 1961).

Wells, O. (1973) Untitled obituary of W. R. Ashby, *Artorga*, "Tribute to Ross Ashby," Cmmnctn 163, 164 (July and August 1973), no page numbers, S. N. 11/6/26; B. N. 17, University of Illinois, Urbana-Champaign, archives.

Whittaker, D. (2003) *Stafford Beer: A Personal Memoir* (Charlbury, Oxfordshire: Wavestone Press).

Whittaker, D. (2009) *Stafford Beer: A New Anthology of Writings* (Charlbury, Oxfordshire: Wavestone Press).

Wiener, N. (1948) *Cybernetics; or, Control and Communication in the Animal and the Machine* (Cambridge, MA: MIT Press).

Wiener, N. (1967 [1950]) *The Human Use of Human Beings: Cybernetics and Society* (New York: Avon Books).

Wiener, N. (1961) *Cybernetics; or, Control and Communication in the Animal and the Machine*, 2nd ed. (Cambridge, MA: MIT Press).

Wildes, K., and N. Lindgren (1985) *A Century of Electrical Engineering and Computer Science at MIT, 1882–1982* (Cambridge, MA: MIT Press).

Willats, S. (1976) *Art and Social Function* (London: Ellipsis). 2nd ed., 2000.

Wilson, C. (1964) *Necessary Doubt* (New York: Trident Press).

Wilson, C. (1971) *The Black Room* (New York: Pyramid).

Wise, M. N., and D. C. Brock (1998) "The Culture of Quantum Chaos," *Studies in the History and Philosophy of Modern Physics*, 29, 369–89.

Wolfram, S. (2002) *A New Kind of Science* (Champaign, IL: Wolfram Media).

Wolfram, S. (2005) "The Generation of Form in *A New Kind of Science*," talk presented at several schools of architecture in 2005, www.stephenwolfram.com/publications/talks/architecture/.

Zangwill, O. L. (1980) "Kenneth Craik: The Man and His Work," *British Journal of Psychology*, 71, 1–16.

Zeidner, J., D. Scholarios, and C. D. Johnson (2001) "Classification Techniques for Person-Job Matching: An Illustration Using the US Army," *Kybernetes*, 30, 984–1005.

Zemanek, H. (1958) "La Tortue de Vienne et les Autres Travaux Cybernétiques," in *Proceedings of the First International Congress on Cybernetics, Namur, 26–29 June 1956* (Paris: Gauthier-Villars), pp. 770–80.

INDEX

Page numbers in italics indicate illustrations.